A LABORATORY GUIDE
TO THE MAMMALIAN EMBRYO

A LABORATORY GUIDE
TO THE MAMMALIAN EMBRYO

Edited by

David K. Gardner

Michelle Lane

Andrew J. Watson

UNIVERSITY PRESS

2004

OXFORD
UNIVERSITY PRESS

Oxford   New York
Auckland   Bangkok   Buenos Aires   Cape Town   Chennai
Dar es Salaam   Delhi   Hong Kong   Istanbul   Karachi   Kolkata
Kuala Lumpur   Madrid   Melbourne   Mexico City   Mumbai   Nairobi
São Paulo   Shanghai   Taipei   Tokyo   Toronto

Published by Oxford University Press, Inc.,
198 Madison Avenue, New York, New York 10016

www.oup.com

Oxford is a registered trademark of Oxford University Press

Library of Congress Cataloging-in-Publication Data
A laboratory guide to the Mammalian Embryo / edited by David K. Gardner,
Michelle Lane, Andrew J. Watson.
p. ; cm.
Includes bibliographical references and index.
ISBN 0-19-517151-9; 0-19-514226-8 (pbk)
1. Embryology, Experimental—Laboratory manuals.   2. Embryology,
Human—Laboratory manuals.
[DNLM: 1. Embryo—Laboratory Manuals.   2. Cells, Cultured—Laboratory
Manuals.   3. Reproductive Techniques, Assisted—Laboratory Manuals. QS
625 L123 2003]   I. Gardner, David K.   II. Lane, Michelle.   III. Watson,
Andrew J. (Andrew John), 1959–
QL961 .L33 2003
571.8'619—dc21   2003000329

9 8 7 6 5 4 3 2 1

Printed in the United States of America
on acid-free paper

Preface

The 1990s saw an exponential increase in research on the preimplantation mammalian embryo. Basic research on laboratory animals has been paralleled with increases in the use of assisted breeding techniques in both domestic animals and humans. Concomitantly, there have been improvements in the technology used to study the early embryo. New biochemical, microscopic, and genetic techniques have been developed and applied at the single-cell level. It is therefore the scope of this book to describe in detail such new developments and to aid the researcher in applying such technology in the laboratory by providing detailed protocols that augment the theoretical aspects of each chapter.

Rather than focusing on a single species, the chapters cover embryos from rodents, domestic animals, and primates, including humans. The techniques and procedures described include updates of established procedures such as in vitro maturation, in vitro fertilization, and embryo culture, together with newer research areas such as the establishment of human embryonic stem cells, gamete transplantation, cloning, and genetic analysis.

Although no single text can cover the entire field of mammalian embryology, we have endeavored to provide today's laboratory workers with a practical guide to many of the techniques used in modern embryology. This book should therefore serve as a useful tool for both students and experienced researchers alike.

ACKNOWLEDGMENTS

The editors thank all of the authors for their valuable contributions and Angela Hunsinger for her assistance with the preparation of this text.

This book is dedicated to Jackie, Glen, and Patricia for their tireless support and understanding.

Contents

Contributors

Jay M. Baltz
Departments of Obstetrics and Gynecology
 (Division of Reproductive Medicine),
and Cellular and Molecular Medicine
University of Ottawa, Canada

D. Page Baluch
Molecular and Cellular Biology Program
School of Life Sciences
Arizona State University
Tempe, Arizona, USA

Lisa C. Barcroft
Department of OB/GYN and Physiology
The University of Western Ontario
London, Ontario, Canada

Parvathi K. Basrur
Department of Biomedical Sciences
Ontario Veterinary College
University of Guelph
Guelph, Ontario, Canada

Dean H. Betts
Department of Biomedical Sciences
Ontario Veterinary College
University of Guelph
Guelph, Ontario, Canada

Debra J. Bloor
School of Biological Sciences
University of Manchester
Manchester, UK

Catherine A. Boyd
The Jones Institute
Eastern Virginia Medical School
Norfolk, Virginia USA

Gerard Brady
School of Biological Sciences
University of Manchester
Manchester, UK

Daniel R. Brison
Department of Reproductive Medicine
St Mary's Hospital, Manchester
and School of Biological Sciences
University of Manchester
Manchester, UK

Wafa Bureau
Department of Biomedical Sciences
Ontario Veterinary College
University of Guelph
Guelph, Ontario, Canada

Michele Calder
Departments of Obstetrics and Gynaecology
 and Physiology
The University of Western Ontario
London, Ontario, Canada

David G. Capco
Molecular and Cellular Biology Program
School of Life Sciences
Arizona State University
Tempe, Arizona, USA

Paul A. De Sousa
Roslin Institute
Roslin, Midlothian, UK

Andras Dinnyes
Roslin Institute
Roslin, Midlothian, UK

Alpesh Doshi
Department of Obstetrics and Gynaecology
University College London
and Assisted Conception Unit
University College Hospital
London, UK

John J. Eppig
The Jackson Laboratory
Bar Harbor, Maine, USA

David K. Gardner
Colorado Center for Reproductive Medicine
Englewood, Colorado, USA

Joyce C. Harper
Department of Obstetrics and Gynaecology
University College London
and Assisted Conception Unit
University College Hospital
London, UK

Laura Hewitson
Department of Obstetrics, Gynecology and
    Reproductive Sciences
University of Pittsburgh School of
    Medicine,
and The Pittsburgh Development Center
Pittsburgh, Pennsylvania, USA

Helen R. Hunter
Department of Reproductive Medicine
St Mary's Hospital
Manchester, UK

D. Holstead Jones-Taggart
Ontario Cancer Institute
Princess Margaret Hospital
University of Toronto
Toronto, Ontario, Canada

Magosaburo Kasai
Laboratory of Animal Science
College of Agriculture
Kochi University
Nankoku, Kochi, Japan

Susan J. Kimber
School of Biological Sciences
University of Manchester
Manchester, UK

Tim King
Roslin Institute
Roslin, Midlothian, UK

W. Allan King
Department of Biomedical Sciences
Ontario Veterinary College
University of Guelph
Guelph, Ontario, Canada

Michelle Lane
Colorado Center for Reproductive Medicine
Englewood, Colorado, USA

Susan E. Lanzendorf
Department of Obstetrics and Gynecology
Washington University in St. Louis
Missouri, USA

Keith E. Latham
The Fels Institute for Cancer Research
    and Molecular Biology
and the Department of Biochemistry
Temple University School of Medicine
Philadelphia, Pennsylvania, USA

Anthony D. Metcalfe
School of Biological Sciences
University of Manchester
Manchester, UK

Ricardo Moreno
Unit of Reproduction and Development
    Physiology Department
Pontifical Catholic University of Chile
Santiago, Chile

Tetsunori Mukaida
Hiroshima HART Clinic
Ohtemachi, Nakaku, Hiroshima, Japan

Alison A. Murray
Department of Biomedical Sciences
University of Edinburgh
Edinburgh, UK

Makoto C. Nagano
Department of Obstetrics and Gynecology
Royal Victoria Hospital
McGill University
Montreal, Quebec, Canada

David R. Natale
Department of Biochemistry and Molecular
    Biology
University of Calgary
Calgary, Alberta, Canada

Queenie V. Neri
The Center for Reproductive Medicine
    and Infertility
Weill Medical College of Cornell
    University
New York, New York, USA

Gula Nourjanova
Cold Spring Harbor Gel Laboratory
Cold Spring Harbor, New York, USA

Gianpiero D. Palermo
The Center for Reproductive Medicine and
    Infertility
Weill Medical College of Cornell
    University
New York, New York, USA

Lesley Paterson
Roslin Institute
Roslin, Midlothian, UK

Christine M. Pauken
Molecular and Cellular Biology
    Program
Department of Biology
Arizona State University
Tempe, Arizona, USA

João Ramalho-Santos
Department of Zoology
University of Coimbra
Coimbra, Portugal

Don Rieger
Animal Biotechnology Embryo
    Laboratory
Department of Biomedical Sciences
University of Guelph
Guelph, Ontario, Canada

William Ritchie
Roslin Institute
Roslin, Midlothian, UK

Claude Robert
Centre de Recherche en Biologic de la
    Reproduction
Département des Sciences Animales
Université Laval
Québec, Canada

Zev Rosenwaks
The Center for Reproductive Medicine
    and Infertility
Weill Medical College of Cornell
    University
New York, New York, USA

Denny Sakkas
Department of Obstetrics and Gynecology
Yale University School of Medicine
New Haven, Connecticut, USA

Gerald Schatten
Development of Obstetrics, Gynecology and
    Reproductive Sciences
University of Pittsburgh School of Medicine,
and The Pittsburgh Development Center,
and The Magee-Women's Research Institute
Pittsburgh, Pennsylvania, USA

Calvin Simerly
Department of Obstetrics, Gynecology and
    Reproductive Sciences
University of Pittsburgh School of Medicine,
and The Pittsburgh Development Center
Pittsburgh, Pennsylvania, USA

Marc-André Sirard
Centre de Recherche en Biologic de la
    Reproduction
Département des Sciences Animales
Université Laval
Québec, Canada

Norah Spears
Department of Biomedical Sciences
University of Edinburgh
Edinburgh, UK

Takumi Takeuchi
The Center for Reproductive Medicine
    and Infertility
Weill Medical College of Cornell
    University
New York, New York, USA

Evelyn E. Teller
Institute of Cell and Molecular Biology
University of Edinburgh
Edinburgh, UK

Lucinda L. Veeck
The Center for Reproductive Medicine
    and Infertility
Weill Medical College of Cornell
    University
New York, New York, USA

Andrew J. Watson
Department of Obstetrics and Gynecology
and Physiology
University of Western Ontario
London, Ontario, Canada

Karen Wigglesworth
The Jackson Laboratory
Bar Harbor, Maine, USA

Ian Wilmut
Roslin Institue
Roslin, Midlothian, UK

Leeanda Wilton
Melbourne IVF
Victoria, Australia

Christine Wrenzycki
Department of Biotechnology
Institut fur Tierzucht und
Tierverhalten (FAL)
Mariensee, Neustadt, Germany

Diane L. Wright
Vincent Reproductive Medicine
and IVF
Massachusetts General Hospital
Boston, Massachusetts, USA

Abbreviations

acetyl-CoA	acetyl coenzyme A
ART	assisted reproductive techniques
bp	basepair
BSA	bovine serum albumin
cGMP	cyclic guanidine 5'-monophosphate
CHAPS	[(cholamidopropyl)dimethylammonio]-l-propanesulfonate
DAPI	4',6-diamidino-2-phenylindole dihydrochloride
DMEM	Dulbecco's modified Eagle's medium
DMSO	dimethylsulfoxide
DNP	dinitrophenol
dpm	disintegrations per minute
DTT	dithiothreitol
EDTA	ethylenediaminetetraacetic acid
EtOH	ethanol
FBS	fetal bovine serum
FISH	flurorescence in situ hybridization
FITC	fluorescein isothiocyanate
FSH	follicle-stimulating hormone
GnRH	gonadotropin-releasing hormone
hCG	human chorionic gonadotropin
HEPES	N-(2-hydroxyethyl)piperazine-N'-ethanesulfonic acid
hMG	human menopausal gonadotropin
HSA	human serum albumin
HTF	human tubal fluid medium
IBMX	3-Isobutyl-l-methylxanthine
ICM	inner cell mass
ICSI	intracytoplasmic sperm injection
i.d.	inner diameter
IVF	in vitro fertilization
IVM	in vitro maturation
kb	kilobase
KSOM	modified simplex optimized medium with elevated potassium

LH	luteinizing hormone
MEM	minimum essential medium
MOPS	4-morpholine propanesulfonic acid
mw	molecular weight
o.d.	outer diameter
PAGE	polyacrylamide gel electrophoresis
PBS	phosphate-buffered saline
PCR	polymerase chain reaction
PMSF	phenylmethylsulfonyl fluoride
PMSG	pregnant mares serum gonadotropin
PVA	polyvinyl alcohol
PVP	polyvinyl pyrrolidone
SDS	sodium dodecyl sulfate
SSC	standard saline citrate
SSCP	single stranded conformational polymorphism
RT-PCR	reverse transcription-polymerase chain reaction
TAE	Tris acetate-EDTA buffer solution
TBE	Tris borate-EDTA buffer solution
TBS	Tris buffered saline
TC	tissue culture
TE	trophectoderm
TEMED	N, N, N^1, N^1-tetramethylethylenediamine
TUNEL	Tdt-mediated dUTP nick-end labeling
X-gal	5-bromo-4-chloro-3-indolyl-β-D-galactoside

A LABORATORY GUIDE
TO THE MAMMALIAN EMBRYO

1

NORAH SPEARS
EVELYN E. TELFER
ALISON A. MURRAY

Follicle Development in Vitro

1. FOLLICULAR AND OOCYTE DEVELOPMENT

The mammalian ovary is endowed by birth or shortly after with a lifetime supply of oocytes. These oocytes are surrounded by somatic cells (forming ovarian follicles) and separated from the interstitial tissue and from each other by a basement membrane. As the ovarian follicles form, the first stage of development is called the primordial follicle. These primordial follicles are in a resting stage, with oocytes surrounded by flattened somatic cells. Between 1 and 2 million primordial follicles are contained within the human ovary at birth. A small proportion of these is continually leaving the resting pool and starting to grow, although only a tiny percentage of these will grow to maturity and release their oocytes for fertilization: the fate of most follicles is to undergo follicular atresia and die (usually during the later stages of development). Growing follicles are characterized by an increase in oocyte size, an increase in the layers of somatic granulosa cells surrounding the oocyte and, later, by the formation of a fluid-filled antral cavity. Preantral growth is slow, whereas the antral phase proceeds more rapidly.

Our ability to support follicular and oocyte development in vitro has advanced greatly over recent years. Although technologies have been developed to culture the follicles of large mammalian species, including primates, the majority of this work has used rodent ovaries. Rodent culture models of follicles at various stages of development have contributed much to our knowledge of the processes that drive follicular development and can also produce viable gametes using various culture methods. In both domestic species and primates (including humans), techniques have been developed in part to investigate the regulation of follicle development, and also (more commonly) to produce a regulated supply of follicles at particular stages of development. The ultimate aim of follicle culture is to be able to grow follicles from the resting primordial stage, in which follicles are laid down during ovary formation, to produce competent oocytes (capable of supporting the development of live young after fertilization). To date, this has been achieved only in the mouse [1], and we are still a long way from being able to use follicle culture to obtain competent oocytes from larger species such as humans. However, much progress has been made in recent years in developing culture techniques both for

3

rodents and for large domestic species and primates at several stages of follicle growth, in particular initiation of primordial follicle growth, development of oocyte–granulosa cell complexes, and growth of antral follicles (Figure 1.1). Different culture systems have been developed with different endpoints in mind. In this chapter, we describe many of the different methods currently carried out and outline their various uses.

It is envisaged that future use of in vitro systems will begin to unravel some of the mechanisms of processes as yet poorly understood, such as what regulates growth ini-

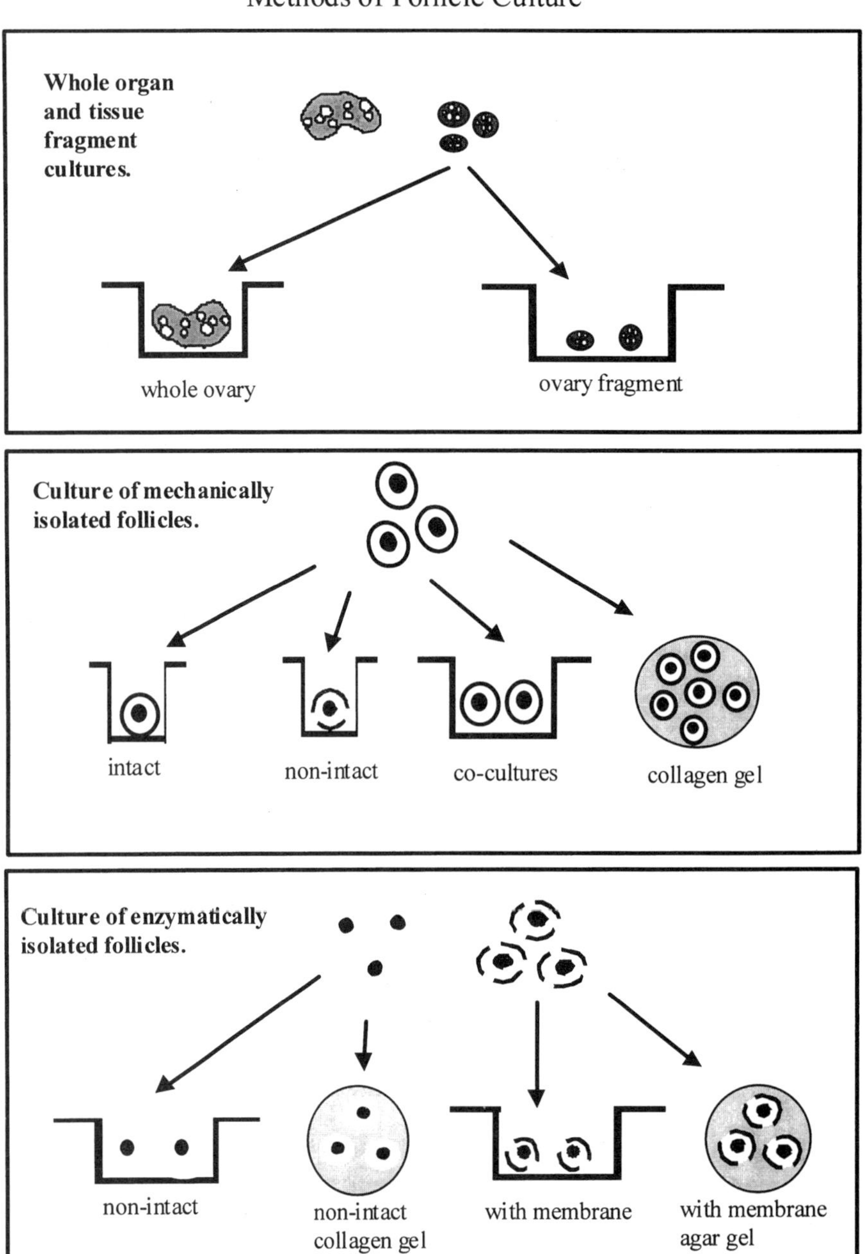

Figure 1.1. Methods of isolation and types of culture systems used. Adapted from Murray and Spears (44), with permission from Thieme Medical Publishers.

tiation in primordial follicles. As development of the oocyte and of the somatic granulosa cells are interdependent, production of mature oocytes is likely to require culture of both oocytes and somatic cells. Knowledge of how oocytes become fully competent will be of particular use if this technology is to be practical in clinical settings. The process of oocyte maturation in the human is very long, and the technical difficulties that need to be overcome are many. The hope is that, with the knowledge gained from other models, it may be possible to accelerate this process in humans in vitro.

1.1 Organ Cultures

Techniques for the culture of ovaries have been described since about 1930. Earlier in vitro techniques tended to use the culture of whole ovaries (2). Although this type of culture preserves normal tissue interactions, it is severely limited in that long-term maintenance of organ explants is difficult. Although it is possible to maintain the cortical regions of the ovary to some extent, the inner medullary region is subject to anoxia, depletion of nutrients and accumulation of metabolites leading to necrosis. Because of these limitations, and due to the time required for full follicular development (especially in larger species), the culture of adult ovaries is probably of limited use and perhaps best confined to the study of a particular ovarian event such as blood flow or ovulation. Ovarian fragments or tissue slices are sometimes used to overcome some of the problems associated with mass. In addition, a number of methods such as suspending or floating organ or tissue slices in culture have been developed, thus maximizing gas and nutrient diffusion and reducing necrosis. Although follicles can grow under these culture conditions, at least up to the antral stage, it is difficult to follow their individual progress. As a result, recent cultures of later stages of follicle development have tended to use isolated follicles. Techniques using whole or slices of ovaries have been reserved for the culture of primordial follicles (at which stage isolated follicles tend to do poorly).

Follicles can remain in the female mammal in the primordial state for the entire reproductive life span of the animal. The processes determining when a particular primordial follicle leaves this resting pool and undergoes growth initiation are, in the main, still a mystery. The reserves of primordial follicles within the ovary represent a vast store of gametes that could be available for in vitro development. Thus, cultures of primordial follicles are useful both as a tool to help investigate the regulation of growth initiation and as a potential source of follicles to supply mature, fertilizable oocytes in vitro.

1.2 Individual Follicle Culture

Short-term cultures of mid- to late-antral follicles and their oocytes have been successfully used to investigate the final stages of follicular growth and oocyte maturation. However, in the 1990s a number of in vitro systems evolved that support the growth and development of follicular units from preantral through to preovulatory stages. Some of these culture systems have produced mature oocytes capable of fertilization, and live young have been born, albeit with limited success rates (3, 4). The majority of these in vitro systems have been devised using rodents as models. The ovaries of rodents are readily available and lack the fibrous tissue associated with the ovaries of larger species. In addition, the time-course of complete follicular development is relatively short in rodents, making it possible to grow follicles in a manner that closely resembles the in vivo situation (Table 1.1). Many studies have used follicles obtained from juvenile mice and rats; these ovaries have few antral follicles, lack corpora lutea, and contain large numbers of follicles at similar stages of preantral development. While rodents have been used extensively in developing culture systems, some progress has also been made in domestic species and humans. Human material is difficult to obtain, and few studies on the in vitro growth of follicles have been reported, but other primate models are also

Table 1.1. Size of Preantral and Mature Follicles of Three Species (Murine, Porcine, and Bovine) and the Estimated Time it Takes in vivo to reach the Preovulatory Stage from Preantral Stages

Species	Preantral Size	Mature Size	Growth Period (Days)
Mouse	150–200	400–500 µm	5–7
Pig	150–300	1.5–3 mm	40–50
Cow	100–150	3.8– >8.5 mm	40–50

Adapted Telfer et al. (40), with permission from Elsevier Science.

available (5). Work on domestic species is sometimes carried out with a veterinary or agricultural aim and sometimes as a model of human ovarian development.

1.2.1 Methods of Isolation

Follicular units can be isolated either mechanically or by enzymatic digestion. Enzymatic digestion using collagenase and DNase has been successfully used to isolate follicles from mice, rats, hamsters, and humans. Small pieces of tissue are incubated with the enzymatic mixture, and by mechanically agitating the pieces through progressively smaller pipette tips, individual follicles are released. The degree of damage to the follicular units depends on the stringency of the treatment and on the type of tissue being used. Digestion of ovary samples from hamsters and humans results in follicles with an intact basement membrane but with no attached thecal cells, compared with mice and rats that lose both thecal cells and basement membrane integrity. Effectively, in the latter species, the follicles collected are oocyte–somatic cell complexes. Although these complexes appear devoid of a basement membrane and attached thecal cells, it is likely that some of these components are transferred into the culture to a greater or lesser degree depending on the methods used.

Mechanical isolation of follicles is a more labor-intensive method of isolating follicles, but it ensures that the basement membrane of the follicle remains intact and preserves follicular architecture. Small pieces of tissue are teased apart, and follicles are dissected from these tissues using fine needles. The main criteria used in selection of follicles by this method are size—a centrally placed oocyte surrounded by tightly packed granulosa cell layers and some attached thecal/stromal cells. Conditions for the isolation of immature follicles from mice are different from that required for follicles from domestic species. Microdissection of follicles from rodents is relatively easy, as there is little stromal tissue present. Collagenase isolation of rodent follicles can be carried out rapidly with no detrimental effects to the oocyte, but collagenase dissociation of bovine preantral follicles has a detrimental effect on bovine oocytes. Collagenase does not seem to have as drastic an effect on pig oocytes, but short incubations in collagenase are not very effective at loosening preantral follicles; as such, collagenase is essentially an ineffective method for harvesting large numbers of preantral follicles. A number of methods for isolating preantral follicles from domestic species have been developed (6), but we have found that microdissection of preantral follicles from cortical slices provides high-quality preantral follicles from bovine, porcine, and ovine ovaries.

1.2.2 Culture Conditions

To a certain extent, the method of isolation determines the culture conditions into which follicles are placed (Figure 1.1). Follicle units that have been isolated by enzymatic means and lack a basement membrane need to be cultured in conditions that prevent the dissociation of the granulosa cell from the oocytes, as contact between the two cell types is necessary if the oocyte is to grow and mature. This has been achieved by placing these isolated complexes onto collagen-impregnated membranes by embedding them in collagen gels (7) or by allowing them to plate down onto substrate adhesive surfaces (8, 9). Follicles with the basement membrane intact (isolated either mechanically or enzymatically) have been cultured by embedding them into agar or collagen gels (10–12) or by placing follicles on various surfaces bathed in culture medium (4, 13, 14). In the latter culture techniques, follicles are maintained as intact units due to culture in conditions that prevent adhesion to culture dishes. Where follicles are placed on surfaces allowing adhesion, follicles tend to rupture during culture, but all cell types remain present in the culture (8, 15). It has become increasingly clear that the oocyte contributes to and controls certain aspects of granulosa cell differentiation, and culture of oocyte–somatic cell complexes have been used extensively and very successfully to elucidate some of these processes (Figure 1.2; see, e.g., Eppig et al. [16]).

Each type of culture system has advantages and disadvantages, and selection of one over the other depends both on the experimental endpoint and the type of information desired. The culture of intact follicles ensures that each cell type (theca, granulosa, and oocyte) is present and allows the investigation of how each of these contributes to both follicular development and growth and maturation of the oocyte. However, this type of culture may not support development of antral follicles from larger species where transfer of nutrients and oxygen diffusion may be problematic. These problems could be overcome, and larger numbers of oocytes maintained in vitro, by using a culture system such as that developed for oocyte–somatic cells that are "open" to the culture media, although the contribution of the extracellular matrices (including those in antral fluid) may then be excluded. Culture systems where all elements of the follicular unit are included but the three-dimensional structure is lost provide some of the advantages of both these systems, although here growth may be impeded and basement membrane/thecal cell components may not differentiate in a manner that resembles the in vivo situation as clearly, nor is it feasible to work with quite as large numbers as enzymatic digestion can provide. Embedding follicles in a matrix may be preferable for gaining information on growth kinetics, but it is not as suitable for examining growth and maturation of the oocytes, as these are difficult to

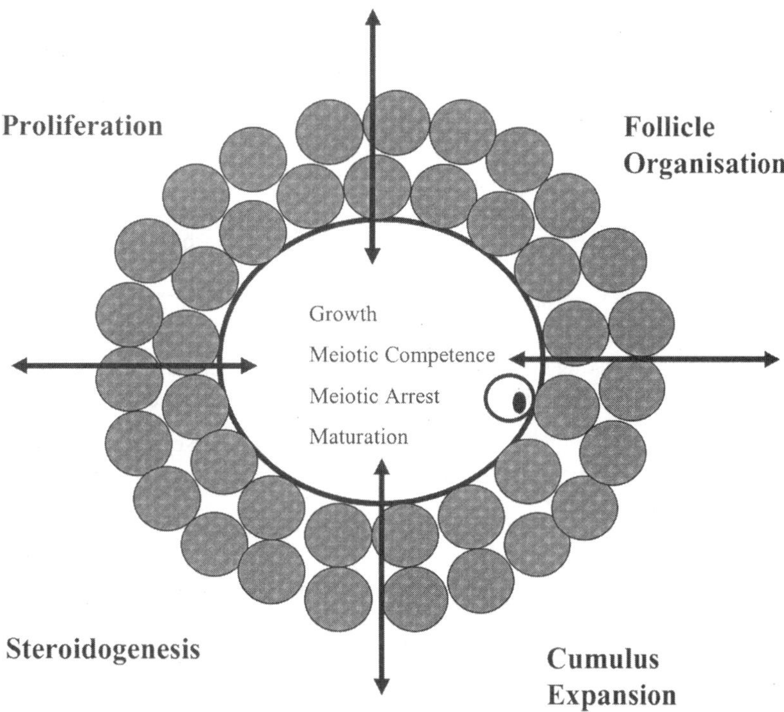

Figure 1.2. Interactions between the oocyte and surrounding somatic cells.

recover from the gels. Given the correct stimulus, some of the culture systems support the formation of antral (or, if non-intact, antral-like) cavities and steroid production, notably in cultures where the basement membrane and thecal/stromal cells are included (8, 15, 17, 18). Follicles maintained as intact units develop thecal layers around the whole follicle. In the rat in vitro system, where all components are present and a substrate adhesion method of culture is used, thecal cells appear to colonize around the edges of the granulosa cell layers and organize as theca interna and externa types (8). The degree of growth and development of follicles depends on a number of factors such as media composition, the age or size of the starting material, stimulation by gonadotrophins and growth factors, and the length of the culture period.

In this chapter, we describe various techniques commonly used to culture primordial, preantral, and early-antral follicles, both for small species such as rodents and for large domestic and primate species. There are, of course, many variations on the techniques described below—too many for us to cover in a chapter such as this. Our aim here is to detail many of the more frequently used culture systems, particularly where we consider that they are likely to be used for much future work. We give examples of how the cultures are carried out and what information each kind can provide. We first consider organ culture primordial follicles and then the culture of late preantral and antral follicles.

2. TECHNIQUES

2.1 Organ Culture of Primordial Follicles

Some of the earliest experiments growing oocytes in vitro involved culturing the whole ovary (2, 19). Organ culture has the advantage of maintaining normal cell contacts, but there are serious disadvantages in maintaining this type of culture over a long period because of perfusion problems. Some of the physical problems can be overcome by using thin cortical sections of the ovary, as this allows the maintenance of contact but provides material easier to deal with in vitro. Organ cultures or cortical slice cultures have successfully allowed the initiation of primordial follicle growth in vitro. It has recently been demonstrated that it is possible to produce competent mouse oocytes after culture from the primordial follicle stage (1). This achievement depended on a two-step procedure involving organ culture for the follicles to initiate growth, followed by isolation of the growing follicles (as organ culture could not further sustain full development). Primordial follicles can be isolated from porcine ovaries (20), but these do not start to grow in vitro. As well as in the rodent, primordial follicle initiation has been observed in cultured bovine, ovine, and primate ovarian cortical strips (5).

2.1.1 Free-floating Organ Cultures
(Rodent and Domestic Species)

In this system, either whole ovaries (from mice [1]) or cortical sections (bovine, primate) are placed in culture medium, free floating with or without a collagen membrane.

1. Dissect mouse ovaries on the day of birth (day 0) and cut in half.
2. Alternatively, remove thin strips of ovarian cortex from the ovaries of domestic species using a scalpel blade and watchmaker's forceps. Trimmed strips to 0.5 × 0.5 × 0.2 mm.
3. Transferred ovaries in a drop of culture medium (Protocol 1.1) using a Pasteur pipette onto a Costar Transwell membrane (non–tissue-culture treated, Nucleopore polycarbonate membrane, 3.0 μm pore size, 24 mm diameter). The Costar membranes sit raised up from the base of the culture dish.

4. Add 1.5 ml medium to the well below the membrane. This means that when the drop of medium containing the ovary is placed on the surface of the membrane, excess medium from the drop is drawn into the well below, leaving the ovary covered by only a thin film of medium.

5. Up to seven ovaries can be placed on each membrane. Incubate at 37°C in 5% CO_2 : 95% air.

6. Replace medium every 2 days by adding 2 ml of medium to the well below the membrane and withdrawing the same volume.

7. Alternatively, boil polycarbonate membranes (Isopore from Millipore) in deionized water to remove PVP coating, autoclave, and fix to the bottom of wells in 96-well plates with a silicon-based sealant (RTV32, Dow Corning). Sterilize plates by UV light. Fill wells with 100 µl medium overlaid with 100 µl silicon fluid and place ovary pieces on top of the membrane. In this instance, one ovary piece is culture per well. Exchange half the medium for fresh medium every other day.

8. After a few days of culture, follicle growth initiation can be clearly seen in the ovary (Figures 1.3A, B).

2.1.2 Culture of Human Follicles

Although the techniques to grow follicles from animal models have been established for some time, the culture of follicles from human ovaries is still in its infancy. One of the main problems in establishing human culture techniques is lack of material. Furthermore, the material used is from a wide age range of patients undergoing gynecological surgery. Such material may well not be optimal, depending on the reason for surgery. The majority of studies on the growth of human follicles have been carried out using cortical slices from ovarian biopsies. These slices contain only primordial and primary follicles. Most of these studies have concentrated on establishing culture conditions that allow the initiation of primordial follicles into the growth phase (21–23). Some attention has been paid to the effects of cryopreservation of the cortical slices, as this would have clinical implications (24–26). From these studies, primordial follicles begin to grow and survive for periods of 14 days when cultured as groups (within the slices), supported by a layer of extracellular matrix. This method is outlined below, but readers should consult recent literature regarding culture media and growth factors (22, 23), as these factors may contribute to growth.

1. Cut ovarian tissue into slices of 0.5–2 mm using a scalpel blade.

2. Place one to three slices onto Millicell CM inserts (6 mm diameter, 1.0 µm pore size; Becton Dickson), which have been precoated with 150 µl Matrigel (Becton Dickson) and placed into 24-well plates (Becton Dickson).

3. Media is as in Protocol 1.2.

4. Maintain cultures in a humidified incubator in 5% CO_2 at 37°C.

2.2 Culture of Preantral or Early-Antral Follicles from Small Mammalian Species

2.2.1 Granulosa–Oocyte Cultures

Animals

Granulosa–oocyte cell complexes can be isolated from preantral follicles of either 10- or 12-day-old mice using collagenase and DNase. The starting age of the animal will determine the total culture period, that is, complexes from 10-day-old mice will be cultured for 12 days and those from 12-day-old mice for 10 days. Mice normally used are (C57BL × CBA) F1. The minimum number of animals required to set up cultures with a good yield is 10.

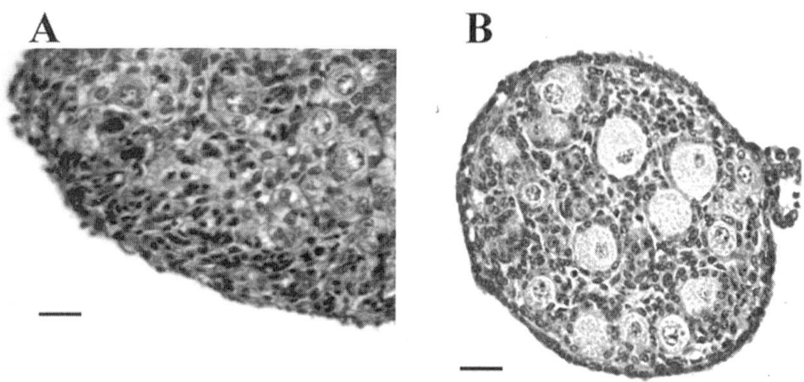
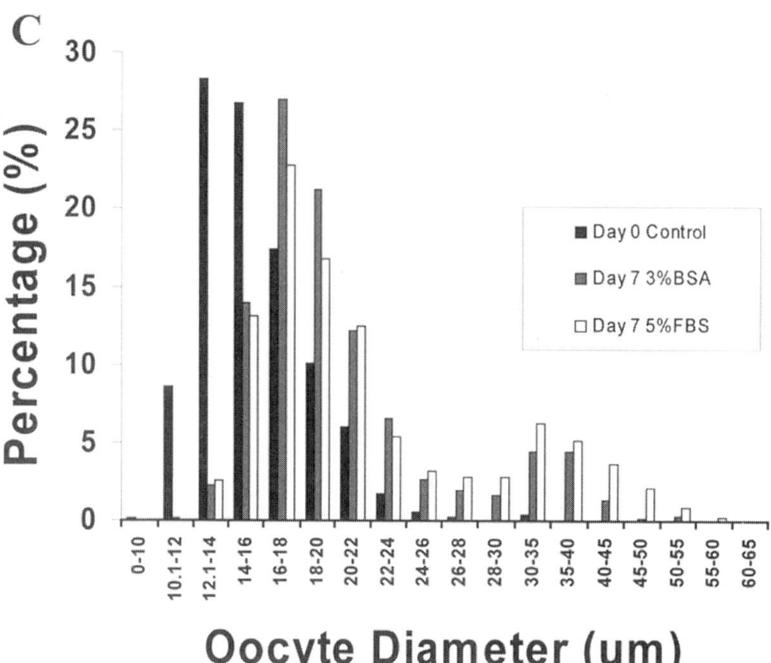

Figure 1.3. Newborn mouse ovaries, quartered and cultured for 7 days. Ovary pieces were cultured in α-MEM medium with 28 μM ascorbic acid and either 3 mg/ml BSA or 5% FBS. After 7 days of culture, ovary pieces were fixed and processed for histological examination. (A) Section of a control ovary fixed on the day of dissection. Most follicles present are at the primordial stage. Bar = 20 μm. (B) Section of ovary cultured in FBS for 7 days, with many growing follicles now present. Bar = 20 μm. (C) Histogram showing the percentage of follicles with oocytes of different diameters in uncultured control ovaries and in both treatment groups after 7 days of culture. Growth of the follicles is supported over the culture period. Statistical analysis (using the nonparametric Kolmogorov-Smirnov test) showed that ovary pieces cultured in the presence of FBS contained a higher proportion of follicles with larger oocytes than those cultured in the presence of BSA ($p < .05$). (From M. Molinek, N. Spears, and E. E. Telfer, unpublished data.)

Follicle Isolation

1. Prepare seven petri dishes as follows:
   - 5 × 3.5 ml Leibovitz + BSA + IBMX (dissection dish + wash dishes)
   - 1 × Leibovitz + BSA + IBMX + 1 mg/ml collagenase
   - 1 × Waymouth media (final wash dish).
   All dishes should be kept at 37°C, and the Waymouth media should be kept in an atmosphere of air + 5% CO_2.
2. Dissect out ovaries and place them in Leibovitz media (Protocol 1.3). All adhering tissue and bursa are removed.
3. Pull apart cleaned ovaries and transfer them to a petri dish containing 1 mg/ml of collagenase; pipette them several times with a P1000 pipette or a sterile Pasteur pipette.
4. Keep the dish at 37°C and pipette the pieces of ovaries at regular intervals (every 2 min).
5. Once the pieces are smaller, use a P200 tip to pipette the pieces up and down.
6. When follicles are isolated, continually remove them to a fresh wash dish using a P200 pipette set at 150 µl.
7. Keep ovaries in collagenase no longer than 30 min.
8. Once all the loose follicles have been transferred to a wash dish, swirl the dish around and take off and discard the single cells at the top of the media using a P200 pipette tip. Transfer the follicles to a fresh dish and repeat this washing process a further three times in wash media to ensure that all debris is removed.
9. After these wash steps transfer the follicles to Waymouth media as in Protocol 1.3 (or whatever culture media is being used) before being set up on culture wells.

Preparation of Culture Dishes for Oocyte Granulosa Cell Complexes

For optimum growth of granulosa–oocyte cell complexes, Transwell collagen membranes from Costar are used as a culture substrate (Transwell –COL Inserts, 12 mm diameter 3 µm pore size; Costar).

1. Place 1.5 ml Waymouth media in the culture wells and place the inserts on top.
2. Place 1 ml media within the inserts and then divide the complexes between the wells. Once the complexes are in the inserts, gently pipette the media to evenly distribute the complexes across the membrane. It is important to ensure even distribution of complexes across the membrane. If the complexes clump, the culture will fail.
3. Transfer culture plates to a modular incubator chamber (Billups-Rothenberg, Del Mar, CA) in an atmosphere of 5% O_2, 5% CO_2, 90% N_2 (5-5-90 gas mix) at 37°C. This 5-5-90 gas mix has been found to be optimum for oocyte growth, although good results can also be obtained by culturing these complexes in 5% CO_2 in air.
4. Feed cultures every 48 h by exchanging 1 ml of media at the bottom with 1 ml of fresh media. Care must be taken during feeding to avoid jolting the cultures, as this can easily dislodge the complexes from the membranes.

During the culture period, some follicles will be dislodged from the membranes, and these will generally be discarded with media changes. Great care has to be taken throughout the culture period to avoid jolting the dishes. If the dishes are disrupted, this will lead to clumping and/or dislodging of the complexes, and both are undesirable. After culturing the granulosa–oocyte cell complexes for 10–12 days, remove them from the surface of the membrane or collagen gel matrix. This is done by sharply snapping the side of the dish with a fingernail to jolt the complexes free of the substratum. The complexes are collected with a drawn glass pipette (drawn from 4-mm Pyrex glass tubing), using a mouth pipette. Once collected the complexes can now be taken for oocyte maturation (see section 2.2.3).

2.2.2 Individual Follicle Cultures

Individual follicles can be manually dissected from the ovaries of prepubertal mice. The main difference between manual dissection and enzymatic dissociation of the ovaries is that the follicular units obtained by manual dissection have undisrupted basal lamina and have thecal cells attached to the follicles, which are included in the culture. Whether the follicles are maintained as intact units or allowed to disrupt depends on the different culture conditions used. However, in both cases the isolation procedure is essentially the same. The type of culture plate used and volume of media required depends on whether follicles are to be grown as intact or non-intact units.

Animals

Follicles can be manually dissected from the ovaries of juvenile mice and rats. The age of the animals used depends on the starting size of the follicles required. Mice and rats between 12 and 14 days of age yield high numbers of follicles 100–130 μm in diameter, whereas animals 21–25 days of age are a better source of follicles at a later stage of preantral development (170–190 μm in diameter). If the endpoint of the culture is to obtain viable oocytes, then the length of the culture period depends on the starting size of the follicles. Early preantral follicles (100–130 μm) are cultured for 12 days, while late preantral follicles (170–190 μm) are cultured for 5–6 days. We routinely culture in the absence of antibiotics or antifungal agents to ensure that neither oocyte nor follicle growth and development is compromised by the presence of these substances. This requires that greater than usual care is taken with sterility during culture procedures. Alternatively, penicillin and streptomycin could be added, as in section 2.2.1.

2.2.3 Culture Conditions for Follicles

Intact Follicles

1. Individual murine follicles are grown in the "U" wells of 96-well non–tissue-culture treated plates (Iwaki, UK Japan). Each well contains 30 μl of culture media (Protocol 1.4) overlaid with 75 μl of sterile silicon fluid (Merck, UK). Sterile silicone fluid is used to overlay the culture media to avoid evaporation. Place follicles individually into the wells and then move into wells of fresh media daily, using finely-drawn and curved glass pipettes. After 5 or 6 days of culture, follicles will have developed to the preovulatory stage, with fertilizable oocytes (Figure 1.4).
2. Alternatively, follicles can be grown under static conditions. In this instance follicles need to be prevented from adhering to the culture plate if they are to be maintained as intact units. Placing the follicles on squares of polycarbonate membrane (Isopore, Millipore) glued to the base of flat bottom culture plates is one technique that has been successfully used in static culture systems (27, 28). Most researchers exchange half of the culture media every second day.

Non-intact Follicles

Culture the follicles under static conditions either in flat-bottom wells of 96-well tissue culture treated plates (e.g., Bibby Sterilin) containing 75 μl culture media or in 20-μl droplets of media placed onto suitable dishes (e.g., 60-mm Falcon petri dishes). In both instances the culture media is overlaid with silicon fluid. Exchange half the media for fresh media every 48 h. Again, oocytes can be successfully fertilized at the end of the culture period.

2.2.4 In vitro Maturation of Rodent Oocytes

Although a number of types of media have been used for isolation and maturation of oocytes, Waymouth medium (Protocol 1.5) has been shown to yield the highest per-

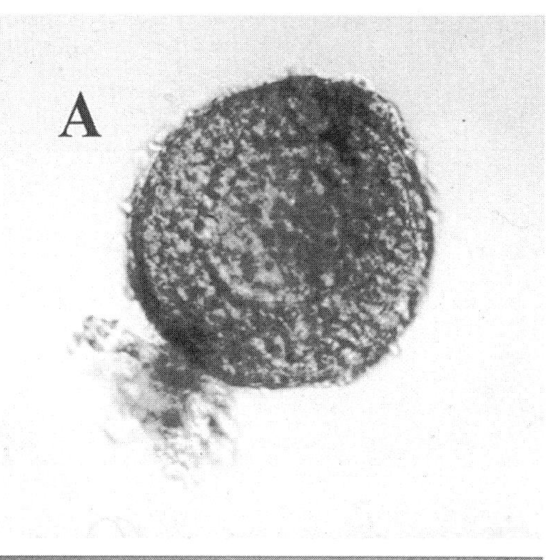

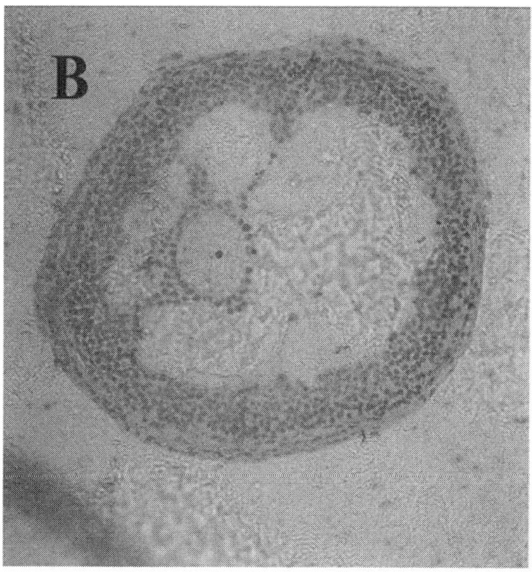

Figure 1.4. (A) Photomicrograph of an intact preantral follicle (180 μm in diameter) freshly dissected from mouse ovary. Note the clump of thecal tissue attached to the follicle. (B) Histological section of a preantral follicle cultured for 6 days as described in text and maintained as an intact unit throughout the culture period. The follicle has grown from around 180 μm in diameter with no antral cavity to 400 μm in diameter with a large antral cavity over the culture period. The thecal tissue is now evenly distributed around the follicle.

centage of mouse oocytes capable of development to the blastocyst stage (29). Other media might be preferable for the oocytes of different species. If fertilization of the in vitro-matured oocyte is part of the proposed experiment, then the medium used for both collection and maturation should be supplemented with 5% FBS or 1 mg/ml fetuin (3) to prevent hardening of the zona pellucida. Oocytes can be matured on a petri dish in drops of medium under washed oil placed in 5% CO_2 incubator overnight, or in 1 ml of medium in snap-cap tubes placed in a 37°C waterbath overnight. Some investigators have found that a gas mixture of 5% CO_2, 5% O_2, and 90% N_2 is an advantageous component of in vitro maturation protocols. After growing follicles using isolated culture techniques, the addition of epidermal growth factors to the maturation medium has been found to be beneficial (30, 31), as is also the case in other species (porcine [32]; bovine [33]; ovine [34]).

2.2.5 Cultures in a Three-Dimensional Matrix

In this system follicles are isolated enzymatically and placed within a collagen gel or agar matrix to retain three-dimensional integrity. This system has been used to culture preantral follicles from mice, hamsters, and humans.

Collagen-gel Solution

Collagen gel is extracted from the tail tendons of rats by the methods of Ehrmann and Gey (35) and Chambard et al. (36).

1. Transfer the tendons to 70% alcohol and rinse in sterile distilled water. Then add 1 g of tendon to 100 ml 1:1000 acetic acid and stir at 40°C for 48 h.
2. Centrifuge the solution at 2000 g on a benchtop centrifuge for 1 h. This solution can be stored at 4°C for up to 8 weeks.

Setting Gel Solution

1. Prepare gels immediately before use by mixing 200 µl serum with 200 µl concentrated (X10) medium 199 at 4°C. (*Note*: Place all dishes on ice.)
2. Adjust the pH by adding 500 mM NaOH until the indicator (phenol red) turns the appropriate color for physiological pH. (The amount of NaOH required has to be calculated by titration for each batch of collagen.)
3. Transfer groups of enzymatically isolated follicles to the wells of a Terasaki plate (Flow Laboratories) and add 10 µl of the collagen gel solution to each well and pipette once to suspend the follicles within the gel.
4. Incubate at 37°C for 2–3 min until the gel sets. Pipette 20 µl of gel solution into empty wells and place the 10 µl gel containing the follicles within it to form a double gel. This double gel is necessary to avoid follicle losses resulting from collagen gel contraction during culture and processing.

Culture Wells

Four collagen droplets are cultured in 1 ml of culture medium (Protocol 1.6) in 24-well culture plates. At the end of the culture period, the collagen gel must be treated with 3 mg/ml collagenase to isolate follicles from the gel.

2.3 Culture of Preantral or Early-Antral Follicles from Large Mammalian Species

2.3.1 Culture of Cattle and Sheep Follicles

Preantral Follicle Isolation

1. Transport bovine and ovine ovaries (obtained from an abattoir) in HEPES-buffered M199, at 25–30°C. In a laminar flow hood, rinse ovaries with 70% alcohol.
2. Take slices of ovarian cortex using a scalpel and place them in dissection medium as in Protocol 1.7 (Figure 1.5).
3. Under the dissecting microscope, isolated preantral follicles (100–200 µm) in a petri dish, from the cortical slices using fine 25-G needles attached to syringe barrels.
4. Select follicles with an intact basement membrane and even granulosa and theca layers for culture.
5. Culture isolated bovine preantral follicles individually in 96-well plates (Bibby Sterilin Ltd., Stone, Staffs, UK) in 250 µl of culture medium (Protocol 1.7).
6. Incubate plates in a sterile humidified atmosphere with 5% CO_2 at 37°C. Replace half of the medium every second day.

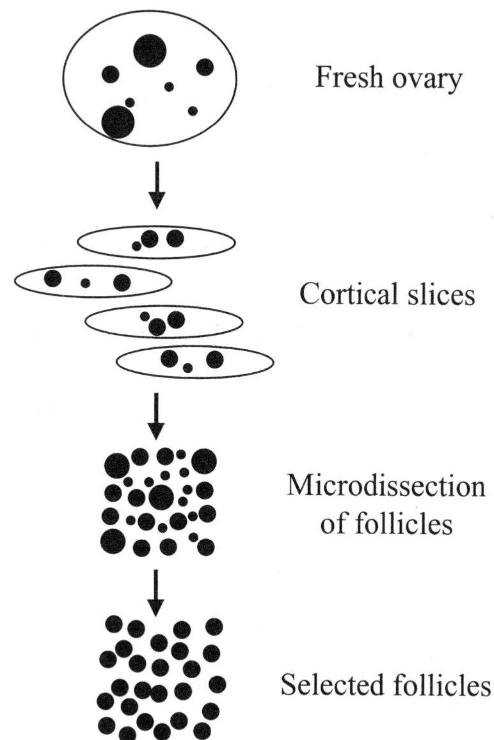

Figure 1.5. Diagram illustrating the method used to obtain individual bovine follicles.

2.3.2 Culture of Porcine Follicles

Systems to support development of porcine oocytes are more advanced than those for other large animal species. The greatest success has been achieved by using preantral follicles from prepubertal pigs. Oocytes from pig preantral follicles grown within a collagen gel matrix have been found to be meiotically competent if they reach a diameter of at least 100 μm (37), but fertilization of in vitro-grown oocytes has not been achieved using this system (37). The follicles isolated in this study consisted of an oocyte of 70–90 μm and granulosa cells with no theca, so that these structures were granulosa–oocyte complexes. Although 30% of the cultured complexes reach the preovulatory size after 16 days of culture, only 3.5% developed an oocyte with a diameter > 110 μm. After removal of the oocyte from its granulosa layers and maturation for 2 days, only 1.3% of oocytes from the initially cultured follicular structures were capable of reaching metaphase II stage of meiosis. This low rate of success to resume and progress in meiotic maturation will be the result of many factors, but the main ones will be the selection and stage of follicles that start the culture and the culture conditions.

The Hirao study (37) grew follicles within a collagen gel matrix, and we have found that this is suitable for porcine preantral follicles < 100 μm in diameter, but once follicles reach a size of 150–200 μm in diameter, transfer to the surface of a collagen matrix improves oocyte growth. It may be that a three-dimensional support and closer contact with other follicles is required at an early stage of development but that as development proceeds this environment becomes inhibitory.

A recent study reported in vitro development of porcine oocytes from preantral follicles that were capable of maturation and fertilization after 4 days of development in vitro (38). This study used North Carolina State University 23 medium (NCSU23) supplemented with 3 mg/ml BSA (Sigma, Poole, Dorset, UK; fraction V), 3.5 µg/ml insulin, 10 µg/ml transferrin; 100 µg/ml ascorbic acid, and 10% porcine serum. Optimum conditions were found if three follicles were cultured per well. The study (38) reported a remarkable growth rate of the oocyte during a short culture period, although this laboratory has been unable to replicate these results.

2.3.3 Culture of Human Follicles

A few studies have attempted to mature human follicles that have begun their growth phase (10, 39). These follicles were isolated and placed into either agar or collagen gels and underwent limited development. Although the follicles underwent some development, little information was obtained on oocyte quality, as these are difficult to recover from the gel matrix.

3. DATA ANALYSIS

3.1 Cultures to Obtain Mature Oocytes

Much of the work using follicle culture is carried out with the technical aim of maturing follicles in vitro to the point at which oocytes can be fertilized (in some cases, particularly with larger species, there will first be a further period of oocyte–cumulus cell culture to promote oocyte maturation). In these instances, data will usually be collected on the percentage of oocytes capable of developing to a further stage after culture. For example, Telfer et al. (40) examined how both the size of the follicle at the start of culture and the presence or absence of serum in the culture medium could affect the ability of the oocytes to progress to stage II of meiosis at the end of the culture period. As can be seen from Table 1.2, only when follicles with a starting diameter of > 200 µm were cultured in the presence of serum were any oocytes able to progress to metaphase II at the end of the culture period.

In other instances (mainly when small-mammal follicles are being cultured), the endpoint of the culture will be a comparison of the percentage of oocytes able to undergo fertilization, or, ideally, to achieve full developmental competence. Eppig and O'Brien (41) have examined how culture conditions can affect both these processes, in comparison to oocytes obtained from in vivo-grown follicles (Figure 1.6).

Table 1.2. Number of Porcine Oocytes Harvested from In Vitro-Grown Follicles after a 20-day Culture Period

Follicle Size (µm)	Culture (N)	No. Recovered	GV (%)	MII (%)	Dead (%)
160–210	Control (80)	39	28 (72)	0	10 (25)
210–260	Control (80)	37	20 (54)	0	14 (38)
160–210	Serum (64)	63	49 (78)	0	13 (21)
210–260	Serum (64)	52	31 (60)	10 (19)	7 (13)

Two populations of follicles were selected by size (160–210 µm and 210–260 µm) at the start of the culture period. At the end of the culture period, oocyte–cumulus cell complexes were liberated from the follicles and placed in maturation media for a total of 44 h. At the end of the maturation period, oocytes were stripped of cumulus cells and then fixed and stained with aceto-orcein to determine their meiotic status (GV= germinal vesicle stage; MII = metaphase II). Reprinted from Telfer et al. (40), with permission from Elsevier Science.

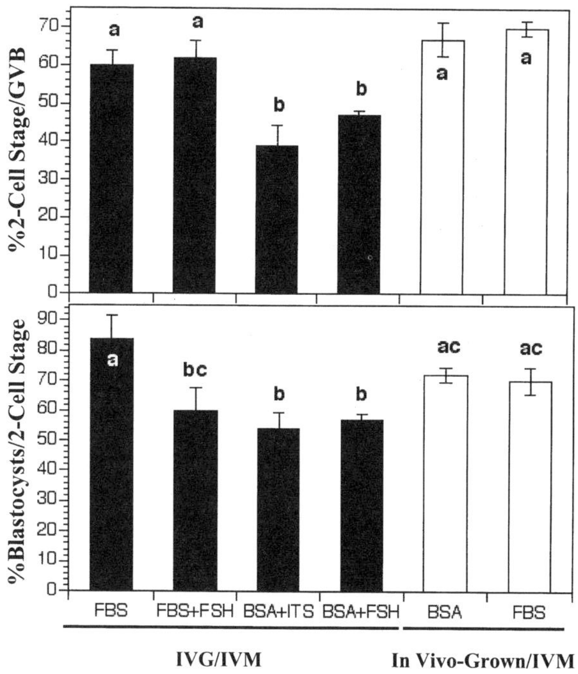

Figure 1.6. Developmental competence of oocytes grown and matured in vitro (IVG/IVM) and those grown in vivo and matured in vitro (in vivo-grown/IVM). The top panel shows the percentage of mature oocytes that cleaved to the two-cell stage after insemination, and the bottom panel shows the percentage of two-cell stage embryos that developed to the blastocyst stage. No common letters on or over the bars indicate a difference of >95% confidence level. Fewer oocytes grown in serum-free medium underwent cleavage to the two-cell stage than those grown in the presence of FBS or those grown in vivo. Likewise, the percentage of embryos derived from oocytes grown in serum-free medium that were competent to complete the two-cell stage to blastocyst transition was significantly lower than embryos derived from oocytes grown in medium supplemented with FBS (but not FBS plus FSH). Reprinted from Eppig and O'Brien (41), with permission from Elsevier Science.

3.2 Cultures to Examine the Regulation of Follicle Development

Where cultures are carried out as a technique to allow investigations into the regulation of oocyte development (rather than to increase the number of oocytes available for fertilization), the type of data collected is more variable. In many instances, information will simply be gathered on the growth of the follicles during culture, such as in Figure 1.3C, which demonstrates that culture of newborn mouse ovaries (which contain mainly primordial follicles at the start of culture) supports follicle growth over the culture period. Furthermore, it shows that ovaries cultured in the presence of FBS contain more developed follicles with larger oocytes than those cultured in the presence of BSA.

In other cases, cultured material might be analyzed to determine molecular or cellular aspects of follicle development. Figure 1.7 shows the results of McGee et al. (27), who analyzed rat follicles cultured in serum-free conditions to determine whether cGMP and FSH in combination could stimulate both follicle growth and also production of inhibin α during follicle culture. Finally, culture systems such as this, which can support development in a physiological manner, can be used to identify new factors regulating follicle development. One recent example of this is work by Vanderhyden and MacDonald (42), using culture of oocyte–granulosa cell complexes to show that oocytes secrete a factor affecting steroid production by the granulosa cells. Figure 1.8 shows that, whereas

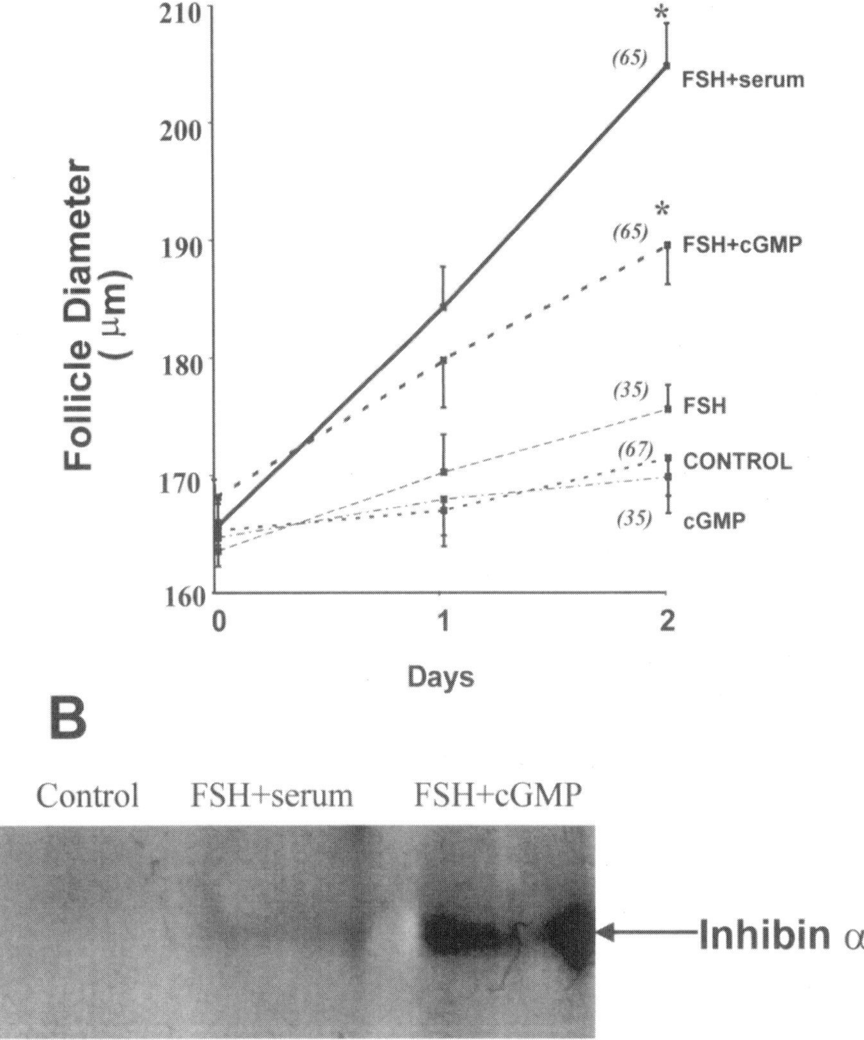

Figure 1.7. Growth of rat preantral follicles in culture. (A) Diameters of individual preantral follicles were measured over 48 h in culture. Treatment groups include serum-free medium alone (control) and with or without 8-br-cGMP (5 mM) and/or FSH (100 ng/ml). For comparison, some cultures were also treated with FSH plus 5% rat serum. Numbers in parentheses represent the number of follicles cultured per group. (B) Immunoblot analysis of inhibin-α-antigen content in cultured preantral follicles. Protein was extracted from 20 follicles/treatment group after culture for 48 h in control conditions, FSH (100 ng/ml) plus serum (5%), or FSH plus 8-br-cGMP (5 mM). Note that follicles grown in FSH plus 8-br-cGMP but in the absence of serum stimulated both growth of the follicles and also production of inhibin-α. Adapted from McGee et al. (27), with permission from The Endocrine Society.

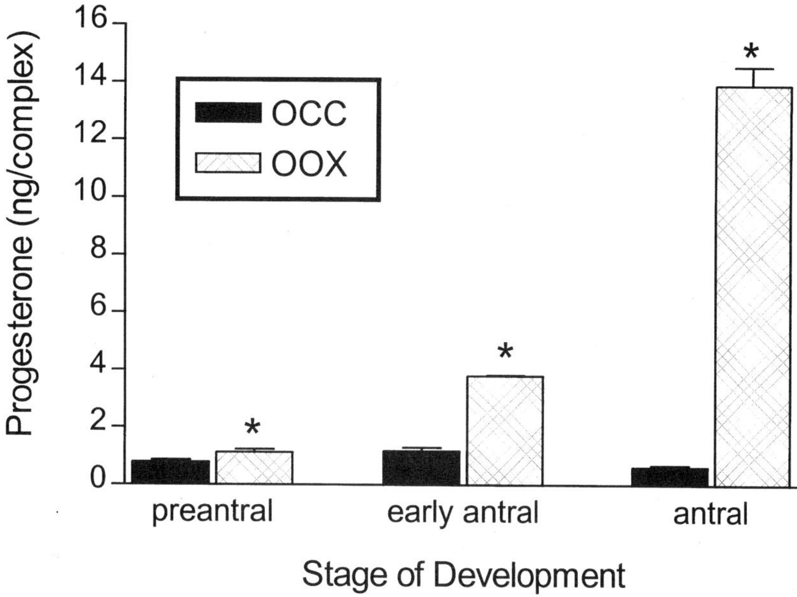

Figure 1.8. Accumulation of progesterone in cultures of intact (OOC) and oocytectomized (OOX) oocyte–granulosa cell complexes from preantral, early antral, and antral follicles. Complexes were stimulated with FSH (150 ng/ml) and testosterone (500 nM) for 48 h. Values are means ± SEMs for a minimum of three experiments. *Significantly different ($p < .05$) from the same-stage OCC. Reprinted from Vanderhyden and MacDonald (42), with permission from the Society for the Study of Reproduction.

oocytectomized complexes secrete increasing quantities of progesterone as the culture progresses, progesterone secretion from intact complexes remains low throughout.

4. CONCLUSION

Many recent studies of follicle and oocyte development have used various methods of follicle cultures. The majority of these studies have used rodents as models because material is easy to obtain and there is a short time span for follicle development, both in vivo and in vitro. In the future, it is likely that increasing effort will be placed in further developing systems to support in vitro growth of the follicles of humans and domestic species, with the more practical endpoint of producing a large supply of competent oocytes. We have detailed some of the more commonly used methods today for use with both small and large mammals, although there are many more variations in use—too many to comprehensively outline here. In general, the system of choice will depend both on the aim of the culture (the endpoint and information wanted) and on how the follicles are obtained from the ovary (whether they are intact or not).

At present, culture techniques that allow fertilizable oocytes to be obtained from primordial follicles are possible only in the mouse. It is hoped that this will also become possible in larger species before too long, probably, as with the mouse, using a succession of different in vitro techniques.

In a recent study, oocytes have been derived from cultured mouse embryonic stem cells, which formed follicle-like structures and even undergone what appears to be parthenogenic activation (43). The viability of these oocytes is not yet known, but this breakthrough opens exciting possibilities for the future production of female germ cells in vitro.

Protocol 1.1. Culture medium for whole organ culture (rodent and domestic species)

Waymouth medium MB752/1 supplemented with

0.23 mM pyruvic acid
50 mg/l streptomycin sulfate
75 mg/l penicillin G
3 mg/ml BSA or 10% FBS

Protocol 1.2. Culture medium for human whole-organ culture

Earle's Balanced Salt Solution, supplemented with

5% inactivated human serum
0.47 mM sodium pyruvate,
0.5 IU rh follicle-stimulating hormone
Penicillin (50 IU/ml) and streptomycin (50 μg/ml)

Protocol 1.3. Media for granulosa–oocyte rodent cultures

Dissection media

Leibovitz (Invitrogen BRL, UK) with

2 mM sodium pyruvate
3 mg/ml bovine serum albumin
75 μg/ml penicillin G + 50 μg/ml sterptomycin

Culture media

To 1 l of Waymouth (MB752/1) medium, add

2.240 g of sodium bicarbonate
75 mg penicillin G potassium salt
50 mg streptomycin
25 mg pyruvic acid sodium salt

Protocol 1.4. Media for individual rodent follicle cultures

Dissection media

Leibovitz L-15 media (Life Technologies)
3 mg/ml BSA (fraction V, Sigma) or FBS

Culture media

α-MEM (Life Technologies, UK)
5% serum (either fetal bovine serum, rat or mouse serum, dependent on type of
culture)
1 IU/ml recombinant FSH
25 ng/ml ascorbic acid

Preparation for Culture

1. For each pair of ovaries, 6 ml of L-15 media is required. Place 1-ml aliquots into
   glass dissection dishes and keep at 37°C.
2. Prepare suitable culture plates (see below) and preincubate to allow media and
   the silicon fluid to equilibrate prior to use.

Follicle isolation

1. Dissect ovaries from the animal and place them in a dish containing L15 medium. Remove all adhering tissue and the bursa and transfer the cleaned ovaries to a fresh dish of L15 medium.
2. Bisect the ovaries and transfer pieces to individual dishes of L15. Keep at 37°C.
3. Using 2 × 25-G needles, gently tear the tissue into smaller pieces. Follicles for culture can then be identified from these pieces and dissected free of surrounding tissue. Acupuncture needles (30 G) mounted into a holder are useful tools for dissection, as these have a fine point and no barb.
4. Transfer isolated follicles (using fine drawn glass pipettes coated with 0.1% BSA solution to prevent sticking) to a fresh L15 dish before selection. Follicles are selected for culture on the basis of size and morphological characteristics. The oocyte needs to be centrally located surrounded by tightly layered granulosa cells with no obvious gaps between them. To maintain intact follicles that will grow well over the culture period, a small clump of thecal-interstitial tissue needs to be attached to the basement membrane. At dissection, this thecal layer tends to be very ragged (Figure 1.4A), but within a day or two of culture, thecal cells from the very edge die, and a smooth, even layer of theca remains round the follicle.

Protocol 1.5. Media for in vitro maturation of rodent oocytes

Media with serum

To1 l of Waymouth (MB752/1) medium add

2.240 g sodium bicarbonate
75 mg penicillin G potassium salt
50 mg streptomycin
25 mg pyruvic acid sodium salt
5% FBS

Serum-free media

Waymouth (MB752/1) media containing

3 mg/ml BSA
5 μg/ml insulin
5 μg/ml iron-saturated transferrin
50 μM 3-isobutyl methylzanthine

Protocol 1.6. Me]dium for three-dimensional matrix cultures

M199 (Invitrogen) supplemented with

75 μg/ml penicillin
50 μg/ml streptomycin
0.1 mg/ml L-glutamine
3.5 μg/ml sodium pyruvate
10% heat-inactivated donor calf serum

Protocol 1.7. Media for cattle and sheep follicle cultures

Dissection media

100 ml Liebovitz's medium (Invitrogen) with

2 mM sodium pyruvate
2 mM glutamine

3 mg/ml BSA
75 µg/ml penicillin G
50 µg/ml streptomycin

(All chemicals from Sigma.)

Culture media

McCoy's 5a medium with bicarbonate, supplemented with

20 nM HEPES
0.1% BSA
3 mM L-glutamine
100 IU/ml penicillin
0.1 mg/ml streptomycin
2.5 µg/ml transferrin
4 ng/ml selenium
10^{-7} M androstenedione
10 ng/ml insulin

(All chemicals from Sigma.)

References

1. O'Brien, M. J., Pendola, J. K., and Eppig, J. J. (2003). *Biol. Reprod.,* **68**, 1682.
2. Martinovitch, P. N. (1938). *Proc. R. Soc. Lond. B.,* **125**, 232.
3. Eppig, J. J., and Schroeder, A. C. (1989). *Biol. Reprod.,* **41**, 268.
4. Spears, N., Boland, N. I., Murray, A. A., and Gosden, R. G. (1994). *Hum. Reprod.,* **9**, 527.
5. Fortune, J. E., Kito, S., and Byrd, D. D. (1999). *J. Reprod. Fertil. Suppl.,* **54**, 439.
6. Telfer, E. E., Webb, R., Moor, R. M., and Gosden, R. G. (1999). *Anim. Sci.,* **68**, 285.
7. Torrance, C., Telfer, E. E., and Gosden, R. G. (1989). *J. Reprod. Fertil.,* **87**, 367.
8. Cain, L., Chatterjee, S., and Collins, T. J. (1995). *Endocrinology,* **136**, 3369.
9. Gore-Langton, R. E., and Daniel, S. A. (1990). *Biol. Reprod.,* **43**, 65.
10. Roy, S. K., and Treacy, B. J. (1993). *Fertil. Steril.,* **59**, 783.
11. Roy, S. K., and Greenwald, G. S. (1996). *Hum. Reprod. Update,* **2**, 236.
12. Gomes, J. E., Correia, S. C., Gouveia-Oliveira, A., Cidadao, A. J., and Plancha, C. E. (1999). *Mol. Reprod. Dev.,* **54**, 163.
13. Nayudu, P. L., and Osborn, S. M. (1992). *J. Reprod. Fertil.,* **95**, 349.
14. Rose, U. M., Hanssen, R. G., and Kloosterboer, H. J. (1999). *Biol. Reprod.,* **61**, 503.
15. Cortvrindt, R., Smitz, J., and Van Steirteghem, A. C. (1996). *Hum. Reprod.,* **11**, 2656.
16. Eppig, J. J., Wigglesworth, K., Dendola, F., and Hirao, Y. (1997). *Biol. Reprod.,* **56**, 976.
17. Boland, N. I., Humpherson, D. G., Leese, H. J., and Gosden, R. G. (1993). *Biol. Reprod.,* **48**, 798.
18. Murray, A. A., Gosden, R. G., Allison, V., and Spears, N. (1998). *J. Reprod. Fertil.,* **113**, 27.
19. Ryle, M. (1969). *J. Reprod. Fertil.,* **20**, 307.
20. Greenwald, G. S., and Moor, R. M. (1989). *J. Reprod. Fertil.,* **87**, 561.
21. Hovatta, O., Wright, C., Krausz, T., Hardy, K., and Winston, R. M. (1999). *Hum. Reprod.,* **14**, 2519.
22. Wright, C. S., Hovatta, O., Margara, R., Trew, G., Winston, R. M., Franks, S., and Hardy, K. (1999). *Hum. Reprod.,* **14**, 1555.
23. Louhio, H., Hovatta, O., Sjoberg, J., and Tuuri, T. (2001). *Mol. Hum. Reprod.,* **6**, 694.
24. Hovatta, O. (2001). *Mol. Cell Endocrinol.,* **169**, 95.
25. Oktay, K., Newton, H., Aubard, Y., Salha, O., and Gosden, R. G. (1998). *Fertil. Steril.,* **69**, 1.
26. Picton, H. M., and Gosden, R. G. (2000). *Mol. Cell Endocrinol.,* **166**, 27.
27. McGee, E., Spears, N., Minami, S., Hsu, S. Y., Chun, S. Y., Billig, H., and Hsueh, A. J. (1997). *Endocrinology,* **138**, 2417.

28. McGee, E. A., Smith, R., Spears, N., Nachtigal, M. W., Ingraham, H., and Hsueh, A. J. (2001). *Biol. Reprod., **64***, 293.
29. van de Sandt, J. J., Schroeder, A. C., and Eppig, J. J. (1990). *Mol. Reprod. Dev., **25***, 164.
30. Boland, N. I., and Gosden, R. G. (1994). *J. Reprod. Fertil., **101***, 369.
31. de la Fuente, R., O'Brien, M. J., and Eppig, J. J. (1999). *Hum. Reprod., **14***, 3060.
32. Abeydeera, L. R., Wang, W. H., Cantley, T. C., Rieke, A., Murphy, C. N., Prather, R. S., and Day, B. N. (2000). *Theriogenology, **54***, 787.
33. Goff, A. K., Yang, Z., Cortvrindt, R., Smitz, J., and Miron, P. (2001). *Reprod. Domest. Anim., **36***, 19.
34. Guler, A., Poulin, N., Mermillod, P., Terqui, M., and Cognie, Y. (2001). *Theriogenology, **54***, 209.
35. Ehrmann, R. L., and Gey, G. O. (1956). *J. Natl. Cancer Inst., **16***, 1375.
36. Chambard, M., Gabrion, J., and Mauchamp, J. (1981). *J. Cell Biol., **91***, 157.
37. Hirao, Y., Nagai, T., Kubo, M., Miyano, T., Miyake, M., and Kato, S. (1994). *J. Reprod. Fertil., **100***, 333.
38. Wu, J., Emery, B. R., and Carrell, D. T. (2001). *Biol. Reprod., **64***, 375.
39. Abir, R., Franks, S., Mobberley, M. A., Moore, P. A., Margara, R. A., and Winston, R. M. (1997). *Fertil. Steril., **68***, 682.
40. Telfer, E. E., Binnie, J. P., McCaffery, F. H., and Campbell, B. K. (2000). *Mol. Cell Endocrinol., **163***, 117.
41. Eppig, J. J., and O'Brien, M. J. (1998). *Theriogenology, **49***, 415.
42. Vanderhyden, B. C., and MacDonald, E. A. (1998). *Biol. Reprod., **59***, 1296.
43. Hubner, K., Fuhrmann, G., Christenson, L. K., Kehler, J., Reinbold, R., de la Fuente, R., Wood, J., Strauss, J. F. 3rd, Boiani, M., and Scholer, H. R. (2003). *Science, **300***, 1251.
44. Murray, A., and Spears, N. (2001). *Semin. Reprod. Med., **18***, 109.

2

MICHELLE LANE
DAVID K. GARDNER

Preparation of Gametes, in Vitro Maturation, in Vitro Fertilization, and Embryo Recovery and Transfer

1. IMPORTANCE OF OPTIMAL CONDITIONS

The conditions used for the collection and manipulation of gametes and/or embryos has a significant impact on the subsequent developmental potential of the embryo. It is therefore essential when handling gametes/embryos that factors such as temperature, pH, medium composition, and so forth, are always optimized during collection and handling of embryos to minimize stress. For example, mammalian oocytes and embryos should never be exposed to a medium lacking amino acids (see below). This care to minimize stress is necessary if the handling period is for a very short time (less than 5 min) or chronic (several hours), as the developmental competence of the embryos can be affected. Whether the final outcome of an experiment is to produce live young or measure the physiology or genetic make up of an embryo, the conditions that the embryo is exposed to during the manipulation period are crucial.

2. HANDLING OF GAMETES AND EMBRYOS

2.1 Composition of Handling Media

The formulation of the handling media is an important consideration and can impact subsequent developmental competence of the embryo. Handling media require a buffering system that maintains pH on the bench. We recommend media that are HEPES or MOPS based rather than phosphate based. The use of PBS should be avoided, as there is a large body of literature demonstrating the inhibitory effect of phosphate on embryos even at very low levels in the micromolar to nanomolar range (1–8). To further reduce any possibility of inducing stress to the embryo, the handling media formulations should be HEPES/MOPS-buffered modifications of the appropriate culture media (i.e., the formulations are tailored to support the physiology of the particular stage of development, such as precompaction or postcompaction; (see chapter 3).

Regardless of the stage of development or the species of gametes and embryos, Eagle's nonessential amino acids (alanine, asparagine, aspartate, glutamate, glycine, proline, serine) are an essential component of handling media. These amino acids have an essential role in maintaining the physiology of gametes and embryos by acting as osmolytes, pH buffers, energy substrates, and chelators (9–11). Even a short-term exposure (5 min) of embryos to a medium lacking these seven amino acids has a significant detrimental effect on subsequent development (Figure 2.1). Therefore, these 7 amino acids should always be present in all formulations for the handling and manipulation of gametes and embryos for all species (see Tables 2.1–2.3) whether the collection and handling period is short term (few minutes) or longer term (several hours).

Handling media should be used at a pH of 7.2–7.3, as this is the physiological intracellular pH for mammalian embryos. This is particularly important for the manipulation of oocytes, as they appear to lack any active pH transport systems (12–14). Handling media are usually supplemented with serum albumin (BSA, HSA) or a synthetic macromolecule such as PVA. The use of serum should be avoided due to its detrimental actions on embryo physiology (10, 15–18).

2.2 Quality Control

The quality of the media used for handling gametes and embryos depends on the quality of the components used to prepare the media. Therefore, for all chemicals and contact supplies it is advisable to employ a rigorous quality assurance/quality control (QA/QC) program before introducing a component into the embryo culture system (see chapter 3).

2.3. Temperature

For best results, gametes and embryos should be handled at 37°C for mice and humans or 38.5°C for cows. The temperature of the warm plates and stages should be calibrated to reflect the temperature in the dish and not taken directly from the stage, as the temperature within the dish is usually several degrees cooler. All warm stages should be checked on a weekly basis. For more detailed description of QC/QA, see chapter 3.

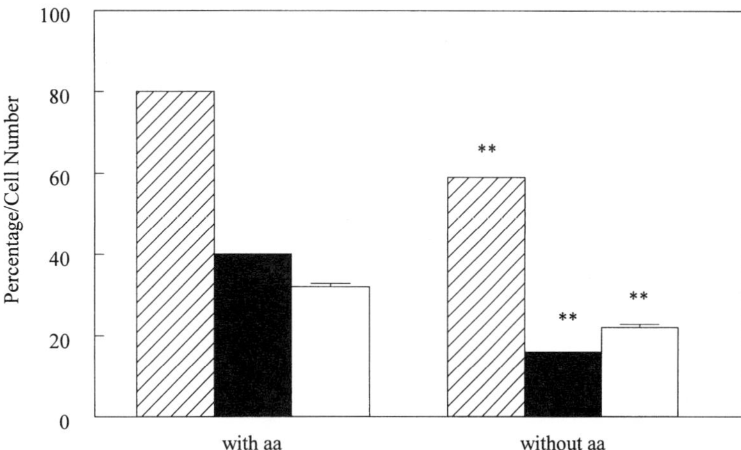

Figure 2.1. Effect of the absence of amino acids (aa) in the collection medium on subsequent embryo development. Hatched bars represent morula development after 72 h of culture; filled bars represent blastocyst development after 96 h of culture; open bars represent blastocyst cell numbers (mean ± SEM). **Significantly different from medium with amino acids ($p < .01$).

Table 2.1. Timetable for Embryo Collection

Day of PMSG Injection	Day of hCG Injection	Zygote, 10 AM, Day of Plug	2-Cell, 11 AM, Day 2	8-Cell, 6–8 AM, Day 3	Morula, 3–4 PM, Day 3	Blastocyst, 8–9 AM, Day 4
Monday	Wednesday	Thursday	Friday	Saturday	Saturday	Sunday
Tuesday	Thursday	Friday	Saturday	Sunday	Sunday	Monday
Wednesday	Friday	Saturday	Sunday	Monday	Monday	Tuesday
Thursday	Saturday	Sunday	Monday	Tuesday	Tuesday	Wednesday
Friday	Sunday	Monday	Tuesday	Wednesday	Wednesday	Thursday
Saturday	Monday	Tuesday	Wednesday	Thursday	Thursday	Friday
Sunday	Tuesday	Wednesday	Thursday	Friday	Friday	Saturday

The light cycle in the mouse room is 14 h daylight and 10 h dark. Daylight is from 5 AM to 7 PM. The room temperature is 21° ± 2°C. Hormone injections are administered between 12 and 1 PM.

3. PROTOCOLS FOR IVM, IVF, AND EMBRYO COLLECTION

This section is divided into three parts: (1) collection of mouse gametes and embryos, (2) IVM and IVF protocols, and (3) media formulations and preparation. Detailed embryo culture protocols are provided in chapter 3.

3.1 Superovulation Protocols for Mice

To increase the numbers of embryos available for an experiment and to reduce the numbers of animals that are required, it is common to induce multiple ovulations in mice by administering hormones (Table 2.1). Superovulation protocols also assist in the synchronization of embryo timing between females. Best results from superovulation protocols are achieved by using prepubertal females (3–5 weeks old). Two hormones are administered by intraperitoneal injection 48 h apart. The first injection of pregnant mare's serum gonadotrophin (PMSG) provides FSH that recruits follicles for ovulation. This is

Table 2.2. Formulations of Media for Manipulation of Mouse Embryos

Component	Handling Medium (mM)	Mouse Fertilization Medium (mM)
NaCl	90.5	100.5
KCl	5.5	5.5
NaHPO$_4$	0.5	0.5
MgSO$_4$	1.0	1.0
NaHCO$_3$	2.0	25.0
CaCl$_2$	1.8	1.8
Lactate	10.5	10.5
Pyruvate	0.35	0.35
Glucose	0.5	3.15
Alanyl-glutamine	0.5	0.5
Alanine	0.1	0.1
Aspartate	0.1	0.1
Asparagine	0.1	0.1
Glutamate	0.1	0.1
Glycine	0.1	0.1
Proline	0.1	0.1
Serine	0.1	0.1
Taurine	0.1	0.1
MOPS/HEPES	23.0	—
Glutathione	—	1.0 g/l
Phenol red	0.005g/l	0.005 g/l

For protocols on how to prepare media, see instructions in chapter 3.

Table 2.3. Formulations of Media used in Bovine IVM/IVF

Component	Modified Fert-Talp	Sperm-Talp	HEPES/MOPS-Fert-Talp
NaCl	114.0	100.0	114.0
KCl	3.2	3.1	3.2
NaHPO$_4$	0.4	0.3	0.4
MgSO$_4$	0.5	1.2	0.5
NaHCO$_3$	25.0	25.0	5.0
CaCl$_2$	2.6	2.6	2.6
Lactate	10.0	21.0	10.0
Pyruvate	0.25	1.0	0.25
Glucose	—	—	—
Alanine	0.1	—	0.1
Aspartate	0.1	—	0.1
Asparagine	0.1	—	0.1
Glutamate	0.1	—	0.1
Glycine	0.1	—	0.1
Proline	0.1	—	0.1
Serine	0.1	—	0.1
MOPS/HEPES	—	10.0	20.0
Phenol red	0.001	0.001	0.001

followed 48 h later by an injection of human choronic gonadotrophin (hCG), which mimics the LH surge and results in ovulation. The injection interval should be kept between 46 and 48 h. It is essential that the hCG injection is given before the endogenous LH surge, which occurs around 15–20 h after the middle of the second dark cycle. Females are then placed with males and allowed to remain overnight. Mating can be detected by the presence of a vaginal plug the next morning. Fertilized embryos can then be collected (see below). Response to superovulation procedures vary significantly between strains of mice. We typically give a dose of 5 IU per mouse for mice 3–4 weeks old.

Hormone solutions can be prepared by reconstituting lyophilized powder in sterile saline at a concentration of 50 IU/ml. For most prepubertal mice this results in a injection of 0.1 ml per mouse. Hormones can be stored for 2–3 weeks at –20°C or at –80°C for 2–3 months. There are significant differences in the ability of different batches of hormones, so each new lot should be prescreened before use.

3.2 Collection of Embryos

In mice (and other rodents), early cleavage stage embryos can be collected from the oviduct of mated animals by flushing the oviduct through the infundibulum with handling media. In addition, for mouse embryos, ovulated oocytes and zygotes (up to around 15 h post-hCG) can be collected by tearing the oviduct in the ampullary region. At this stage the ampulla is swollen and the embryos are under positive pressure and are therefore expelled from the oviduct when the wall of the ampulla is breached.

A timetable for the collection of different stage mouse embryos is provided in Table 2.1. Details for the dissection of the reproductive tracts and the collection of zygotes, two-cell, eight-cell/morula, and blastocyst stage embryos is provided in Protocols 2.1–2.4.

3.3 Isolation and Maturation of Germinal Vesicle Stage Mouse Oocytes

Immature germinal vesicle stage oocytes are obtained when mice are around 22–26 days of age. It has been reported that using mice between 4 and 5 weeks old results in a low yield of quality oocytes (19). Oocytes surrounded by granulosa cells are aspirated from ovaries 44–46 h after PMSG injection. Ovaries are removed from the female in warmed handling media. The cumulus-enclosed oocytes can then aspirated from the follicles, washed, and placed into maturation media to mature overnight (see Protocols 2.5 and 2.6).

Once mouse oocytes are removed from the follicles, the zona hardens and becomes impenetrable by sperm. Therefore it is necessary to either include serum in the media for collection or add fetuin (1 mg/ml) to media without serum.

3.4 In Vitro Fertilization of Mouse Oocytes

Mouse oocytes are fertilized by sperm at around 13–15 h post-hCG or after 16–17 h of in vitro maturation (Protocol 2.7). The success of IVF in the mouse is highly dependent on the strain of male mouse. For example, sperm from F_1 hybrids fertilize oocytes at high rates (range of 80–90%), but sperm from CF1 males fertilize oocytes at very low rates (range of 10–30%), regardless of the strain of mouse that the oocytes originated from.

3.5 Collection, Aspiration, Maturation, and Fertilization of Bovine Oocytes

It is common to obtain slaughterhouse material for the generation of IVM/IVF/IVC (in vitro maturation/in vitro fertilization/in vitro culture) bovine embryos. The protocols used in our laboratory for the collection, maturation, and fertilization of bovine oocytes are provided in Protocols 2.8–2.10.

3.6 Embryo Transfer in Mice

In mice, embryos are transferred to pseudopregnant recipients. Pseudopregnant recipients are obtained by mating females to vasectomized males. Cleavage-stage embryos are usually transferred to the oviduct and blastocyst-stage embryos to the uterus. Depending on the desired result of the embryo transfer, the day of the recipient (day of pregnancy) can vary. Usually, for experiments involving determining the relative health of an embryo after culture, it is advisable to use synchronous recipients, as this will be the strictest test of a comparison of the embryo's development compared to in vivo-developed embryos. However, if the endpoint of the embryo transfer is to produce off-spring (e.g., after transgenic manipulation), it is preferable to use –1 (for cleavage stages and blastocysts) or –2 (for blastocysts) asynchronous recipients.

3.6.1 Vasectomies

Males are vasectomized by removing the vas deferens. Males are anesthetized and a portion of the vas deferens removed (Protocol 2.11). Males are allowed to recover from the surgery for 2–3 weeks and then they are mated to superovulated females, and on day 2 after plugging the oviducts are flushed. The presence of oocytes rather than two-cell embryos indicates that the surgery was successful.

3.6.2 Preparation of Pseudopregnant Recipients

Females at 8–12 weeks of age are placed with the vasectomized males, and the next morning mating is checked by the presence of a vaginal plug. To increase the chances of the females cycling and being mated, the status of the vagina can be examined to select females (20). The day of plug is designated day 1 of pseudopregnancy.

3.6.3 Embryo Transfer

Protocols for oviduct and embryo transfers are provided in Protocol 2.12. Due to the separated cervix in the mouse, it is possible to transfer embryos from different treatments to each uterine horn. This has the advantage of removing any variation in results that may result from different recipient females. When embryos from two different treatments are transferred to the recipient female, it is common to assess implantation and fetal development between days 14 and 18.

The strain of mice used as the recipient female can have a bearing on the outcome of the transfer procedure. If the endpoint of the embryo transfer is to produce live young, it is very important to use strains of mice that will make good foster mothers. In our experience F_1 hybrids (C57BL/6 × CBA) and CF1 females make good recipients and foster mothers.

CONCLUSION

The conditions under which embryos/gametes are collected and manipulated can have a significant effect on the subsequent developmental competence of the embryo. Therefore, conditions should aim to minimize stress to the gametes and embryos.

Protocol 2.1. Dissection of reproductive tracts

Procedure

1. Bring the mated female mice into the dissection area.
2. Place a 35-mm petri dish containing 2–3 ml of handling medium on a warm stage at 37°C adjacent to the dissection area.
3. Disinfect instruments with 70% ethanol.
4. Sacrifice the mouse humanely by cervical dislocation and place it ventral side up on absorbent paper.
5. Disinfect the abdomen with 70% ethanol.
6. Cut the peritoneum to expose the body cavity.
7. Move the coils of gut to one side to expose the reproductive tract.
8. Hold the utero-tubal junction with watchmaker's forceps, and, using the point of a pair of scissors, blunt dissect strip the surrounding connective tissue from the uterus.
9. For collection of the oviduct, make a cut between the oviduct and the ovary (avoid cutting the ovary). For collection of zygotes and two-cell embryos, make a second cut below the utero-tubal junction to isolate the oviduct from the uterus. For collection of eight-cell embryos, make a second cut around one-third of the way down the uterine horn. Place the dissected tissue into the collection dish of handling medium (see Table 2.2).
10. For collection of the uterus, make a cut below the utero-tubal junction of both uterine horns to isolate the uterus from the oviduct. Make a second cut through the cervix and place the connected uterine horns into the collection dish of handling medium.

Protocol 2.2. Collection of mouse ovulated oocytes and zygotes

Preparation

Before collection of the oviducts, warm 10 ml of handling medium and a 1-ml aliquot of hyaluronidase (1mg/ml) to 37°C.

Prepare pulled Pasteur pipettes with an internal volume of >100 μm for manipulation of the embryos. A pipette that is just larger than the embryos is essential to ensure adequate washing and minimal transfer of medium between drops.

Procedure

1. Place approximately 500 µl of handling medium into a new collection dish, either a 35-mm petri dish or organ-well dish.
2. Transfer an oviduct to the new collection dish, one at a time.
3. Stabilize the oviduct with fine forceps and locate the swollen ampulla.
4. Tear the ampulla of the oviduct to release the cumulus mass. Discard the oviduct and transfer a new oviduct to the collection dish. Repeat procedures 3–4 until the cumulus masses from each oviduct are released (Figure p2.1).
5. Add the 500 µl of warmed hyaluronidase solution (1 mg/ml, 1100 U/ml) to the dish to remove cumulus cells. Expose for 20–60 s and collect embryos as soon as they become denuded.
6. During the exposure of the cumulus complex to hyaluronidase, set up three washing drops of handling medium (approximately 50 µl) in a collection dish or the lid of a collection dish.
7. Wash denuded embryos through the three drops of handling medium.
8. Embryos are now ready to be placed into culture.

Embryos from some strains are very susceptible to their environment. Therefore, it is important to minimize the time from extraction of the cumulus masses and placement in culture. It is advisable to keep this time period to a maximum of 10 min. Therefore, it is preferable to do multiple smaller embryo collections rather than one large collection.

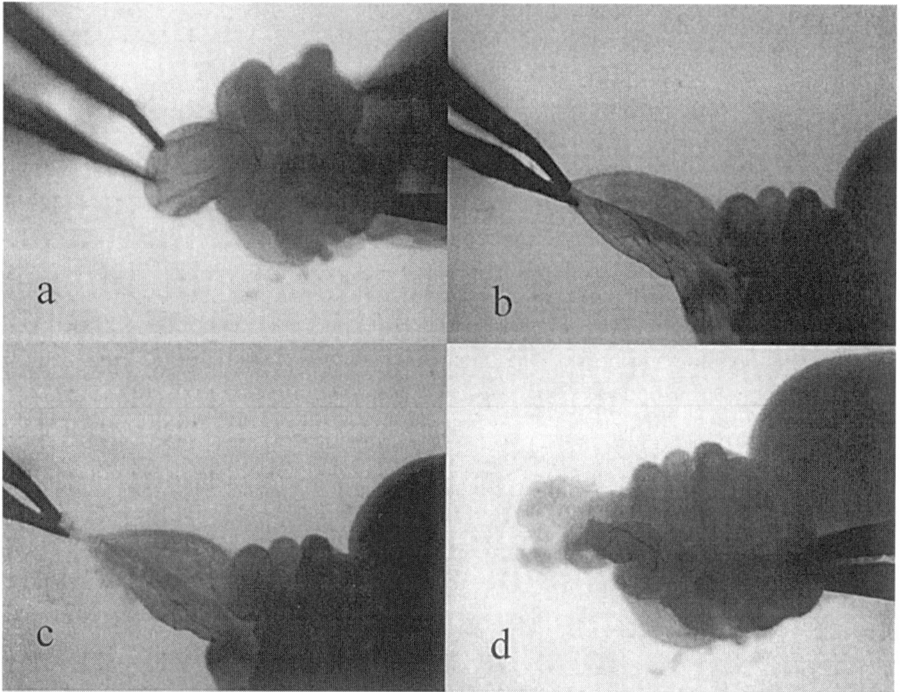

Figure p2.1. Collection of mouse zygotes from the ampullary region of the oviduct: (a) The swollen ampulla containing the cumulus-oocyte complex is located. (b) Using a pair of watchmaker's forceps the ampullary region of the oviduct is grabbed. (c) The ampulla is then torn with the watchmaker's forceps. (d) The cumulus-oocyte complex is released into the collection medium.

Protocol 2.3. Collection of mouse two-cell and eight-cell embryos

Preparation

Before collection of the oviducts, warm 10 ml of handling medium to 37°C.

Prepare pulled pasteur pipettes with an internal volume of >100 μm for manipulation of the embryos. A pipette that is just larger than the embryos is essential to ensure adequate washing and minimal transfer of medium between drops.

Attach a sterile 1-ml syringe containing handling medium to a 32–35 gauge blunted needle.

Procedure

1. Place approximately 1 ml of handling medium into a new collection dish (either a 35-mm petri dish or organ well dish).
2. Transfer an oviduct to the new collection dish, one at a time.
3. Stabilize the oviduct with fine forceps and locate the infundibulum. Gently grasp the neck of the infundibulum with a pair of watchmaker's forceps.
4. Insert the blunted needle into the infundibulum (see Figure p2.2).
5. Gently flush medium through the infundibulum. The coils of the oviduct will swell, and the embryos will flush from the cut uterine horn.
6. Collect and wash embryos in another well of handling medium and place into culture.

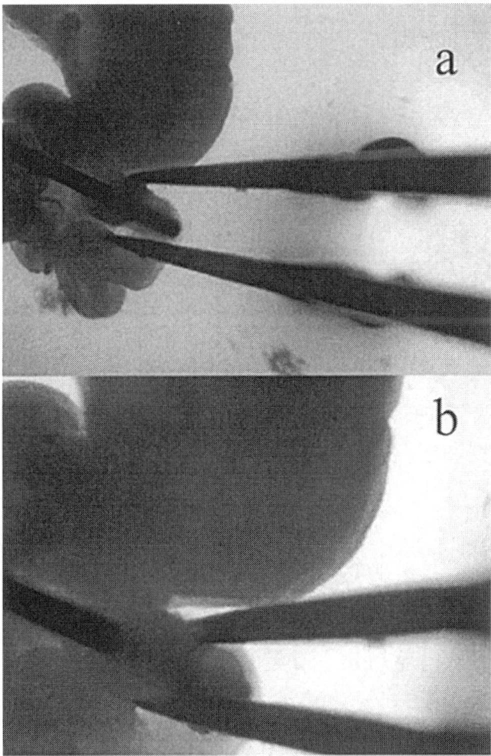

Figure p2.2. Flushing 2-cell embryos from the oviduct. (a) and (b) picture of a 35 gauge needle inserted into the infundibulum of the oviduct. Media is then flushed through the needle and the embryos are flushed from the oviduct into the collection medium.

Protocol 2.4. Collection of mouse blastocysts

Preparation

Before collection of the uteri, warm 10 ml of handling medium to 37°C.

Prepare pulled Pasteur pipettes with an internal volume of >150 nm for manipulation of the embryos. A pipette that is just larger than the embryos is essential to ensure adequate washing and minimal transfer of medium between drops.

Attach a sterile 1-ml syringe containing handling medium to a 32–35 G blunted needle.

Procedure

1. Place approximately 1 ml of handling medium into a new collection dish (either a 35-mm petri dish or organ-well dish).
2. Transfer a pair of connected uterine horns to the new collection dish one at a time.
3. Insert the blunted needle into the cervix and up into one of the uterine horns.
4. Gently flush medium through the uterine horn. The embryos will flush from the cut end of the uterine horn.
5. Repeat by inserting the needle into the other uterine horn and flush medium through this horn.
6. Collect and wash embryos in another well of handling medium.

Protocol 2.5. Collection and aspiration of mouse germinal vesicle stage oocytes

Preparation for processing ovaries

Before collection of the ovaries, warm 20 ml of HEPES/MOPS-maturation medium (table 2.3) to 37°C.

Attach a sterile 27-gauge needle to 1-ml syringe.

Prepare pulled Pasteur pipettes with an internal volume of about 500 μm for manipulation of the cumulus-enclosed oocytes complexes. A pipette that is just larger than the complexes is essential to ensure adequate washing and minimal transfer of medium between drops.

Procedure

1. Place 2–3 ml of HEPES/MOPS-maturation medium into a 35-mm petri dish on a heated stage at 37°C.
2. Using watchmaker's forceps, isolate the ovary and remove ovary from bursa.
3. Place the dissected ovary into the warm HEPES/MOPS maturation medium.
4. After all ovaries are collected, aspirate the cumulus-oocyte complexes under a stereo microscope.
5. Place ovaries into a fresh dish containing 2–3 ml of warmed HEPES maturation medium. Release cumulus oocytes complexes from the antral follicles by puncturing the follicles with the needle.
6. When all of the antral follicles have been aspirated, collect the cumulus-oocyte complexes using the pulled Pasteur pipette and place them into a fresh dish of HEPES maturation medium.

Protocol 2.6. Maturation of mouse germinal vesicle stage oocytes

Preparation of Maturation dishes

Germinal vesicle stage oocytes are matured overnight for approximately 16–17 h at 37°C in 6% CO_2:5%O_2:89%N_2.

The morning of culture, dishes containing maturation medium are prepared. Maturation can occur either in 35-mm petri dishes containing 2–3 ml of maturation medium or in organ-well dishes containing 800 μl–1 ml of maturation medium covered with oil. The benefits of using the organ-well dishes are that changes in pH and evaporation can be minimized by the use of the oil overlay. However, due to the smaller surface area, a smaller number of cumulus oocytes complexes can be matured in the organ well dishes. Dishes should be equilibrated in the incubator for a minimum of 3 h.

For each maturation culture dish, prepare a separate wash dish. An organ-well dish containing maturation medium and overlaid with oil is preferable, as this reduces the changes in pH that may occur during washing of the cumulus-enclosed oocytes complexes.

Procedure

1. Using a pulled Pasteur pipette, transfer cumulus-enclosed oocyte complexes from the HEPES/MOPS handling medium to a wash dish containing maturation medium.
2. Wash cumulus-enclosed oocytes complexes by swirling the wash dish.
3. Transfer the washed cumulus-enclosed oocytes complexes to the maturation dish. It is essential that the cumulus-enclosed oocytes complexes are separated in the maturation dish. Therefore, before placing in the incubator, the dish should be shaken gently to enable the oocytes to settle evenly throughout the dish.
4. Place maturation dishes containing the cumulus-enclosed oocytes complexes in the incubator for 16–17 h.

Removal of the oocytes from the ovary results in hardening of the zona pellucida of the oocytes. Therefore, it is essential that either serum or fetuin be included in all of the media used for the isolation and maturation of the oocytes (see Table 2.3).

Protocol 2.7. In vitro fertilization of mouse oocytes

Preparation of animals

Administer PMSG solution to female mice via intraperitoneal injection (5 IU; i.e., 0.1 ml solution per mouse) at 1800 h ± 30 min, 63 h before sperm collection.

Administer hCG solution to female mice via intraperitoneal injection (5 IU; i.e., 0.1 ml solution per mouse) at 1800 h ± 30 min, 48 h after the PMSG injection was administered, 15 h before sperm collection.

Preparation

The day before fertilization, prepare necessary media and dishes. It is advisable for sperm capacitation and washing to use medium that is covered with oil to minimize changes in pH and to avoid evaporation.

For sperm: (per male) 1 × 1000 μl mouse fertilization medium (Table 2.2) under oil in an organ-well dish, equilibrated at 5% O_2, 6% CO_2, 89% N_2.

For washing: (per 5 females) 1 × 800 μl mouse fertilization medium under oil, equilibrated at 5% O_2, 6% CO_2, 89% N_2.

For insemination: (per female) 1 × 90 μl mouse fertilization medium under oil, equilibrated at 5% O_2, 6% CO_2, 89% N_2.

For culture: appropriate culture medium (G1; see Chapter 3) in 20 μl drops under oil, equilibrated at 5% O_2, 6% CO_2, 89% N_2. Set up 2 drops per treatment as well as an additional wash drop.

Procedure

1. On the day of fertilization, sacrifice males 13–15 h after administering the hCG injection to females.

2. Place the male on its back. Spray the abdomen with 70% ethanol and make a cut through the skin and the peritoneum to expose the body cavity.

3. Pull the fat pad up with forceps to expose the testes and the epididymus. Locate the vas deferens and gently blunt dissect the connective tissue away from the vas deferens.

4. Cut the vas deferens around 0.5 cm away from the epididymus. While holding the end of the vas deferens, dissect the epididymus away from the fat pad and the testes and place into a dish of warmed handling medium.

5. Remove the sperm dish containing the mouse fertilization medium from the incubator and place an epididymus with attached vas deferens into the dish. Slide forceps along the vas deferens to remove the sperm through the cut end. Puncture three or four holes in the epididymus to release further sperm from the epididymus. Repeat for the other epididymus and vas deferens. Return to the dish to the incubator for 1 h.

6. After 1 h the sperm should be dispersed and appear hyperactive. When taking sperm from the dish, take only dispersed, motile sperm.

7. To count the sperm for insemination, place 95 µl of water into a 5-ml tube. Using a different tip (as water will immobilize sperm), remove 5 µl of sperm and mix well with the water. Using a fresh tip, load 10 µl of this immobilized sperm into the top and bottom of the hemocytometer chamber. Count 5 squares (usually 4 corners and middle) of both the top and bottom. Average the counts. There should be between 30–40 sperm. This will result in an insemination volume of around 1×10^4 sperm/ml. If this is not the case, the insemination volume can be adjusted to compensate for low sperm numbers.

8. Add 10 µl of sperm (taken from dispersed, motile sperm in the dish) to each 90 µl drop of mouse fertilization medium. When all drops have been inseminated, return the fertilization dishes to the incubator.

9. Ovulated oocytes should be collected using the procedures outlined in Protocols 2.1 and 2.2. For fertilization, cumulus-enclosed oocytes either freshly ovulated or matured in vitro are washed well in warmed handling medium.

10. Wash oocytes once in mouse fertilization medium and place into the fertilization drops containing the sperm, transferring a minimal amount of medium. Place 1–2 cumulus masses per drop or 20–30 matured oocytes. Return the dish to the incubator. Gametes are coincubated for 4 h.

11. After 4 h, oocytes should be denuded and washed well in handling medium. Oocytes can then be placed into culture in medium G1 (see Chapter 3).

Protocol 2.8. Collection, aspiration, and maturation of bovine oocytes

Preparation

Ovaries should be collected from the abattoir in saline. The temperature at which the ovaries are collected and stored for transport should be kept constant. Therefore, depending on the distance and circumstances for the collection of the ovaries, the saline can be warmed to 35–39°C, or, if this is not possible, the saline should be kept at room temperature.

For oocytes collection and maturation, the following media should be prepared:

- 2 l of saline for washing of the ovaries. This should be a similar temperature to the saline used for transport.
- 100 ml of HEPES-TCM-199 (Protocol 2.10) for searching and washing.
- Dishes for maturation. Maturation can occur either in large volumes of 500 µl covered with oil or in 50 µl microdrops under oil. It is preferable to always cover the dishes with oil to minimize changes in pH and osmolality that may occur. Incubate dishes at 38.5°C, in 6% CO_2 in air for a minimum of 4 h.

Procedure

1. Before aspiration, wash ovaries well with saline to remove any blood or other contaminants. Ovaries can be washed with 1% Nolvasan solution (Aveco Co. Inc., Fort Dodge, IA) and 1% 7X (ICN Biochemicals Inc., Costa Mesa, CA) to ensure that there is no transfer of contaminants to the culture.

2. Aspirate cumulus-enclosed oocytes from follicles 2–10 mm in size using a 17–19 gauge needle. The needle can be connected to either a 5-ml syringe (soaked overnight in sterile water to remove any lubricant from the plunger) or to a vacuum pump. It is essential that the pressure on the pump be around 50 mm Hg so that the oocytes are not denuded by the aspiration process.

3. Collect follicular fluid containing the cumulus-enclosed oocytes in 10-ml tubes and place the tubes in the tube warmer at 38.5°C. Leave the tubes for 5 min to allow the cumulus-enclosed oocyte complexes to settle in the bottom of the tube.

4. Locate cumulus-enclosed oocytes from the follicular aspirate. A grid is scratched onto the outer bottom of a 10-mm petri dish using a scalpel blade or needle to assist in the searching. Using a sterile Pasteur pipette, remove the bottom 1 ml of the follicular aspirate containing the sediment from the tube and transfer it to the searching plate. To this sediment add 1–2 ml of HEPES-TCM-199 medium. Add a further 2–3 ml of HEPES-TCM-199 medium to a 35-mm petri dish for collection of the cumulus-enclosed oocyte complexes.

5. Using a stereomicroscope, collect the cumulus-enclosed oocyte complexes and place them into the collection dish. Select only oocytes with 3 or more layers of cumulus cells, which are tightly compacted. Oocytes with expanded cumulus masses or with no or few cumulus layers result in the production of few viable embryos and are therefore not routinely used. Place all of the cumulus-enclosed oocyte complexes that meet these criteria into the second 35-mm dish containing HEPES-TCM-199.

6. Wash cumulus-enclosed oocyte complexes in a well/drop of maturation medium and place them in the wells/drops of maturation medium. For 500 µl wells of maturation medium 30–50 cumulus-enclosed oocytes are cultured in each well. For 50-µl drops 5–10 cumulus-enclosed oocytes are cultured in each drop. Return cumulus-enclosed oocytes to the incubator and mature for 22–24 h at 38.5°C, in 6% CO_2 in air.

Protocol 2.9. In vitro fertilization of bovine oocytes and preparation for culture

Preparation

1. Media required for in vitro fertilization of bovine oocytes: 10 ml of Sperm Talp (modified Tyrodes solution), 10 ml of modified Fertilization (fert) Talp, and 30 ml of fert HEPES medium. (Formulations for media are shown below.)

2. Dishes for fertilization. Fertilization can occur either in large volumes of 500 µl covered with oil or in 50-µl microdrops under oil. It is preferable to always cover the dishes with oil to minimize changes in pH and osmolality that may occur. Dishes are incubated at 38.5°C, in 6% CO_2 in air for a minimum of 4 h. An additional 1 ml of modified fert Talp is placed in the incubator to equilibrate for resuspension of the final sperm pellet.

3. Place sperm Talp and fert HEPES in a tube warmer to warm to 38.5°C.

Procedure

1. Oocytes are fertilized 22–24 h after being placed into maturation medium. Preparation of the washed sperm by swim-up should begin around 1.5 h before the insemination time. Place 1 ml of warmed sperm Talp into 4.5-ml tubes. Thaw

out the appropriate number of semen straws. Remove straws from the liquid nitrogen storage tank and place them into a 37°C waterbath for 1 min. Dry straws with a Kimwipe and cut one end of the straw with sterile scissors. Hold the cut end of the straw over an empty tube and cut the plugged end of the straw and collect the semen into the tube. Using a pipette, gently layer 200 µl of the semen under each 1 ml of sperm Talp. Cap the tubes and place them into the 37°C waterbath for 1–1.5 h.

2. During this time, transfer cumulus-enclosed oocytes from the maturation dishes to the fertilization dishes. Transfer cumulus-enclosed oocytes from the maturation medium into a 35-mm petri dish containing 2–3ml of fert HEPES medium warmed to 38.5°C. Gently swirl the dish to wash the oocytes. Repeat the wash.

3. Remove fertilization dishes from the incubator and wash the cumulus-enclosed oocytes once in modified fert Talp and then transfer them to the fertilization drops. For 500-µl wells of modified fert Talp medium, 30–50 cumulus-enclosed oocytes are cultured in each well. For 50-µl drops, 5–10 cumulus-enclosed oocytes are cultured in each drop. Return fertilization dishes to the incubator.

4. After 1 h, remove the tubes containing the sperm swim-up from the waterbath and remove the upper 800 µl from each swim-up tube and transfer it to a centrifuge tube. Spin the tube in a centrifuge at 700 G for 5 min.

5. Remove the tube from the centrifuge and remove the supernatant down to the pellet. Add 3 ml warm sperm Talp and spin for an additional 5 min at 700 G. After the spin, again remove the supernatant, leaving the pellet in a small volume.

6. Resuspend the pellet using a pipette. Remove 5 µl from the pellet and mix with 95 µl of water in a clean tube. Load 10 µl of this dilution into a hemocytometer. Count the number of sperm in 5 squares. Divide 7500 by this count. Add this to the volume of the sperm pellet to a fresh tube and make up the volume to 300 µl using the equilibrated modified fert Talp. For fertilization drops of 50 µl, the insemination volume is 2 µl for the 50-µl drops and 20 µl for the 500-µl wells of medium.

7. Remove fertilization dishes from the incubator. To the fertilization drops containing the cumulus-enclosed oocytes, add sperm and heparin at the optimal concentration for the bull to be used. Heparin concentrations usually range between 1 and 5 µg/ml of heparin and sperm concentrations range between 0.5 and 2.0×10^6/ml. These should be optimized for each bull. Return dishes to the incubator and coincubate gametes at 38.5°C, in 6% CO_2 in air for 18 h.

Preparation for Culture

1. The presumptive zygotes should be placed into culture approximately 18 h after insemination. Embryo culture is done at 38.5°C in 6% CO_2 and 5% O_2, balance N_2 (see chapter 3 for details on preparation of culture media and culture dishes). Warm a flask containing 50 ml of handling medium to 38.5°C.

2. Place 2 ml handling medium into a 15-ml centrifuge tube. Remove the fertilization plates from the incubator. Transfer embryos from the fertilization dishes to the handling medium. Vortex embryos for 2 min.

3. Place the centrifuge tube into tube warmer and let it sit for 2 min to allow the embryos to settle to the bottom of the tube.

4. Using a Pasteur pipette, collect the bottom 1–2 ml from the bottom of the tube and place it into a 35-mm dish. Transfer denuded zygotes to a fresh dish containing 2–3 ml of handling medium using a pipette only slightly larger than the embryos. Embryos are now ready to be placed into culture (see chapter 3).

Protocol 2.10. Preparation of maturation media

Preparation of HEPES TCM-199/MEM

1. Measure out 1 l of sterile 18Ωohm water.
2. Add 1 bottle of HEPES-TCM-199/MEM to the 1 l of water; rinse the bottle three times.
3. Add 2.102 g $NaHCO_3$.
4. Add 0.0275 g sodium pyruvate.
5. Add 0.06 g penicillin.
6. Add 4 g BSA.
7. Add 10 ml of this medium to a tube containing 0.8 g of NaOH. Using this 2 M NaOH solution, adjust pH of the medium to 7.30 ± 0.05.
8. Filter through a 0.2-μm filter.

Maturation TCM-199/MEM stock solution

1. Measure out 1 l of sterile 18Ωohm water.
2. Add 1 bottle of TCM-199/MEM to the 1 l of water; rinse the bottle three times.
3. Add 0.420 g $NaHCO_3$.
4. Add 0.0275 g sodium pyruvate.
5. Add 0.06 g penicillin.
6. Filter through a 0.2-μm filter.

Preparation of maturation medium

9.0 ml Mat-TCM-199/MEM
1.0 ml/0.04 g FCS/BSA
10 μl Epidermal growth factor (10–100 ng/ml)
10 μl Hormones (FSH)

Filter through a 0.2-μm low protein binding filter.

Protocol 2.11. Vasectomized males for producing pseudopregnant recipients

Media

Stock solution: 0.5 ml butan-2-ol, 0.5 g 2,2,2-tribromoethanol. Store in dark at room temperature.
Working solution: 10 ml saline, 120 μl stock solution. May need to warm to 37°C to dissolve. Store in dark at room 4°C.
Dose: 0.02 ml/g body weight.

Vasectomy

1. Anesthetize the male and place him ventral side up on the microscope stage. Avertin can be used for mice at a dose of 0.02 ml/g body weight.
2. Spray the area between the anus and penis with 70% alcohol.
3. Make an incision about 2–3 mm long ventrally along the scrotal sac below the penis through the body wall and peritoneum.
4. Locate the bursa sac containing the testes and gently tear a hole at the bottom of the sac near the epididymus.
5. Locate the epididymus and pull through the hole in the bursa.
6. Locate the vas deferens, which runs from the epididymus up the length of the testes. The vas deferens can be distinguished because it runs along side a blood vessel.
7. Using a pair of forceps the vas deferens can be blunt dissected away from the testes and the blood vessel. Dissect away a length of vas deferens at least 1–2 cm long.

8. Cut the vas deferens away from the epididymus and then cut the vas deferens again at least 1 cm from the original cut. Remove the piece of vas deferens.
9. Repeat steps 4–8 for the other testes.
10. Using absorbable suture, stitch the body wall together, and then, using either absorbable suture or silk, stitch the skin together.
11. Place the mouse on the warm stage until he begins to wake up.
12. Observe the mouse until fully recovered from the anesthetic and give analgesics.

Checking for sterility

1. Allow males 2–3 weeks to recover from the operation.
2. Superovulate two females for each male. After the hCG injection, place two females with each male.
3. The following morning check for vaginal plugs.
4. On day 2 flush the oviducts of the mated females using procedure in Protocol 2.3. The presence of unfertilized eggs in females which had definite vaginal plugs indicates that the males are sterile and can be used to obtain pseudopregnancies.

Protocol 2.12. Embryo transfer

Preparation of pseudopregnant recipients

Ideally females should be 8–12 weeks of age.

1. Place two mature females with the vasectomized males the evening before the desired day 1 of pregnancy. The number of plugs can be increased by assessing the stage of the estrous cycle by the appearance of the vagina (20).
2. The following morning assess mating by the presence of a vaginal plug. This is designated day 1 of pseudopregnancy.

Transfer media

Embryos should be transferred in a HEPES/MOPS modification of the culture medium, as the time taken to load and transfer the embryos would result in a significant loss in CO_2 from bicarbonate-buffered medium and a rise in pH. Immediately before transfer, place the embryos into 800-μl drops (in an organ well or 4-well plate) of HEPES/MOPS media that has been warmed to 37°C. Place the dish containing the embryos onto a warm stage.

Isolation of reproductive tract

1. Anesthetize females and lay them face down on the microscope stage. Avertin can be used for mice at a dose of 0.02 ml/g body weight.
2. Spray the back of the mouse with 70% ethanol.
3. Using forceps, make a part in the hair of the mouse ventrally down the midline of the back. This part should be around 3–4 cm long and adjacent to the top of the hips of the mouse.
4. Make an incision along the part through the skin to expose the body cavity. The middle of this incision should be adjacent to the hips of the mouse. This will ensure that the reproductive tract is easy to find. Wipe the incision with 70% ethanol to remove any loose hairs.
5. Using a pair of watchmaker's forceps, blunt dissect the body cavity from the skin on the left side of the mouse. Continue until the opening in the skin can be moved over the area of the ovary and fat pad.
6. Move the opening in the skin over the ovary. The ovary can often be visualized as a pink area through the peritoneum. If the ovary is visible through the peritoneum, then move the opening to the left of the back muscles and make the incision adjacent to the hips on the mouse. Make an incision into the peritoneum.

7. Using watchmake's forceps, grab the fat pad and pull the ovary, oviduct, and beginning of the uterus out of the body wall. Clip the fat pad to the back of the mouse. Do not touch the ovary during this procedure, as this can cause the ovary to bleed and makes the subsequent surgery very difficult (if an oviduct transfer) and can affect the outcome of the transfer procedure.

8. Cover the reproductive tract with a sterile gauze soaked in sterile saline.

9. Load embryos into the transfer pipette. This is a key factor in the success of the transfer procedure. The pipette should have a lumen just bigger than the embryos to be transferred and be sturdy enough to insert into the reproductive tract without bending. Additionally, the pipette should have a blunt end that is polished by placing the end of the pipette briefly into a yellow flame to smooth the edges. Break the capillary force of the pipette by filling the pipette intermittently with air bubbles and medium. Place a large air-bubble in the pipette and load the embryos in a small volume immediately after the air bubble. Once all of the embryos are in the pipette take up a further air bubble into the end of the pipette. The embryos are then loaded in <1 μl between two air bubbles. Usually 5–6 embryos are transferred to each side of the reproductive tract if the outcome of the experiment is to examine the fetuses on days 14–18 and 3–4 embryos per side if the outcome is to generate offspring, as this size litter is more easily fed and cared for by the foster mother (this depends on the strain of mouse used as the recipient).

Oviduct transfer

For cleavage-stage embryo transfer, the embryos are transferred to the oviduct on day 1 pseudopregnant recipient. The embryos are transferred through the infindibulum of the oviduct.

10. Once the reproductive tract has been isolated as above, locate the infindibulum of the oviduct through the bursa covering the ovary and oviduct. Using watchmaker's forceps, gently tear the bursa over the infindibulum. Take care to avoid tearing any blood vessels that run through the bursa, as this will result in the bursa, filling with blood. If this happens, flush the area with sterile saline and pat dry with sterile gauze.

11. Using watchmaker's forceps, grab the neck of the infindibulum.

12. Insert the pipette into the infindibulum and blow in the embryos. The amount of fluid to blow into the ampulla can be followed by watching the air bubbles in the pipette. (When this procedure is first being practiced, it is useful to blow a colored dye into the ampulla to ensure that the pipette has been correctly inserted into the infindibulum).

13. Check the pipette to ensure that all of the embryos have been expelled.

14. Grab the left edge of the incision in the peritoneum and hold open with forceps. Using watchmaker's forceps, hold on to the fat pad only and ease the tract back into the body cavity.

15. To transfer to the right side, repeat steps 5–9. Turn the female around so that she faces toward you and then repeat steps 10–14.

16. At the end of the procedure, close the incision in the skin using either wound clips or suture.

13. Place the mouse on the warm stage until she begins to wake up.

17. Observe the mouse until fully recovered from the anesthetic and give analgesics.

Uterine embryo transfer

Later stage embryos can be transferred to the uterus of pseudopregnant recipients. Transfers are usually performed on the afternoon of day 3 (day 3 pseudopregnant recipients) or the morning of day 4 (day 4 pseudopregnant recipients). The embryos are transferred directly to the uterus of the recipient.

10. Using watchmaker's forceps, grab the reproductive tract at the utero-tubal junction. Pick up a 27-G needle attached to a 1-ml syringe and also the pipette with the embryos.
11. Pull the uterus taut.
12. Using the 27-G needle, make a hole in the uterus around one-third of the way down from the oviduct. Ensure that the hole goes into the lumen of the uterus and not into the muscle walls. Remove the needle and put aside while keeping your eye on the hole.
13. Insert the pipette into the hole made with the needle. Ensure that the pipette is inside the lumen by gently moving the pipette up and down inside the uterus. If there is no resistance, then the pipette is in the lumen.
14. Expel the embryos into the uterus. The amount of fluid to blow into the ampulla can be followed by watching the air bubbles in the pipette.
15. Check the pipette to ensure that all of the embryos have been expelled.
16. Grab the left edge of the incision in the peritoneum and hold open with forceps. Using watchmaker's forceps, hold on to the fat pad only and ease the tract back into the body cavity.
17. To transfer to the right side, repeat steps 5–9. Turn the female around so that she faces toward you and then repeat steps 10–16.
18. At the end of the procedure, close the incision in the skin using either wound clips or suture.
19. Place the mouse on the warm stage until she begins to wake up.
20. Observe the mouse until fully recovered from the anesthetic and give analgesia.

References

1. Barnett, D. K., and Bavister, B. D. (1996). *Hum. Reprod.*, **11**, 177.
2. Barnett, D. K., Clayton, M. K., Kimura, J., and Bavister, B. D. (1997). *Mol. Reprod. Dev.*, **48**, 227.
3. Lane, M., Ludwig, T. E., and Bavister, B. D. (1999). *Mol. Reprod. Dev.*, **54**, 410.
4. Ludwig, T. E., Squirrell, J. M., Palmenberg, A. C., and Bavister, B. D. (2001). *Biol. Reprod.*, **65**, 1648.
5. Schini, S. A., and Bavister, B. D. (1988). *Biol. Reprod.*, **39**, 1183.
6. Seshagiri, P. B., and Bavister, B. D. (1991). *Mol. Reprod. Dev.*, **30**, 105.
7. Quinn, P. (1995). *J. Assist. Reprod. Genet.*, **12**, 97.
8. Quinn, P., Moinipanah, R., Steinberg, J. M., and Weathersbee, P. S. (1995). *Fertil. Steril.*, **63**, 922.
9. Gardner, D. K., and Lane, M. (1997). *Hum. Reprod. Update*, **3**, 367.
10. Gardner, D. K., and Lane, M. (1998). *Hum. Reprod.*, **13 Suppl. 3**, 148.
11. Gardner, D. K., Lane, M., and Schoolcraft, W.B. (2000). *Hum. Reprod.*, **15 Suppl. 6**, 9.
12. Lane, M., Baltz, J. M., and Bavister, B. D. (1999). *Biol. Reprod.*, **61**, 452.
13. Lane, M., Baltz, J. M., and Bavister, B. D. (1999). *Dev. Biol.*, **208**, 244.
14. Phillips, K. P., and Baltz, J. M. (1999). *Dev. Biol.*, **208**, 392.
15. Thompson, J. G., Gardner, D. K., Pugh, P. A., McMillan, W. H., and Tervit, H. R. (1995). *Biol. Reprod.*, **53**, 1385.
16. Gardner, D. K., Lane, M., and Schoolcraft, W. B. (2000). *Hum. Reprod.*, **15 Suppl 6**, 9.
17. Gardner, D. K., and Lane, M. (2003). In: *Biotechnology of Human Reproduction*, A. Reveilli, I. Tur-Kaspa, J. G. Holte, and M. Massobrio, eds., p. 181. The Parthenon Publishing Group, New York.
18. Dorland, M., Gardner, D. K., and Trounson, A. (1994). *J. Reprod. Fertil. Abstr. Ser.*, **13**, 70.
19. Eppig, J. J., O'Brien, M., and Wigglesworth, K. (1996). *Mol. Reprod. Dev.*, **44**, 260.
20. Champlin, A. K., Dorr, D. L., and Gates, A. H. (1973). *Biol. Reprod.*, **8**, 491.

3

DAVID K. GARDNER

MICHELLE LANE

Culture of the Mammalian Preimplantation Embryo

1. EMBRYO CULTURE MEDIA

There are several treatises on the composition of embryo culture media and the role of specific medium components in supporting mammalian embryo development in vitro (1–3). It is the aim of this chapter to describe in detail how such media should be used in the laboratory and to identify the potential pitfalls that can lead to impaired embryo development in culture. Embryos considered here include those from the mouse, human, and domestic animal species.

This chapter will not address the use of co-culture in the production of mammalian preimplantation embryos. There are two reasons for this: first and most important, it has been demonstrated that the use of novel culture media, such as those described within this work, are superior to co-culture systems (4, 5). Second, co-culture typically uses serum as the protein supplement. A developing embryo should never come into contact with serum, as it induces significant abnormalities in the physiology and gene expression of the embryo (6, 7). Finally, the concept of co-culture depends on being able to maintain two totally different cells types in vitro. Due to the idiosyncratic and dynamic physiology of the embryo, this is not feasible (8).

1.1 Media Formulations

Historically, embryo culture media fall into two categories: simple and complex (8). Media classified as simple media, such M16, CZB, and HTF, are derivatives of those designed in the 1960s for the development of the embryos of inbred mice strains (9). The formulations of these media are shown in Table 3.1. Typically, simple media are characterized by their lack of amino acids. A modification of such media is P1 (10), whose composition is also shown in Table 3.1.

Media classified as complex are those used for embryo culture that were actually designed with the requirements of somatic cells in mind (i.e., they are tissue culture media). Examples of such media include Ham's F10, TCM 199, and α-MEM. Such media and their modifications, such as blastocyst medium (Ham's F10 with added

41

Table 3.1. Composition of Simple Embryo Culture Media

Component (mM)	M16	CZB	HTF	P1
NaCl	94.59	81.26	101.6	101.6
KCl	4.78	4.69	4.69	4.69
KH_2PO_4	1.19	1.18	0.37	0.00
$CaCl_2 \cdot 2H_2O$	1.71	1.70	2.04	2.04
$MgSO_4 \cdot 7H_2O$	1.19	1.18	0.20	0.20
$NaHCO_3$	25.0	25.0	25.0	25.0
Na pyruvate	0.33	0.27	0.33	0.33
Na lactate	23.28	31.30	21.4	21.4
Glucose	5.56	0.00	2.78	0.00
Glutamine	0.00	1.0	0.00	0.00
Taurine	0.00	0.00	0.00	0.05
Citrate	0.00	0.00	0.00	0.5
EDTA	0.00	0.11	0.00	0.00

Phenol red is usually added to media at concentrations of 0.005–0.1 g/l. Albumin is usually added in the form of HSA or BSA. BSA preparations should be low lipid.

glutathione [11]), can be used to support the development of the embryo from the late eight-cell stage onward but are not ideally suited for the development of the zygote and cleavage stages.

Since 1990, new embryo culture media have been formulated that do not fit into either category. These media developed from two different design strategies. The first involved the use of computer-based simplex optimization to generate media with increasing effectiveness (12). The medium eventually developed from this approach was KSOM. This medium was subsequently improved by the addition of amino acids, (13), ultimately leading to the formulation of KSOMAA (14). The second strategy was fundamentally different, basing media formulations on three factors: the levels of nutrients within the female reproductive tract, the changing requirements of the embryo, and minimizing in vitro-induced stress as determined by studies on embryo physiology and metabolism. To accommodate the changing levels of nutrients within the female reproductive tract and the dynamics of embryo physiology, two media were formulated. The first medium, G1 (6), was developed to support the zygote and cleavage stage embryo, and medium G2 (15) was developed with the requirements of the postcompaction embryo in mind. The compositions of media G1.2 and KSOMAA are listed in Table 3.2. Table 3.3 details the composition of the media G2.2 and blastocyst medium.

It is essential to appreciate that the culture media are only one part of the overall culture system. Specifically, factors in the laboratory have a significant impact on the performance of culture media, aside from the actual formulations used. Key factors affecting culture medium efficacy are macromolecules, carbon dioxide concentration, oxygen concentration, medium renewal, temperature, pH, type of medium overlay, incubation volume, embryo density per volume, culture vessel, air quality, technical diligence, and quality control. Each of these factors is covered in detail below.

2. COMPONENTS OF AN EMBRYO CULTURE SYSTEM

2.1 Macromolecules

Embryo culture media are typically supplemented with a macromolecule, such as serum or albumin, which can serve both physiological and practical functions. The physiological functions, typically restricted to albumin, include the ability to bind growth factors

Table 3.2. Composition of Media G1.2 and KSOMaa

Component (mM)	G1.2	KSOMaa
NaCl	90.08	95.0
KCl	5.5	2.5
KH_2PO_4	0.00	0.35
$NaH_2PO_4 \cdot 2H_2O$	0.25	0.00
$CaCl_2 \cdot 2H_2O$	1.8	1.7
$MgSO_4 \cdot 7H_2O$	1.0	0.2
$NaHCO_3$	25.0	25.0
Na pyruvate	0.32	0.20
Na lactate	10.5	10.0
Glucose	0.50	0.20
EDTA	0.01	0.01
Alanine	0.1	0.05
Arginine	—	0.3
Asparagine	0.1	0.05
Aspartate	0.1	0.05
Cystine	—	0.05
Glutamate	0.1	0.05
Glutamine	0.5	1.0
Glycine	0.1	0.05
Histidine	—	0.1
Isoleucine	—	0.2
Leucine	—	0.2
Lysine	—	0.2
Methionine	—	0.05
Phenylalnine	—	0.1
Proline	0.1	0.05
Serine	0.1	0.05
Taurine	0.1	0.00
Threonine	—	0.2
Tryptophan	—	0.025
Tyrosine	—	0.1
Valine	—	0.2

Phenol red is usually added to media at concentrations of 0.005–0.1 g/l.
Albumin is usually added in the form of HSA or BSA. BSA preparations
should be low lipid.

and chelate heavy metals (8). The practical function is as a surfactant to facilitate embryo manipulations by preventing them from sticking to glass or tissue culture ware.

Serum has been used for many years in embryo culture systems, both clinically and in domestic animal embryology. However, from a physiological perspective, the mammalian embryo is never exposed to serum in vivo. The fluids of the female reproductive tract are not simple serum transudates (16), but rather specialized environments for the development of the embryo (17). Serum can best be considered a pathological fluid. The use of serum should be avoided due to its detrimental effects on the embryos such as perturbed energy metabolism, ultrastructural damage to the mitochondria (6, 18, 19), and abnormal accumulation of lipids (4, 20, 21). Embryos cultured with serum also have a reduced tolerance to cryopreservation (19). Additionally, in animal models the use of serum in the culture media has been associated with increases in complications during pregnancy and in neonates (18, 22).

Albumin is the most commonly used macromolecule in embryo culture media. Although serum albumin is a relatively pure fraction, it is still contaminated with fatty acids and other small molecules. The latter include an embryotrophic factor, citrate, which stimulates cleavage and growth in rabbit morulae and blastocysts (23). Furthermore, there are significant differences between each serum albumin, therefore, it is essential

Table 3.3. Composition of Media for Later (around the Eight-cell Stage) Preimplantation Embryos

Component (mM)	G2.2	Blastocyst Medium[a]
Alanine	0.1	0.1
Alanyl-glutamine	1.0	—
Arginine	0.6	1.0
Asparagine	0.1	0.1
Aspartate	0.1	0.1
Biotin	—	0.0001
Calcium chloride	1.8	0.3
Calcium lactate	—	1.0
Choline chloride	0.0072	0.005
Copper sulfate	—	0.00001
Cysteine	—	0.2
Cystine	0.1	—
Folic acid	0.0023	0.003
Glucose	3.15	6.1
Glutamate	0.1	0.1
Glutamine	—	1.0
Glutathione	—	1.0
Glycine	0.1	0.1
Histidine	0.2	0.1
i-Inositol	0.01	0.003
Iron sulfate	—	0.003
Isoleucine	0.4	0.02
Leucine	0.4	0.1
Lysine	0.4	0.1
Magnesium sulfate	1.0	1.1
Methionine	0.1	0.03
Nicotinamide	0.0082	0.005
Pantothenate	0.0042	0.003
Phenylalanine	0.2	0.03
Potassium bicarbonate	—	5.0
Potassium chloride	5.5	3.8
Potassium phosphate	—	0.6
Proline	0.1	0.1
Pyridoxine	0.0049	0.001
Riboflavin	0.00027	0.001
Serine	0.1	0.1
Sodium bicarbonate	25.0	20.0
Sodium chloride	90.08	116.6
Sodium lactate	5.87	—
Sodium phosphate	0.25	1.1
Sodium pyruvate	0.10	1.0
Thiamine	0.003	0.003
Thioctic acid	—	0.001
Threonine	0.4	0.03
Thymidine	—	0.003
Tryptophan	0.5	0.003
Tyrosine	0.2	0.01
Valine	0.4	0.03
Vitamin B-12	—	0.001
Zinc sulfate	—	0.0001

[a]Blastocyst medium is a modification of Ham's F10 containing glutathione.

that each batch is screened for its ability to adequately support embryo development. This can be achieved using a mouse embryo bioassay and embryo transfers (3). Albumin has been shown to better maintain embryo physiology and metabolism in vitro compared to embryos cultured in the presence of a synthetic macromolecule such as polyvinyl alcohol (PVA) (24, 25).

Recombinant human serum albumin has become available. Not only can recombinant albumin readily replace blood-derived albumin in an embryo culture system (26, 27), but both mouse and cow embryos cultured in the presence of recombinant albumin exhibit an increased tolerance to cryopreservation (28). The development of recombinant human serum albumin has therefore helped eliminate the problems inherent with using blood-derived products and should eliminate variability.

Albumin is the most abundant protein in the female reproductive tract (16); however, there are other macromolecules, such as mucins, that are present at higher levels than albumin (29). The role of mucins during the preimplantation period has yet to be elucidated, but their role during implantation has been better characterized (30). The female reproductive tract contains significant levels of glycosaminoglycans (31), the levels of which can vary during the menstrual/estrous cycle (32). One glycosaminoglycan of interest is hyaluronan. Hyaluronan is a high molecular-mass polysaccharide and can be obtained endotoxin- and prion-free from a yeast fermentation procedure. In the mouse hyaluronan levels in the uterus increase at the time of implantation (33). It has been shown in the pig (34) and cow (35) that hyaluronan can act in synergy with albumin in an embryo culture system. Gardner et al. (36) showed this to be true also for the mouse and showed that hyaluronan could completely replace albumin. Gardner et al. (36) also showed that the presence of hyaluronan in the medium for embryo transfer results in a significant increase in embryo implantation rates. Furthermore, similar to results with recombinant albumin, the presence of hyaluronan in the culture medium increases the cryosurvivability of blastocysts (36, 37). Recombinant albumin and hyaluronan confer a synergistic benefit to the embryo (28, 36, 38).

Alternatives to physiological macromolecules are synthetic macromolecules such as PVA (39) and PVP. Such an approach has worked for the in vitro development of embryos from several mammalian species (14, 39, 40). However, PVA is not able to maintain the physiology and metabolism of the embryo, and bovine embryos cultured in the presence of PVA do not survive cryopreservation as well as those cultured in the presence of albumin (24). Furthermore, the use of PVA has been associated with the development of birth defects in mouse fetuses (36).

2.2 pH

When considering medium pH, it is important to understand that the actual pH of the surrounding medium (pH_o; typically 7.4) is different from that inside the embryo (pH_i; 7.2) (41–43). Rather than the pH of the medium itself controlling pH_i, it is also the specific components of the culture media, such as lactic and amino acids, that affect and buffer pH_i. While high concentrations of lactate in the culture medium can drive pH_i down (41), amino acids increase the intracellular buffering capacity and help maintain the pH_i at around 7.2 (44). As the embryo has to maintain pH_i against a gradient when incubated at pH 7.4, it would seem prudent to culture embryos at lower pH_o. Typically a pH_o of 7.2–7.3 works well.

The pH of a CO_2/bicarbonate-buffered medium is not easy to quantitate. A pH electrode can be used, but one must be quick, and the same technician must take all readings to ensure consistency. An alternative approach is to take samples of medium in an airtight syringe and measure the pH with a blood-gas analyzer. A final method necessitates the presence of phenol red in the culture medium and the use of Sorensons's phosphate buffer standards (3). This method allows visual inspection of a medium's pH with a tube in the incubator and is accurate to 0.2 pH units (Table 3.4).

Table 3.4. Preparation of Color Standards for pH of Media

pH at 18°C	Solution A (ml)	Solution B (ml)
6.6	62.7	37.3
6.8	50.8	49.2
7.0	39.2	60.8
7.2	28.5	71.5
7.4	19.6	80.4
7.6	13.2	86.8

Stock A: 9.08g KH_2PO_4 (0.067M), 10 mg phenol red in 1 l water. Stock B: 9.46g Na_2HPO_4 (0.067M), 10 mg phenol red in 1 l water. Measure the pH with meter and adjust pH as required (i.e., add solution A to lower the pH; add solution B to increase the pH). The pH standards should be filter sterilized and can then be kept for up to 6 months.

2.3 Carbon Dioxide

The concentration of CO_2 has a direct impact on the pH of a bicarbonate-buffered medium. Although most culture media work over a wide range of pH (7.2–7.4), it is preferable to ensure that pH does not go above 7.4. It is therefore advisable to use a CO_2 concentration of 6–7%. The amount of CO_2 in the incubation chamber can be calibrated with a Fyrite, although this calibration is only accurate to ±1%. An alternative method is to use an infrared metering system that can be calibrated and is accurate to around 0.2%.

Medium buffered with CO_2/bicarbonate require at least 4 h in a CO_2 environment to equilibrate prior to use. Equilibrate the volume of medium to be used and do not keep bottles of medium gassed in the incubator, especially media containing amino acids (see below). When using a CO_2/bicarbonate-buffered medium, it is essential to minimize the amount of time the culture dish is out of the CO_2 incubator to prevent increases in pH. To help reduce pH fluctuations, modified pediatric isolettes designed to maintain temperature, humidity, and CO_2 concentration can be used.

A more practical alternative is to use a different buffering system to CO_2/bicarbonate. Alternative buffers include HEPES or MOPS (45), which are typically used at 20–23 mM together with 5–2 mM bicarbonate. Such buffering systems do not require a CO_2 environment. The use of PBS for any oocyte or embryo manipulations is not recommended. Both the presence of high levels of phosphate and the absence of amino acids can have a significant detrimental impact on the embryo (46).

2.4 Oxygen

The levels of oxygen to which embryos are exposed in vivo are significantly below levels present in air (20%). The concentration of oxygen in the lumen of the rabbit oviduct is 2–6% (47, 48), whereas the oxygen concentration in the oviduct of hamster, rabbit, and rhesus monkey is 8% (49). The oxygen concentration in the uterus is lower than that in the oviduct, ranging from 5% in the hamster and rabbit to 1.5% in the rhesus monkey (49).

Several studies on different species have clearly demonstrated that culture at a reduced oxygen concentration (5–7%) results in enhanced embryo development in vitro (mouse [50, 51]; sheep and cow [52]; goat [53]; human [54]). More significantly, exposure to 5–7% oxygen in vitro leads to higher rates of fetal development (54, 55). However, the embryos of humans and certain F_1 mice can develop well in culture in the presence of atmospheric oxygen (20%). This has therefore produced some confusion regarding the optimal concentration for these species. Human embryos cultured in a low oxygen environment (5%) produce blastocysts with significantly more cells than those embryos cultured in a high oxygen environment (20%) (54). Embryos derived from outbred mice, such as CF1, develop significantly better in culture in a reduced oxygen environment (51). Considering the physiology of the reproductive tract and the beneficial effects of using a reduced oxygen concentration as determined in controlled studies, it is advisable to culture embryos of all species at low oxygen concentrations (5–7%).

2.5 Medium Renewal

When simple media were used to culture embryos, there was no need to renew the culture medium because such media lacked labile components. However, with the routine inclusion of amino acids in embryo culture media, there is a need to renew the culture medium every 48 h. This is due to the spontaneous deamination of amino acids at 37°C and to their metabolism by the embryos, both of which result in the release of ammonium into the culture medium. Spontaneous deamination is responsible for the majority of ammonium released. The significance of medium renewal cannot be overstated: chronic exposure of embryos to ammonium not only impairs development in culture (56, 57), but in the case of the mouse retards fetal growth and can lead to abnormal fetal development (58). Ammonium has also been implicated in the induction of oversized fetuses in sheep (59, 60).

The buildup of ammonium in the culture system is independent of the incubation volume used. Rather, the appearance of ammonium is linked to the concentration of amino acids in the medium, especially that of glutamine. So whether one uses a 20-μl drop of medium in a dish or 800 μl of medium in a well, the concentration of ammonium after a certain time will be the same in both, as deamination of amino acids in the incubator is not volume related. Furthermore, whether an oil overlay is used has no effect on the final ammonium concentration. Ideally, therefore, medium should be renewed every 48 h.

2.6 Temperature

Although oocytes and embryos can develop after incubation at room temperature, this should be avoided. Initially there were concerns about the stability of the cytoskeleton at room temperature (61), which in turn would affect embryo development. However, it is evident that procedures such as ICSI can be performed at room temperature without compromising human embryo development. However, exposure of embryos to temperatures below 37°C will reduce the speed of development. Therefore, maintaining the temperature of oocytes and embryos at the incubation temperature during all stages of manipulation should be considered important and can be facilitated by the use of warming stages. When using warming stages, along with test tube warmers, it is important to calibrate the temperature to that of the medium inside the culture dish. For example, to maintain the temperature of a 100-μl drop of medium in a tissue culture dish at 37°C, the temperature of the heating stage may read 38.5°C rather than 37°C due to the thermal absorption of the dish. Therefore, one should calibrate each warming stage, heating block, and so on, according to its use.

2.7 Embryo Density per Volume

The culture of mammalian embryos in reduced volumes of medium and/or in groups significantly increases blastocyst development (62–65) and increases blastocyst cell number (64). Culturing embryos in reduced volumes increases subsequent viability after transfer (64). It has been proposed that the benefit of growing embryos in small volumes and/or in groups is due to the production of specific embryo-derived autocrine/paracrine factor(s) that stimulate development (63, 64). The culture of embryos in large volumes will result in a dilution of such factors so that they become ineffectual (6). This phenomenon is not confined to the mouse, in which several embryos reside in the female tract at one time, but has also been reported for the sheep and cow, which like the human are monovular (57, 66). It has been shown in both the mouse and cow that increasing the embryo:incubation volume ratio specifically stimulates the development of the inner cell mass. This explains the increased viability of embryos cultured in reduced volumes in groups.

2.8 Medium Overlay

To prevent changes in osmolality, it is desirable to overlay the medium used with an inert layer, typically paraffin, mineral oil, or silicon oil. The use of such an overlay is a prerequisite to using microdrop volumes. Which ever oil is chosen, it should be pre-screened before use (see quality control), as there is batch-to-batch variation of raw materials, and certain lots can be toxic to the embryo (67). An oil overlay also reduces the speed of CO_2 loss and the associated increase in pH.

2.9 Culture Vessel

It is imperative that the culture vessel undergoes rigorous quality control (see below), as not all tissue culture materials are consistent with embryo viability. Choose a brand such as Falcon that offers high-quality dishes and tubes that can sustain both sperm motility and embryo development.

2.10 Air Quality

The significance of air quality in an embryology laboratory has been the focus of attention (68). Should air quality be questionable, there are several alternatives available to deal with this situation. Stand-alone air purification systems that include carbon filters may be useful.

2.11 Technical Diligence

Adequate training of embryologists is fundamental to success. As the success of a given procedure can be due to the skill of the individual, it would be fair to classify assisted reproductive technology as art. However, with appropriate training one can minimize variability between embryologists. Key aspects of embryo manipulation that directly impact embryo culture outcome are speed and the ability to move embryos in the smallest volumes possible. This greatly enhances washing procedures and results in good embryo development. Embryologists also need to understand the fundamentals of embryo development so that they know how their actions can affect embryo development and pregnancy outcome.

2.12 Quality Control

Quality control (QC) systems are present in clinical embryology laboratories due to the requirements of various regulatory bodies. In contrast, rigorous QC systems are often forsaken in research laboratories, primarily due to limited resources.

Establishing an appropriate QC system for the IVF laboratory is a prerequisite for establishing a successful laboratory. The types of bioassays conducted for QC have been the focus of much discussion (3). It is important to understand the limitations of the assays performed and to use data obtained from bioassays in an appropriate fashion. Quality control should not be limited to the culture media used, but should include all contact supplies and gases used in an IVF procedure.

A practical and quantifiable bioassay is the culture of pronucleate mouse embryos in protein-free media, as protein has the ability to mask the effects of any potential toxins present (69, 70), along with scoring the embryos at key times and determining blastocyst cell number. The most common system used by our laboratory for QC is outlined in Protocol 3.1. The stage at which the embryos are cultured has an impact on development. Embryos collected at the pronucleate stage do not tend to fair as well in culture as those collected at the two-cell stage. Reports that mouse embryos can develop in culture in medium prepared using tap water (71, 72) should be interpreted carefully after taking into account the types of media used and the supplementation of medium with protein. Silverman et al. (72) used Ham's F-10. This medium contains amino acids, which can chelate any possible toxins present in the tap water, such as heavy metals. George et al. (71) included high levels of BSA in their zygote cultures to the blastocyst stage. Furthermore, all studies used blastocyst development as the sole criterion for assessing embryo development. Blastocyst development is a poor indicator of embryo quality and does not accurately reflect developmental potential (73). Therefore, rates of development should be determined by scoring the embryos at specific times during culture (2). Key times to examine the embryos include the morning of day 3 to determine the extent of compaction, the morning and afternoon of day 4 to determine the degree of blastocyst formation, and the morning of day 5 to assess the initiation of hatching. Figure 3.1 can be used to help define specific stages of embryo development. Finally, for the embryos that form blastocysts in a given time, typically on the morning of day 5, cell numbers should be determined by staining and counting (Protocol 3.3). When new components of culture media can affect the development of the inner cell mass directly, such as essential amino acids, a differential nuclear stain should be performed to determine the extent of ICM development (Protocol 3.4). Using such an approach, it is possible to identify potential problems in culture media before the media are used.

The minimum QC requirements for incubators include the use of an infrared monitor to calibrate CO_2, certified thermometers and a pronucleate MEA.

3. CULTURE OF EMBRYOS

This section outlines the basics for mammalian embryo culture. The basic culture protocol for embryo culture is outlined in Protocol 3.2. There are evident differences between species, such as temperature, which should be 37°C for embryos of the mouse and human but 38.5°C for the embryos of domestic animals. The duration of culture will also depend on the species. Incubation volumes also differ due to the different metabolic activities of the embryos from different species. Typically 10 mouse embryos can be cultured in a 20-µl drop; up to 4 human or domestic animal embryos can be cultured in a 50-µl drop. (However, different media will not be the same due to different levels of nutrients present in the medium). As discussed, the medium should be renewed every 48 h regardless of the species and incubation volume.

Embryo cultures should ideally be performed in an atmosphere of 5% O_2 and 6% CO_2. This gas environment can be created using either a tri-gas incubator or a modular incubator chamber/desiccator and a cylinder of special gas mix. It is advisable to minimize the number of times the incubators is accessed during a given culture. The more incubator chambers available, the better. However, when dealing with large numbers of embryos, a tri-gas incubator is more practical and is a sensible investment. It is advisable to minimize the number of times the incubators or chambers are accessed during a given culture. The detrimental effects that frequent opening of the incubator can have on embryo development have been clearly demonstrated (51).

All embryo manipulations should be performed using a pulled Pasteur pipette or a displacement pipette such as those originally used to load gels. The latter can now be obtained commercially for use with embryos and come in different tip-diameters. Should a pulled Pasteur pipette be used, then control of fluid can be achieved using a syringe attached through tubing or via a mouth pipette and tubing (though this cannot

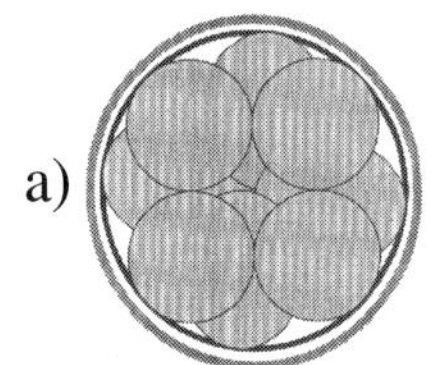

8-cell Embryo

Individual cells can clearly be distinguished

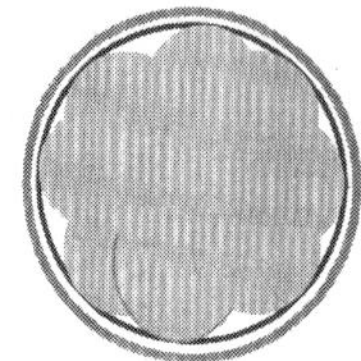

Compacted Embryo

The cells of the embryo have begun to form an epithelium and it is hard to distinguish individual cells

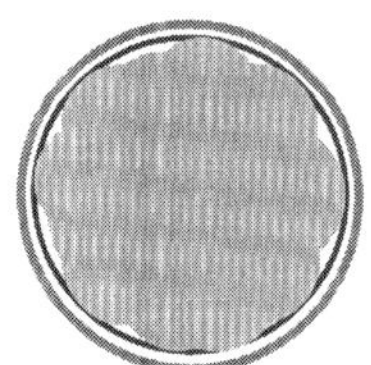

Morula

The embryo now has 16 to 32 cells and individual cells can no longer be identified

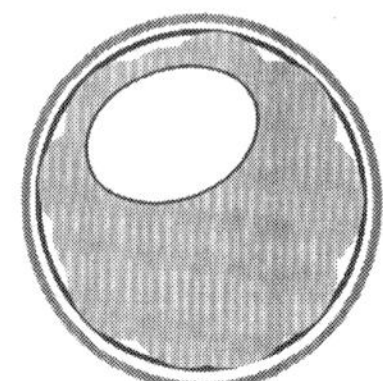

Early Blastocyst

The blastocoel cavity is <2/3 of the volume of the embryo

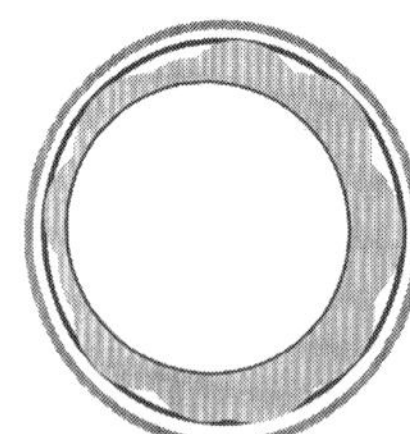

Blastocyst

The blastocoel cavity is ≥ 2/3 of the volume of the embryo

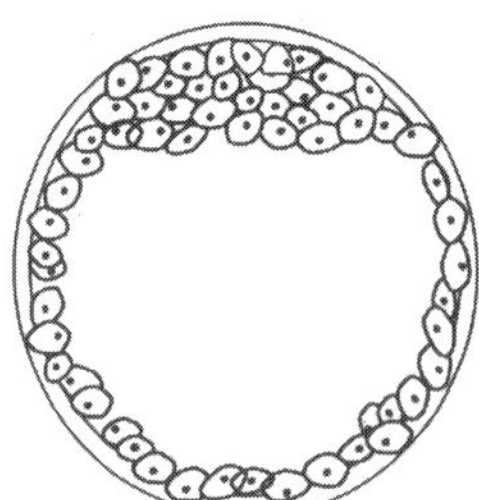

Expanded Blastocyst

The blastocoel is full and the embryo has increased in volume and the zona pellucida has started to thin

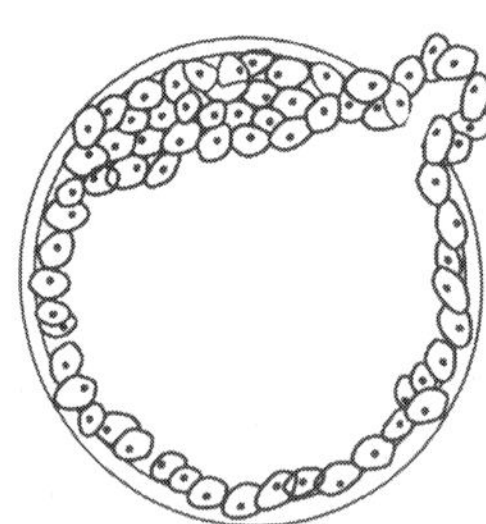

Hatching Blastocyst

The embryo has started to herniate through the zona pellucida

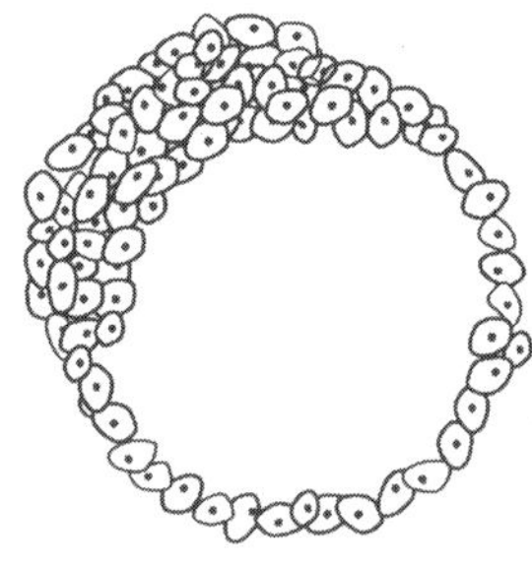

Fully Hatched Blastocyst

The embryo has completely escaped from the zona pellucida

a)

b)

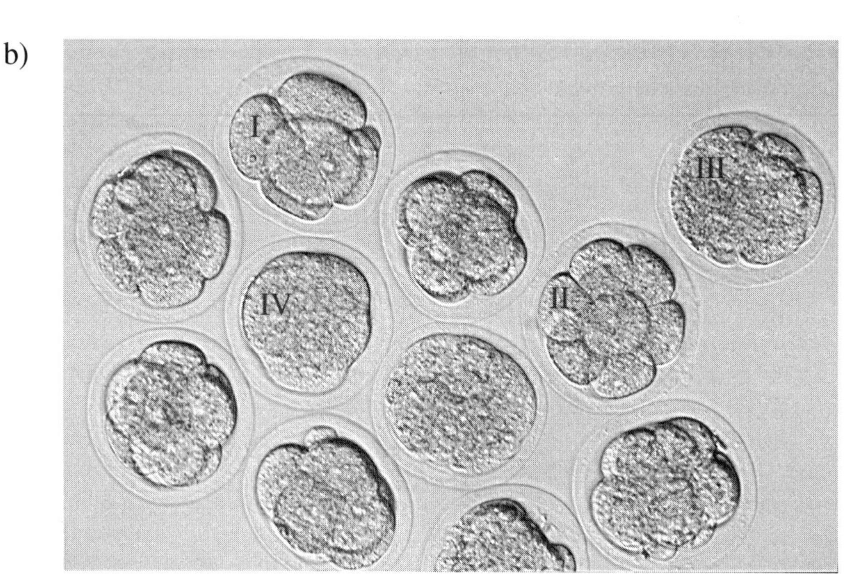

Figure 3.1 (Facing page). Grading of embryo development. As embryo development is a continuum, it is arbitrary where the lines are drawn to distinguish specific stages of development. In this figure, the development of the eight-cell stage to the fully hatched blastocyst is represented schematically. The key is that all embryologists in the same laboratory or clinic give embryos of the same development stage the same score for consistency. (a) Schematic of embryo development. (b) Photomicrograph of embryos on day 3 of culture: I, six-cell embryo; II, eight-cell embryo; III, eight-cell embryo starting to compact; IV, compacted eight-cell embryo.

be advocated for clinical use). It is important to use a pipette with the appropriate size tip (i.e., just larger than the diameter of the embryo) and to break the capillary action of the pipette by trapping air bubbles in the fluid column (Figure 3.2). Using the appropriate-size tip minimizes the volumes of culture medium moved with each embryo, which typically should be less than 1 μl. Such volume manipulation is a prerequisite for successful embryo washing and subsequent culture. Ideally, glass Pasteur pipettes should be washed before use with tissue-culture grade water and then heat sterilized. All oocyte and embryo manipulations should be in media containing amino acids (51, 74).

4. ASSESSMENT OF EMBRYO DEVELOPMENT

Assessment of embryo development is discussed in full in chapter 4. It is important to use a sequential approach for scoring embryo development and not simply rely on endpoint determinations such as blastocyst development. Blastocyst development alone is a relatively poor indicator of developmental competence (73). Therefore, when the endpoint of an experiment is not an embryo transfer, to further establish the health of an embryo, in vitro markers of developmental competence such as ICM differentiation (see Protocol 3.4) and metabolism (see chapters 7 and 9) should be determined.

5. PREPARATION OF CULTURE MEDIA

The in-house preparation of culture media is a time-consuming procedure, and steps are required to ensure that all components are of the highest quality. Therefore, every item in the culture system should undergo a prescreen with a sensitive mouse embryo assay to

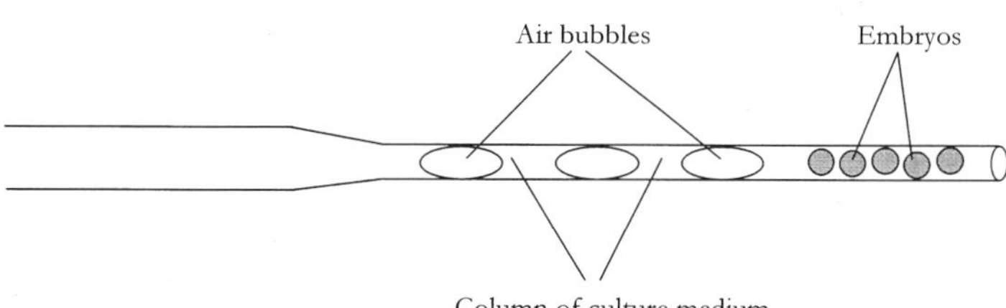

Figure 3.2. Schematic of a pulled glass pipette used to control embryos. The capillary action of the pipette is broken by trapping air bubbles in the column of medium in the pulled region of the pipette. This is achieved by moving the pipette in and out of the medium rapidly while applying light suction to the pipette. Embryos can then be aspirated into the pipette with the minimum of culture medium.

ensure that embryo development and physiology are not compromised due to the inclusion of a suboptimal component. Although media are available commercially, for experiments on the composition of embryo culture media etc., preparation of media in-house is essential. This section details how to make culture media and handling media from stock solutions. These instructions can be used for the preparation of any culture media.

5.1 Preparation of Stock Solutions

As most culture media used in research are used in small volumes, it is not necessary to make large volumes of media. Only small volumes of 10–50 ml are typically required. Therefore, it is easiest to prepare media from concentrated stock solutions. Media components are divided into like components with similar in-solution shelf-lives. For example, a stock A would usually contain salts and carbohydrates that are stable in solution for several months. However, for all media it is necessary to add the calcium chloride as a separate stock solution because this chemical can precipitate with other salts.

Individual stock solutions are prepared and can be filtered through a 0.2-μm filter and stored at 4°C. An example of stock groupings and shelf-lives are shown in Protocol 3.5.

5.2 Preparation of Media from Stock Solutions

Media are formulated using the concentrated stock solutions described above. For each case the water is added to the vial/tube and the solutions added one at a time. Before using media, characteristics such as osmolality and pH at 6% CO_2 should be checked. A protocol for making a generic culture medium (GCM) from stock solutions is outlined in Protocol 3.5.

6. CONCLUSION

The formulations of culture media are a key aspect of embryo culture. However, as described in detail, the way in which a medium is used has a significant impact on its performance. Therefore, it is important to consider the culture system as a whole and monitor all aspects to ensure consistent and successful embryo development.

Protocol 3.1. Outline of mouse bioassay for screening media components

Media are prepared containing the component to be tested. The medium should be protein-free.

Embryos are collected from an F_1 hybrid strain of mice (CBA × C57) at around 22 h post-hCG (see Chapter 2).

Zygotes are placed into culture in a control medium and the test medium and cultured to the blastocyst stage. Embryo development is assessed at specific time points to determine on-time development. On the morning of day 5 blastocyst cell number is determined.

The test item is determined to be embryo safe if all of the parameters measured (development on days 2, 3, 4, and 5 and blastocyst cell numbers) are not significantly worse than those grown in the control media.

Day 0: Place dishes into the incubator (6% CO_2 in air) after 4 PM.
Day 1 (10 AM): Collect zygotes and place into culture in the control and test treatments.
Day 2 (4 PM): Determine development to the four-cell stage.

Day 3 (9 AM): Determine number of embryos at the eight-cell stage and number beginning compaction.

Day 4 (4 PM): Determine development to the blastocyst stage.

Day 5 (9 AM): Determine development to the expanded/hatching blastocyst stage. Determine blastocyst cell numbers (see Protocol 3.3).

Protocol 3.2. Basic culture protocol for culture of mammalian embryos

Preparation of culture system

Embryo cultures should ideally be performed in 6% CO_2 and 5% O_2. This gas environment can be created using either a tri-gas incubator or a modular incubator chamber/dessicator with a certified gas mix (6% CO_2, 5% O_2 and 89% N_2).

All contact supplies used for embryo culture should be of tissue-culture grade and should be prescreened with the mouse bioassay before introduction into the culture system.

Day before culture

1. After 4 PM on the day before embryo culture, G1 culture dishes should be prepared. Rinse the pipette tip by taking up the required volume and then expelling into a discard dish. Using this prerinsed sterile tip, place drops of culture media into the petri dish (for mouse 10 µl, cow 15 µl, and human 25 µl). Always set up several wash drops in each dish. Optimally there should be one wash drop for every culture drop (see Figure p3.1). Immediately cover drops with paraffin oil to avoid evaporation. Prepare no more than 2–3 dishes at one time. Using a new tip for each drop, first rinse the tip and then round each drop up to the final volume (final culture volumes are 20 µl for mouse, 30 µl for cow, and 50 µl for human).

2. Immediately place the dish in the incubator at 6% CO_2. Gently remove the lid of the dish and set it at an angle on the side of the plate. Dishes must equilibrate in the incubator with a semi-opened lid for a minimum of 4 h (this is the minimal measured time for the media to reach correct pH under oil) and for a maximum of 18 h to prevent excessive ammonium production.

Day 1

1. Wash zygotes in handling media (see chapter 2) and then in the wash drops of medium G1. Washing entails picking up the embryos 2–3 times in a minimal volume and moving them around within the drop. Embryos should then be washed

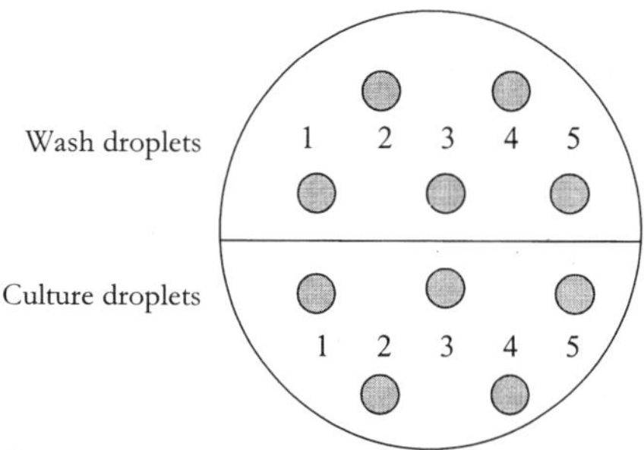

Figure p3.1. Schematic of arrangement of media drops in a culture dish.

in the wash drops in the culture dish and then transferred to the culture drops. Optimal development is achieved when embryos are cultured in groups (57, 64). For mice 10 embryos are cultured per 20-μl drop, for cows 4 embryos per 30-μl drop, and for humans 4 embryos in each 50-μl drop. Increasing the number of embryos can result in a significant depletion of the nutrient pool by the embryos and may result in the nutrients being depleted before the end of the culture period.

2. Dishes should be returned to the incubator and cultured for 48 h (mouse and human) or 72 h (cow) to around the eight-cell stage.

Day 2 (mouse and human), day 3 (cow)

1. After 4 PM on the day before embryo changeover into G2, culture dishes should be prepared. Rinse the pipette tip by taking up the required volume and then expelling into a discard dish. Using this prerinsed sterile tip, place drops of culture media into the petri dish (for mouse 10 μl, cow 15 μl, and human 25 μl). Always set up several wash drops in each dish. Immediately cover drops with paraffin oil to avoid evaporation. Prepare no more than 2–3 dishes at one time. Using a new tip for each drop, first rinse the tip and then round each drop up to the final volume (final culture volumes are 20 μl for mouse, 30 μl for cow, and 50 μl for human).

2. Immediately place the dish in the incubator at 6% CO_2. Gently remove the lid of the dish and set at an angle on the side of the plate. Dishes must equilibrate in the incubator with a semi-opened lid for a minimum of 4 h (this is the minimal measured time for the media to reach correct pH under oil) and for a maximum of 18 h to prevent excessive ammonium production.

Day 3 (mouse and human), day 4 (cows)

1. Change embryos over from medium G1 into G2. It is essential that embryos be washed well from G1 into G2, as components in the G1 such as EDTA have detrimental effects on the development of the blastocyst. Therefore, it is essential to wash embryos in a handling medium (containing amino acids; see chapter 2) before culture in medium G2.

2. Remove embryos from the G1 culture dishes and wash twice in a handling medium containing amino acids (see Chapter 2) that has been warmed to 37°C. Washing entails picking up the embryos 2–3 times in a minimal volume and moving them around within the dish. For mice 10 embryos are cultured per 20-μl drop, for cows 4 embryos per 30-μl drop, and for human 4 embryos per 50-μl drop. Increasing the number of embryos can result in a significant depletion of the nutrient pool by the embryos and may result in the nutrients being depleted before the end of the culture period.

3. Dishes should be returned to the incubator and cultured for a further 48 h (mouse and human) or 72 h (cow) to the blastocyst stage.

Protocol 3.3. Determination of embryo cell number

Solutions

Prepare 500 ml of MOPS/HEPES buffered medium with no protein.

Triton solution

1. Add 20 μl Triton X-100 to 100 ml sterile MOPS/HEPES buffered medium (0.02%).
2. Shake until Triton is fully dissolved (vortex if necessary).

3. Store at 4°C.
4. Discard 8 weeks after preparation.

Propidium iodide solution

1. Weigh out 4 mg of propidium iodide.
2. Add to 200 ml sterile MOPS/HEPES buffered medium.
3. Shake until completely dissolved.
4. Store in the dark at 4°C.
5. Discard 8 weeks after preparation.

Preparation of staining dishes

1. Label and date each petri dish.
2. Number the bottom of each dish according to the number of treatments (i.e., a dish labeled 1, 2, 3 corresponds to a culture with one control and two test droplets).
3. Dispense 10 µl drops of propidium iodide into dish. Cover droplets with 3.5 ml paraffin oil. Dispense an additional 10 µl of propidium iodide into each droplet (thus allowing the staining drops to "round up").
4. Cover dishes with aluminum foil (propidium iodide is light sensitive).

Preparation of washing dishes

1. Label the bottom of each wash dish (organ well or 4-well plate) "W" and label the dish with the treatment number(s).
2. Dispense 800 µl of MOPS/HEPES buffered medium into the well of the organ culture dish or into each well of the four-well multidish.

Preparation of Triton dishes

1. Label the bottom of each Triton dish (organ well or 4-well plate) "T" and label the dish with the treatment number(s).
2. Dispense 800 µl Triton medium into the well of the organ culture dish or into each well of the four-well multidish.

Staining procedure

The staining procedure can be performed at room temperature or at 37°C.

1. Place expanded blastocysts and hatching blastocysts into the Triton solution for 45 s.
2. Move the embryos to the wash dish containing MOPS/HEPES buffered medium; transfer as little media as possible.
3. Transfer the embryos to the propidium iodide drops.
4. Embryos are stained sufficiently after around 10 min. If embryos are not to be counted immediately, then the propidium iodide dish may be wrapped with foil and stored at 4°C for several days.

Mounting

1. Label a glass slide with a treatment number using a diamond pen.
2. Apply three small drops of glycerol to the slide.
3. Transfer one or two embryos from the propidium iodide to each drop of glycerol.
4. Gently apply three cover slips to the slide.
5. Store in the dark at 4°C until reading.
6. Read slides immediately.

Counting

The cell number of each embryo is determined by counting the cell nuclei using a fluorescence microscope with an ultraviolet light under a green filter. If the embryos appear to be three-dimensional, apply a small amount of pressure to the cover slip with the tip of a pen. This should cause the embryo to flatten into a two-dimensional image, thus enabling an accurate cell count.

Protocol 3.4. Differential nuclear staining

Solutions

Incidental solutions

> Absolute ethanol: ~5 ml for final wash step.
> Wash medium: MOPS/HEPES buffered medium with no protein should be used.
> *Glycerol:* used for mounting embryos.

Anti-DNP (dinitrophenol) solution

1. Divide anti-DNP into 10 µl aliquots.
2. Store at –20°C.
3. Discard 8 weeks after preparation.

Bisbenzimide solution

1. Weigh out 5 mg of bisbenzimide stain.
2. Add to 200 ml absolute ethanol to yield a 25 µg/ml solution.
3. Wrap in foil.
4. Store at 4°C.
5. Discard 16 weeks after preparation.

Guinea pig serum

1. Add 2 ml of MOPS/HEPES buffered medium to dry serum.
2. Store at 4°C.
3. It is critical to discard 1 month after preparation.

Pronase solution

1. Add 0.25 g of Pronase to 50 ml of sterile MOPS/HEPES buffered medium to yield a 0.5% solution.
2. Freeze in 200 µl aliquots
3. Store at –20°C.
4. Discard 8 weeks after preparation.

Propidium iodide solution

1. Weigh out 4 mg of propidium iodide.
2. Add to 200 ml sterile MOPS/HEPES buffered medium with no protein.
3. Shake until completely dissolved.
4. Store in the dark at 4°C.
5. Discard 8 weeks after preparation.

PVP medium

1. Add 4 mg/ml PVP to 10 ml culture medium.
2. Store at 4°C.
3. Discard 8 weeks after preparation.

Day of staining

On day of staining, prepare TNBS (2,4,6-trinitrobenzenesulfonic acid) solution by adding 10 μl TNBS to 90 μl of PVP medium. Cover with foil and store at 4°C.

Prepare anti-DNP solution by adding 90 μl MOPS/HEPES-buffered medium to a 10-μl aliquot of anti-DNP. Prepare complement by adding 50 μl of propidium iodide stain to 50 μl of guinea pig serum; cover with foil.

Preparation of staining dishes

1. Label and date each Petri dish.
2. Number the bottom of each dish according to the number of treatments (i.e., a dish labeled 1, 2, 3 corresponds to a culture with one control and two test droplets).
3. All dishes should be prepared with 20 μl drops under 3.5 ml paraffin oil. The following dishes should be prepared:

   1 dish with Pronase
   1 dish with TNBS solution
   1 dish with anti-DNP solution
   1 dish with complement
   3 dishes of wash medium with a minimum of 2 wash drops per treatment

4. Cover the TNBS dish with foil and store at 4°C.
5. Cover the complement dish with foil.
6. Place all dishes (except the TNBS dish) on a warm stage at 37°C for at least 1 h.

Staining procedure

All staining procedures should be performed at 37°C, except where specified. A timer should be used. Dishes should be prewarmed for 1 h at 37°C.

Day 1

1. Place blastocyst and hatching blastocyst embryos into the Pronase solution to dissolve the zona at 37°C. *Note*: This is the key step to ensure the staining procedure works, any remaining zona will render the remaining steps useless. Watch the zona dissolve under the microscope and immediately remove the embryos and place them into wash drops. This should take between 2 and 5 min for mice and up to 10 min for cows. Wash the embryos through an additional wash drop before proceeding to step 2.
2. Incubate the embryos in the TNBS dish, covered with foil at 4°C for 10 min, then move them into fresh wash drops. Wash through additional wash drops until there is no visible yellow color of the TNBS solution in the wash drop. Move as little fluid as possible to the wash; use a pipette that is just slightly bigger than the diameter of the embryos. Be careful as the embryos are now sticky. If too much media has been transferred (which is apparent when there is visible yellow color of the TNBS in the wash drop), extra wash steps must be added.
3. Incubate the embryos in anti-DNP BSA for 10 min at 37°C, then move them to fresh wash drops. Wash through an additional wash drop before proceeding to step 4.
4. Place the embryos in the complement dish for 5–10 min at 37°C. For best results, the evenness of propidium iodide stain may be checked under a green filter after 5 min. *Note*: Embryos are now very fragile.
5. While the embryos are in the complement, set up a four-well Nunclon dish with 800 μl of bisbenzimide and cover with foil. Pipette embryos into bisbenzimide, cover dish with foil, and store at 4°C overnight.

Day 2

6. Transfer embryos into 100% alcohol to wash away excess bisbenzimide.

Siliconized slide preparation

Siliconized slides should be prepared in a fume hood.

1. Put on nitrile gloves before handling any of the equipment.
2. Fill one glass bath with siliconizing solution (Repelcote, BDH, Poole, Dorset, UK).
3. Fill slide holders with slides.
4. Dip slides that are in the holder into the bath; make sure that they are completely covered by the Repelcote. If not, quickly top off the bath until they are covered.
5. Leave slides in the bath for 60 s.
6. Remove the slide holder and drain the slides still in the slide holder on a flat surface for 3 min.
7. Place the slides into a container.
8. When all of the boxes have been siliconized, rinse the slides well 5 times with RO water.
9. Lay out the slides on absorbent towel and dry overnight.
10. Replace the slides in the boxes and label the boxes as siliconized; initial and date.

Mounting

Embryos should be mounted immediately after the staining procedure. If necessary, the 100% alcohol dish may be wrapped in foil and stored at 4°C for around 1 week.

1. Siliconized slides should be used.
2. Label a glass slide with a treatment number using a diamond pen.
3. Apply three small drops of glycerol to the slide.
4. Transfer one or two embryos from the alcohol dish to each drop of glycerol. Transfer as little alcohol as possible, otherwise embryos will adhere to the slide as a three-dimensional object, and counting will be difficult.
5. Gently apply three cover slips to the slide.
6. Store in the dark at 4°C until reading.
7. Slides should be read immediately.

Counting

The cell number of each embryo is determined by counting the cell nuclei using a fluorescence microscope with an ultraviolet filter. Under the UV filter, cells of the trophectoderm appear pink, and cells of the inner cell mass appear blue. If the embryos appear to be three-dimensional, apply a small amount of pressure to the cover slip with the tip of a pen. This should cause the embryo to flatten into a two-dimensional image, thus enabling an accurate cell count.

Trophectoderm cells are stained with both propidium iodide and bisbenzimide during this procedure and thus appear pink. However, propidium iodide is very light sensitive and photobleaches over time. If the procedure takes a long time, then the blue bisbenzimide stain may overpower the propidium iodide under the UV filter. Therefore, under the UV filter count total cells (both pink and blue), then count only trophoectoderm cells (orange) under the green filter and subtract from the total to give the inner cell mass number.

Protocol 3.5. Preparing media for research use

All of the media listed in tables 3.1–3.3 are commercially available. This guide to media preparation is based on the assumption that all chemicals and contact supplies have previously been screened using a mouse embryo bioassay. Rather than using a specific medium as an example, we here present how to formulate a Generic Culture Medium, containing amino acids, designated GCM. This protocol can be adapted to make any embryo culture media such as those listed in tables 3.1–3.4.

Groupings and shelf-life of stock solutions

Stock A (salts and carbohydrates): X10 stock; use for 3 months
  Sodium chloride
  Potassium chloride
  Potassium/sodium phosphate
  Magnesium sulfate/chloride
  Glucose
  Lactate (L-isomer only)
  Penicillin
Stock B: X10 stock; use for 2 weeks
  Sodium bicarbonate
  Phenol red
Stock C: X100 stock; use for 1 month
  Calcium chloride/lactate
Stock D: X100 stock; use for 2 weeks
  Sodium pyruvate
Stock E: X100 stock; use for 3 months
  EDTA
Stock F: X100 stock; use for 3 months
  Amino acids
Stock G: X100 stock; use for 3 months
  Vitamins
Stock H: X10 stock; use for 3 months
  HEPES / MOPS
  This stock is adjusted to pH 7.3 ± 0.5 by using a concentrated stock solution of NaOH (2–5 M).

Stock solutions that are ×10 are prepared by weighing the grams for 1 l and dissolving in 100 ml. Stock solutions that are ×100 are prepared by weighing the grams for 1 l and dissolving in 10 ml.

All stock solutions should be filtered through a 0.2-μm filter immediately after they are prepared. Stock solutions can then be stored at 4°C. Vitamins can be frozen, thawed once, and stored in the dark at –20°C.

Preparation of culture medium GCM and handling-GCM from stock solutions

Stock	GCM (10 ml)	Handling-GCM (10 ml)
H_2O	7.5	7.5
A	1.0	1.0
B	1.0	0.16
C	0.1	0.1
D	0.1	0.1
E	0.1	0.1
F	0.1	0.1
G	0.1	0.1
H	—	0.84

Water should first be added to the culture flask/tube using a sterile pipette. Each component is then added using a displacement pipette. Rinse each tip after the addition of each stock solution. Immediately after they are prepared, media should be filtered through a 0.2-μm filter and then be stored at 4°C for 2–4 weeks.

Media can then be supplemented with serum albumin. If serum albumin is to be added as a solution, the amount of water should be decreased for the amount of albumin solution to be added.

References

1. Biggers, J. D. (1987). In: *The Mammalian Preimplantation Embryo Regulation of Growth and Differentiation in Vitro*, B. D. Bavister, ed., p. 1. Plenum Press, New York.
2. Bavister, B. D. (1995). *Hum. Reprod. Update*, **1**, 91.
3. Gardner, D. K., and Lane, M. (1999). In: *Handbook of in Vitro Fertilization*, 2nd ed., A. O. Trounson and D. K. Gardner, eds., p. 205. CRC Press, Boca Raton, FL.
4. Gardner, D. K. (1998). In: *Animal Breeding: Technology for the 21st Century*, J. R. Clark, ed., p. 13. Harwood Academic Publishers, London.
5. Gardner, D. K., and Lane, M. (2002). In: *Principles of Cloning*, J. Cibelli, R. Lanza, K. Campbell, and M. D. West, eds., p. 187. Elsevier Science, New York.
6. Gardner, D. K. (1994). *Cell Biol. Int.*, **18**, 1163.
7. Wrenzycki, C., Herrmann, D., Carnwath, J. W., and Niemann, H. (1999). *Mol. Reprod. Dev.*, **53**, 8.
8. Gardner, D. K., and Lane, M. (1993). In: *Handbook of in Vitro Fertilization*, D. K. Gardner and A. O. Trounson, eds., p. 85. CRC Press, Boca Raton, FL.
9. Whittingham, D. G. (1971). *J. Reprod. Fertil. Suppl.*, **14**, 7.
10. Pool, T. B., Atiee, S. H., and Martin, J. E. (1998). *Infertil. Reprod. Med. Clinics N. Am.*, **9**, 181.
11. Milki, A. A., Hinckley, M. D., Fisch, J. D., Dasig, D., and Behr, B. (2000). *Fertil. Steril.*, **73**, 126.
12. Lawitts, J. A. and Biggers, J. D. (1993). *Methods Enzymol.*, **225**, 153.
13. Ho, Y., Wigglesworth, K., Eppig, J. J., and Schultz, R. M. (1995). *Mol. Reprod. Dev.*, **41**, 232.
14. Biggers, J. D., McGinnis, L. K., and Raffin, M. (2000). *Biol. Reprod.*, **63**, 281.
15. Barnes, F. L., Crombie, A., Gardner, D. K., Kausche, A., Lacham-Kaplan, O., Suikkari, A. M., Tiglias, J., Wood, C., and Trounson, A. O. (1995). *Hum. Reprod.*, **10**, 3243.
16. Leese, H. J. (1988). *J. Reprod. Fertil.*, **82**, 843.
17. Gardner, D. K., Lane, M., Calderon, I., and Leeton, J. (1996). *Fertil. Steril.*, **65**, 349.
18. Thompson, J. G., Gardner, D. K., Pugh, P. A., McMillan, W. H., and Tervit, H. R. (1995). *Biol. Reprod.*, **53**, 1385.
19. Abe, H., Yamashita, S., Satoh, T., and Hoshi, H. (2002). *Mol. Reprod. Dev.*, **61**, 57.
20. Abe, H., Yamashita, S., Itoh, T., Satoh, T., and Hoshi, H. (1999). *Mol. Reprod Dev.*, **53**, 325.
21. Gardner, D. K. (1999). *J. Reprod. Fertil. Suppl.*, **54**, 461.
22. Walker, S. K., Heard, T. M., and Seamark, R. F. (1992). *Theriogenology*, **37**, 111.
23. Gray, C. W., Morgan, P. M., and Kane, M. T. (1992). *J. Reprod. Fertil.*, **94**, 471.
24. Eckert, J., Pugh, P. A., Thompson, J. G., Niemann, H., and Tervit, H. R. (1998). *Reprod. Fertil. Dev.*, **10**, 327.
25. Thompson, J. G., Sherman, A. N., Allen, N. W., McGowan, L. T., and Tervit, H. R. (1998). *Mol. Reprod. Dev.*, **50**, 139.
26. Gardner, D. K., and Lane, M. (2000). *Fertil. Steril.*, **74 Suppl 3**, O86.
27. Bungum, M. (2002). *RBM Online*, **4**, 233.
28. Gardner, D. K., Maybach, J. M., and Lane, M. (2001). *Proc. World Congr. Fertil. Steril.*, **17**, 226.
29. Pool, T. B. (2001). In: *ART and the Human Blastocyst*, D. K. Gardner and M. Lane, eds., p. 105. Springer-Verlag, New York.
30. Meseguer, M., Pellicer, A., and Simon, C. (1998). *Mol. Hum. Reprod.*, **4**, 1089.
31. Laurent, C., Hellstrom, S., Engstrom-Laurent, A., Wells, A. F., and Bergh, A. (1995). *Cell Tissue Res.*, **279**, 241.
32. Carson, D. D., Dutt, A., and Tang, J. C. (1987). *Dev. Biol.*, **120**, 228.
33. Zorn, T. M., Pinhal, M. A., Nader, H. B., Carvalho, J. J., Abrahamsohn, P. A., and Dietrich, C. P. (1995). *Cell Mol. Biol.*, **41**, 97.

34. Kano, K., Miyano, T., and Kato, S. (1998). *Biol. Reprod.*, **58**, 1226.
35. Furnus, C. C., de Matos, D. G., and Martinez, A. G. (1998). *Theriogenology*, **49**, 1489.
36. Gardner, D. K., Rodriegez-Martinez, H., and Lane, M. (1999). *Hum. Reprod.*, **14**, 2575.
37. Stojkovic, M., Peinl, S., Stojkovic, P., Zakhartchenko, V., Thompson, J. G., and Wolf, E. (2002). *Reproduction*, **124**, 141.
38. Hooper, K., Lane, M., and Gardner, D. K. (2000). *Biol. Reprod.*, **62 Suppl. 1**, 249.
39. Bavister, B. D. (1981). *J. Exp. Zool.*, **217**, 45.
40. Keskintepe, L., Burnley, C. A., and Brackett, B. G. (1995). *Biol. Reprod.*, **52**, 1410.
41. Edwards, L. J., Williams, D. A., and Gardner, D. K. (1998). *Mol. Reprod. Dev.*, **50**, 434.
42. Lane, M., Baltz, J. M., and Bavister, B. D. (1998). *Biol. Reprod.*, **59**, 1483.
43. Phillips, K. P., Leveille, M. C., Claman, P., and Baltz, J. M. (2000). *Hum. Reprod.*, **15**, 896.
44. Edwards, L. J., Williams, D. A., and Gardner, D. K. (1998). *Hum. Reprod.*, **13**, 3441.
45. Good, N. E., Winget, G. D., Winter, W., Connolly, T. N., Izawa, S., and Singh, R. M. (1966). *Biochemistry*, **5**, 467.
46. Gardner, D. K., Pool, T. B., and Lane, M. (2000). *Semin. Reprod. Med.*, **18**, 205.
47. Mastroianni, L. J., and Jones, R. (1965). *J. Reprod. Fertil.*, **9**, 99.
48. Ross, R. N., and Graves, C. N. (1974). *J. Anim. Sci.*, **39**, 994.
49. Fischer, B., and Bavister, B. D. (1993). *J. Reprod. Fertil.*, **99**, 673.
50. Quinn, P., and Harlow, G. M. (1978). *J. Exp. Zool.*, **206**, 73.
51. Gardner, D. K., and Lane, M. (1996). *Hum. Reprod.*, **11**, 2703.
52. Thompson, J. G., Simpson, A. C., Pugh, P. A., Donnelly, P. E., and Tervit, H. R. (1990). *J. Reprod. Fertil.*, **89**, 573.
53. Batt, P. A., Gardner, D. K., and Cameron, A. W. (1991). *Reprod. Fertil. Dev.*, **3**, 601.
54. Gardner, D. K., Lane, M., Johnson, J., Wagley, L., Stevens, J., and Schoolcraft, W. B. (1999). *Fertil. Steril.*, **72**, S30.
55. Harlow, G. M., and Quinn, P. (1979). *Aust. J. Biol. Sci.*, **32**, 363.
56. Gardner, D. K., and Lane, M. (1993). *Biol. Reprod.*, **48**, 377.
57. Gardner, D. K., Lane, M., Spitzer, A., and Batt, P. A. (1994). *Biol. Reprod.*, **50**, 390.
58. Lane, M., and Gardner, D. K. (1994). *J. Reprod. Fertil.*, **102**, 305.
59. McEvoy, T. G., Robinson, J. J., Aitken, R. P., Findlay, P. A., and Robertson, I. S. (1997). *Anim. Reprod. Sci.*, **47**, 71.
60. Sinclair, K. D., McEvoy, T. G., and Carolan, C. (1998). *Theriogenology*, **49**, 218.
61. Pickering, S. J., Braude, P. R., and Johnson, M. H. (1991). *Hum. Reprod.*, **6**, 142.
62. Wiley, L. M., Yamami, S., and Van Muyden, D. (1986). *Fertil. Steril.*, **45**, 111.
63. Paria, B. C., and Dey, S. K. (1990). *Proc. Natl. Acad. Sci. USA*, **87**, 4756.
64. Lane, M., and Gardner, D. K. (1992). *Hum. Reprod.*, **7**, 558.
65. Salahuddin, S., Ookutsu, S., Goto, K., Nakanishi, Y., and Nagata, Y. (1995). *Hum. Reprod.*, **10**, 2382.
66. Ahern, T. J., and Gardner, D. K. (1998). *Theriogenology*, **49**, 194.
67. Erbach, G. T., Bhatnagar, P., Baltz, J. M., and Biggers, J. D. (1995). *Hum. Reprod.*, **10**, 3248.
68. Cohen, J., Gilligan, A., and Garrisi, J. (2001). In: *Textbook of Assisted Reproductive Techniques*, D. K. Gardner, A. Weissman, C. Howles, and Z. Shoham, eds., p. 17. Martin Dunitz, London.
69. Fissore, R. A., Jackson, K. V., and Kiessling, A. A. (1989). *Biol. Reprod.*, **41**, 835.
70. Flood, L. P., and Shirley, B. (1991). *Mol. Reprod. Dev.*, **30**, 226.
71. George, M. A., Braude, P. R., Johnson, M. H., and Sweetnam, D. G. (1989). *Hum. Reprod.*, **4**, 826.
72. Silverman, I. H., Cook, C. L., Sanfilippo, J. S., Yussman, M. A., Schultz, G. S., and Hilton, F. H. (1987). *J. In Vitro Fert. Embryo Transf.*, **4**, 185.
73. Gardner, D. K., and Lane, M. (1998). *Hum. Reprod.*, **13 Suppl. 3**, 148.
74. Escriba, M. J., Silvestre, M. A., Saeed, A. M., and Garcia-Ximenez, F. (2001). *Reprod. Nutr. Dev.*, **41**, 181.

4

DENNY SAKKAS
MICHELLE LANE
DAVID K. GARDNER

Assessment of Preimplantation Embryo Development and Viability

1. IMPORTANCE OF EMBRYO ASSESSMENT

The need to identify the most viable embryo is of fundamental importance. In particular, human IVF clinics have faced this problem because stimulation protocols (1) generate more embryos to choose from and because of the increase in IVF-related multiple births (2, 3).

Many methods have been suggested to evaluate embryo viability in both animal models and in human IVF programs. In human IVF clinics, a limiting factor is that these measurements need to be noninvasive and not time consuming. Routinely, the embryos selected for transfer are chosen on the basis of their morphology and rate of development in culture.

In this chapter we discuss a series of assessment methods, which in combination will facilitate the selection of those embryos in a given cohort with the highest viability. Criteria used are the assessment of morphological markers at different stages of development at critical time points. Developmental indices include the assessment of the pronuclear stage embryo, cleavage rate, and development to the blastocyst stage. Finally, we also elaborate on more complex techniques that have proved beneficial in animal models and in the future may further assist in selection of the most viable embryos. Although the examples given in this chapter are from humans, the embryo assessment and scoring systems described can be readily adapted for different laboratory or domestic species.

2. MORPHOLOGICAL ASSESSMENT

2.1 The Pronuclear-Stage Embryo

The many transformations that take place during the fertilization process make the pronuclear stage a dynamic stage to assess. The human oocyte contains the majority of the developmental materials and maternal mRNA, for ensuring that the embryo reaches the four- to eight-cell stage (4). The quality of the oocyte therefore plays a crucial role for determining embryo development and subsequent viability. A number of studies have postulated that embryo quality can be predicted from the pronuclear-stage embryo (see below). The features assessed include the orientation of pronuclei relative to the polar bodies, alignment of pronuclei and nucleoli, the appearance of the cytoplasm, presence of nuclear precursor bodies (NPB), and the timing of nuclear membrane breakdown (5). Tesarik and Greco (5) postulated that the normal and abnormal morphology of the pronucleus was related to the developmental fate of human embryos. They retrospectively assessed the number and distribution of NPB in each pronucleus of fertilized zygotes that led to embryos that implanted. The characteristics of these zygotes were then compared to those that led to failures in implantation. The features that were shared by zygotes that had the 100% implantation success were (1) the number of NPB in both pronuclei never differed by more than three and, (2) the NPB were always polarized or not polarized in both pronuclei but never polarized in one pronucleus and not in the other. Zygotes not meeting these criteria were more likely to develop into preimplantation embryos that had poor morphology and or that experienced cleavage arrest. The presence of at least one embryo that had met these criteria at the pronuclear stage led to a pregnancy rate of 22/44 (50%) compared to only 2/23 (9%) when none were present.

A further criterion of pronuclear embryos that may affect embryo morphology is the orientation of pronuclei relative to the polar bodies. Oocyte polarity is clearly evident in nonmammalian species. In mammals, the animal pole of the oocyte may be estimated by the location of the first polar body, whereas after fertilization, the second polar body marks the embryonic pole (6). In human oocytes a differential distribution of various factors within the oocyte has been described, and anomalies in the distribution of these factors, in particular the side of the oocyte believed to contain the animal pole, are thought to affect embryo development and possibly fetal growth (6–8). Following from this hypothesis, Garello et al. (9) examined pronuclear orientation, polar body placement, and embryo quality to ascertain if a link existed between a plausible polarity of oocytes at the pronuclear stage and further development. The most interesting observation involved the calculation of the position of the polar body. They found that with a larger displacement in the position of the polar bodies, there was a concurrent decrease in the morphological quality of preimplantation-stage human embryos. They postulated that the misalignment of the polar body might be linked to cytoplasmic turbulence, hence disturbing the delicate polarity of the zygote.

A study by Scott and Smith (7) devised an embryo score on day 1 on the basis of alignment of pronuclei and nucleoli, the appearance of the cytoplasm, nuclear membrane breakdown, and cleavage to the two-cell stage. Patients who had an overall high embryo score ($\geq$15) had a pregnancy and implantation rate of 34/48 (71%) and 49/175 (28%), respectively, compared to only 4/49 (8%) and 4/178 (2%) in the low embryo score group. Scott et al. (10) also showed that zygotes displaying equality between the nuclei had 49.5% blastocyst formation, and those with unequal sizes, numbers, or distribution of nucleoli had 28% blastocyst formation. Figure 4.1 summarizes the key morphological aspects of the pronucleate stage human embryo.

2.2 Cleavage-stage Embryos

The most widely used criteria for selecting the best embryos at the cleavage stages have been based on cell number and morphology. In one such study, Cummins et al. (11) established an embryo quality and development rating and found that good ratings for both were more likely to result in clinical pregnancies. Other studies have also found advantages in transferring embryos on the basis of a morphological and developmental assessment (12, 13).

Bavister (14) highlighted the problem that most clinics have in selecting the best embryos. He stated that the examination of embryos at arbitrary time points during

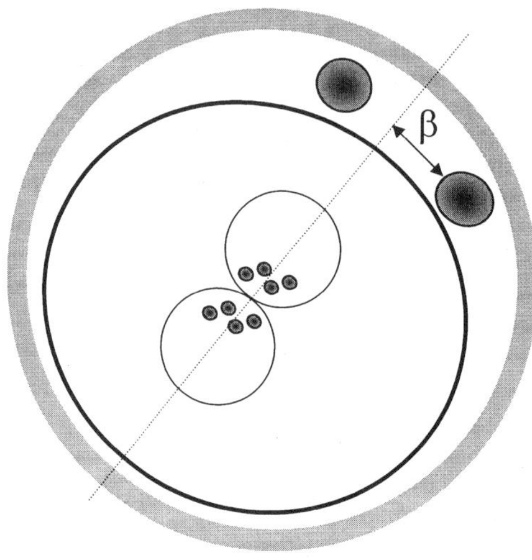

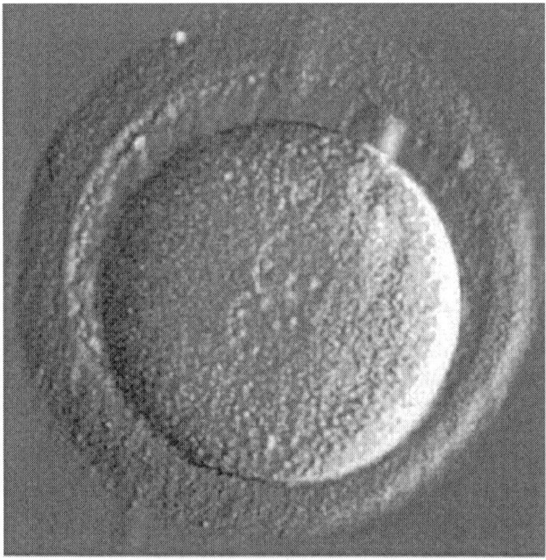

Figure 4.1. Ideal features shared by zygotes that have high viability at 18–19 h postinsemination/
injection. (top) The number of nucleolar precursor bodies (NPB) in both pronuclei never differed by more than 3. The NPB are always polarized or not polarized in both pronuclei but never
polarized in one pronucleus and not in the other. The angle from the axis of the pronuclei and the
farthest polar body is less than 50°. (From Sakkas [46], with permission of Martin Dunitz Press).
(bottom) Photomicrograph of a pronucleate human embryo.

development could be quite misleading with respect to categorizing the stage of development reached and timeliness of development. The selection of a critical time point is
essential so as to maximize the differences between embryos. Observations of embryo
development in culture are sometimes made infrequently so that precise data on cleavage timing are usually not available.

   The majority of studies that have used and reported embryo selection criteria on the
basis of cell number and morphology state that embryos were selected on day 2 or
day 3. As discussed by Bavister (14), one of the most critical factors in determining
selection criteria is to ascertain strict time points to compare the embryos. A four-cell
embryo scored in the morning of day 2 is definitely not the same as one that was scored
as a four-cell in the afternoon. Studies have used cleavage to the two-cell stage at 25 h
after insemination or microinjection as the critical time point for selecting embryos
(15, 16). In a larger series of patients, it was found that 45% of patients undergoing IVF
or ICSI have early cleaving two-cell embryos. Patients who have early cleaving two-

cell stage embryos allocated for transfer on day 2 or 3 have significantly higher implantation and pregnancy rates. In particular, nearly 50% of the patients who have two early cleaving two-cell embryos transferred achieve a clinical pregnancy (17) (Figure 4.2). The embryos that cleave early to the two-cell stage have also been reported to have a significantly higher blastocyst formation rate (18). It is also interesting to note that, in the embryo scoring system described by Scott and Smith (7), embryos that had already cleaved to the two-cell stage by 25–26 h after insemination were assigned an additional score of 10. This score is a sizeable part when the high-quality embryos were judged to be those scoring ≥15.

A vast number of variations on the theme have been published; however, some studies by Gerris et al. (19) and Van Royen et al. (20) used strict embryo criteria to select single embryos for transfer. The necessary characteristics of their top-quality embryos were established by retrospectively examining embryos that had a very high implantation potential. These top-quality embryos had the following characteristics: four or five blastomeres on day 2 (41–44 h postinsemination/injection) and at least seven blastomeres on day 3 (66–71 h postinsemination/injection), absence of multinucleated blastomeres, and <20% of fragments on day 2 and day 3 after fertilization. When these criteria were used in a prospective randomized clinical trial comparing single and double embryo transfers, it was found that in 26 single embryo transfers where a top quality embryo was available, an implantation rate of 42.3% and ongoing pregnancy rate of 38.5% was obtained. In 27 double embryo transfers, an implantation rate of 48.1% and ongoing pregnancy rate of 74% was obtained. The characteristics of cleavage stage embryos are highlighted in Figure 4.3.

2.3 Development to the Blastocyst Stage

An interest in culturing human embryos to the blastocyst stage has always existed, in particular as there were always concerns as to the logic of transferring early-cleavage-stage embryos to the uterine environment. Limited data existed that indicated an improved pregnancy rate when placing blastocysts into the uterus. Buster et al. (21) recovered in vivo-developed human blastocysts by uterine lavage and transferred the same blastocysts to achieve a high implantation rate (3/5, 60%), well above that currently observed in most IVF cycles. Similar implantation rates have now been published using sequential embryo culture media by a number of clinics (22–24).

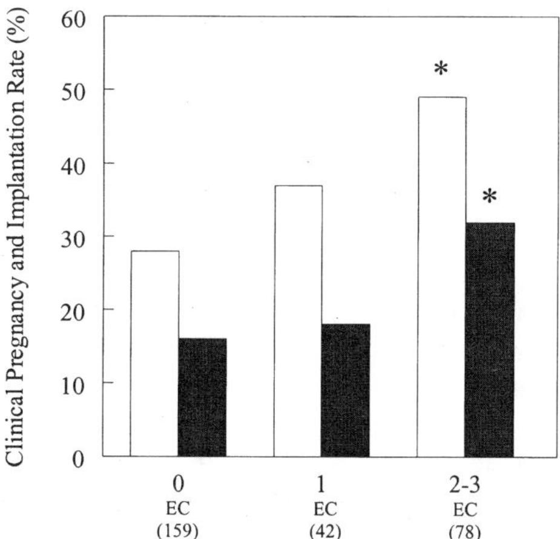

Figure 4.2. Outcome of embryo transfer related to the number of early cleaving (EC) two-cell human embryos. Open bars, pregnancy rate; solid bars, implantation rate. *Significantly different ($p<.01$) from the no early cleavage group. The number of cycles are in parentheses.

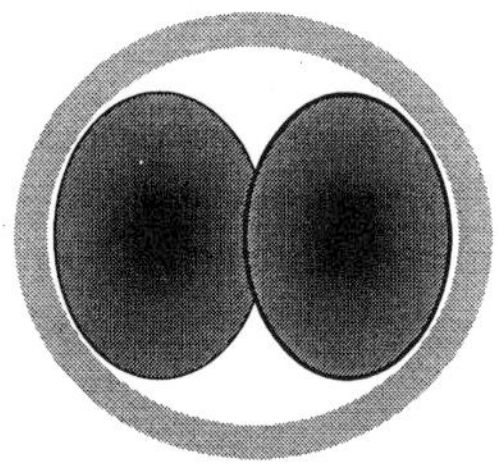

25 - 26h post insemination / injection
embryo should be at the 2-cell stage with
equal blastomeres and no fragmentation

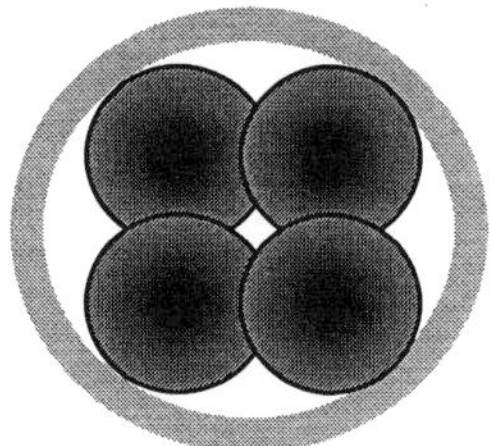

42 - 44h post insemination / injection
embryo should have 4 or more blastomeres
and less than 20% fragmentation

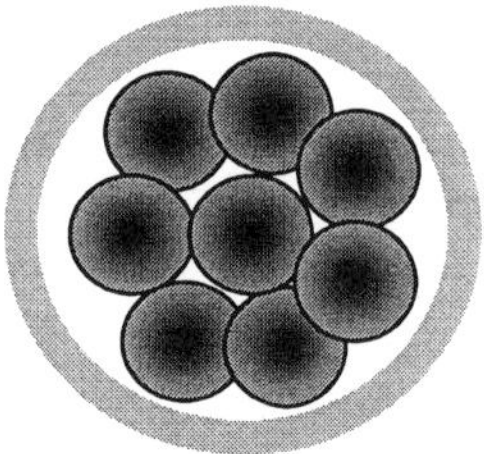

66 - 68h post insemination / injection
embryo should have 8 or more blastomeres
and less than 20% fragmentation

Figure 4.3. Ideal features of the cleavage-stage embryo. (From Sakkas [46], with permission of Martin Dunitz Press.)

The time of blastocyst development is evidently important, but forming a blastocyst per se is not the criterion most strongly associated with pregnancy outcome (Figure 4.4). In the mouse model several aspects of blastocyst development and physiology were quantitated and related to subsequent fetal development (25). It was determined that total cell number, inner cell mass (ICM) cell number, and glycolysis had the strongest correlation with blastocyst viability. Blastocyst formation and hatching were poorly correlated with pregnancy outcome.

This theory has been further substantiated by a number of studies that adopted a scoring method for the blastocyst transferred. This is highlighted in the study by Gardner et al. (26), who showed that blastocysts of high quality led to the highest pregnancy and implantation rates. Blastocysts were scored according to the expansion state of the blastocoel cavity and the number and cohesiveness of the ICM and trophectodermal cells. Blastocysts with a full blastocoel cavity, a well-populated and tightly formed ICM, and a cohesive trophectoderm with many cells were given a score of 3AA or greater and designated as the top-scoring blastocysts (Figure 4.5). When two such blastocysts were transferred, pregnancy rates were >80%, and implantation rates were 70%. The transfer of one top-scoring blastocyst in the cohort of two still led to pregnancy rates >60% and implantation rates of 50%. The importance of blastocyst quality in relation to pregnancy outcome has also been shown by Balaban et al. (27). The other factor important in the assessment of blastocysts is the time of blastocyst formation. In a previous study when cases where only day 5 and 6 frozen blastocysts were transferred were compared to those frozen on or after day 7 and transferred, the pregnancy rates were 7/18 (38.9%) and 1/16 (6.2%), respectively (28). In these cases expanded blastocysts with a definable ICM and trophectoderm were frozen. These results showed that even though blastocysts could

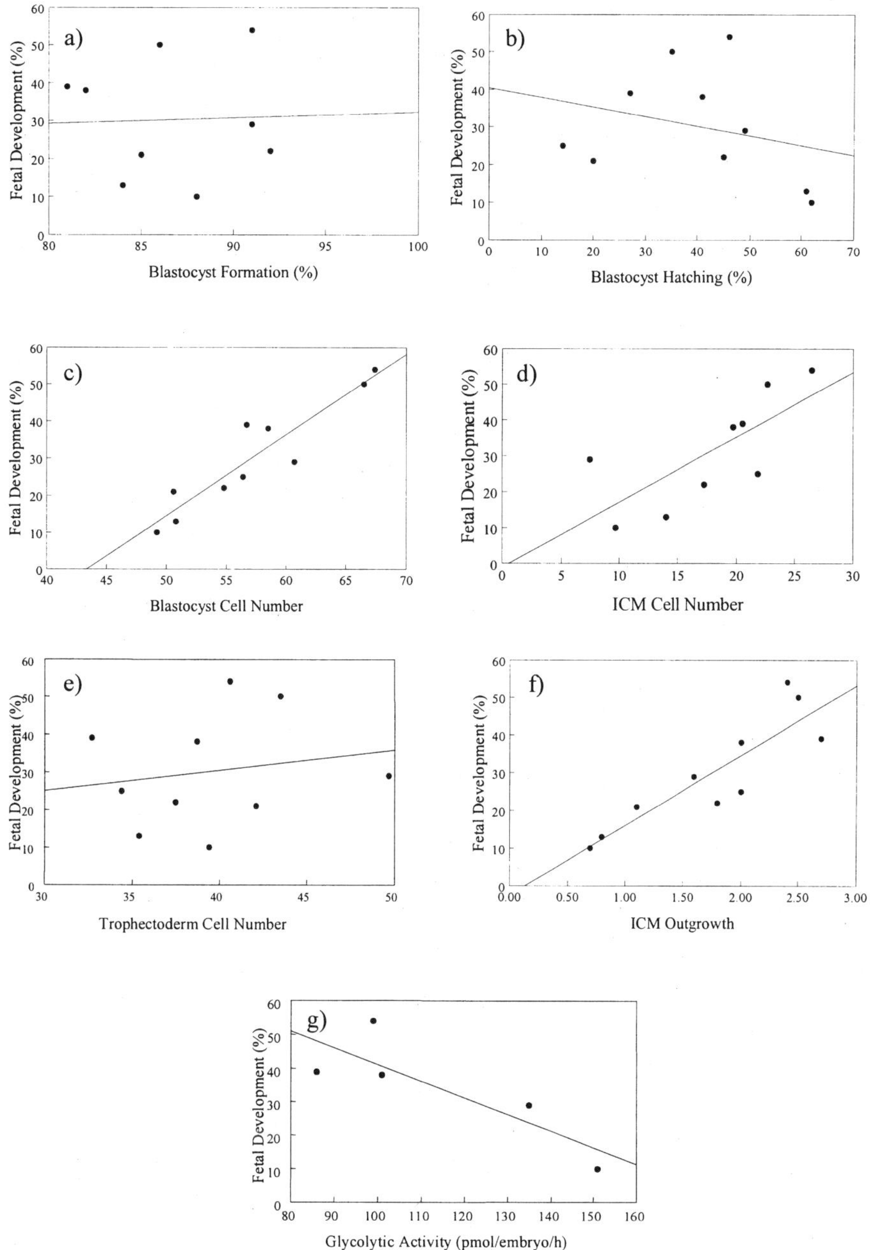

Figure 4.4. Correlation between parameters for embryo assessment and fetal development after transfer. The plots show the relationship between seven parameters used to determine the effectiveness of different culture media to support mouse blastocyst development, and the resultant viability after transfer (25). Dots represent mean values. (a) blastocyst formation and fetal development ($p > .1$); (b) blastocyst hatching and fetal development ($p > .1$); (c) total blastocyst cell number and fetal development ($p < .01$); (d) inner cell mass cell number and fetal development ($p < .05$); (e) trophectoderm cell number and fetal development ($p > .1$); (f) inner cell mass outgrowth and fetal development ($p < .01$); (g) glycolytic activity and fetal development ($p < .07$). The relationship between glycolysis and viability has been further analyzed (36). The techniques used to assess embryo development and metabolism can be found in chapters 3, 8, and 9. (From Lane and Gardner [25], with permission from *Reproduction*.)

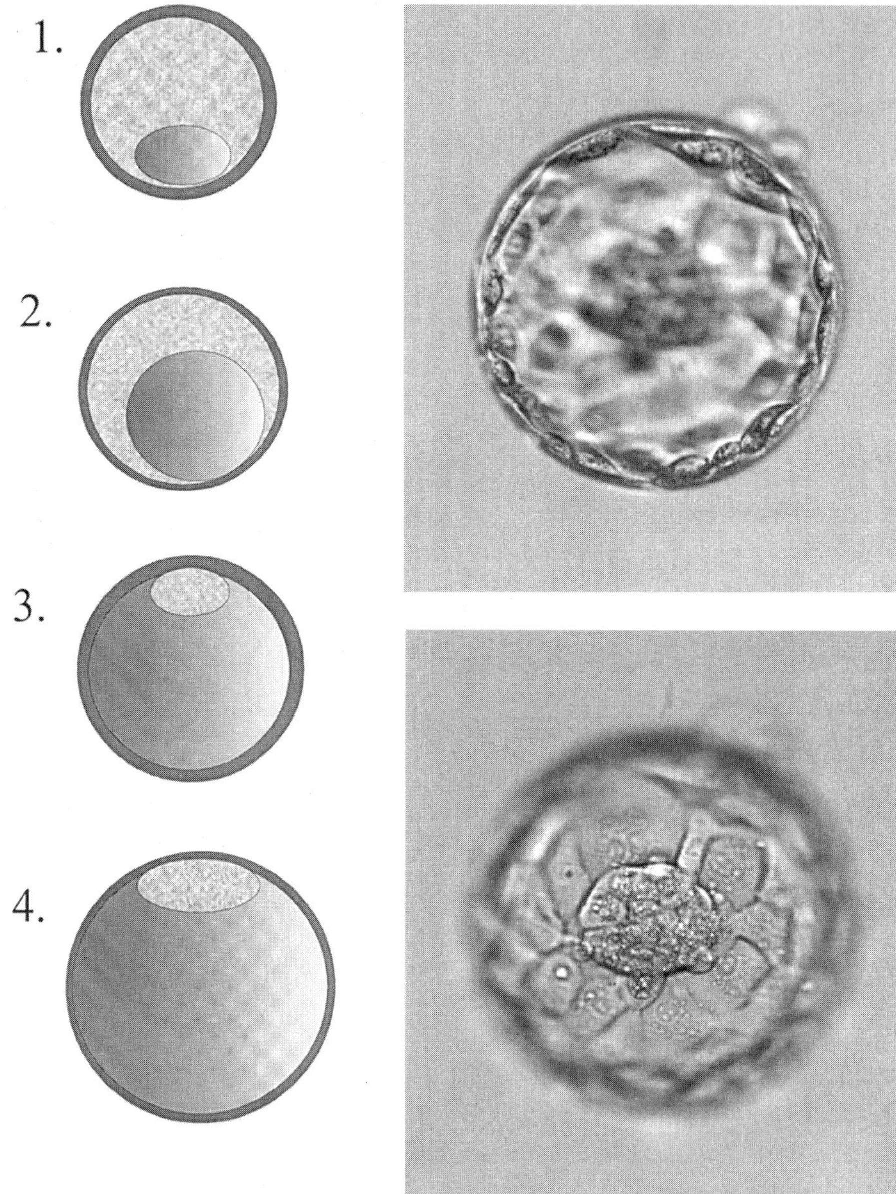

Figure 4.5. Blastocyst scoring system used to select embryos for transfer. (a) Initially blastocysts are given a numerical score from 1 to 6 based on their degree of expansion and hatching status: (1) early blastocyst; the blastocoel being less than half the volume of the embryo; (2) blastocyst; the blastocoel being greater than or equal to half of the volume of the embryo; (3) full blastocyst; the blastocoel completely fills the embryo; (4) expanded blastocyst; the blastocoel volume is now larger than that of the early embryo and the zona is thinning; (5) hatching blastocyst; the trophectoderm has started to herniate though the zona; (6) hatched blastocyst; the blastocyst has completely escaped from the zona. The initial phase of the assessment can be performed on a dissection microscope. The second step in scoring the blastocysts should be performed on an inverted microscope. For blastocysts graded as 3–6 (i.e., full blastocysts onward) the development of the inner cell mass (ICM) and trophectoderm can then be assessed. *ICM grading*: (A) tightly packed, many cells; (B) loosely grouped, several cells; (C) very few cells; (D) degenerate ICM. *Trophectoderm grading*: (A) many cells forming a cohesive epithelium; (B) few cells forming a loose epithelium; (C) very few, large cells; (D) degenerate cells. (b) Photomicrograph of a human blastocyst (score 4AA morning of day 5). The blastocoel cavity is expanding and the zona is thinning (score of 4). The trophectoderm is composed of numerous cells (score of A). Blastocyst diameter is 140 μm. (c) Photomicrograph of the same human blastocyst in panel b taken in a different focal plane. The ICM is visible and is composed of numerous tightly packed cells (score of A). (Adapted from Gardner and Schoolcraft [45] with permission of Parthenon Publishing.)

be obtained, the crucial factor was when they became blastocysts. When taking this into account, the best blastocysts would be those that develop by day 5. In the bovine model it has also been demonstrated that those embryos that form blastocysts earlier are the most viable (29).

A number of studies have attempted to investigate whether the embryos that are growing the best at the earlier stages of development are also those that reach the blastocyst stage. Although this seems logical, the results are not conclusive. For example, Graham et al. (30) reported that the criteria for embryo selection on day 3 seem to be inadequate for selecting blastocysts. In contrast, Shapiro et al. (31) found a predictive value of 72-h blastomere cell number on blastocyst development and success of subsequent transfer based on the degree of blastocyst development. Similarly, Langley et al. (32) showed that there was a relationship between embryo cell number on day 3 and the potential to form a blastocyst (Figure 4.6).

3. ASSESSMENT OF METABOLISM TO SELECT EMBRYOS

Renard et al. (33) observed that day-10 cattle blastocysts which had a glucose uptake > 5 μg/h developed better both in culture and in vivo after transfer than those blastocysts with a glucose uptake below this value. However, due to the insensitivity of the spectrophotometric method used, they could not quantitate glucose uptake by younger embryos. Using the technique of noninvasive microfluorescence, Gardner and Leese (34) measured glucose uptake by individual day-4 mouse blastocysts before transfer to recipient females. Those embryos that went to term had a significantly higher glucose uptake in culture than those embryos that failed to develop after transfer. However, such studies were retrospective and as such could not conclusively demonstrate whether it was possible to identify viable embryos before transfer using metabolism as a marker. Therefore, in a prospective study, Lane and Gardner (35) used glucose uptake and lactate production to determine glycolytic activity in individual day-5 mouse blastocysts before transfer. Blastocysts were classified as viable or nonviable according to their rate

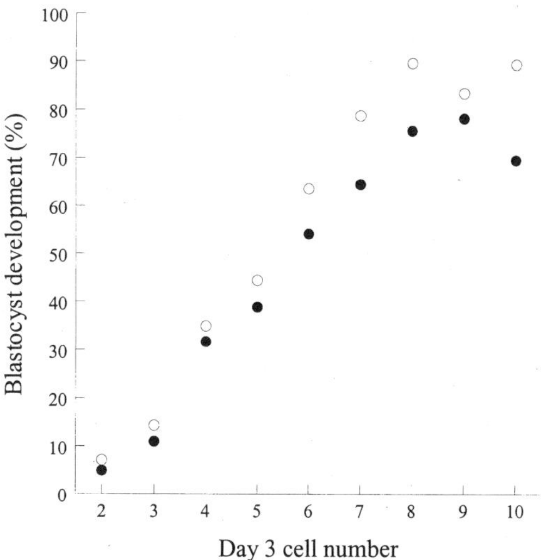

Figure 4.6. Relationship between cell number and blastocyst development on day 3. The filled circles represent IVF patients, the open circles represent the oocyte donors. There is a significant linear trend between cell number and blastocyst development from the two- to eight-cell stage for both groups of patients (linear regression; $r = .96$, $p < .01$). (From Langley et al. [32], with permission of the European Society of Human Reproduction and Oxford University Press.)

of glucose uptake and lactate production. Metabolism of mouse blastocysts of equal dimensions and morphology was quantitated noninvasively and the blastocysts were classified as either viable or nonviable. Those blastocysts that exhibited a pattern of glycolytic utilization similar to that of embryos developed in vivo (high glucose uptake and low lactate production) had a developmental potential of 80%, whereas those blastocysts that exhibited an excessive lactate production (i.e., aberrant glycolytic activity) had a developmental potential of only 6%. Therefore, both the rate of nutrient uptake and its subsequent fate are important determinants of embryo viability.

Similarly, studies on nutrient uptake and subsequent viability have been performed on the human embryo. In a retrospective analysis, Conaghan et al. (36) observed an inverse relationship between pyruvate uptake by two- to eight-cell embryos and subsequent viability. However, it is important to note that such measurements were performed before human embryonic genome activation. It is therefore plausible that the observed differences in pyruvate uptake reported by Conaghan et al. (36) reflected differences inherited from the oocyte and did not represent the true physiology of the later stage embryo. Furthermore, the medium used to assess nutrient consumption was a simple one lacking amino acids and vitamins. In a study on human morulae and blastocysts of different degrees of expansion, no conclusive data were generated on the ability of nutrient consumption to predict pregnancy outcome (37). Again, however, the medium used to assess embryo metabolism was a simple one, lacking pyruvate, lactate, amino acids, and vitamins. Under such severe culture conditions the resultant stress on the embryos would have been enormous, and therefore it is questionable whether any meaningful data could have been obtained. In fact, one would expect embryos undergoing such a treatment to be compromised. In contrast, Van den Bergh et al. (38) showed that in patients who conceived following blastocyst transfer, their embryos had a higher glucose uptake than those blastocysts that failed to establish a pregnancy. Similar to the data reported on the mouse (35), viable human blastocysts had a significantly lower glycolytic activity than those embryos that did not establish a pregnancy. Importantly, in the work of Van den Bergh et al. (38), a compete medium was used for the metabolic assessment, thereby alleviating the culture-induced stress associated with the work of Jones et al. (37).

Two studies have determined the relationship between human embryo nutrition and subsequent development in vitro. Gardner et al. (39) determined that glucose consumption on day 4 by human embryos was twice as high in those embryos that went on to form blastocysts. Furthermore, blastocyst quality affected glucose uptake. Poor-quality blastocysts consumed significantly less glucose than top-scoring embryos. Significantly, within a cohort of human blastocysts from the same patient with the same alphanumeric score (i.e., 4AA), there was a significant spread of metabolic activities. These embryos were cultured in sequential media G1 and G2 and their metabolism assessed in the medium G2 to prevent metabolic transformation. Therefore, assessing metabolic activity and metabolic normality may prove to be a feasible way to determine embryonic health. In a study on amino acids, Houghton et al. (40) determined that alanine release into the surrounding medium on day 2 and day 3 was highest in those embryos that did not form blastocysts. Details of metabolic analysis can be found in chapter 8.

The assessment of embryo metabolism to select viable embryos may be most effective when examining embryos after cryopreservation. A study on the metabolism of day-7 cattle blastocysts before and after cryopreservation showed that it was possible to identify those blastocysts capable of reexpansion in the hours immediately after thaw (41). Those blastocysts that survived the freeze–thaw procedure had a significantly higher glucose uptake and lactate production than those embryos that did not reexpand and subsequently died. Of greater significance, however, was the observation that there was no overlap in the distribution of glucose uptake by the viable and nonviable embryos and very little overlap of lactate production, suggesting that it is feasible to use metabolic criteria for the prospective selection of viable embryos post thaw.

4. APPLICATION OF SEQUENTIAL ANALYSIS OF EMBRYOS

The selection criteria discussed thus far have all shown benefits in identifying individual embryos that have a higher viability. A multiple-step scoring system that encompasses all the above criteria would allow us to allocate a score to those embryos that attain a defined hurdle at each step of assessment. The scoring would therefore be based on positive points at each step (Protocol 4.1). Embryos that do not show the required pattern of development at the specific time would not score. To implement this system, embryos would need to be cultured singularly.

4.1 Examples of Data Collected

The criteria established are seen as ideal hurdles of development. At every step an embryo would be given a set score when it reached the ideal criteria of a certain stage. Only embryos reaching the desired hurdle obtain a score. It would, however, be possible that an embryo may not pass one step, but would pass the hurdle at a following step. The embryo or embryos attaining points at each step would therefore be the ones that would be selected for transfer. For example, if one is attempting to transfer a single embryo to a patient, the following scenario could be envisaged. An embryo may not pass any of the earlier hurdles but still form a high-grade blastocyst on day 5. If this were the most successful of the cohort of embryos, then this would be the embryo selected. If, however, six blastocysts were observed on day 5, all of equally high grade, then the blastocyst that had achieved the most positive scores at each of the previous hurdles could be transferred. Furthermore, patients who have low numbers of embryos and may have transfer on day 2 or 3 could be assessed using the initial criteria, and the embryo that passed the initial hurdles would be selected. A proposed schedule of embryo selection is given in Figure 4.7. Examples of an embryo scoring schedule are also given in Figure 4.8. It is important to note that to date the strongest criteria of selection appear to be a high-quality blastocyst on day 5 of development; hence the scoring system is biased to attribute a higher value to embryos reaching this stage. These systems can be readily adapted for different species.

5. TROUBLESHOOTING

One practical issue for performing such a selection process is that embryos would have to be cultured in individual drops. This may remove any necessary benefits of culturing embryos in groups (33). A further practical issue is that embryos will need to be observed more often; however, using drop culture systems under oil should allay this. The extra observations, if performed under a controlled heated and gassed climate, should not be detrimental to the embryo. In prolonged culture, pronuclear assessment, change over into new media on day 3, and assessment of the blastocyst is already performed. The extra assessment periods would be the check of early-cleaving two-cell embryos and assessment of embryos on day 2 and day 4. One optional observation would be that of the polar body placement, as described by Garello et al. (9). This assessment criterion involves photography, followed by calculation from the photograph, which would involve a further manipulation of the zygote once the polar body displacement has been calculated.

6. FUTURE POSSIBILITIES OF ASSESSMENT

The expanding knowledge in molecular techniques and the ability to assess minute amounts of material will surely benefit single embryo assessment methods. Indeed, we have al-

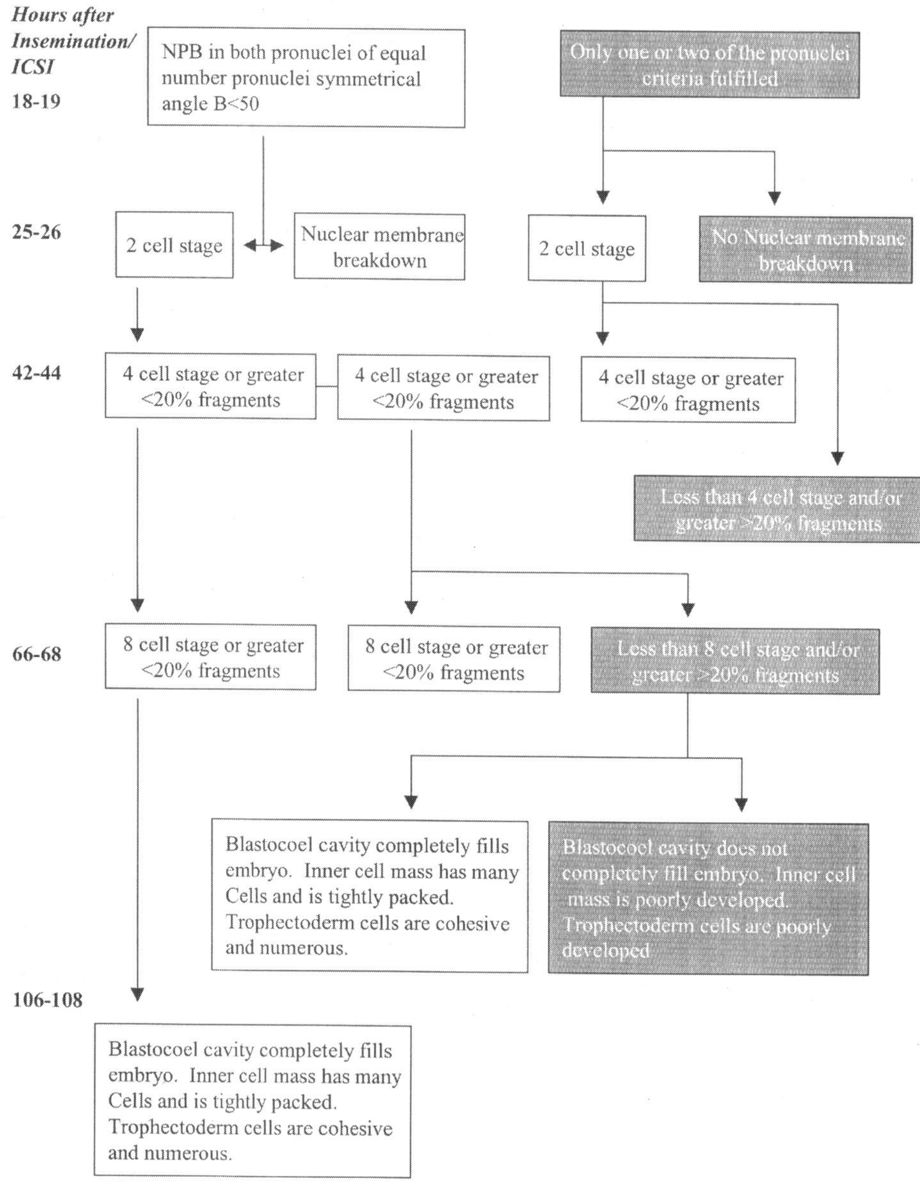

Figure 4.7. A strategy for the sequential analysis of embryo development. NPB, nucleolar precursor bodies.

ready seen certain techniques that have used chromosomal analysis of polar bodies or single blastomeres to improve the chances of older patients achieving pregnancy (42–44).

7. CONCLUSION

For many years the preimplantation embryo was considered a static entity. However, it seems illogical to make a single static assessment on the day of transfer and believe that this is indicative of a whole series of complex developmental processes. The embryo passes numerous hurdles as it develops, and we must assess each of these hurdles so that we can select the most viable embryo. In adopting this strategy we will be able to maintain high implantation and pregnancy rates.

Patient Name: Date of Birth:

Date:

Time (h post insem / ICSI)		Score	1	2	n
DAY 1	Equal size and symmetry of PN	10			
18-19	Alignment between the PN and polar bodies	5			
	Lack of heterogeneity and granularity in cytoplasm	10			
	Presence of PN with both polarized or both not-polarized NPB	5			
	A difference of less than 3 in the number of NPB in the PN	5			
	Polar bodies are not displaced from each other	5			
25-26	2-cell embryo	10			
	Nuclear membrane breakdown	5			
DAY 2	**Cell Number**				
44-46	**Grade**				
	Number of blastomeres ≥ 4	10			
	Fragmentation of less than 20%	10			
	No multinucleated blastomeres	5			
DAY 3	**Cell Number**				
66-68	**Grade**				
	Number of blastomeres ≥ 8	10			
	Fragmentation of less than 20%	10			
	No multinucleated blastomeres	5			
DAY 4	**Cell Number / Stage**				
94-96	**Grade**				
	Blastocoel cavity	15			
	Compaction	10			
DAY 5	**Stage**				
106-108	**Grade**				
	Blastocyst - 4AA	25			
	Blastocyst - 3AA	25			
	Blastocyst - 4 with one A	15			
	Blastocyst - 3 with one A	15			
	Blastocyst <= 2	10			
EMBRYO TOTAL					

Figure 4.8. Sequential scoring system for human embryos.

Protocol 4.1. Sequential assessment of embryos

Day 1

1. Quickly identify any oocyte(s) present and carefully strip away the surrounding cumulus cells using micropipette.
2. Place in wash droplet.

3. Observe and/or photograph denuded embryo(s) and note the number of pronuclei and polar bodies present.
4. Quickly transfer the fertilized embryos to specific drops in day-1 plate and assess features.
5. Note all observations on score sheet.
6. At 25–26 h after insemination/ICSI, assess embryos for cleavage to the two-cell stage and note on worksheet.

Day 2

1. At 42–44 h after insemination/ICSI, assess embryos for cleavage to the four-cell or greater stage and for fragmentation and note on worksheet.
2. Prepare a separate dish by placing culture medium drops onto base of the dish using the next sequential embryo culture media if progressing to blastocyst stage (see chapter 3). Label 1, 2, 3, 4, and so on, and C (center).

Day 3

1. Assess embryos 66–68 h after insemination/ICSI for cleavage to the eight-cell or greater stage and for fragmentation and note on worksheet.
2. Transfer all embryos to drops in the new culture media if using sequential culture media.

Day 4

1. Assess embryos 94–98 h after insemination/ICSI for early signs of blastocoel cavity and compaction.
2. Note on worksheet.

Day 5

1. Assess embryos for the presence of a blastocoel cavity.
2. Grade as to their stage (see figure 4.5).

Select embryos for transfer according to those with the highest total score on the day of transfer.

References

1. Trounson, A. O., Leeton, J. F., Wood, C., Webb, J., and Wood, J. (1981). *Science*, **212**, 681.
2. Martin, P. M., and Welch, H. G. (1998). *Fertil. Steril.*, **70**, 478.
3. Templeton, A. (2000). *Hum. Reprod.*, **15**, 1662.
4. Braude, P., Bolton, V., and Moore, S. (1988). *Nature*, **332**, 459.
5. Tesarik, J., and Greco, E. (1999). *Hum. Reprod.*, **14**, 1318.
6. Gardner, R. L. (1997). *Development*, **124**, 289.
7. Scott, L. A., and Smith, S. (1998). *Hum. Reprod.*, **13**, 1003.
8. Antczak, M., and Van Blerkom, J. (1999). *Hum. Reprod.*, **14**, 429.
9. Garello, C., Baker, H., Rai, J., Montgomery, S., Wilson, P., Kennedy, C. R., and Hartshorne, G. M. (1999). *Hum. Reprod.*, **14**, 2588.
10. Scott, L., Alvero, R., Leondires, M., and Miller, B. (2000). *Hum. Reprod.*, **15**, 2394.
11. Cummins, J. M., Breen, T. M., Harrison, K. L., Shaw, J. M., Wilson, L. M., and Hennessey, J. F. (1986). *J. In Vitro Fertil. Embryo Transf.*, **3**, 284.
12. Edwards, R. G., Fishel, S. B., Cohen, J., Fehilly, C. B., Purdy, J. M., Slater, J. M., Steptoe, P. C., and Webster, J. M. (1984). *J. In Vitro Fertil. Embryo. Transf.*, **1**, 3.
13. Steer, C. V., Mills, C. L., Tan, S. L., Campbell, S., and Edwards, R. G. (1992). *Hum. Reprod.*, **7**, 117.
14. Bavister, B. D. (1995). *Hum. Reprod. Update*, **1**, 91.
15. Shoukir, Y., Campana, A., Farley, T., and Sakkas, D. (1997). *Hum. Reprod.*, **12**, 1531.
16. Sakkas, D., Shoukir, Y., Chardonnens, D., Bianchi, P. G., and Campana, A. (1998). *Hum. Reprod.*, **13**, 182.

17. Sakkas, D., Percival, G., D'Arcy, Y., Sharif, K., and Afnan, M. (2001). *Fertil. Steril.*, **76**, 1150.
18. Fenwick, J., Platteau, P., Murdoch, A. P., and Herbert, M. (2002). *Hum. Reprod.*, **17**, 407.
19. Gerris, J., De Neubourg, D., Mangelschots, K., Van Royen, E., Van de Meerssche, M., and Valkenburg, M. (1999). *Hum. Reprod.*, **14**, 2581.
20. Van Royen, E., Mangelschots, K., De Neubourg, D., Valkenburg, M., Van de Meerssche, M., Ryckaert, G., Eestermans, W., and Gerris, J. (1999). *Hum. Reprod.*, **14**, 2345.
21. Buster, J. E., Bustillo, M., Rodi, I. A., Cohen, S. W., Hamilton, M., Simon, J. A., Thorneycroft, I. H., and Marshall, J. R. (1985). *Am. J. Obstet. Gynecol.*, **153**, 211.
22. Gardner, D. K., Vella, P., Lane, M., Wagley, L., Schlenker, T., and Schoolcraft, W. B. (1998). *Fertil. Steril.*, **69**, 84.
23. Gardner, D. K., Schoolcraft, W. B., Wagley, L., Schlenker, T., Stevens, J., and Hesla, J. (1998). *Hum. Reprod.*, **13**, 3434.
24. Behr, B., Pool, T. B., Milki, A. A., Moore, D., Gebhardt, J., and Dasig, D. (1999). *Hum. Reprod.*, **14**, 454.
25. Lane, M., and Gardner, D. K. (1997). *J. Reprod. Fertil.*, **109**, 153.
26. Gardner, D. K., Lane, M., Stevens, J., Schlenker, T., and Schoolcraft, W. B. (2000). *Fertil. Steril.*, **73**, 1155.
27. Balaban, B., Urman, B., Sertac, A., Alatas, C., Aksoy, S., and Mercan, R. (2000). *Fertil. Steril.*, **74**, 282.
28. Shoukir, Y., Chardonnens, D., Campana, A., Bischof, P., and Sakkas, D. (1998). *Hum. Reprod.*, **13**, 676.
29. Hasler, J. F. (1992). *J. Dairy Sci.*, **75**, 2857.
30. Graham, J., Han, T., Porter, R., Levy, M., Stillman, R., and Tucker, M. J. (2000). *Fertil. Steril.*, **74**, 495.
31. Shapiro, B. S., Harris, D. C., and Richter, K. S. (2000). *Fertil. Steril.*, **73**, 582.
32. Langley, M. T., Marek, D. M., Gardner, D. K., Doody, K. M., and Doody, K. J. (2001). *Hum. Reprod.*, **16**, 902.
33. Renard, J. P., Philippon, A., and Menezo, Y. (1980). *J. Reprod. Fertil.*, **58**, 161.
34. Gardner, D. K., and Leese, H. J. (1987). *J. Exp. Zool.*, **242**, 103.
35. Lane, M., and Gardner, D. K. (1996). *Hum. Reprod.*, **11**, 1975.
36. Conaghan, J., Hardy, K., Handyside, A. H., Winston, R. M., and Leese, H. J. (1993). *J. Assist. Reprod. Genet.*, **10**, 21.
37. Jones, G. M., Trounson, A. O., Vella, P. J., Thouas, G. A., Lolatgis, N., and Wood, C. (2001). *Reprod. BioMed. Online*, **3**, 124.
38. Van den Bergh, M., Devreker, F., Emiliani, S., and Englert, Y. (2001). *Reprod. BioMed. Online*, **3**, **Suppl. 1**, 8.
39. Gardner, D. K., Lane, M., Stevens, J., and Schoolcraft, W. B. (2001). *Fertil. Steril.*, **76**, 1175.
40. Houghton, F. D., Hawkhead, J. A., Humpherson, P. G., Hogg, J. E., Balen, A. H., Rutherford, A. J., and Leese, H. J. (2002). *Hum. Reprod.*, **17**, 999.
41. Gardner, D. K., Pawelczynski, M., and Trounson, A. O. (1996). *Mol. Reprod. Dev.*, **44**, 472.
42. Gianaroli, L., Magli, M. C., Munne, S., Fiorentino, A., Montanaro, N., and Ferraretti, A. P. (1997). *Hum. Reprod.*, **12**, 1762.
43. Gianaroli, L., Magli, M. C., Ferraretti, A. P., and Munne, S. (1999). *Fertil. Steril.*, **72**, 837.
44. Verlinsky, Y., Cieslak, J., Ivakhnenko, V., Evsikov, S., Wolf, G., White, M., Lifchez, A., Kaplan, B., Moise, J., Valle, J., Ginsberg, N., Strom, C., and Kuliev, A. (1997). *Genet. Test.*, **1**, 231.
45. Gardner, D. K., and Schoolcraft, W. B. (1999). In: *Towards Reproductive Certainty: Fertility and Genetics Beyond 1999*, R. Jansen and D. Mortimer, eds., p. 378. Parthenon Press, Carnforth.
46. Sakkas, D. (2001). In: *Textbook of Assisted Reproductive Techniques*, D. K. Gardner, A. Weissman, C. M. Howles, Z. Shoham, eds., p. 223. Martin Dunitz, London.

5

GIANPIERO D. PALERMO

QUEENIE V. NERI

TAKUMI TAKEUCHI

LUCINDA L. VEECK

ZEV ROSENWAKS

Micromanipulation of Gametes

1. MALE-FACTOR INFERTILITY AND ICSI

Since the first human birth following in vitro fertilization (IVF) in 1978, this procedure has been used extensively to treat infertility. However, in many cases of male-factor infertility spermatozoa cannot fertilize, and a number of micromanipulation techniques have been developed to overcome this inability. Micromanipulation of human gametes has not only allowed fertilization in cases of severe oligozoospermia, even by defective spermatozoa, but has provided a powerful tool for exploring the basic elements of oocyte maturation, fertilization, and early development. Micromanipulation techniques also now permit the diagnosis and sometimes even the correction of genetic anomalies, as well as optimization of implantation in certain cases.

Intracytoplasmic sperm injection (ICSI) is now the primary micromanipulation technique used for treating male-factor infertility. It entails the mechanical insertion of a selected spermatozoon directly into the cytoplasm of a human oocyte. This procedure was first applied to human gametes in 1988 (1), but human pregnancies were reported first only 4 years later (2). Since that time, ICSI has achieved consistent fertilization and high pregnancy rates in couples with suboptimal spermatozoa (2–7). The success of this procedure appears not to be influenced by the collection method, origin or quality of the semen, or by the presence of antisperm antibodies. The fact that virtually any spermatozoon can induce oocyte activation and pronucleus formation has extended the use of ICSI to azoospermic patients as well.

Until the establishment of ICSI, azoospermia was considered to be a poorly treatable form of male-factor infertility. In fact, the fertilization rate obtained with standard in vitro insemination in male-factor patients was 10–13%. The application of ICSI to testicular spermatozoa has been used to treat azoospermia successfully (8), with the first ICSI series using epididymal spermatozoa reported soon thereafter (9). Although the fertilization rates obtained with surgically retrieved spermatozoa were satisfactory, they were significantly lower than those achieved with fresh ejaculates. Because surgically retrieved spermatozoa appeared to benefit from a more aggressive immobilization procedure (10), the reason was attributed possibly to a higher lipid content of the sperm membrane. Aggressive membrane permeabilization may exert its effect by facilitating the release of a sperm cytosolic factor that is able to activate the oocyte (11, 12).

Approximately 10% of male-factor infertility has been attributed to complete azoospermia, which presents either in an obstructive or a nonobstructive form. In obstructive cases, spermatozoa are retrieved by microsurgical epididymal sperm aspiration or by percutaneous epididymal sperm aspiration, and only when the epididymal access is lacking is testicular sampling deemed appropriate. Obstructive azoospermia is characterized by a normal sperm production and often by congenital absence of the vas deferens, a condition linked to a cystic fibrosis gene mutation. Nonobstructive azoospermia, in contrast, is characterized by a varying degree of spermatogenic failure and is often associated with an increased number of chromosomal abnormalities (13, 14). The only method for retrieval in the latter case involves direct biopsy of the testis (15). More recently, the testicular fine needle aspiration (TEFNA) technique has been introduced as an alternative to the open biopsy approach (16).

ICSI has also proven to be useful when only a low number of oocytes are retrieved and when oocytes and embryos are to be considered for preimplantation genetic diagnosis (PGD). Because analysis of the polar body requires corona cell removal and opening of the zona pellucida, ICSI becomes the only option for avoidance of polyspermy in such cases. Cytogenetic analysis of embryos for PGD, particularly if performed by PCR, also benefits from the absence of potentially contaminating spermatozoa attached to the zona pellucida. In presenting a realistic option for treatment of sperm-related problems, ICSI has allowed embryologists to concentrate their efforts on another important limiting factor for conception: oocyte/embryo quality. In a significant proportion, failure of conception is linked to embryo abnormalities that can perhaps be attributed to cytosolic defects at the respiratory or metabolic level. In some instances, however, these embryos can be rescued by transferring ooplasm collected from a normal donor oocyte (17).

On the other hand, some genetically abnormal oocytes may develop into morphologically normal embryos, but these either are unable to implant or, if they develop to the fetal stage, do not progress to term. It is well known that many such oocytes have a chromosomal abnormality, an anomaly seen more commonly with advancing maternal age, and attempts are underway to avoid oocyte aneuploidy. This can be accomplished by micromanipulation that involves nuclear transplantation (18), an approach that requires not only removal of cumulus cells but opening of the zona pellucida, so ICSI is the method of insemination.

In this chapter, we review the different techniques of sperm and oocyte manipulation that are used for ICSI and consider the techniques that hold promise in the treatment of oocyte/embryo abnormalities.

2. INDICATIONS FOR ICSI

Despite agreement in some areas, no universal standards for patient selection have emerged in regard to ICSI. There is a general consensus that ICSI should be performed following fertilization failure in standard IVF with putatively normal oocytes, and where an appropriate sperm concentration was used even in microdrops—useful criteria for all male-factor patients, even those who have not been treated before.

Although oocytes that failed to fertilize with standard IVF techniques can sometimes be reinseminated successfully by ICSI, this introduces a risk of creating embryos from aged eggs (19). In our own limited experience, six of eight pregnancies established by micromanipulation of such oocytes miscarried, and cytogenetic studies performed on the aborted fetuses provided evidence of chromosomal abnormalities. Thus, notwithstanding a recent report of normal pregnancies (20), we currently reinseminate unfertilized oocytes only for research purposes.

In regard to sperm numbers, a count of $< 5 \times 10^6$/ml generally reduces the likelihood of fertilization with normal IVF procedures, regardless of etiology (21), and $< 0.5 \times 10^6$/ml with $< 4\%$ normal forms in the initial ejaculate is considered unsuitable for standard IVF. In addition, occasional fertilization failure may occur with an apparently normal ejaculate (22), possibly due to spontaneous hardening of the zona pellucida after in vitro culture (23) or to an inherently impenetrable zona around oocytes that often reveal ooplasmic inclusions (24, 25). The rare sperm abnormality that prevents fusion of the perivitelline spermatozoon with the oolemma (26) also justifies ICSI. In many instances, however, failure of fertilization is due to multiple sperm abnormalities common in cases of severe oligo-, astheno-, or terato-zoospermia as defined by the World Health Organization (27), and there is a clear consensus that in such cases where IVF rates drop to $< 10\%$ (28), ICSI is the only treatment option (29).

2.1 Ejaculated Spermatozoa

Semen samples are collected by masturbation preferably after 3 days of abstinence and are first allowed to liquefy for at least 20 min at 37°C. Viscous samples are first incubated for 3–5 min in 2–3 ml of HEPES-buffered medium (M-HEPES) containing 200–500 IU of chymotrypsin (Sigma Chemical Co., St. Louis, MO). See Tables 3.1 to 3.3.

After evaluating concentration and motility in a counting chamber, 5 µl of a sperm suspension is spread on prestained slides (Testsimplets, Boehringer Mannheim) to assess morphology and the presence of other cells or bacteria. Morphological parameters are evaluated according to Kruger et al. (30), and a sample is considered to be compromised when the sperm density is $< 20 \times 10^6$/ml, total motility is $< 40\%$, and normal morphology is $< 5\%$.

Adequate samples ($\geq 20 \times 10^6$/ml concentration with $\geq 40\%$ motility) are washed by centrifugation at 500 g for 5 min in culture medium supplemented with 5% human serum albumin (Plasmanate), and sperm are then allowed to swim up. Suboptimal samples are washed in HTF by a single centrifugation at 1,800 g for 5 min. The resuspended pellet is layered on a three-layer discontinuous density gradient (90–70–50%) and centrifuged at 300 g for 20 min. A two-layer density gradient is used (95–47.5%) when samples have a sperm density of $< 5 \times 10^6$/ml and $< 20\%$ motility. Spermatozoa are isolated from the high-density fraction by double centrifugation (1,800 g for 5 min) after adding 4–5 ml of culture medium. When necessary, the final suspension can be adjusted to $1–1.5 \times 10^6$ spermatozoa/ml by adding medium. Samples are incubated at 37°C in a gas atmosphere of 5% CO_2 in air until use.

2.2 Surgically Retrieved Spermatozoa

2.2.1 Epididymal Spermatozoa

Spermatozoa from men with irreparable obstructive azoospermia are obtained by microsurgical epididymal sperm aspiration (31, 32) or percutaneous epididymal sperm aspiration (33) and, when epididymal access is lacking, by testicular sampling. Variable volumes of fluid (1–500 µl) generally containing concentrated spermatozoa are aspirated from the epididymal lumen with a glass micropipette or a metal needle. The preparation and selection of epididymal spermatozoa is performed as described above.

2.2.2 Testicular Spermatozoa

The testicular approach, by open biopsy or by fine needle aspiration, is used for azoospermic men with nonobstructive azoospermia as well as obstructive cases when the epididymal approach is not possible. A biopsy of approximately 500 mg is rinsed in medium to remove red blood cells and divided into small pieces under a stereomicroscope (34). Motility or twitching is then assessed at 100–200× magnifaction, and a second biopsy specimen is obtained if spermatozoa are not found. Spermatozoa can be released by shredding the testicular tissue with two glass slides or fine tweezers, producing unraveled and broken tubules. Other methods include vortexing or crushing the biopsy in a tissue homogenizer. If more than 10×10^6/ml red blood cells are present, the sample is also treated with an erythrocyte-lysing buffer solution (155 mM NH_4Cl, 10 mM $KHCO_3$, 2 mM EDTA, pH 7.2) (35), after which the sperm suspensions are layered on a discontinuous two-layer density gradient (95–47.5%). After centrifugation, the pellet from the 95% fraction is prepared as described above.

2.2.3 Cryopreservation of Surgically Retrieved Spermatozoa

For cryopreservation of excess epididymal spermatozoa (35), a suspension (at a concentration of approximately 30×10^6/ml) is diluted volume for volume with an equal amount of cryopreservation medium (Freezing Medium-Test Yolk Buffer with Glycerol; Irvine Scientific, Irvine, CA). Up to 1-ml aliquots are placed in 1-ml cryogenic vials (Nalgene Brand Products, Rochester, NY), kept at –20°C for 35 min, exposed to liquid N_2 vapor at –70°C for 10 min, and then plunged in liquid N_2 at –196°C. When required, vials are thawed at room temperature. In regard to testicular spermatozoa, the suspension of microdissected testicular tissue is cryopreserved similarly by the addition of equal amounts volume for volume of cryopreservation medium (Irvine Scientific) and processed as described above.

2.3 Oocytes

Oocytes are retrieved after pituitary desensitization with a gonadotropin-releasing hormone agonist and follicle stimulation with a combination of human menopausal gonadotropins (hMG) (Pergonal, Serono, Waltham, MA; Humegon, Organon Inc., West Orange, NJ), and FSH (Gonal-F, Serono; Follistim, Organon). Human chorionic gonadotropin (hCG) is administered when criteria (e.g., ultrasound, estradiol levels) for oocyte maturity are met, and retrieval is performed 35 h later by vaginal ultrasound-guided puncture. Under the inverted microscope at 100×, the cumulus–corona-cell complexes are scored as mature, slightly immature, completely immature, or slightly hypermature. Thereafter, the oocytes are incubated up to 4 h depending on their state. Immediately before micromanipulation, the cumulus-corona cells are necessarily removed for oocyte observation and accurately controlled by the use of the holding and/or injecting pipette. Such removal involves oocyte exposure to M-HEPES containing 40 IU/ml of hyaluronidase (type VIII, Sigma), then hand-drawn aspiration in and out using a Pasteur pipette with an inner diameter of approximately 200 μm. In assessing oocyte maturation and integrity, metaphase II (MII; the only suitable stage for ICSI) is judged according to the absence of the germinal vesicle and the presence of an extruded polar body.

2.4 Microinjection

2.4.1 Setting

Immediately before injection, 1 μl of the sperm suspension is diluted with 4 μl of a 10% PVP solution (approximately 300 mOsmol; PVP-K 90, molecular weight 360,000; ICN Biochemicals, Cleveland, OH) in M-HEPES medium placed in the middle of the plastic Petri dish. The viscosity of this solution slows down the aspiration and prevents sperm cells from sticking to the injection pipette. Avoidance of too many spermatozoa in the PVP-containing droplet also reduces the level of contaminants that can be carried into the oocytes during injection. However, when < 500,000 spermatozoa are available *in toto*, the suspension is first concentrated in approximately 3 μl and transferred directly to the injection dish, which contains M-HEPES supplemented with 10% plasmanate). Each oocyte is placed in a 5-μl droplet of this medium surrounding the central drop containing the sperm suspension. The droplets are covered with lightweight paraffin oil. Spermatozoa are aspirated from the central droplet or the concentrated 3 μl sperm suspension drop and transferred into the droplet containing PVP to remove debris and gain aspiration control.

Injection is performed on a heated stage (Easteach Laboratory, Centereach, NY) fitted on a Nikon Diaphot inverted microscope at 400× magnification using Hoffman Modulation Contrast optics. This microscope is equipped with two motor-driven coarse-control manipulators and two hydraulic micromanipulators (MM-188 and MO-109, Narishige Co. Ltd.). The micropipettes are fitted into a tool holder controlled by two IM-6 microinjectors (Narishige).

2.4.2 Sperm Selection and Immobilization

At a magnification of 400×, it is difficult to select spermatozoa on the basis of morphology while they are in motion and without the use of stain. Nonetheless, normally shaped spermatozoa can be selected, to a certain extent, by observing their shape, light refraction, and motion patterns in a viscous medium.

Spermatozoa are aspirated and after release positioned transversely to the tip of the pipette, which is gently lowered to compress the mid-region of the sperm flagellum against the petri dish, thus immobilizing it (Figure 5.1). It is important to note that motile spermatozoa do not fertilize as well as immobilized spermatozoa.

It has been proposed that such immobilization involves a membrane permeabilization that may allow the release of a sperm cytosolic factor that activates the oocyte (11). Interestingly, a more aggressive immobilization is required for surgically retrieved than for ejaculated spermatozoa (10).

2.4.3 Injection

The oocyte is held in place by the suction applied through the holding pipette (Figure 5.2). During the injection procedure, a better grip of the egg is established when the inferior pole of the oocyte is touching the bottom of the dish. The injection pipette is lowered, and the outer right border of the oolemma on the equatorial plane at 3 o'clock is brought

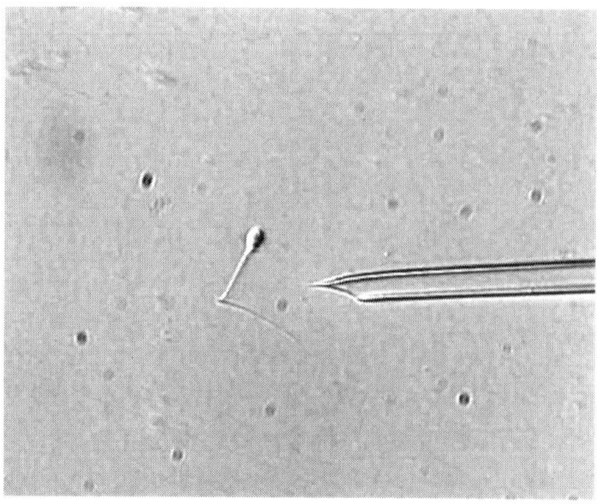

Figure 5.1. Immobilized human spermatozoon. The sperm flagellum is kinked by mechanical action (600×).

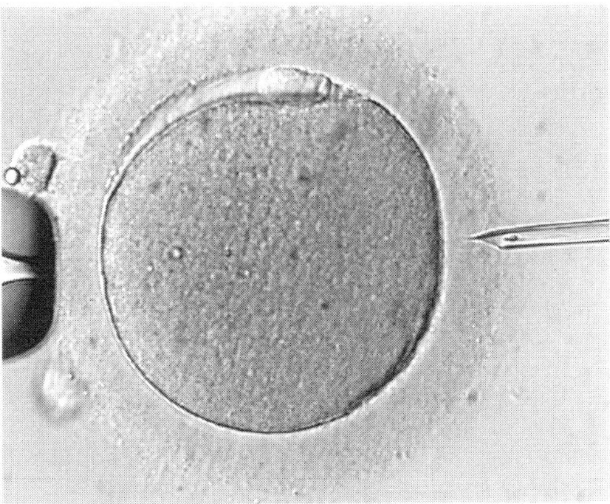

Figure 5.2. Piercing the zona pellucida with the ICSI needle (400×).

into focus. The spermatozoon is then brought close to the beveled opening of the injection pipette (Figure 5.2). This is pushed into the zona to indent the oolemma, at which point a break in the oolemma should occur at the approximate center of the egg. This break is anticipated by a sudden quivering of the convexities of the invaginated oolemma above and below the penetration point, as well as by the uptake of the ooplasmic organelles and a backward movement of the spermatozoon itself up the pipette. The cytoplasmic organelles and the spermatozoon are then slowly released back into the cytoplasm. The aspiration of cytoplasm is thought to provide an additional stimulus for activation of the egg. To optimize its interaction with the oocyte cytosol, the spermatozoon should be ejected beyond the tip of the pipette (Figure 5.3) to maintain an intimate comingling with the organelles during pipette withdrawal. While the pipette remains at the approximate center of the egg, reaspiration of some surplus medium ensures that the cytoplasmic organelles "embrace" the sperm, thereby reducing the size of the breach created in the ooplasm. This also facilitates closure of the terminal part of the funnel-shaped opening at 3 o'clock (Figure 5.4). Once the pipette is removed, the border of the opening should maintain a funnel shape with a vertex into the egg, but if the border becomes inverted the cytoplasmic organelles may leak out, and the oocyte may lyse (36).

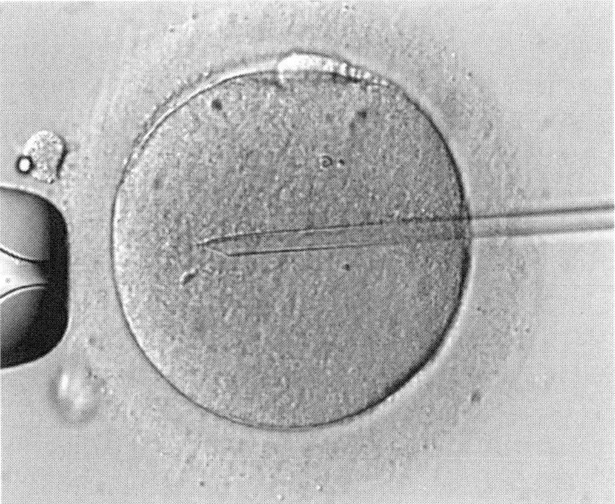

Figure 5.3. Positioning of the spermatozoon into the cytoplasm of a metaphase II oocyte (400×).

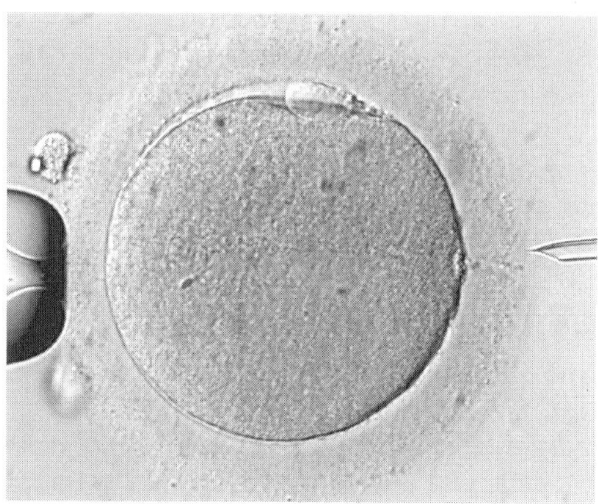

Figure 5.4. Sealing of the oolemma after withdrawal of the injection tool (400×).

3. RESULTS

From September 1993 through June 1999, we performed ICSI in 3360 cycles with ejaculated spermatozoa and in 468 cycles with surgically retrieved spermatozoa. All couples undergoing surgical retrieval were genetically screened and counseled. Mean maternal age was 36.2 ± 5 years for cases involving ejaculated spermatozoa and 34.5 ± 5 years for surgically retrieved spermatozoa. Clinical pregnancy was defined as the presence of a gestational sac as well as at least one fetal heartbeat.

3.1 Ejaculated Spermatozoa

A total of 3360 ICSI cycles were performed with ejaculated spermatozoa: 355 cycles where the semen parameters were normal and 3005 with abnormal parameters according to the World Health Organization and Kruger's strict criteria. After injection, 94.4% of 28,478 oocytes survived, and 75.3% displayed two pronuclei. Though the group sizes varied greatly, the origin of the semen sample when mature spermatozoa were used (fresh or cryopreserved spermatozoa obtained by masturbation, electroejaculation, or bladder catheterization) did influence fertilization rate ($p = .0001$; Table 5.1). When the fertilization and pregnancy rates of ejaculated spermatozoa were compared with surgically retrieved spermatozoa, a significant difference was found ($p = .0001$; Table 5.2).

3.2 Surgically Retrieved Spermatozoa

From a total of 309 cycles performed with epididymal spermatozoa and 159 cycles with testicular spermatozoa, we observed that cryopreservation clearly impaired motility parameters ($p < .0001$) and pregnancy outcome ($p < .001$), though the ICSI fertilization rate was unaffected (Table 5.3). The fertilization and pregnancy rates after ICSI

Table 5.1. Fertilization Rates According to Semen Origin and Collection Method

Semen Origin	Cycles	Oocyte Fertilized/Oocyte Inseminated
Fresh	3,154	20,026/26,539 (75.4%)*
Cryopreserved	161	1,084/1,503 (72.1%)*
Electroejaculate	31	250/312 (80.1%)*
Retrograde ejaculate	6	55/59 (93.2%)*
Cryopreserved electroejaculate	8	40/65 (61.5%)*

*χ^2 2 × 5, 4 *df*; effect of etiology and collection method of spermatozoa on fertilization rate, $P = 0.0001$.

Table 5.2. Fertilization and Pregnancy Rates According to Semen Origin

	Spermatozoa	
	Ejaculated	Surgically Retrieved
Cycles	3,360	468
Fertilizations	21,455/28,478 (75.3%)*	3,163/4,485 (70.5%)*
Clinical pregnancies	1,471 (43.8%)†	253 (54.0%)†

*χ^2 2 × 2, I df; effect of spermatozoal source on fertilization rate, $P = 0.0001$.
†χ^2 2 × 2, 1 df; effect of spermatozoal source on clinical pregnancy rate, $P = 0.0001$.

were lower when testicular spermatozoa were used, but were similar for both fresh and cryopreserved.

3.3 Pregnancy

Among the 3828 ICSI cycles, 56.3% presented with a positive β-hCG (Table 5.4). Of these, 301 were biochemical (7.9%) and 129 were empty sacs (3.4%). A viable fetal heartbeat was observed in 1724 patients of which 192 miscarried or aborted, and 15 had ectopic pregnancies. As of March 1, 2001, the ongoing pregnancy rate was 39.6% per retrieval and 41.6% per replacement. The evolution and outcome of the implanted embryos (Table 5.5) demonstrated an implantation rate of 22.9% sacs with a positive fetal heartbeat. An additional 17.7% of 2786 embryos were lost, as detailed in Table 5.5. The frequency of high-order gestations at 6 weeks was 45.7% (788/1724): 547 twins (31.7%); 211 triplets (12.2%); 28 quadruplets (1.6%); 1 with 5 fetal heartbeats (0.6%) and 1 with 6 fetal heartbeats (0.6%).

Of the 146 miscarriages, 37 cases were karyotyped. All trisomies occurred in women > 35 years old. Furthermore, among 26 patients > 36 years of age (mean 39.0 ± 2), 10 pregnancies were therapeutically terminated because of trisomy 21, two for trisomy 18, one for chromosome 18 translocation, and one case of Klinefelter's syndrome. Other reasons for termination were one neural tube defect, one physical accident, five terminations for unknown reason, and another five were elective. In the age group ≤ 35 years (mean 33.3 ± 2), three patients were aborted due to a mosaic Klinefelter (47,XXY/47,XY), and two because of trisomy 21.

Among a total of 2190 babies born from 1485 deliveries, the gender distribution included 1114 males and 1076 females (ratio 1.04:1), with 516 (34.7%) twin pregnancies, 93 (6.3%) triplet, and 1 (0.07%) quadruplet. ICSI newborns experienced a lower rate of congenital malformations than with standard IVF ($p < .05$; Table 5.6). Only 1.8% of the 2190 newborns exhibited congenital abnormalities: 23 were major and 17 were

Table 5.3. Sperm Parameters and ICSI Outcome with Fresh or Frozen/Thawed Surgically Retrieved Spermatozoa

	Epididymal Spermatozoa		Testicular Spermatozoa	
	Fresh	Frozen/Thawed	Fresh	Frozen/Thawed
Cycles	154	155	137	22
Density (10^6/ml SD)	25.6 ± 36	21.1 ± 27	0.4 ± 1	0.2 ± 0.3
Motility(% ± SD)	18.4 ± 16*	3.0 ± 7*	7.6 ± 13	1.8 ± 6
Morphology (% ± SD)	2.1 ± 3	1.4 ± 2	0	0
Fertilization(%)	1,220/1,656 (73.7)	1,015/1,386 (73.2)	824/1,276 (64.6)	104/167 (62.3)
Clinical pregnancies (%)	103 (66.9)†	73 (47.1)†	69 (50.4)	8 (36.4)

*Student's t-test, two independent samples; effect of cryopreservation on epididymal spermatozoa, $P < 0.0001$.
†χ^2, 2 × 2, 1 df; effect of cryopreservation on clinical pregnancy rate, $P < 0.001$.

Table 5.4. Pregnancy Outcome of 3,573 ICSI Cycles

	No.	No. of Negative Outcomes
ICSI cycles	3,828	—
Embryo transfer	3,650	—
Positive hCG* test results	2,154 (56.3%)	301 biochemical pregnancies; 129 blighted ova
Positive fetal heartbeats	1,724 (45.0%)	192 miscarriages and therapeutic abortions; 15 ectopic pregnancies
Deliveries and ongoing pregnancies	1,517 (39.6%)	—

*hCG indicates human chorionic gonadotropin.

minor (Table 5.7). As indicated, 22 (55.0%) of the malformations occurred in the multifetal gestations.

No differences were found in the miscarriage rate nor in the frequency of congenital malformations as a function of sperm maturity. For example, in cases involving testicular spermatozoa, 98 of 157 patients presented a positive hCG titer, with 7 pregnancies being subsequently aborted. One such baby exhibited a ventricular septal defect. In attempt to discover any relationship between congenital abnormalities and poor sperm quality, couples were divided into three groups: (1) cryptozoospermic (i.e., sperm present only after processing); (2) oligozoospermic ($\leq 1 \times 10^6$ spermatozoa in the ejaculate); and (3) no morphologically normal spermatozoa. Our evidence provides no correlation between congenital anomalies and any abnormal semen parameter (data not shown). In fact, there were no statistical differences in the frequency of congenital malformations between couples considered in these three groups and the remaining couples with less severely compromised semen.

4. CONCERNS ABOUT ICSI

A concern in considering the use of suboptimal spermatozoa for ICSI is the potential for transmitting the genetic abnormalities responsible for male infertility (4, 5, 7, 10, 29, 36–41). Despite this, ICSI is accepted as the only therapeutic option for patients with congenital absence of the vas deferens, associated with a gene deletion labeling these individuals as carriers of cystic fibrosis (42, 43). Similarly, infertile patients with Kartagener's syndrome or other ciliary dyskinesia that renders spermatozoa immotile may transmit this condition to their offspring. In these cases, genetic screening and patient counseling is crucial, and preimplantation genetic testing as well as prenatal evaluation should be considered.

Liebaers et al. (44) reported only 1% of sex chromosomal anomalies after ICSI, and in our ICSI program ($n = 2190$) there was a 1.8% incidence of congenital abnormalities compared to 3.0% in babies born as a result of standard IVF and compared to 3.6% in

Table 5.5. In Vivo Evolution and Pregnancy Outcome of Embryos Generated by ICSI

	No.	No. of Complications in Pregnancy Evolution
Embryos replaced	12,176	—
Embryonic sacs implanted	3,116 (25.6%)	144 anembryonic sacs
Implanted embryos displaying a fetal heartbeat	2,786 (22.9%)	150 vanishing; 15 ectopics; 199 miscarriages; 100 selective reductions; 30 therapeutic abortions
Remaining fetal heartbeats at 20 wks gestation	2,230 (18.3%)	23 fetal demises
Live offspring delivered	2,207 (18.1%)	17 neonatal mortalities
Surviving offspring	2,190 (18.0%)	—

*Percentages indicate proportion of embryos.

Table 5.6. Incidence of Congenital Abnormality in Relation
to the Insemination Technique

	ICSI	IVF
Cycles	3,828	3,399
Offspring delivered	2,190	1,865
Newborns with major malformations	23 (1.0%)	30 (1.6%)
Newborns with minor malformations	17 (0.8%)	24 (1.3%)
Total malformations	40 (1.8)*	54 (2.9)*

*χ^2, 2 × 2, 1 df; Difference in congenital malformations between ICSI and IVF,
$P < 0.05$.

the general population (45). Thus, our data agree with conclusions from other programs
(46–48) that the incidence of fetal abnormalities after ICSI is no higher than that after
natural conception and standard IVF, despite the fact that it often involves suboptimal
spermatozoa (39). As already mentioned, there is, of course, a reason to be concerned
that the problems that necessitated recourse to ICSI in some cases (e.g., astheno-, oligo-,
and azoospermia) could emerge in the male children (49, 50).

Finally, more needs to be known about the ongoing development of children conceived by ICSI, who are being studied to assess both their physical and psychological
progress. Bowen et al. (51) have reported that although most children conceived by ICSI
are normal, there may be a small delay in development of ICSI children at 1 year compared to those conceived by routine IVF or naturally. However, neither ICSI ($n = 201$)
nor IVF ($n = 131$) children had any slower mental development than 1283 Dutch children of the same age conceived naturally, nor did they have a slower mental development than the general population (52, 53). However, larger studies need to be conducted
to be able to draw definitive conclusions.

5. FUTURE DEVELOPMENTS

5.1 Sex Preselection and Prefertilization Genetics

Certain trends are now emerging in attempts to solve other infertility disorders in human
reproduction. Many are linked to genetic issues such as complete spermatogenetic arrest and oocyte aging, while others relate to sex-linked diseases.

Table 5.7. Relationship of Congenital Malformation
or Chromosomal Defect with Pregnancy Order

	Gestation Type	
Category/Abnormality	Singleton	Multiple*
Major		
Cardiovascular	2	2
Chromosomal	3	2
Facial	0	1
Gastrointestinal	4	3
Neurological	1	4
Urological	0	1
Minor		
Cardiovascular	0	1
Neurological	1	0
Skeletal	1	3
Urogenital	6	5

*Multiple gestations: 13 twins, 8 triplets.

In humans, there are at least 6000 heritable defects, of which about 370 are known to be X-linked. Because X-linked diseases are generally expressed by the male child of carrier mothers, controlling the gender of the offspring would obviously prevent this class of disease. Preimplantation genetic diagnosis already provides a partial solution to the problem, allowing the transfer of healthy embryos of a selected sex, and so far represents the only alternative to prenatal diagnosis and therapeutic abortion.

Identification of the spermatozoon's gender (X or Y) would provide a less invasive approach in the prevention of sex-linked diseases. Currently, 50% of all embryos undergoing preimplantation gender assessment in relation to sex-linked diseases must be discarded because they are or may be affected. Such wastage could be eliminated by fertilization with an X or a Y spermatozoon, as desired. Such separation techniques have been the subject of active research and of controversy (54–56). Several techniques relying on physical or immunological differences between X- and Y-chromosome-bearing spermatozoa have been devised (e.g., serum albumin gradients, discontinuous density gradients, sperm surface antigen, and flow cytometry). The developmental history and applications of these approaches have been extensively reviewed (57). Thus far, none of these methods has achieved a complement of 100% of X- or Y-bearing spermatozoa, and concerns regarding the safety of these methods still exist.

The only validated technology is based on the well-known difference in the DNA present in X and Y sperm in humans, representing a difference in mass of 2.8%. This incorporates modified flow cytometry to sort X- and Y-bearing sperm, and published results clearly prove the effectiveness and efficiency of the current sexing process in a broad range of applications (58–60). Resulting populations of X or Y sperm are generally used in conjunction with IVF or intrauterine insemination. Improvements in the production rate of sexed sperm, high-speed sorting is one of the newer technological advances to increase sorted sperm throughput. Initially this approach allowed the sorting of 350,000 sperm/h, but now 6×10^6 of each sex/h are sorted, under routine conditions. Sorting only the X population results in about 18×10^6 sperm/h. Improvements in the technology will no doubt lead to much greater usage of sexed sperm, depending on the species involved.

Based on his finding that the X chromosome of most mammals carries more DNA than the Y chromosome, Morruzi (61) was the first to suggest that this might provide the basis for separating X- and Y-chromosome-bearing sperm. While the autosomes have an essentially identical DNA content, the Y chromosome carries less DNA than the larger X chromosome. Taking advantage of this real difference for purposes of X- and Y-bearing sperm separation has only been possible since the advent of flow cytometry. The first attempts using DNA and flow cytometric analysis did not succeed (62). However, with the combination of improved staining methods and a realization that aspherical cells such as sperm must be orientated to the excitation source, the relative DNA difference could be measured (63). Sorting sperm into separate and nearly pure X and Y populations based on the DNA difference was first accomplished with the modified flow cytometry/cell sorter instrumentation (64, 65). However, the sperm discussed in these reports were dead because they had been sonicated to remove the tails (versus intact motile sperm).

Improved cellular staining and the utilization of vital fluorochromes to label the DNA of living sperm led to the sorting of X- and Y-chromosome-bearing sperm for the production of offspring of predetermined sex (65, 66). The sorting of sperm into separate populations of X and Y based on DNA requires a flow cytometer/cell sorter with a modified configuration. The newer generation of high-speed cell sorters have become available over the past 4 years and operate at sample pressures that range from 0.84 kg/cm^2 to 4.22 kg/cm^2. At the present time, this sorting system modified for sperm can produce 6×10^6 X sperm and 6×10^6 Y sperm/h.

An improved orienting system (nozzle) was first fitted to the standard speed sorter. This improvement in orientation on the modified cell sorter through replacement of the beveled needle system with an orienting nozzle was reported by Rens et al. (67). This modification has increased the percentage of sperm oriented from 25% to 70%. The modified flow cytometer and cell sorter is essential for attaining separate populations of viable X and Y sperm on a repeatable basis (66). Detection of the small differences in the amount of DNA between the X and Y sperm can be achieved with virtually all commercial cell sorters that have been manufactured in the past 25 years if they have the basic modifications for sperm (68).

The most critical aspect of the preparation for sperm sexing is the insult imposed on the sperm, which is reflected in their fertilizing capacity (58,66). Examples of such potential insults are the addition of stain, rewarming of the stained sperm, subjecting the sperm to pressure changes in the sorting system, and, finally, centrifugation. Using high-speed sorting, an aliquot of 1 ml containing 150×10^6 sperm or 10 times the amount with the standard-speed sorter is mixed with the same volume of Hoechst 33342 (bisbenzimide; Calbiochem-Behring, La Jolla, CA), added at the rate of 8 μl/ml. The suspension is then incubated at 35°C for approximately 1 h to allow a uniform penetration of the stain, which is essential to minimize stain variation and helps reduce the coefficient of variation of the sperm separation.

The need to maintain sperm viability throughout the sorting process is critical. Once the fluorescently stained sperm pass into the flow cell, the stream is surrounded by a sheath fluid of PBS containing 0.1% BSA. This contributes to an increasing dilution effect, and the BSA helps minimize agglutination of the sperm.

The process described above is most effective for sperm used for insemination, but for sperm used with IVF to produce sexed embryos, the yolk in the TEST-yolk can be reduced to less than 5% so as not to interfere with fertilization. A concentration of 2% yolk is appropriate as an environment for sorted sperm used for IVF (69).

In some species where the DNA difference between X and Y sperm is greater than in humans (e.g., *Chinchilla langier*), 100% purity in the sort may be feasible, with the highest purities being attained when the DNA difference is greater than 3.5%, when sperm are efficiently oriented to the laser beam, and when there is uniform staining. Sorting human sperm presents more of a challenge because the differences in DNA mass between X and Y sperm is about 2.8%. However, the percentages of purity can range routinely from 75% Y to 90% X for sorted human sperm (70, 71). However, in some samples the proportion of dead sperm and those of abnormal morphology may impact the sorting process.

With regard to the procedure, at some point during or immediately after the completion of a sort to be used for insemination or IVF, a presiliconized and/or BSA presoaked tube is placed in position to receive the sample—generally some 100,000 sperm per tube. In the other method of determining the sperm sex ratio, FISH, the fluorescent signals indicate the proportion of Y-chromosome-bearing sperm carrying the Y microsatellite DNA probe. PCR also has been useful in determining the proportions of X and Y sperm in a sorted sample. Sort reanalysis for DNA has an advantage, however, because PCR may take from 3 to 4 h, as may FISH (71), whereas sort reanalysis can be done in 30 min (72). This is important because the sexed sperm should be used for IVF insemination soon after sorting for maximal fertility. And in that regard, high-speed cell sorting is a significant technological improvement over standard speed cell sorters.

A method able to separate X and Y sperm on a large scale, at least on a scale that could be applied to semen production practice at an assisted reproductive technology center, would be one based on the hypothesis that sex-specific proteins are more highly conserved than non–sex-specific proteins. The approach most often considered in this context is that of isolating a protein from the sperm surface of the X- or Y-bearing sperm and preparing an antibody to that protein. The assumption is that affinity chromatography or a magnetic bead technique for separation of the antibody-labeled population would be adaptable for use in assisted reproduction. Approximately 1000 proteins have been mapped on the sperm surface. However, no difference could be detected between proteins present on X and Y sperm, respectively suggesting that sex-specific surface proteins are not expressed (73–75).

The current technology for sorting sperm by flow cytometric sorting into X and Y populations at 90% purity of X or Y sperm is adaptable to virtually all mammalian species including humans. Currently, sperm sorting by this basic method is also being conducted in human clinical trials under the trademark MicroSort (76). At the time of this writing, the flow cytometric sorting method of Johnson et al. (66) modified for high-speed sorting with an orienting nozzle (59) provides the only fully validated means of preselecting the sex of spermatozoon and thus of mammalian offspring.

5.2 Transplantation of Spermatogonia

In the female, a finite number of primary oocytes are present at birth, and this pool progressively dwindles during the lifetime. Thus, stem cell spermatogonia are the only self-renewing cell type in the adult capable of providing a genetic contribution to the next generation (77).

Spermatogonia exist in three states: stem cell, proliferative, and differentiating spermatogonia (78). The first two are frequently designated as undifferentiated spermatogonia, and the stem cell spermatogonium is most resistant to a variety of insults, often surviving where other germ cells are destroyed. The less frequent division of this stem cell population is believed to be one reason for their relative ability to survive (78, 79). Although it is difficult to be certain about which of the spermatogonia are capable of self-renewal, both stem cell and proliferative spermatogonia appear poised for further differentiation (78–81).

In 1994, Brinster and Zimmermann (77) reported that spermatogonial stem cells isolated from normal male mice repopulated sterile testes when injected directly into seminiferous tubules, later showed normal morphological characteristics, and ultimately differentiated into normal spermatozoa. In 1996, Clouthier et al. (82) examined the feasibility of transplanting spermatogonia from other species to the mouse. Marked testicular cells from transgenic rats were transplanted to the testes of 10 immunodeficient mice, and rat spermatogenesis occurred in all of them. Among epididymides of eight mice, the three belonging to those with the longest transplants (> 110 days) contained morphological rat spermatozoa (82). The success of rat spermatogenesis in mouse testes would appear to open the possibility of xenogeneic spermatogenesis for other species combinations.

With this in mind, we have transplanted germ cells from azoospermic men to the testes of mutant aspermatogenic (W/W^v) and severe combined immunodeficient mice (SCID) (83). Spermatogenic cells were obtained from testicular biopsy specimens from 24 men (average age of 34.3 ± 9 years) with obstructive ($n = 16$; OA) and nonobstructive ($n = 8$; NOA) azoospermia by first digesting testis biopsies with collagenase to promote separation of individual cells. The concentration of spermatogenic cells in the OA group was 6.6×10^6 cells/ml and 1.3×10^6 cells/ml in the NOA group ($P < .01$). The germ cells were injected into mouse seminiferous tubules using a microneedle (40 µm inner diameter) on a 10-ml syringe, with trypan blue used as an indicator of the accuracy of this transfer. Mice were maintained from 50 to 150 days to allow time for germ cell colonization and development before sacrifice. Testes were then fixed for histological evaluation, and approximately 100 tubule cross-sections were examined for human spermatogenic cells.

The different spermatogenic cell types were distributed equally in the OA samples, ranging from spermatogonia to fully developed spermatozoa, but in the NOA group the majority of cells were spermatogonia and spermatocytes. A total of 23 testes from 14 W/W^v mice and 24 testes from 12 SCID mice were injected successfully, as judged by the presence of spermatogenic cells in histological sections of testes removed immediately after the injection (Figures 5.5–5.7). However, sections from the remaining testes

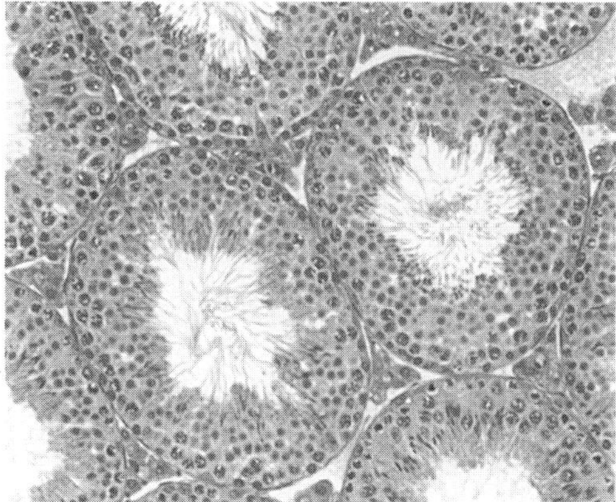

Figure 5.5. Section of seminiferous tubules displaying complete spermatogenesis (200×). Sperm flagella are protruding into the tubular cavity.

examined up to 150 days after injection showed tubules lined only with Sertoli cells with no xenogeneic germ cells surviving. The reason the two recipient mouse strains did not allow the implantation of human germ cells was probably due to signaling molecule divergence between primates and rodents that occurred more than 100 million years ago (84).

5.3 Nuclear Transplantation into Immature Oocytes

In some cases of infertility, a limiting step for the assisted reproductive technologies (ARTs) is the availability of normal oocytes, and this has stimulated research into cryopreservation of oocytes at different maturational stages and freezing of the ovarian cortex for young women undergoing chemotherapy and/or radiotherapy (85).

One problematic cause of infertility is reflected in the higher incidence of oocyte aneuploidy in older women. Independently of the initial indication or the ART technique used to solve it, the pregnancy rate follows a downward slope with advancing maternal age starting at 35 years (86). Although such patients can only be treated by oocyte donation at the present time, more and more women are becoming interested in bearing their own

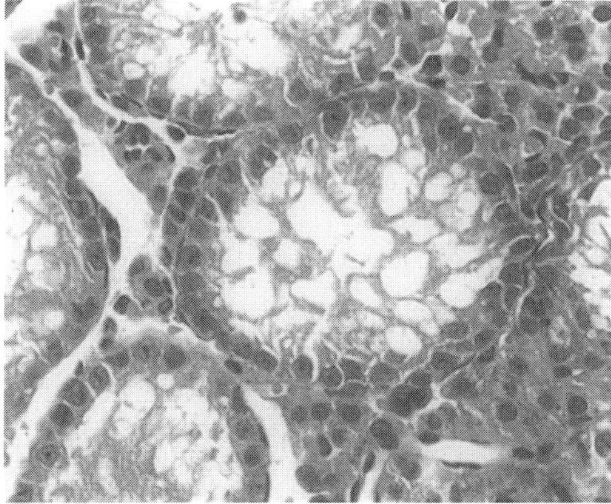

Figure 5.6. Seminiferous tubules of SCID mouse after busulfan treatment (chemical sterilization) (400×). The spermatogenic cells are absent, and only Sertoli cells are lining the inner surface of the tubular wall.

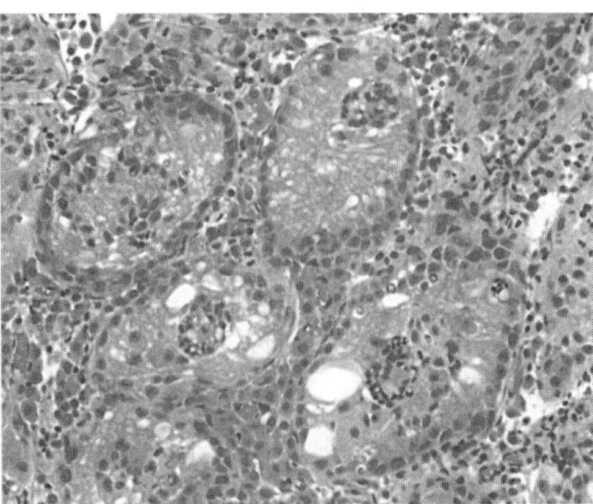

Figure 5.7. Busulfan-treated tubules immediately after human germ cell injection (200×). Spermatogenic cells are present in clumps in the tubular lumens.

genetic offspring. Recently, therefore, transfer of nuclei from "old" oocytes into "young" ooplasm has been proposed as a way of ameliorating the effects of age on oocytes.

In regard to nuclear transplantation, there is a concern related to the introduction of foreign mitochondrial DNA, which creates heteroplasmy. This has stimulated investigation as to how mitochondria are distributed in oocytes of different maturational stages, in isolated karyoplasts/cytoplasts, and in reconstituted oocytes after nuclear transplantation. Finally, the importance of understanding the role of the mitochondrial genome in early embryonic development has become undeniable.

Overall, the various aspects of this work have aimed to clarify the realistic place of nuclear transplantation among the tools becoming available to ARTs, to identify the most efficient nuclear transplantation and electrofusion procedures, and to explore the possibility of "manufacturing" viable oocytes. Finally, we have sought to clarify the impact of the mitochondrial genome in the early development of embryos coming from oocytes generated by nuclear transplantation procedures.

5.3.1 Implications of Nuclear Transplantation

In a large majority of infertile couples, oocyte abnormalities may be metabolically or genetically based, the former sometimes being a consequence of the latter (87). Some infertile couples consistently produce morphologically abnormal embryos that supposedly arise from oocytes with metabolic defects, some of which may be treated by transferring ooplasm aspirated from a fertile donor oocyte (17).

In contrast, some genetically abnormal oocytes can develop into embryos of normal appearance that would be considered suitable for uterine transfer (88). However, the large majority of chromosomally abnormal embryos probably do not implant, and even when they do, the fetus generally does not survive to term. The decline in the ability to conceive with maternal age (86) appears to be primarily related to an increased incidence in oocyte aneuploidy (89). In accordance with this, 37.2% of morphologically normal eight-cell embryos in the 40–45 maternal age group were found to express chromosomal aberrations (88). The risk of conceiving an aneuploid fetus has been reported to increase from 6.8% for women 35–39 years old to about 50% in women 45 years or older (90). Nevertheless, much older women have become pregnant after replacement of embryos derived from young donor oocytes.

The abnormalities of oocytes' chromosomes associated with aging are due primarily to non-disjunction of bivalent chromosomes during meiosis I (89). Aging of its organelles has an effect on the meiotic spindle's ability to direct a balanced segregation of the chromosomes (91, 92). Why aging affects this step adversely is still unknown, but damage from free oxygen radicals (93, 94) and/or a compromised perifollicular microcirculation may be responsible (95, 96).

Cytoplasmic transfusion into eggs or embryos of other mammals has been shown to improve their developmental potential (97–100), and the transfer of cytoplasm from fertile donor oocytes to putatively low-grade human oocytes has resulted in successful pregnancy and delivery of a normal baby (17, 101). However, because this procedure was performed on mature oocytes, it has no bearing on the prior events of meiosis I that may lead to a chromosomal imbalance. The latter may be overcome only by substituting—by nuclear transplantation—the nucleus of a germinal vesicle (GV)-stage oocyte to an enucleated immature oocyte from a younger woman (102).

5.3.2 Preliminary Results with Nuclear Transplantation

Mouse GV oocytes were retrieved by puncturing ovarian follicles of unstimulated B6D2F$_1$ females, and cumulus-free oocytes were cultured in medium (M199) supplemented with a phosphodiesterase inhibitor (0.2 mM 3-isobutyl-1-methylxanthine; Sigma) to prevent spontaneous GV breakdown.

All the micromanipulation and electrofusion procedures were performed in a shallow plastic Petri dish on a heated stage, placed on an inverted microscope equipped with hydraulic micromanipulators (103). The zona pellucida was breached by a glass microneedle, then oocytes were exposed to 25 µg/ml of cytochalasin B. The GV surrounded by a small amount of cytoplasm (GV karyoplast) was removed by a micropipette with a 20 µm inner diameter. Next the GV karyoplast was inserted into the perivitelline space of a previously enucleated oocyte (GV cytoplast). Each grafted oocyte was aligned with a micromanipulator between two microelectrodes perpendicular to their axes. To induce fusion, a single or double 1.0 kV/cm direct current fusion pulse was delivered for 100 µs in an electrolyte medium (M2) by an Electro Cell Manipulator (BTX 200 and 2001, Genetronics, Inc., San Diego, CA). Then, after washing and culture for 30 min in a cytochalasin B-free medium, these oocytes were examined to confirm cell survival and fusion, cultured further, and then examined at 12–16 h after the fusion treatment. Finally, their nuclear maturation was evaluated, as evidenced by extrusion of the first polar body (PB).

For similar human studies, GV oocytes were obtained, as described earlier, from consenting patients undergoing ICSI (4, 10, 29, 36). Immediately before ICSI, corona cells were removed by enzymatic and mechanical treatments, and the denuded oocytes were examined under an inverted microscope to assess their integrity and maturation stages (4, 10, 29, 36). The nuclear transplantation procedure is shown in Figures 5.8–5.11. The reconstituted immature oocytes were cultured and were examined at 24 and 48 h after electrofusion to evaluate nuclear maturation, characterized by disappearance of the GV and extrusion of the first PB.

In the mouse, nuclear transplantation into GV-stage oocytes followed by extrusion of a PB revealed an overall efficiency of 80%. Interestingly, this aggressive technique appears not to increase the incidence of chromosomal abnormalities (103). Human oocytes were reconstituted with an efficiency of 73%, with a lower maturation rate of 62% following reconstitution, comparable to that of control human GV oocytes, as was the 20% incidence of aneuploidy among the constructed oocytes (104). Those reconstituted human oocytes were successfully fertilized by ICSI at a rate of 52%, but although they underwent early embryonic cleavage after ICSI, their survival rate and embryo quality appeared suboptimal compared to control oocytes matured *in vivo* (105).

In a limited number of transfers of aged GV oocyte into a younger ooplast, there was a normal haploidization during the first meiotic division accompanied by the extrusion of the first PB. In other studies, lower maturation rates and impaired embryo development have been the rule, probably attributable to the suboptimal in vitro culture conditions currently available for human oocyte maturation (102–107).

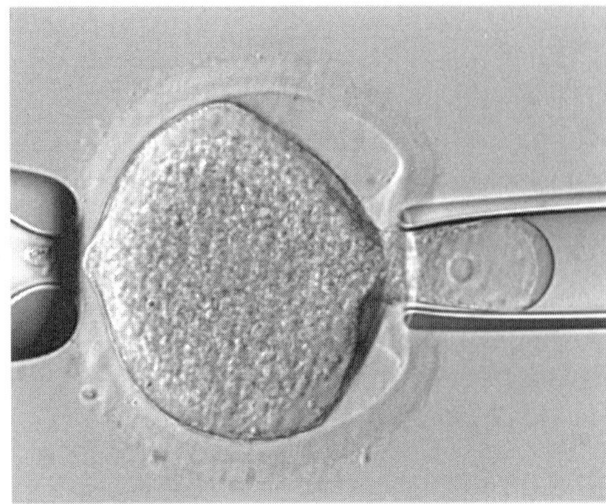

Figure 5.8. Aspiration of the nucleus from a germinal vesicle stage human oocyte (400×). The cytoplasmic bridge is still visible.

Although further cytogenetic information is needed about oocytes reconstituted by transfer of an old nucleus into a younger ooplast, this approach appears to be the treatment option for age-related oocyte aneuploidy. Although further technical improvements are needed, especially in regard to the in vitro maturation process, this approach now provides an interesting avenue to explore in attempting to prevent chromosomal defects associated with oocyte aging.

5.3.3 An Alternative Source of Oocytes

Although better culture conditions might enhance the limited ability of human GV oocytes to mature in vitro after nuclear transplantation, obtaining a sufficient number from older women for such reconstitution is a limiting factor. It is possible, however, that sufficient number of oocytes might be created by a form of cloning—by transplantation of a nucleus taken from a patient's somatic cell into an enucleated ooplast obtained from a younger donor. Such a construction of viable oocytes from somatic cells would benefit older women, women with premature ovarian failure, or those considered "poor responders."

Because GV oocytes can complete the first meiotic division spontaneously in vitro, it is possible that GV-stage ooplasm might suppport the haploidization of diploid somatic nu-

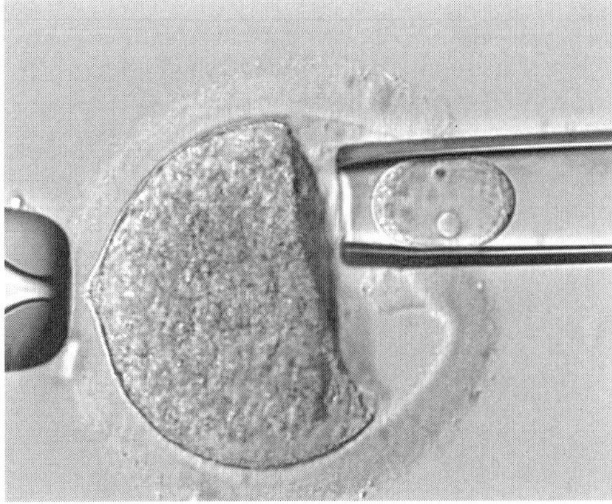

Figure 5.9. Transfer of the isolated karyoplast into the perivitelline space of an enucleated oocyte (400×).

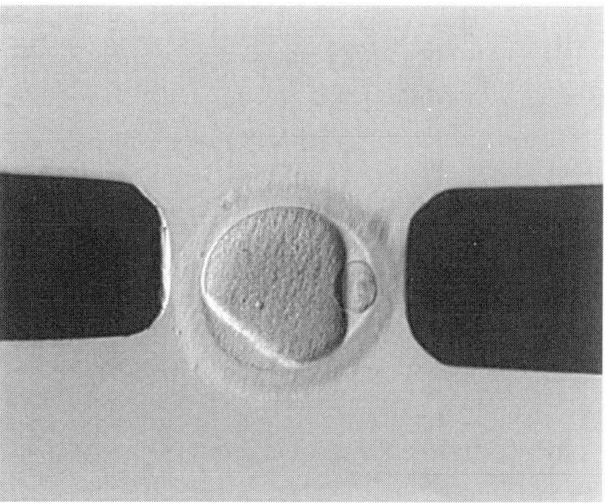

Figure 5.10. Cell alignment for electrofusion by microelectrodes under micromanipulator control (200×).

clei to metaphase II (108). Immature ooplasm is capable of inducing haploidization-like reduction division of transplanted somatic cell nuclei (107, 109), and we have proposed that the transfer of somatic nuclei to GV ooplasts and their ensuing haploidization may provide a source of viable mammalian oocytes. If oocytes can indeed be reconstituted by such techniques, this would greatly benefit patients who are candidates for oocyte donation.

To obtain somatic cell nuclei, endometrial stromal cells were collected from consenting patients undergoing endometrial cell coculture during IVF. Endometrial stromal and glandular cells were isolated by enzymatic digestion using 0.2% collagenase type II, separated by differential sedimentation (110), and cultured in a long-term culture medium supplemented with 10% FBS. In the case of the mouse, cumulus–oocyte complexes were obtained from B6D2F$_1$ mice after ovarian stimulation with PMSG and hCG, and cumulus cells isolated by brief exposure to hyaluronidase were cultured for up to 30 days. GV oocytes were retrieved from the same strain of mouse by puncturing follicles of unstimulated ovaries and were denuded by mechanical removal of cumulus cells.

All the micromanipulation and electrofusion procedures were carried out in a shallow plastic petri dish on a heated stage under an inverted microscope. A hole was made in the zona pellucida of the GV oocyte with a glass needle, and after short exposure to cytochalasin B, the GV surrounded by a small amount of ooplasm was removed with a glass micropipette (103). Cultured human endometrial stromal and mouse cumulus cells

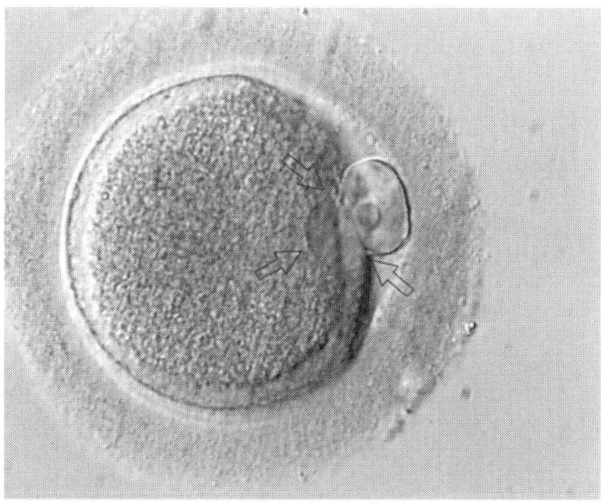

Figure 5.11. Grafted oocyte 30 min after electrofusion (400×). Drifting of the transplanted nucleus (arrows) into the host ooplasm.

were isolated with trypsin-EDTA. Either a stromal or a cumulus cell was then inserted subzonally into an enucleated mouse GV oocyte. Each grafted oocyte was manually aligned between two microelectrodes, and cell fusion was induced by applying direct current. The resulting reconstituted oocytes were allowed to mature for 14–16 h until extrusion of the first PB (103). The distribution of nuclear chromatin between the ooplasm and the PB was evaluated by specific DNA staining. For this, some mature oocytes were stained with DAPI solution and evaluated under a fluorescent microscope, while others were anchored between a microslide and coverslip, fixed with methanol/acetic acid (3:1; v/v), and stained with 1% aceto-orcein solution.

A total of 45 GV oocytes were enucleated, then fused with somatic cells. Staining showed the presence of metaphase chromosomes in the ooplasm and in the PB. The overall efficiency of the sequence from an intact GV oocyte to a reconstituted oocyte with an extruded PB was 60.0% for human endometrial cells and 42.8% for mouse cumulus cells. A longer period of in vitro culture did not induce any additional PB extrusions.

Other studies were performed to evaluate the karyotypes of reconstituted mature oocytes. Among a total of 78 intact GV oocytes, 77 were successfully enucleated and grafted with a single cumulus cell, 56 being reconstituted finally by electrofusion. Subsequently, 29 extruded the first PB during the 14–16 h of culture. Of the 13 oocytes that matured successfully and whose karyotypes were analyzable, 5 had 20 chromosomes, 5 were aneuploid, and 3 were diploid.

The observation that immature mouse ooplasts can accomplish haploidization of human or murine somatic cell nuclei suggests that this approach may provide an alternative source of viable human oocytes. However, more detailed cytogenetic information is needed, and the presence of the somatic cell centrosome in these manufactured oocytes means that the fertilization process and later embryo cleavage must be carefully investigated. It also remains to be established that genetic imprinting of the oocyte reconstructed from a cultured somatic cell is comparable to that in natural haploid.

Although this approach to haploidization of somatic cell nuclei may at first bring cloning to mind, the resulting haploid oocytes obviously still need the contribution of the paternal genome for further development. Nonetheless, these preliminary findings suggest that nuclear transplantation may provide a means of manufacturing viable oocytes from defective oocytes.

6. SUMMARY

ICSI is by far the most effective of the techniques designed to overcome an inability of spermatozoa to fertilize. It leads to high rates of fertilization and pregnancy despite very low sperm numbers or extreme defects of sperm motility and morphology. The only requirement is that spermatozoa should be viable, as reflected often in some form of tail beat (though the spermatozoon should be immotile when injected). Because of its nature, the technique has raised some justifiable concerns, but babies born as a consequence of ICSI have a neonatal malformation rate no different from babies resulting from standard IVF (39, 46) and even from natural conception. Because it permits high rates of fertilization and pregnancy and is apparently safe, ICSI is being used increasingly for non–male-factor situations where any aspect of gamete function is in doubt. Because ICSI may allow procreation by subfertile men, including some with no spermatogenesis (where immature germ cells are used), the genetic abnormalities at the root of these problems can be transmitted to and be reflected postpubertally in some male babies. Therefore, it is important that couples treated by ICSI are subjected to karyotype screening and, when oligo- or azoospermic, are tested for cystic fibrosis carrier status or deletions present in the Y chromosome (14).

The positive outcome of ICSI is largely independent of the motility, morphology, and concentration of the sperm sample, even in couples where these characteristics are severely impaired, and even when no spermatozoa are present in the initial ejaculate (111, 112). Its successful use with surgically retrieved spermatozoa proves that ICSI is able to achieve fertilization regardless of the maturation status of the gametes. It also allows a bypass of the steps in spermiogenesis and epididymal sperm maturation, of the acrosome reaction, of binding to the zona pellucida and fusion with the oolemma. Our experience suggests that the evolution of pregnancies and the occurrence of congenital malformations after treatment by ICSI are similar to the outcome in other assisted reproduction procedures. As reported in several studies (39, 46–48, 52, 113–115), the outcome for these offspring is in the normal range, and any congenital anomalies seen thus far seem to be nongenetic in nature and are not clearly linked to sperm defects as currently classified. However, because subfertile men do have a higher frequency of chromosomal abnormalities (39, 44, 116), the earlier concerns generated by the nature of ICSI procedure have shifted to the possible transmission to male offspring of the genetic defects causing a man's infertility. Thus, genetic screening, including preimplantation and prenatal diagnosis, as well as genetic counseling should be offered to these patients, particularly those undergoing ICSI with immature gametes. Careful monitoring of the technique together with continued, meticulous obstetric and pediatric follow-up will certainly help to alleviate any concerns (52, 117).

Although great progress has been made in treatment of infertility with ICSI and other techniques, a considerable number of patients still fail to conceive and spermatogenic failure and oocyte aging appears to be responsible in a large proportion of such cases. In some instances, donor gametes can be used, but where not, emerging reproductive technology and preliminary experimental results provide some hope for alternative sources of gametes through which these infertile patients can finally conceive their own genetic child. In conjunction with ICSI, fertilization of human oocytes with immature sperm precursors such as spermatids and even secondary spermatocytes occasionally has resulted in healthy babies. In view of the possibility for transplantation and even xenotransplantation of spermatogonia to a mouse testis, human male stem cells treated similarly might provide an attractive source in, for example, the treatment of maturation-arrested male germ cells and male cancer patients. Finally, current efforts aimed at refinement of techniques for transplantation of somatic cell nuclei and their haploidization within oocytes may prove to be a practical way of eradicating age-related aneuploidy and so constitute a novel source of healthy oocytes.

References

1. Lanzendorf, S. E., Maloney, M. K., Veeck, L. L., Slusser, J., Hodgen, G. D., and Rosenwaks, Z. (1988). *Fertil. Steril.*, **49**, 835.
2. Palermo, G., Joris, H., Devroey, P., and Van Steirteghem, A. C. (1992). *Lancet*, **340**, 17.
3. Palermo, G., Joris, H., Derde, M. P., Camus, M., Devroey, P., and Van Steirteghem, A. C. (1993). *Fertil. Steril.*, **59**, 826.
4. Palermo, G. D., Cohen, J., Alikani, M., Adler, A., and Rosenwaks, Z. (1995). *Fertil. Steril.*, **63**, 1231.
5. Palermo, G. D., Schlegel, P. N., Hariprashad, J. J., Ergün, B., Mielnik, A., Zaninovic, N., Veeck, L. L., and Rosenwaks, Z. (1999). *Hum. Reprod.*, **14**, 741.
6. Van Steirteghem, A. C., Liu, J., Joris, H., Nagy, Z., Janssenwillen, C., Tournaye, H., Derde, M. P., Van Assche, E., and Devroey, P. (1993). *Hum. Reprod.*, **8**, 1055.
7. Colombero, L. T., Hariprashad, J. J., Tsai, M. C., Rosenwaks, Z., and Palermo, G. D., (1999). *Fertil. Steril.*, **72**, 90.
8. Schoysman, R., Vanderzwalmen, P., Nijs, M., Segal, L., Segal-Bertin, G., Geerts, L., van Roosendaal, E., and Schoysman, D. (1993). *Lancet*, **342**, 1237.
9. Tournaye, H., Devroey, P., Liu, J., Nagy, Z., Lissens, W., and Van Steirteghem, A. (1994). *Fertil. Steril.*, **61**, 1045.
10. Palermo, G. D., Schlegel, P. N., Colombero, L. T., Zaninovic, N., Moy, F., and Rosenwaks, Z. (1996). *Hum. Reprod.*, **11**, 1023.
11. Dozortsev, D., Rybouchkin, A., De Sutter, P., Dhont, M., (1995). *Hum. Reprod.*, **11**, 2960.

12. Palermo, G. D., Avrech, O. M., Colombero, L. T., Wu, H., Wolny, Y. M., Fissore, R. A., Rosenwaks, Z. (1997). *Mol. Hum. Reprod.*, **3**, 367.

13. Reijo, R., Alagappan, R. K., Patrizio, P., and Page, D. C. (1996). *Lancet*, **347**, 1290.

14. Girardi, S. K., Mielnik, A., and Schlegel. P. N. (1997). *Hum. Reprod.*, **12**, 1635.

15. Schlegel, P. N., Palermo, G.D., Goldstein, M., Menendez, S., Zaninovic, N., Veeck, L. L., and Rosenwaks, Z. (1997). *Urology*, **49**, 435.

16. Friedler, S., Raziel, A., Strassberger, D., Soffer,Y., Komarovsky, D., and Ron-el, R. (1997). *Hum. Reprod.*, **12**, 1488.

17. Cohen, J., Scott, R., Alikani, M., Schimmel, T., Munné, S., Levron, J., Wu, L., Brenner, C., Warner, C., and Willadsen, S. (1998). *Mol. Hum. Reprod.*, **4**, 269.

18. Zhang, J., Blaszczyk, A., Grifo, J., Li, L., Licciardi, F., Noyes, N., and Krey, L. (1997). *Fertil. Steril.*, **Suppl. S1**, S154.

19. Nagy, Z. P., Staessen, C., Liu, J., Joris, H., Devroey, P., and Van Steirteghem, A. C. (1995). *Fertil. Steril.*, **64**, 1130.

20. Morton, P. C., Yoder, C. S., Tucker, M. J., Wright, G., Brockman, W. D. W., and Kort, H. I. (1997). *Fertil. Steril.*, **68**, 488.

21. Yovich, J. L., and Stanger, J. D. (1984). *J. In Vitro Fertil. Embryo Transf.*, **1**, 172.

22. Chia, C. M., Sathananthan, H., Ng, S. C., Law, H. Y., and Edirisinghe, W. R. (1984). *Proceedings of the 18th Singapore-Malaysia Congress of Medicine, Singapore*, p. 52. Academy of Medicine, Singapore.

23. De Felici, M., and Siracusa, G. (1982). *Gamete Res.*, **6**, 107.

24. Bedford, J. M., and Kim, H. H. (1993). *Hum. Reprod.*, **8**, 453.

25. Van Blerkom, J., and Henry, G. (1992). *Hum. Reprod.*, **3**, 379.

26. Lalonde, L., Langalis, J., Antaki, P., Chapdelaine, A., Roberts, K. D., and Bleau, G. (1988). *Fertil. Steril.*, **49**, 316.

27. World Health Organization. (1987). *Intl. J. Androl.*, **7**, 1.

28. deKretser, D. M., Yates, C. A., McDonald, J., Leeton, J. F., Southwick, G., Temple-Smith, P. D., Trounson, A. O., and Wood, E. C. (1985). In: *Gamete Quality and Fertility Regulation*, R. Rolland, M. J. Heineman, S. G. Hillier, and M. Vemer, eds., p. 213. Elsevier Science, Amsterdam.

29. Palermo, G. D., Cohen, J., and Rosenwaks, Z. (1996). *Fertil. Steril.*, **65**, 899.

30. Kruger, T. F., Menkveld, R., Stander, F. S. H., Lombard, C. J., Van der Merwe, J. P., van Zyl, J. A., and Smith, K. (1986). *Fertil. Steril.*, **46**, 1118.

31. Schlegel, P. N., Berkley, A., Goldstein, M., Cohen, J., Alikani, M., Adler, A., Gilbert, B. R., Rosenwaks, Z., (1994). *Fertil. Steril.*, **61**, 895.

32. Schlegel, P. N., Cohen, J., Goldstein, M., Alikani, M., Adler, A., Gilbert, B. R., Palermo, G. D., and Rosenwaks, Z. (1995). *Fertil. Steril.*, **64**, 421.

33. Tsirigotis, M., Pelankos, M., Yazdani, N., Boulos, A., Foster, C., and Craft, I. L. (1995). *Br. J. Urol.*, **76**, 765.

34. Silber, S. J., Van Steirteghem, A. C., Liu, J., Nagy, Z., Tournaye, H., and Devroy, P. (1995). *Hum. Reprod.*, **10**, 148.

35. Verheyen, G., De Croo, I., Tournaye, H., Pletinck, I., Devroey, P., and Van Steirteghem, A. C. (1995). *Hum. Reprod.*, **10**, 2956.

36. Palermo, G., Alikani, M., Bertoli, M., Colombero, L., Moy, F., Cohen, J., and Rosenwaks, Z. (1996). *Hum. Reprod.*, **11**, 172.

37. Ma, K., Inglis, J. D., Sharkey, A., Bickmore, W. A., Hill, R. E., Prosser, E. J., Speed, R. M., Thomson, E. J., Jobling, M., and Taylor, K. (1993). *Cell*, **75**, 1287.

38. Kobayashi, K., Mizuno, K., Hida, A., Komaki, R., Tomita, K., Matsushita, L., Namiki, M., Iwamoto, T., Tamura, S., and Minowada, S. (1994). *Hum. Mol. Genet.*, **3**, 1965.

39. Palermo, G. D., Colombero, L. T., Schattman, G. L., Davis, O. K., and Rosenwaks, Z. (1996). *J. Am. Med. Assoc.*, **276**, 1893.

40. Brandell, R. A., Mielnik, A., Liotta, D., Ye, Z., Veeck, L. L., Palermo, G. D., and Schlegel, P. N. (1998). *Hum. Reprod.*, **13**, 2812.

41. Palermo, G. D., Schlegel, P. N., Sills, E. S., Veeck, L. L., Zaninovic, N., Menendez, S., and Rosenwaks, Z. (1998). *N. Engl. J. Med.*, **338**, 588.

42. Patrizio, P., Asch, R. H., Handelin, B., and Silber, S. J. (1993). *Hum. Reprod.*, **8**, 215.

43. Schlegel, P. N., Cohen, J., Goldstein, M., Alikani, M., Adler, A., Gilbert, B. R., Palermo, G. D., and Rosenwaks, Z. (1995). *Fertil. Steril.*, **64**, 421.

44. Liebaers, I., Bonduelle, M., Van Assche, E., Devroey, P., and Van Steirteghem, A. (1995). *Lancet*, **346**, 773.
45. New York State Department of Health (1990).
46. Bonduelle, M., Hamberger, L., Joris, H., Tarlatzis, B. C., and Van Steirteghem, A. C. (1995). *Hum. Reprod. Update*, **1**, 3.
47. Bonduelle, M., Legein, J., Derde, M. P., Buysse, A., Schietecatte, J., Wisanto, A., Devroey, P., Van Steirteghem, A. C., and Liebaers, I. (1995). *Hum. Reprod.*, **10**, 3327.
48. Bonduelle, M., Aytoz, A., Van Assche, E., Devroey, P., Liebaers, I, and Van Steirteghem, A. (1998). *Hum. Reprod.* **13**, 781.
49. De Jonge, C., and Pierce, J. (1995). *Hum. Reprod.*, **10**, 2518.
50. Patrizio, P. (1995). *Hum. Reprod.*, **10**, 2520.
51. Bowen, J., Gibson, F., Leslie, G., and Saunders, D. (1998). *Lancet*, **351**, 1529.
52. Bonduelle, M., Joris, H., Hofmans, K., Liebaers, I., and Van Steirteghem, A. (1998). *Lancet*, **22**, 1553.
53. Bonduelle, M., Heurckmans, N., De Ketelaere, V., Derde, M-P., Liebaers, I., and Van Steirteghem, A. (2000). *Hum. Reprod.*, **15 Suppl. 1**, 63.
54. Ericsson, R. J., Langerrio, C. N., and Nishimo, M. (1973). *Nature* **246**, 421.
55. Brandriff, B. F., Gordon, L. A., Haendel, S., Singer, S., Moore, D. H., and Gledhill, B. L. (1986). *Fertil. Steril.*, **46**, 678.
56. Johnson, L. A. (1995). *Reprod. Fertil. Dev.*, **7**, 893.
57. Sills, E. S., Kirman, I., Thatcher, S. S., and Palermo, G. D. (1998). *Arch. Gynecol. Obstet.*, **261**, 109.
58. Johnson, L. A., and Welch, G. R. (1999). *Theriogenology*, **52**, 1323.
59. Johnson, L. A., Welch, G. R., and Rens, W. (1999). *J. Anim. Sci.*, **77 Suppl. 2**, 213.
60. Seidel, G. E. Jr, Schenk, J. L., Kerickoff, L. A., Doyle, S. P., Brink, Z., Green, R. D., and Cran, D. G. (1999). *Theriogenology*, **52**, 1407.
61. Moruzzi, J. F. (1979). *J. Reprod. Fertil.*, **57**, 319.
62. Gledhill, B. L., Lake, S., Steinmetz, L. L., Gray, J. W., Crawford, J. R., Dean, P. N., and Van Dilla, M. A. (1976). *J. Cell Physiol.*, **87**, 367.
63. Pinkel, D., Gledhill, B. L., Van Dilla, M. A., Stephenson, D., and Watchmaker, G. (1982). *Cytometry*, **3**, 1.
64. Johnson, L. A., and Clarke, R. N. (1988). *Gamete Res.*, **21**, 335.
65. Johnson, L. A., Flook, J. P., and Look, M. V. (1987). *Gamete Res.*, **17**, 203.
66. Johnson, L. A., Flook, J. P., and Hawk, H. W. (1989). *Biol. Reprod.*, **41**, 199.
67. Rens, W., Welch, G. R., and Johnson, L. A. (1998). *Cytometry*, **33**, 476.
68. Johnson, L. A., and Pinkel, D. (1986). *Cytometry*, **7**, 268.
69. Rath, D., Johnson, L. A., Dobrinsky, J. R., and Welch, G. R. (1997). *Theriogenology*, **47**, 795.
70. Johnson, L. A., Welch, G. R., Keyvanfar, K., Dorfmann, A., Fugger, E. F., and Schulman, J. D. (1993). *Hum. Reprod.*, **8**, 1733.
71. Fugger, E. F., Black, S. H., Keyvanfar, K., and Schulman, J. D. (1998). *Hum. Reprod.*, **13**, 2367.
72. Welch, G. R., and Johnson, L.A. (1999). *Theriogenology*, **52**, 1343.
73. Hendriksen, P. J. M. (1999). *Theriogenology*, **52**, 1295.
74. Hendriksen, P. J. M., Welch, G. R., Grootegoed, J. A., Van Der Lende, T., and Johnson, L. A. (1996). *Mol. Reprod. Dev.*, **45**, 342.
75. Sills, E. S. S., Kirman, I., Colombero, L. T., Hariprashad, J., Rosenwaks, Z., and Palermo, G. D. (1998). *Am. J. Reprod. Immunol.*, **40**, 43.
76. Fugger, E. F. (1999). *Theriogenology*, **52**, 1435.
77. Brinster, R. L., and Zimmermann, J. W. (1994). *Proc. Natl. Acad. Sci. USA*, **91**, 112980.
78. Russell, L. D., Ettlin, R. A., Hikim, A. P., and Clegg, E. D. (1990). In: *Histological and Histopathological Evaluation of the Testis*, L. D. Russel, and R. Ettlin, eds. p. 1. Cache River Press, Clearwater, FL.
79. Ewing, L., Davis, J. C., and Sirkin, B. R. (1980). *Int. Rev. Physiol.*, **22**, 41.
80. Clermont, Y., and Bustos-Obregon, E. (1968). *Am. J. Anat.*, **122**, 237.
81. Huckins, C. (1971). *Anat. Rec.*, **169**, 533.
82. Clouthier, D. E., Avarbock, M. R., Maika, S. D., Hammer, R. E., and Brinster, R. L. (1996). *Nature*, **381**, 30.

83. Reis, M. M., Tsai, M. C., Schlegel, P. N., Feliciano, M., Raffaelli, R., Rosenwaks, Z., and Palermo, G. D. (2000). *Zygote*, **8**, 97.

84. Nagano, M., McCarrey, J. R., and Brinster, R. L. (2001). *Biol. Reprod.*, **64**, 1409.

85. Oktay, K., Newton, H., Aubard, Y., Salha, O., and Gosden, R. (1998). *Fertil. Steril.*, **69**, 1.

86. Tietze, C. (1957). *Fertil. Steril.*, **8**, 89.

87. Van Blerkom, J. (1996). *J. Soc. Gyn. Invest.*, **3**, 3.

88. Munné, S., Alikani, M., Tomkin, G., Griffo, J., and Cohen, J. (1995). *Fertil. Steril.*, **64**, 382.

89. Dailey, T., Dale, B., Cohen J., et al. (1996). *Am. J. Hum. Genet.*, **59**, 176.

90. Hassold, T., and Chiu, D. (1985). *Hum. Genet.*, **70**, 11.

91. Battaglia, D. E., Goodwin, P., Klein, N. A., and Soules, M. R. (1996). *Hum. Reprod.*, **11**, 2217.

92. Volarcik, K., Sheean, L., Goldfarb, J., Woods, L., Abdul-Karim, F. W., and Hunt, P. (1998). *Hum. Reprod.*, **13**, 154.

93. Tarín, J. J., Vendrell, F. J., Ten, I., Blanes, R., van Blerkom, J., and Cano, A. (1996). *Mol. Hum. Reprod.*, **2**, 895.

94. Tarín, J. J. (1995). *Hum. Reprod.*, **10**, 1563.

95. Gaulden, M. (1992). *Mutat. Res.*, **269**, 69.

96. Van Blerkom, J., Antczak, M., and Schrader, R. (1997). *Hum. Reprod.*, **12**, 1047.

97. Muggleton-Harris, A., Whittingham, D. G., and Wilson, L. (1982). *Nature*, **299**, 460.

98. Pratt, H. P., and Muggleton-Harris, A. L. (1988). *Development*, **104**, 115.

99. Flood, J. T., Chillik, C. F., Van Uem, J. F., Iratini, I., and Hogden, J. D. (1990). *Fertil. Steril.*, **53**, 1049.

100. Levron, J., Willadsen, S., Bertoli, M., and Cohen, J. (1996). *Hum. Reprod.*, **11**, 1287.

101. Cohen, J., Scott, R., Schimmel, T., Levron, J., Willadsen, S. (1997). *Lancet*, **350**, 186.

102. Zhang, J., Wang, C. W., Krey, L., Lui, H., Meng, L., Blaszczyk, A., Alder, A., and Griffo, J. (1999). *Fertil. Steril.*, **71**, 726.

103. Takeuchi, T., Ergün, B., Huang, T. H., Rosenwaks, Z., and Palermo, G.D. (1999). *Hum. Reprod.*, **14**, 1312.

104. Takeuchi, T., Tsai, M. C., Gong, J., Veeck, L. L., Davis, O. K., Rosenwaks, Z., and Palermo, G. D. (1999). *Hum. Reprod.*, **Suppl.**, O-057.

105. Takeuchi, T., Tsai, M. C., Gong, J., Veeck, L. L., Rosenwaks, Z., and Palermo, G. D. (1999). *Fertil. Steril.*, **72 Suppl. 1**, S124.

106. Takeuchi, T., Ergun, B., Huang, T. H., Rosenwaks, Z., and Palermo, G. D. (1998). *Fertil. Steril.*, **70 Suppl. 1**, S86.

107. Takeuchi, T., Tsai, M. C., Spandorfer, S. D., Rosenwaks, Z., and Palermo, G. D. (1999). *Hum. Reprod.*, **Suppl.**, O-013.

108. Kubelka, M., and Moor, R. M. (1997). *Zygote*, **5**, 219.

109. Tsai, M. C., Takeuchi, T., Bedford, J. M., Reis, M. M., Rosenwaks, Z., and Palermo, G. D. (2000). *Hum. Reprod.*, **15**, 988.

110. Barmat, L. I., Liu, H. C., Spandorfer, S. D., Xu, K., Veeck, L., Damario, M.A., and Rosenwaks, Z. (1998). *Fertil. Steril.*, **70**, 1109.

111. Bonduelle, M., Desmyttere, S., Buysse, A., Van Assche, E., Schietecatte, J., Devroey, P., Van Steirteghem, A. C., and Liebaers, I. (1994). *Hum. Reprod.*, **9**, 1765.

112. Aytoz, A., Camus, M., Tournaye, H., Bonduelle, M., Van Steirteghem, A., and Devroey, P. (1998). *Fertil. Steril.*, **70**, 500.

113. Baron, L., Basso, G., Blanco, L., and Valzacchi, R. (1998). *Fertil. Steril.*, **70 Suppl. 1**, S29.

114. ESHRE Task Force on Intracytoplasmic Sperm Injection. (1998). *Hum. Reprod.*, **13**, 1737.

115. Sutcliffe, A. G., Taylor, B., Li, J., Thornton, S., Grudzinskas, J. G., and Lieberman, B. A. (1999). *Br. Med. J.*, **318**, 704.

116. Liebaers, I., Bonduelle, M., Van Assche, E., Devroey, P., and Van Steirteghem, A. (1995). *Lancet*, **346**, 1095.

117. Neri, Q. V., Squires, J., Lo, J., Rosenwaks, Z., and Palermo, G. D. (2001). *Hum. Reprod.* **Suppl. 1**, O-058.

6

JOYCE C. HARPER
ALPESH DOSHI

Micromanipulation: Biopsy

1. PREIMPLANTATION GENETIC DIAGNOSIS

IVF and the biopsy of polar bodies from oocytes, blastomeres from cleavage-stage embryos, or trophectoderm cells from blastocysts followed by single cell diagnosis has enabled us to help couples who are at risk of transmitting inherited diseases to their children and to improve pregnancy rates for certain groups of IVF patients. These procedures are used in a technique called preimplantation genetic diagnosis (PGD) (1). PGD was developed in 1990 to offer an alternative treatment for couples at risk of transmitting an inherited disease to their offspring. The main option for such couples is to become pregnant naturally and undergo prenatal diagnosis between 12 and 16 weeks of pregnancy (either by chorionic villus sampling or amniocentesis). The main drawback of prenatal diagnosis is that if the fetus is affected by the disease, the couple have to decide whether to continue the pregnancy or terminate the pregnancy. This is obviously not an ideal option for any couple trying to have a healthy family. With PGD, the couples go through IVF so that embryos are generated in vitro, and one or two cells are removed and used for the genetic diagnosis. Only unaffected embryos are transferred, so that the pregnancy is started with an unaffected fetus. The embryo biopsy is relatively easy to perform for an experienced embryologist. However, the single cell diagnosis is a highly specialized technique and is technically challenging, even for experienced molecular or cytogenetic biologists.

The patients that opt for PGD may be fertile. They are known to be at risk of a particular genetic disorder, either due to already having an affected child or because other family members are affected. In the case of patients carrying chromosome abnormalities, these couples often experience repeated miscarriages due to unbalanced chromosomes in the embryo and fetus, which most often are lethal. Couples may have already undergone prenatal diagnosis and repeated termination of pregnancies. They may have moral or religious objections to termination, or they may be infertile and also be carrying a genetic disease (which may or may not cause their infertility), so PGD is a sensible step to add to their fertility treatment (2, 3).

1.1 Embryo Biopsy

Theoretically, there are three stages at which cells can be removed from an embryo for PGD. The first is removal of the first and second polar bodies. Originally this involved removal of just the first polar body from the oocyte and was therefore called preconception diagnosis (4). However, for the diagnosis to be accurate, the second polar body is also required, and this is not extruded until fertilization. Therefore, the second polar body is removed from the zygote, after conception. It has also been proposed to remove 10–20 trophectoderm cells from the blastocyst (5, 6). This technique will give a large number of cells, but it has many drawbacks. The main biopsy method used for PGD is removal of 1–2 blastomeres from the cleavage-stage embryo on day 3 of development, when the embryo should be at the 6- to 10-cell stage (7). A full description of these techniques and the advantages and disadvantages are presented in sections 1.2 and 1.3.

1.2 Single Cell Diagnosis

Two techniques are used for single cell diagnosis in PGD. PCR is used for the diagnosis of single gene defects, including specific diagnosis of X-linked disease, autosomal recessive and dominant disorders, and the triplet repeat diseases (8). FISH is used to examine chromosomes and so is used for sexing embryos for X-linked disease (9, 10) and for chromosome analysis, in the case of translocations, insertions, and so on (11–13). More recently, FISH has also been used to check for the chromosomes commonly involved in aneuploidy (chromosomes 13, 18, 21, X and Y) in embryos from patients undergoing routine IVF procedures as a method to improve IVF pregnancy rates (14).

1.3 Polymerase Chain Reaction

In PCR, a specific fragment of DNA is amplified thousands of times. This is achieved by using specific primers that are usually designed to amplify the area of DNA containing the mutation. Once the DNA is amplified, a number of techniques can be used to analyze the DNA fragments (8). The original method used was heteroduplex analysis (15), but more recently techniques such as single-stranded conformational polymorphism (SSCP), restriction enzyme digestion, and fluorescent PCR have been used (8). For most single-gene defects, there are hot spots of mutation. For example, the most common cystic fibrosis mutation in the Caucasian population is deltaF508, a three basepair deletion. However, there are more than 700 other cystic fibrosis mutations reported. Therefore, a different set of PCR primers may be needed for each different mutation.

To add to the difficulty in amplifying DNA from single cells, PCR for PGD is complicated by contamination and a phenomenon known as allele dropout (ADO) or preferential amplification (8). Contamination can occur in any PCR reaction, not just in single cell PCR. However, it is more important in single cell PCR because it may lead to a misdiagnosis. Contamination may be caused by cumulus cells or sperm attached to the zona pellucida, which may become dislodged during the embryo biopsy procedure. Therefore it is essential to remove all cumulus cells before the biopsy and to use ICSI for fertilization to ensure there are no sperm embedded in the zona. Contamination can also be caused by DNA in the atmosphere or from cells from laboratory staff. A number of PGD misdiagnosis have been reported, and in most cases this was probably caused by contamination (1). Therefore, polymorphic markers can be used to determine if the DNA being amplified is from the parents (16, 17). A similar method is used in forensics for DNA fingerprinting. For the markers to be informative, they need to show a different pattern at both alleles for both the mother and the father, so that the alleles that the embryo has inherited can be easily identified. Because the mutation analysis is also required, a multiplex PCR is performed. Fluorescent PCR has been used for such analysis (16).

ADO or preferential amplification is where one of the alleles does not amplify during the PCR reaction. For example, if DNA from a carrier was examined, two DNA fragments should be produced: one with the normal gene and one carrying the mutation. In ADO, only the normal or mutated gene is seen instead of both. ADO must therefore be kept to a minimum in any PCR PGD, but is especially important for dominant disorders (where the affected embryo will carry one normal and one abnormal gene) and recessive disorders where the couple carry different mutations and only one mutation is being examined.

1.4 Fluorescent in Situ Hybridization

FISH has been used to examine chromosomes and is often used to complement cytogenetic techniques such as karyotyping. Because it is very difficult to obtain metaphase chromosomes from embryos that could be used for karyotyping, interphase FISH has been the method of choice for examining chromosomes for PGD. Two types of FISH probes can be used in interphase FISH: repeat sequence probes, which often bind to the centromere or alpha satellite regions of chromosomes, and, more recently, locus-specific probes, which bind to a specific sequence of a chromosome. FISH probes are labeled with fluorochromes and hybridize to their specific chromosome. When examined with a fluorescence microscope, dots of fluorescence can be seen where the probes have bound to their complementary sequence. Since 1991, FISH has been used to sex embryos for patients at risk of X-linked disease (9), originally just using repeat probes for the X and Y chromosomes. More recently, three-color FISH has been used for embryo sexing, usually with the addition of a repeat probe for a chromosome (10, 18). Ideally the extra probe should be for chromosome 21 to reduce the risk of Down's syndrome, but the repeat probes for the acrocentric chromosomes (chromosomes 13, 14, 15, 21, 22) cross-hybridize and are not efficient. Therefore, locus-specific probes have to be used to examine these chromosomes (11, 13, 19, 20).

For patients carrying chromosome abnormalities, such as Robertsonian or reciprocal translocations, locus specific or repeat sequence probes are used for the chromosomes in question (11, 13, 19, 20). For Robertsonian translocations, which involve the joining of two acrocentric chromosomes, locus-specific probes are used. Only one probe is needed for each chromosome involved in the translocation, but two different probes could be used for the most important chromosome to ensure against hybridization failure of one of the probes. However, for reciprocal translocations, any chromosomes can be involved and any break points. Therefore, each translocation is usually unique to that family. In some cases, commercial probes may not be available, which would make PGD difficult to develop. PGD for inversions (20), gonadal mosaicism (19), and Klinefelter's syndrome (21) have been reported.

In recent years, several groups have reported the use of embryo biopsy and FISH to examine oocytes or embryos from couples undergoing IVF, with the view of trying to improve the chance of pregnancy. For older patients undergoing IVF, one group has used polar body analysis (22), and the other group has used cleavage-stage biopsy (14). This procedure may also be useful for patients experiencing recurrent miscarriages, even though their chromosomes are normal (23).

Through research into early development using FISH, it has been found that chromosomal mosaicism is high in human preimplantation embryos at cleavage (18, 24, 25) and blastocyst stages (26, 27). The most common abnormality observed is tetraploid/diploid mosaicism, which is probably due to errors in mitotic cell-cycle checkpoints during early embryonic divisions. This mosaicism may not have a significant effect on embryo development, but it can cause problems in PGD, as the cell removed may not be representative of the rest of the embryo. This is most important in the PGD of chromosome abnormalities, as chromosomally normal and abnormal cells have been seen in the same embryo (J. Harper, personal observation). Therefore, due to chromosomal mosaicism, misdiagnosis of chromosome abnormalities can occur.

1.5 Future and Ethics

Immediate research in PGD is concerned with ensuring that the diagnosis is accurate. For PCR accuracy can be improved by the use of informative markers (8, 17). In PCR and FISH, methods are being developed to examine more than one gene or more than a few chromosomes at one time. For PCR, this may involve the use of DNA chips or arrays to allow examination of many genes and mutations simultaneously (1, 17). For FISH, the most promising technique is comparative genomic hybridization (CGH). In this technique the blastomere DNA is labeled with a fluorescent dye (e.g., red), test DNA is labeled with a different color (e.g., green), and the two are cohybridized onto a control metaphase spread. The chromosomes from at least six metaphase spreads are analyzed, and the ratio of red to green is analyzed. If the blastomere has an extra chromosome, that chromosome will appear more red in the metaphase. Conversely, if the blastomere is lacking a chromosome, that chromosome will appear more green. However, for CGH to work in a single blastomere, a whole genome amplification method, such as DOP (degenerative oligonucleotide primer) PCR has to be used to obtain enough DNA (28). CGH has recently been successfully applied to human blastomeres (29, 30). Besides being technically challenging, the main limitation of CGH is the time required (currently about 72 h). Until the procedure can be more reproducible and less time consuming, this technique will not be suitable for PGD.

In most countries, PGD is being used for the diagnosis of serious genetic or chromosomal diseases or for the diagnosis of aneuploidy in older IVF patients to increase the pregnancy rates. However, the use of PGD for sex selection for social reasons has been undertaken in Australia and Jordan. In Jordan, 250 cycles were undertaken, and in only 1 case did the couple wish to select for a girl. Another controversial use of PGD was undertaken in 2000 in the United States. A family with a child affected with Fanconi's anemia wanted another child who would be unaffected with this disease but who could also be HLA matched so the cord blood could be used to treat the first child. This use of PGD has received a mixed response from the PGD community and the media. Most of the concern is that in the future PGD may be used to select for characteristics besides sex, such as hair color, height, and intelligence, and so design the perfect baby. This has resulted in legislation regulating PGD in most countries (31).

1.6 PGD Consortium

The European Society of Human Reproduction and Embryology PGD Consortium was set up in 1997 to

- Survey the availability of PGD for different conditions facilitating cross-referral of patients
- Collect prospective and retrospective data on accuracy, reliability, and effectiveness of PGD
- Initiate follow-up studies of pregnancies and children born
- Produce guidelines and recommended PGD protocols to promote best practice
- Formulate a consensus on the use of PGD.

The largest task of the consortium has been to collate, analyze, and publish data concerning PGD. This has included referral, cycle, pregnancy, birth data, and details of protocols. Two reports have been published (2, 3). The second data collection reported on 1319 cycles of PGD and included clinics from Australia, Belgium, Denmark, France, Greece, Italy, South Korea, Spain, Israel, The Netherlands, Sweden, the United Kingdom, and the United States. The biggest disappointment of this data collection was the low pregnancy rate (18% per oocyte retrieval). It is hoped that with the help of the consortium, PGD will become a more reliable and accepted treatment for couples at risk of transmitting an inherited disease to their offspring.

2. TECHNIQUES AND PRINCIPLES

PGD is based on the progress of IVF, micromanipulation and biopsy of gametes and embryos, and on the progressive achievements in the genetic analysis of single cells. Micromanipulation techniques have been widely used to biopsy mammalian embryos to successfully remove genetic material and analyze it without affecting the viability of the embryo (32). This technique was applied for the first time in humans in 1989 by Handyside and colleagues for diagnosing X-linked diseases (15). In 1990 the first children born as a result of biopsying human embryos showed no detectable birth defects, which indicated that embryo biopsy was a safe and efficacious tool to use for PGD in humans (15).

Embryo biopsy involves two main steps: zona puncture and removal of genetic material. The two methods that are currently being used clinically to obtain genetic material for PGD are: (1) biopsy of the first and second polar bodies from metaphase II oocytes pre- and postfertilization (or simultaneously after fertilization) and (2) biopsy of blastomeres from cleavage-stage embryos on day 3 postfertilization. Blastocyst biopsy has also been proposed, but it has not been applied clinically to date.

2.1 Polar Body Biopsy

Genetic analysis of the first polar body was first reported by Verlinsky et al. (4) (Figure 6.1). The first and second polar bodies are a byproduct of meiosis and are not required for normal embryonic development. Hence their removal to determine the genetic status of the oocyte/zygote may not pose any deleterious risk on the developing embryo (33). This technique has been used for diagnosing age-related aneuploidy (34–36), translocations (37), and single-gene defects (38, 39). Polar body biopsy can be performed sequentially (40) or simultaneously in one manipulation (22, 35). Sequential biopsy may be advantageous as the first polar body degenerates.

Polar body biopsy does pose certain problems that are not encountered with cleavage-stage embryo biopsy. Two of the groups using polar body biopsy have used mechanical means of opening the zona pellucida (partial zona dissection) because it has been shown that further embryonic development is hindered in oocytes subjected to acid tyrodes (41). This could be due to the direct effect of the acid to the oocyte as a result of lowering the pH during exposure. The use of a 1.48-μm laser to ablate the zona in mice zygotes has been reported to facilitate polar body biopsy (42).

2.1.1 Advantages and Disadvantages of Polar Body Biopsy

The advantages of polar body biopsy are that, first, it is ethically more acceptable to certain countries (e.g., Malta and Germany) where research and discarding embryos is prohibited. Second, the first and second polar bodies may not be required for further embryonic development and hence the viability of the oocytes and resulting embryos may be equally maintained.

One disadvantage of polar body biopsy is that it can only be applied to diagnose maternally inherited diseases. Recombination and post zygotic events leading to chromosomal abnormalities cannot be detected by this method. Diseases that are detected by assessing changes in gene product are also impossible, as well as gender determination (sexing).

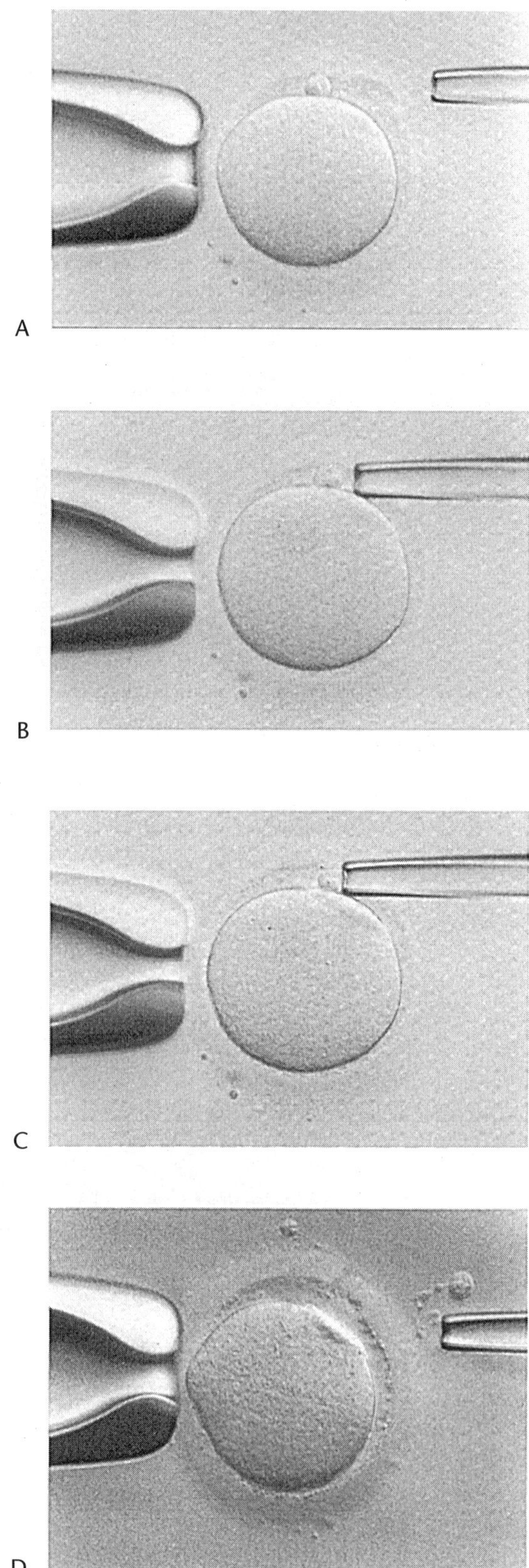

Figure 6.1. Polar body biopsy. (a) A hole is made in the zona pellucida using a laser; (b) the aspiration pipette is placed near the hole; (c) the polar body is aspirated; (d) polar body and biopsied oocyte.

Another disadvantage is that the oocyte/zygote is subjected to two manipulations in cases of sequential biopsy of the first and second polar body rather than to a single manipulation at the cleavage stage. In addition, ICSI has to be carried out in between regardless of the type of diagnosis, as insemination by IVF could lead to polyspermy and sperm contamination for PCR diagnosis. Cleavage-stage biopsy of embryos derived from these manipulated zygotes has also been reported to confirm diagnosis (38). This is highly labor intensive and can be detrimental to the embryo.

A third disadvantage is that, in cases where the first and second polar bodies are removed at the same time in one manipulation, the first polar body degenerates between the time of oocyte retrieval and biopsy following fertilization (34). This may lead to failure of diagnosis in the first polar body due to fragmenting or degenerating DNA. The biopsy of only the first polar body would lead to an overall lower reliability of diagnosis, as only one cell is available for analysis, in contrast to cleavage stages, where up to two cells can be taken for independent testing. In addition, it is thought that the risk of misdiagnosis is greater from polar bodies than from blastomeres (43). Finally, sometimes the first polar body is not very well detached from the oolema, which can lead to lysis during the biopsy procedure.

2.2 Cleavage-stage Biopsy

Biopsy of human embryos at early cleavage stages (two to four cells) has been thought to be unsuitable for PGD because of the excessive reduction in cellular mass, leading to a lower pregnancy rate (44, 45). This has been believed to be as a result of impaired cellular differentiation of the inner cell mass (ICM) and trophectoderm (TE) cells in blastocysts (46). The eight-cell stage has been suggested to be most suitable for biopsy because of its higher mitotic index and totipotency (47). In addition, removal of up to two cells from an eight-cell embryo does not have any detrimental effect on its metabolism or on the ratio of ICM and TE cells in blastocysts derived from biopsied embryos when compared with controls (48). Similarly, after cryopreservation of human embryos, normal and healthy live births have been reported from embryos that had only 50% of their blastomeres intact on thawing, thus emphasizing the totipotent nature of early embryonic cells. Cleavage-stage embryo biopsy to date remains the most favorable choice for obtaining genetic material by most PGD centers worldwide; from 759 cycles, 755 used cleavage-stage biopsy and blastomere aspiration, 3 used polar body biopsy, and 1 used both polar body and cleavage-stage biopsy (3). From these data, a successful biopsy was achieved in 96% of the cases.

Many different methods have been used in animal models to remove embryonic cells before compaction (49–51). Moreover, micromanipulation and biopsy of cattle embryos is commonly used to create chimeras for agricultural use. However, only a few of these methods have been successfully applied to human embryos (Figure 6.2).

2.3 Methods of Zona Drilling

2.3.1 Acid Tyrodes Drilling

Acid Tyrodes has been used in embryology since 1986 (52) and has been the best choice to date for cleavage-stage embryo biopsy (3) (Figures 6.2a; see also Figure 6.4). Acid Tyrodes solution is commercially available (Medicult, Denmark, or Vitrolife, Sweden) or can be made in house. The pH of acid tyrodes should not be lower than 2.2, as this can be detrimental in lowering the pH of the biopsy drop when drilling. Cleavage-stage embryo biopsy using acid tyrodes usually incorporates the use of a double tool holder on the micromanipulator. A detailed account of acid tyrodes drilling is given in the cleavage-stage biopsy protocol (Protocol 6.1). The main drawback of using acid tyrodes is its toxicity and cytoplasmic acidification, which changes the intracellular pH, leading

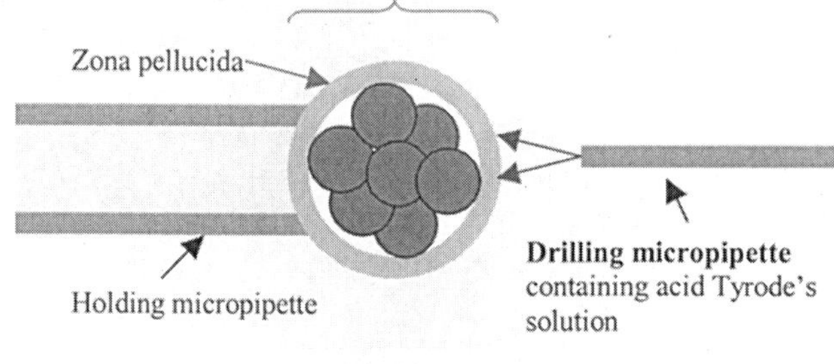

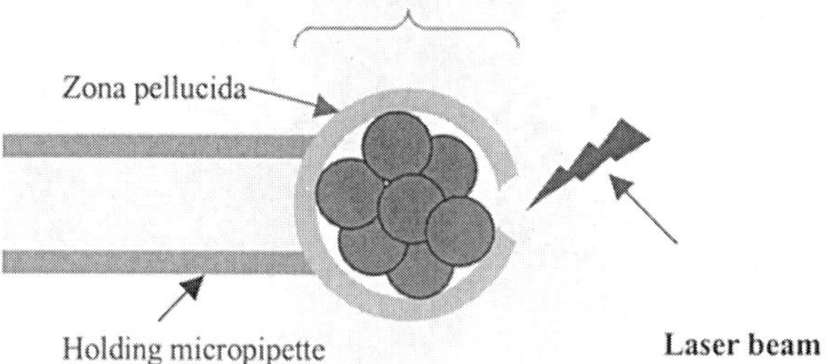

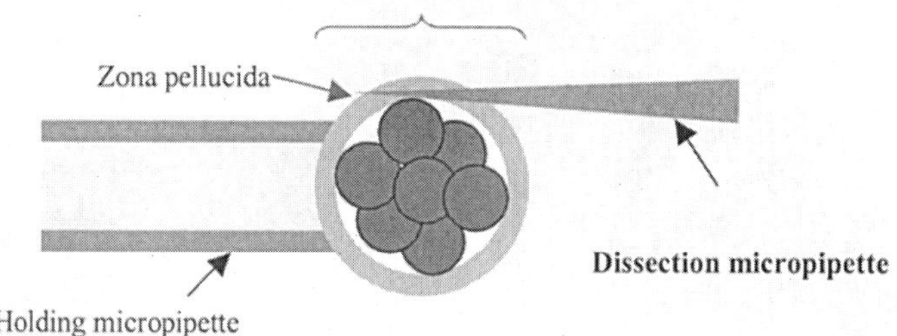

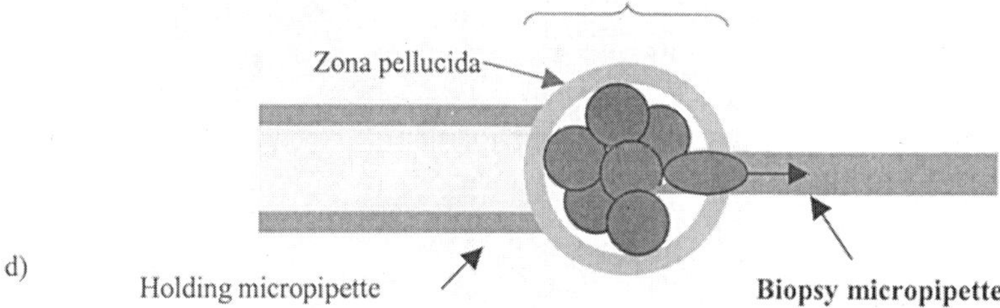

Figure 6.2. Methods of zona pellucida opening. (a) Acid tyrode's drilling, (b) laser drilling, (c) mechanical drilling (partial zona dissection), (d) zona pellucida piercing. (Figure courtesy of P. Phopong, Siriraj Hospital, Bangkok.)

to cytoplasmic degeneration (53). Several studies have reported that chemical drilling is more detrimental to oocytes than mechanical means (54–56). In addition, exposure of four-cell human embryos to acid tyrodes has also been suggested to reduce their viability and preimplantation development (41, 44, 54). In contrast, several published reports indicate that there is no significant adverse effect from acid tyrodes in mice and human embryos (7, 57–59). In humans it has been shown that assisted hatching using acid tyrodes at early-cleavage stages increases the implantation rate.

2.3.2 Laser Drilling

Lasers have been successfully used over the last decade and their safety has been proven in both mouse and human gametes and embryos (42, 60–62) (Figure 6.2b). Feichtinger et al. reported the first healthy live birth after attempting laser-assisted partial zona dissection to assist hatching in human embryos (63, 64). Over the years several reports have highlighted the use of different types of lasers (60, 65–67), some of which have shown adverse effects such as affecting the genetic structure of oocytes (68), abnormal embryonic development (66), and inducing sister chromatid exchange in Chinese hamster ovary cells (69). Recently the preferred choice has been the near infrared (NIR) solid-state compact diode 1.48-μm laser (7, 61, 70, 71). The wavelength of light emitted is highly absorbed by the glycoproteins of the zona pellucida and not by blastomeres or the culture media. However, if the exposure time (pulse irradiation time [PIR] range 3–100 ms) is too long, then the photothermal energy can cause selective damage to further embryonic development (J. Harper and A. Doshi, personal observations). The usual exposure time to drill a hole of 10–15 μm is 10–12 ms. It is advisable to use a couple of shots at low exposure times rather that a single shot of PIRs > 15 ms to give the appropriate size of hole needed. Nevertheless, when used within short exposure times, human and mouse embryos showed no signs of any extraneous thermal damage under light or electron microscopy (61). The 1.48-μm diode laser has been used in polar body biopsy in mouse zygotes (42) as well as for mouse and human blastocyst biopsy (6, 72, 73). The efficiency of the procedure and blastocyst recovery rates were encouraging, suggesting that laser-assisted blastocyst biopsy could be a viable tool for clinical PGD.

The advantages of laser drilling are

1. There is no need for disposable micropipettes for drilling, and risk of introducing any extraneous contamination is avoided (when drilling with acid tyrodes).
2. Consistent results are obtained because controlled hole sizes can be obtained, whereas with acid tyrodes there is no regulation of hole size, and it can vary between embryos. Very large holes drilled as a result of acid tyrodes can result in the unwanted loss of blastomeres and affect implantation (S. Munne, personal communication).
3. Because there are variations in zona pellucidae both between and within cohorts of human oocytes and embryos (e.g., hardening, thickness), laser drilling can achieve more consistent results rather than subjecting the embryo to large amounts of acid tyrodes.
4. It has been observed that laser drilling results in a lower degree of blastomere lysis during biopsy compared to acid tyrodes (74).
5. The rapid delivery of photothermal energy is easily dissipated and the method offers a high degree of precision.

The disadvantages of laser drilling are

1. It is very expensive, and for certain centers performing small number of cases it is not cost effective.

2. Laser drilling creates the aperture in the form of a groove rather than a spherical hole. This can in certain cases cause constriction of the blastomere through it and result in lysis.

Many centers are making the transition from acid tyrodes to laser drilling for cleavage stage embryo biopsy (7). However, the former still remains the most commonly used technique (2, 3).

2.3.3 Mechanical Drilling

Also known as partial zona dissection (PZD), this technique of creating an opening in oocytes and embryos has been used clinically by two centers worldwide for PGD (2) (Figure 6.2c). Mechanical drilling involves the use of a sharpened micropipette to mechanically pierce the zona and make a hole or a slit. Malter and Cohen (41) suggested that PZD of the human oocyte was a nontraumatic method compared to zona drilling with acid tyrodes. However, the precision in positioning the hole is jeopardized when compared to acid tyrodes or laser. Advantages of mechanical drilling are that there are no effects of chemicals or heat during the procedure. In addition, reports indicate that the fertilization rate (for polar body biopsy) and further embryonic development is not compromised in comparison with their respective controls using this method (38). It has been suggested that with PZD the openings can be very narrow in comparison with spherical apertures using other drilling methods. This can subsequently induce more trauma to the cells during biopsy as well as lowering the blastocyst hatching rates (75, 76).

2.3.4 Zona Piercing

Zona piercing has not been used clinically for PGD, but it has been tested in the mouse model. Wilton et al. (51) and Thompson et al. (77) used this approach to biopsy blastomeres from four-cell and eight-cell embryos, respectively (Figure 6.2d). A sharpened, beveled micropipette was used to puncture the zona and aspirate the cell(s) with the same pipette. Results showed that the blastocyst rates were comparable between the manipulated and control embryos (51,77). This method is not feasible for application to human embryos due to the thickness of the zona pellucida.

2.4 Methods of Blastomere Removal

2.4.1 Aspiration

Aspiration is the most commonly used method to remove blastomeres for clinical embryo biopsy after drilling with either acid tyrodes, laser, or PZD (3) (Figure 6.3a and 6.4). A finely fire polished biopsy pipette (30–40 μm internal diameter) containing biopsy medium is inserted close to the aperture, and the blastomere in contact with the pipette is aspirated using gentle suction with either hydraulic, pneumatic, or mouth-controlled syringes. The most important factor determining a successful biopsy is careful and gentle control during aspiration because the blastomeres can be very fragile and prone to lysis. If the hole is drilled in between two blastomeres, then the biopsy pipette is brought close to the hole, after which it can be moved up and down near the cell to be biopsied, followed by aspiration. This is more problematic in comparison with holes drilled directly adjacent to the blastomere. However, when opting for two cells, it is advisable to only drill one hole and aspirate both cells from the same hole rather that making another one, which can cause embryonic hatching from two sites, leading to monozygotic twinning (M. Montag, personal communication). Once at least

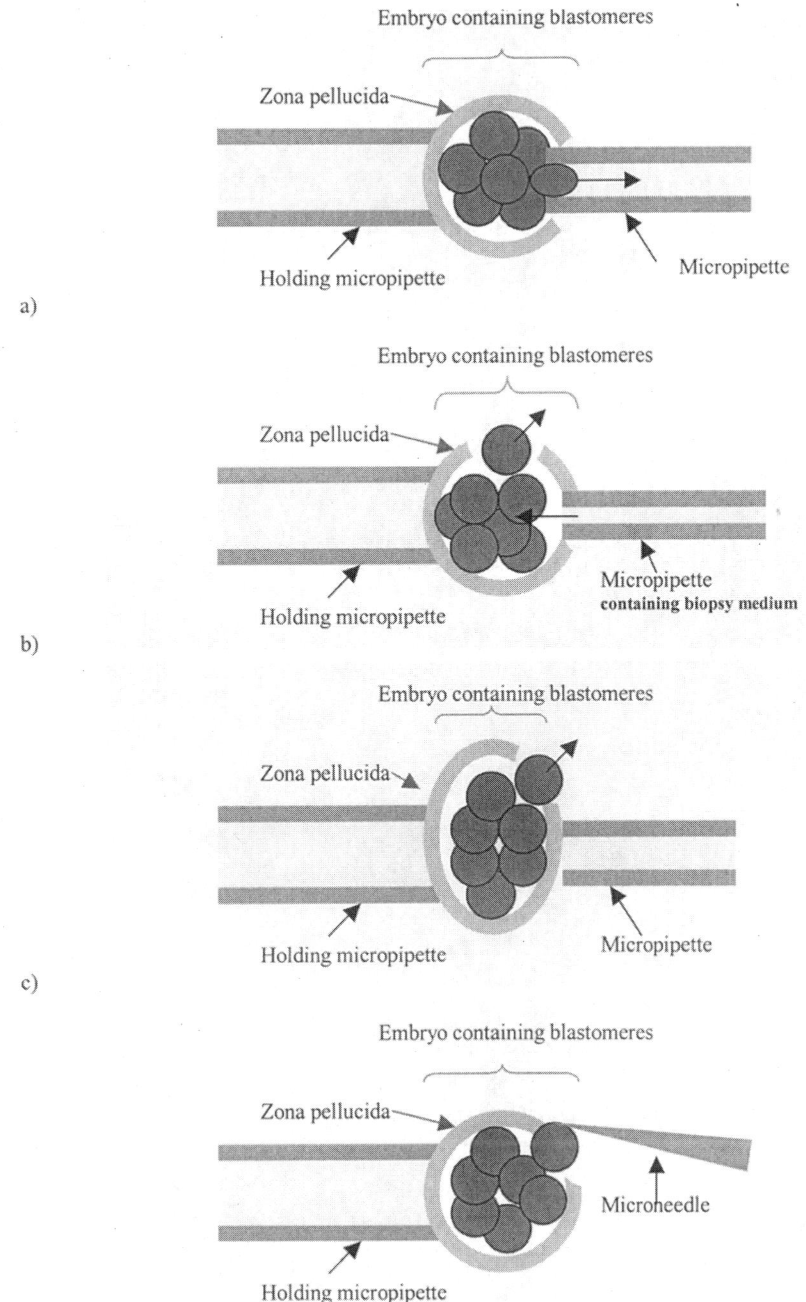

Figure 6.3. Methods of blastomere removal. (a) Aspiration, (b) displacement, (c) extrusion, (d) stitch and pull. (Figure courtesy of P. Phopong, Sivivaj Hospital, Bangkok.)

half of the blastomere is aspirated within the biopsy pipette, it should be gently withdrawn to dislodge the cell from the rest of the embryo, followed by gradual removal of the cell from the pipette. Engulfing the entire cell in the pipette can be more traumatic and can result in cell blebbing and thus lysis. However, in highly compact embryos engulfing the cell might be necessary to dislodge the cell from the rest of the embryo. It is important that the right size pipette is used for aspiration. If too small a

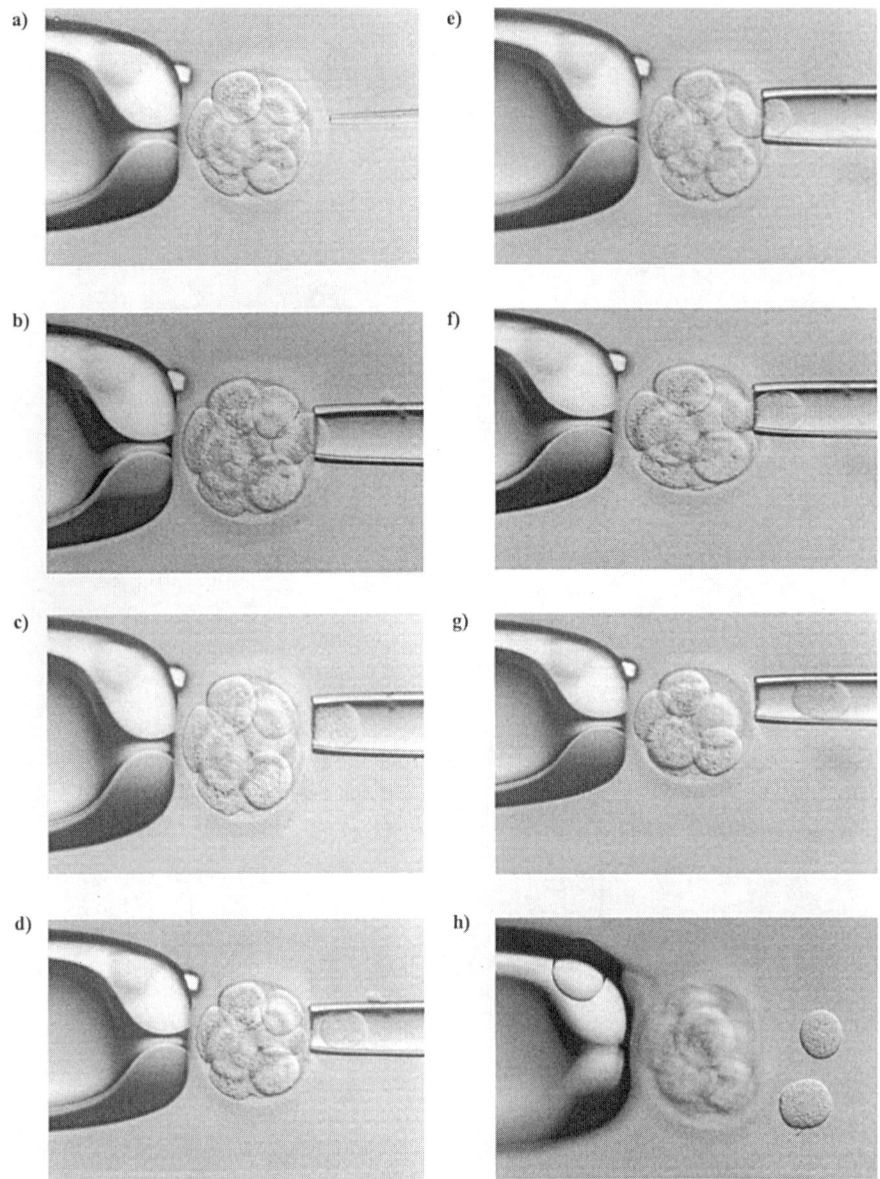

Figure 6.4. (a) Zona drilling with acid tyrodes; (b) aspirating the first blastomere; (c, d) aspirating and dislodging the blastomere from the rest of the embryo; (e) aspirating the second blastomere from the same hole; (f) maneuvering the blastomere to dislodge from the embryo (compact); (g) engulfing the second blastomere into the pipette; (h) two biopsied blastomeres regain shape.

pipette is used, then there is increased risk of cell constriction, eventually leading to lysis. Conversely, using a very large pipette can subject nearby blastomeres to be aspirated unnecessarily.

2.4.2 Displacement

Displacement was first tried by Roudebush et al. (78) to biopsy mouse embryos at the four- to eight-cell stage (Figure 6.3b). Two openings were made by zona slitting, and a cell was protruded from the first puncture site by injecting culture medium through the second hole. Cell survival after biopsy was much higher than with direct aspiration (55%

vs. 22%). A similar technique was used by Pierce et al. (79) to biopsy eight-cell mouse and human embryos. No significant difference was seen in blastocyst rate between biopsied and control mouse embryos, and 96% of the biopsies cells underwent further cleavage in culture. Blastocyst formation was observed in 69% of the biopsied human embryos. Pierce et al. (80) applied the displacement method in clinical practice to detect chromosomal abnormalities in cleavage-stage embryos.

2.4.3 Extrusion

Once an aperture has been made in the zona, a cell is biopsied by applying gentle pressure using a blunt pipette at another site close to the cell to be biopsied, thus protruding it through the opening. Gordon and Gang (81) biopsied mouse embryos using the extrusion method, and Levinson et al. (82) applied this principle to human embryos in clinical PGD for sexing (Figure 6.3c).

2.4.4 Stitch and Pull

The stitch-and-pull method has been described in the review by Tarin and Handyside (45) (Figure 6.3d). The biopsy is carried out by drilling an opening in the zona, followed by removal of the cell by stitching movements using a fine, sharp glass needle. Once the needle is inserted through the zona, the blastomere is pushed upward toward the opening. This method was first used by Muggleton-Harris et al. (83) to analyze cells biopsied at both the cleavage and blastocyst stage. Cell lysis is more prevalent with this method of biopsy.

2.4.5 Single-Pipette Biopsy

The most commonly used three-pipette technique, in which one pipette is used to hold the embryo, another is used for drilling (acid tyrodes), and the third is used to aspirate the blastomere, was modified in favor of a single pipette which performs both functions (i.e., drilling and aspirating). The internal diameter of the pipette ranges from 34 to 51 μm depending on the number of cells in the embryo. Once a hole is drilled by acid tyrodes, negative suction is applied to aspirate all acidic medium around the site of drilling. The pipette is emptied of all acidic tyrodes, and fresh biopsy medium is loaded, after which the one or two cells are removed by aspiration. Inzunza et al. (84) and Chen et al. (85) reported healthy live births from embryos biopsied in this way.

2.5 Technical Considerations

2.5.1 Biopsy Time

Cleavage-stage embryo biopsy is carried out on day 3 after insemination when the embryo contains between 6 and 10 cells (Figure 6.4). Most centers prefer to do the biopsy in the morning of day 3 if most of the embryos have attained the right stage of development. However, it is advisable to wait until the afternoon if embryos are at the five-cell stage or less. Moreover, the decision to wait depends on how many cells are required for diagnosis. In addition, it is important to take into account timings of human chorionic gonadotropin (hCG) injection, egg collection, hours after insemination/injection (86), and times at which the embryos are scored on day 2 and 3.

It is still unclear in humans what impact biopsying embryos with four to five cells has in humans. In a study carried out by Tarin et al. (44), embryos biopsied on day 2 after insemination showed a great degree of retarded cleavage rates after biopsy as well as low cell numbers in resulting blastocysts (44). In contrast, pregnancies have been reported from embryos biopsied at the four-cell stage (A. Thornhill, unpublished data).

In the mouse model biopsying embryos on day 2 results in abnormal differentiation of the ICM and the TE, resulting in abnormal postimplantation development.

2.5.2 Decompaction and the Use of Ca^{2+}/Mg^{2+}-free Medium

The embryo starts to undergo the events of compaction on day 3 at the eight-cell stage (87). Compaction involves the production of desmosomes and gap and tight junction proteins, which causes the blastomeres to adhere to one another to maximize intercellular contacts. The integrity of tight junctions is calcium and magnesium dependent. In a study carried out by Dumoulin et al. (88), incubation of embryos in Ca^{2+}/Mg^{2+}-free medium for short periods of time resulted in less cell lysis and shorter biopsy time. Moreover, there was no difference in the blastocyst rate and the number of cells in the blastocysts in the control and experimental groups. This led the authors to conclude that decompaction of the embryo was a reversible process (88). Similar conclusions were made in the mouse model when embryos were biopsied at the four-cell stage (89), and no detrimental effect was seen on further embryonic development after biopsy. However, in a study reported by Van Blerk et al. (90), in which mouse morulae were decompacted using Ca^{2+}/Mg^{2+}-free medium, a significant drop was seen in the number of live offspring born in comparison to controls when transferring these embryos to foster mice. This may suggest that decompaction can be reversed after biopsy only if performed before a certain developmental stage. A survey carried out in 12 centers performing embryo biopsy for PGD revealed that more than half use Ca^{2+}/Mg^{2+}-free medium routinely (2).

2.5.3 Cell Selection

Cleavage-stage embryo biopsy offers the advantage of a choice of cells to biopsy, whereas polar body and blastocyst biopsy do not. It is essential that a few guidelines are followed when selecting cells for biopsy before attempting to drill a hole:

1. Visualizing the nucleus is quite difficult when all the cells are overlapping within the embryo. Moreover, the granularity of the cytoplasm sometimes makes the nucleus less clear. However, visualization is not impossible, and the nucleus can sometimes be clearly seen under high magnification. It has been suggested that in the future the use of fluorescence might make it easier to locate a nucleus in a biopsied cell (91, 92). Large fragments are often mistaken as blastomeres, and hence visualization of the nucleus prevents biopsying any anucleate fragments.
2. It is always advisable to select the smaller blastomeres for biopsy within an embryo. This indicates that the cell has undergone its mitotic division. Large, undivided cells usually suggest that the cell has either not undergone mitosis or that it is still in the process of division (karyokinesis and cytokinesis). Biopsying large cells not only reduces embryonic mass significantly but also risks the cell being in metaphase, making the chromosomes more prone to loss during fixation.
3. Multinucleation of blastomeres is not an uncommon phenomenon observed in cleavage-stage embryos. Binucleate and multinucleate cells should strongly be avoided during biopsy, as they can result in unclear diagnosis. In addition, such cells could have derived as a result of failure of cytokinesis after karyokinesis.

2.5.4 Number of Cells Biopsied

With the discovery of mosaicism within preimplantation embryos, it has become common practice in some centers to routinely biopsy two cells and analyze them independently before transferring. This gives greater diagnostic accuracy and a higher indication of normality. In regard to this practice, it is important to bear in mind that the potential to implant decreases with greater amount of embryonic mass removed (91, 93). However, recent studies have shown that there is no difference in pregnancy and implantation and live birth rates when two cells are biopsied compared to one (7). This ties in with data obtained by Hardy et al. (48), who showed that there was no significant difference in the ratio of TE and ICM nuclei when up to two cells were biopsied at the eight-cell stage. The decision to take one or two cells is based on the stage of development of the embryo and the type of diagnosis. In cases of dominant disorders discovered by PCR it has become essential to analyze two cells because allele dropout can cause serious misdiagnosis if diagnosis is based on one cell analysis. However, for sexing by FISH as well as for diagnosis of autosomal recessive disorders, one-cell diagnosis can be considered because transferring carrier embryos is not lethal (94).

2.5.5 Cryopreservation of Biopsied Embryos

Cryopreservation of biopsied embryos is becoming a necessity with patients who have a greater number of normal embryos than those transferred. Experiments in mice have shown promising results in achieving successful cryopreservation of biopsied embryos (51, 53, 93). Cryopreservation of biopsied human embryos would not be as simple as methods applied for unbiopsied embryos due to the presence of a hole in the zona (which acts as a protective barrier to regulate the inflow of toxic cryoprotectant). To date, few groups have attempted freezing biopsied embryos. One of the first reports by Joris et al. (74) tried freezing biopsied day 3 embryos using a slow freeze/thaw procedure with DMSO. Survival in the biopsied group was significantly lower than in intact controls. Nevertheless, in biopsied embryos that did survive the freeze/thaw procedure, blastocyst formation was seen in a small percentage. Similar results were confirmed by Magli et al. (95). This suggests that although survival rates are poor with current freezing protocols, freezing biopsied embryos can be attempted. Lee and Munne (96) reported a live birth from embryos derived from polar body biopsy and frozen using a one-step method as previously described (97). Unpublished reports have claimed survival (>50% of cells intact) in 75% of biopsied, frozen and thawed embryos, with clinical pregnancy rates of 25% per embryo transferred. However, the modification to previous methods reported was culturing biopsied embryos to day 5 before attempting freezing using glycerol and sucrose as cryoprotectants (J. Catt, personal communication).

2.5.6 Contamination

Aseptic techniques are of utmost importance in an embryology laboratory. Moreover, in cases of PCR, contamination can be very crucial because it can result in amplification failure or preferential amplification. Extraneous sources of contamination are listed below:

1. Always ensure that new batches of media that have been tested for quality control are used for PGD cycles, and avoid using bottles that have been previously opened. It is good practice to batch test all media used for embryo culture and PGD by PCR to confirm purity.
2. Biopsy medium containing HSA should be tested by PCR to ensure that no amplification results from any human sources of DNA present in the HSA, which could lead to inconclusive results.
3. All cumulus cells should be carefully removed before the biopsy procedure. Any remnants of these cells can contain DNA that will amplify during PCR.
4. ICSI should be used for all PCR cases. Any extra sperm bound to the zona as a result of IVF can lead to amplification of sperm DNA.
5. Nontoxic, powder-free gloves, hats, masks, and sterile clothing should be worn at all times to avoid any cells (e.g., skin) from the personnel contaminating the

test material. It is advisable that all embryologists and biopsy practitioners have their buccal cells analyzed by PCR so that any contamination present during a clinical case can be traced.

2.6 Blastocyst Biopsy

The human blastocyst consists of two differentiated cell lines: TE and ICM. Two-thirds of the cells are TE and one-third are ICM. As the TE gives rise predominantly to the placenta and does not participate in the development of the fetus, several of its cells may be selectively removed without posing any risk to the resulting embryo proper (45). To some this may be ethically more acceptable because the cells biopsied do not contribute to fetal development. Morphologically normal human blastocysts consist of 60–125 cells on day 5–7 postinsemination (48). This would potentially offer a higher number of cells to be biopsied and available for genetic analysis. The higher number of cells available may decrease the likelihood of misdiagnosis, which can occur with the limited number of cells obtained from cleavage-stage biopsy. This would also allow more diagnostic techniques to be performed (5, 6). Blastocyst biopsy and livebirths have been reported in mice and primates (98, 99). The mouse model has since then been used to develop techniques of embryo biopsy and examine their safety (56, 78, 100). Studies in spare human blastocysts have shown that biopsy of up to 10 TE cells has proved to be safe in terms of hCG production in vitro and further blastocyst development compared to controls (101). If this technique were to be extrapolated to clinical practice, then the availability of more cells would reduce problems such as amplification failure or allele dropout for PCR (102). In addition, for FISH the presence of more than two cells would virtually guarantee a result for each embryo biopsied, as problems of split signals, signal overlap, or probe hybridization failure would be significantly less misleading. The availability of more cells would also increase the diagnostic possibilities for PGD, such as screening more chromosomes for FISH and analyzing more specific sequences with PCR (5).

2.6.1 Disadvantages of Blastocyst Biopsy

The main limitation of blastocyst biopsy is that a low number of embryos make it to the blastocyst stage in vitro. Even with sequential culture systems and co-culture, reports indicate around 50% blastocyst formation of normally fertilized zygotes (103). This would result in very few embryos available for PGD and would restrict the chances of patients having more embryos diagnosed. Moreover, since a good number of oocytes and embryos are needed for a successful PGD, blastocyst biopsy reduces the chances of having an embryo transferred in patients with fewer than nine oocytes collected (104).

The accuracy of diagnosis will be jeopardized if the TE cells show genetic variation from the ICM. This has known to happen in 1% of conceptions in which the chromosomal status of the embryo is different from the placenta (confined placental mosaicism). This may be a mechanism of early development in which abnormal cells are preferentially allocated to the trophectoderm lineage (105). Mosaicism is commonly seen in blastocysts, but the degree is significantly lower than in cleavage-stage embryos (27). In addition, mosaic cells have been seen even in the ICM (28, 106). It is believed that only about four cells from the ICM go on to make the embryo proper. This may suggest that mosaic cells within the ICM may be eliminated at this stage. These findings challenge previous beliefs that chromosomally imbalanced embryos will be eliminated as a result of extended culture in vitro (103, 107–109).

There is limited time to perform diagnosis because the blastocyst has limited viability in vitro. Hence there is pressure to put the blastocysts back before they collapse. The biopsy procedure can be very time consuming, as first a slit has to be made in the zona, followed by incubation for 6–24 h until some TE cells herniate through the opening, after which the extruded cells are excised. It has been suggested that the incision has to be made at the pole opposite to the ICM. However, this is not always straightforward because some blastocysts have very small ICMs, making it less distinct.

The only advantage in performing blastocyst biopsy clinically is that more cells can be obtained for analysis. However, it has been suggested that an alternative to blastocyst biopsy is to proliferate cleavage-stage blastomeres in vitro to yield more cells for diagnosis (110). Blastocyst biopsy using lasers has been proposed in human embryos for clinical PGD (6), but to date no center has reported using blastocyst biopsy clinically, and hence its widespread use awaits large-scale clinical assessment. In the future, the safety of blastocyst biopsy might be readily evaluated from multicenter data collected by the ESHRE PGD Consortium.

3. CONCLUSIONS

We have outlined the three methods of embryo biopsy in this chapter. Worldwide, the majority of clinics are performing cleavage-stage biopsy at the six- to eight-cell stage, using acid Tyrodes to drill the zona and aspiration to remove the blastomeres (3). There has been an increase in the use of the laser for zona drilling (2, 3), but the cost of the laser is still high, and its use has not been approved in every country. For those performing cleavage-stage biopsy, the decision whether to take one or two cells is still being debated (7). Performing PGD reliably from one cell is technically challenging and is especially difficult in the cases of chromosomal abnormalities and dominant disorders. However, taking two cells may reduce implantation. The use of Ca^{2+}/Mg^{2+}-free medium has made the biopsy technically easier, but the use of all these techniques needs to be monitored to ensure there are no detrimental effects.

Protocol 6.1. Cleavage-stage embryo biopsy

Materials

Solutions and Equipment	Sources
Micromanipulators	Research Instruments
	Eppendorf
	Narishige
Biopsy pipettes	Cook IVF
	Humagen
	Conception Technologies
Sequential culture medium	Cook IVF
	Medicult
	Vitrolife
Acid tyrodes solution	Medicult
	Vitrolife
Ca^{2+}/Mg^{2+}-free biopsy medium	Medicult
	Vitrolife

ICSI, oocyte, and embryo culture

1. Carry out all oocyte and embryo culture in sequential IVF media (e.g., Cook IVF, G1/G2, Vitrolife).
2. For cases involving PCR, use ICSI for insemination to avoid the risk of any contamination from the DNA of sperm. Routine IVF can be carried out in cases of FISH diagnosis where there is no male-factor problem.
3. Group oocytes together after the oocyte collection until the time of ICSI. Oocytes should be denuded as for the routine ICSI protocol. However, it is extremely

important to remove all cumulus and corona cells from the zona pellucida before commencing injection, as this can lead to contamination. Commercial denuding pipettes (internal diameter of 0.127–0.129 mm) can be used.

4. Score ICSI as normal (i.e., grading the oocyte cytoplasm, maturity, polar body fragmentation, and membrane elasticity).
5. Pick up about 5 sperm and transfer them to a fresh droplet of PVP in each dish before introducing the oocytes for injection. The microinjection needle should be changed after all the sperm have been separated into a fresh drop of PVP in all dishes used for ICSI to ensure that there is no contamination of residual sperm stuck to the pipette, which will be introduced into the oocytes.
6. Videotape the ICSI procedure for future reference, if desired.
7. After microinjection, culture all the oocytes individually in the order that they were injected. This will enable monitoring of an embryo derived from a good oocyte if a selection is available on the day of embryo transfer.
8. Double label all culture dishes clearly (on the lid as well as under the dish) with the oocyte/embryo number and patient details to avoid any confusion.
9. Carry out routine scoring and assessment for pronuclei. Normally fertilized zygotes should be transferred into fresh, pre-equilibrated drops of cleavage medium, correspondingly labeled with the number on the dish.
10. Remove any remnants of cumulus cells before transferring the zygotes. Evaluate the embryos morphologically on days 2 and 3 after insemination. Scoring on day 3 should be done just before commencing biopsy.

Preparation of biopsy dishes for acid tyrodes drilling

1. Use sterile tissue culture 60-mm dishes (Falcon 35-3002) for the biopsy procedure, prepared the evening before the biopsy (Figure P6.1).
2. Make three drops of Ca^{2+}/Mg^{2+}-free embryo biopsy medium with a single drop of acid Tyrodes (pH 2.3–2.4) as shown in Figure P6.1.
3. Overlay the drops with approximately 6 ml of washed oil.
4. Make one dish per embryo to biopsy. Make a spare dish for priming the pipettes.
5. Keep the dishes overnight in an atmosphere of 6% CO_2 and 37°C for equilibration.

Setting the micromanipulator

1. The biopsies can performed under an Olympus inverted microscope with suitable micromanipulators controlled by air (Figure P6.2).
2. A double tool holder is required in most centers using mechanical or acid tyrodes drilling (Figure P6.2), but a method reporting the use of the same pipette for

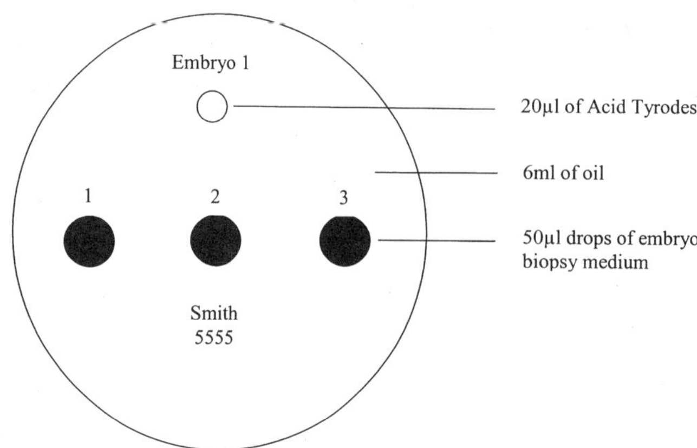

Figure P6.1. Preparation of biopsy dishes for acid tyrodes drilling.

Figure P6.2. Setting the micromanipulator.

zona drilling and blastomere aspiration has been highly successful for one group (84, 111).

3. A number of commercially available micromanipulation pipettes are available.

Cook IVF embryo biopsy pipettes (Hammersmith and UCL design)
Holding pipette: K-HPIP-2335, external diameter 120 μm
Zona drilling pipette: K-AHP-1035, internal diameter 5–10 μm
Blastomere aspiration ppette: K-EBPH-3535, internal diameter 35 μm (for embryos $\geq$6 cells); K-EBPH-4035, internal diameter 40 μm (for embryos $\leq$5 cells.

4. Carefully align all 3 pipettes on the same level using course and fine adjustments.
5. Before the biopsy, prime the holding and aspiration pipettes with Ca^{2+}/Mg^{2+}-free biopsy medium and prime the drilling pipette with acid tyrodes.

Embryo biopsy

Preparation for biopsy

1. Prepare the biopsy dishes the night before (see above).
2. Videotape all embryo biopsies for future reference, if desired.
3. Carefully record details of the biopsy (see sheet at end of protocol) and ensure that the embryo and biopsied blastomere number can be identified.
4. Take the embryo to be biopsied from its culture dish and place it in the first drop of the biopsy medium in the appropriately numbered biopsy dish.
5. Wash the embryo in this drop and transfer it to the second drop of the biopsy medium.
6. Depending on the level of compaction, determine the timing of preincubation. Preincubation should not exceed 10 min in any case. Embryos that do not appear very compact should be biopsied at once. Keep the dish in the incubator during preincubation.
7. It may be easier to start with an average-quality embryo, then move to the best quality embryo and biopsy the poorer embryos last. Embryos that have fewer than 5 cells should be left till the end in case they divide further.

Zona drilling

8. Using the holding and the aspiration pipette, gently secure the embryo and position by rotation so that the blastomere(s) to be biopsied are at the 3 o'clock position.
9. Check the presence of the nucleus under the highest magnification (400×) in the blastomeres to be biopsied.

10. Once secured to the holding pipette, lift the embryo slightly from the bottom of the dish and lower the magnification to 200× for the biopsy.
11. Introduce the zona drilling pipette into the biopsy drop and push it against the zona so that an indentation in the zona is observed at 3 o'clock. It is important to ensure that the blastomere(s) to be biopsied and the hole are in the same plane of focus.
12. The hole can be drilled in between two blastomeres to facilitate removal of two blastomeres through the same hole and to avoid lysis of the blastomere.
13. Once the correct plane has been achieved, the pipette should be held close to the zona and a fine stream of acid tyrodes expelled.
14. As the zona thins, move the pipette closer to the zona and expel acid tyrodes until a clear hole is made or until the zona pops.
15. Once the hole is drilled, immediately press the equilibration button to cancel all negative and positive pressure, which will prevent any more expulsion of acid tyrodes in the biopsy drop.
16. Move the embryo away from the site of drilling to avoid any toxic effects of the acid tyrodes.
17. Remove the zona drilling pipette from the biopsy drop immediately.

Blastomere aspiration

18. Place the aspiration pipette very close to the hole and apply gentle suction.
19. As the blastomere is aspirated, gradually move the pipette away from the embryo. It is sometimes possible to pull the blastomere away from the embryo just by partial aspiration of the blastomere.
20. Care should be taken to ensure that the blastomere does not shoot up into the aspiration pipette, as this can cause the blastomere to lyse.
21. Expel the blastomere gently from the pipette and check for the presence of an intact nucleus.
22. If a further blastomere is required, use the same hole.
23. A second hole may lead to the embryo hatching from both sites, increasing chances of monozygotic twinning.
24. Rotate or move the embryo gently on the same plane with the aspiration pipette to facilitate removal of the second blastomere through an angle. (Getting the second blastomere can be quite time consuming.)
25. If the blastomere is difficult to aspirate due to compaction, move the pipette gently up and down against the zona to try and dislodge the blastomere.
26. As soon as the required blastomeres are aspirated, remove the aspiration pipette from the biopsy drop to avoid suction of the blastomeres back up the pipette.
27. Reprime the drilling pipette between each biopsy to ensure that the acid Tyrodes is concentrated.
28. If a blastomere has lysed during the aspiration process, change the aspiration pipette before starting the next biopsy to reduce the risk of any contamination.

After biopsy

1. As soon as the biopsy is complete, wash the embryo in the third drop of the biopsy medium and then wash it thoroughly in the first well of blastocyst medium before transferring it into the correspondingly numbered microdrop under oil.
2. (These dishes are made the evening before the biopsy and preequilibrated [Figure P6.3].)
3. Promptly place the dish containing the biopsied embryos back inside the incubator.
4. The biopsied cells must remain in the labeled biopsy dish to avoid confusion and handed to the geneticist, who prepares the blastomeres for FISH or PCR.

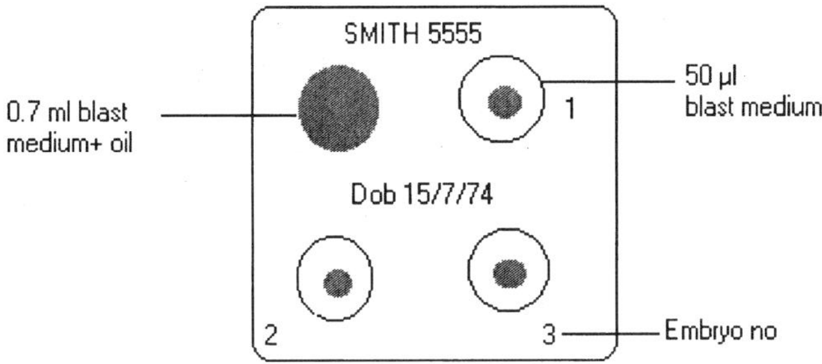

Figure P6.3. Preparation of culture dish for development post-biopsy.

Troubleshooting

Resealing of the zona after drilling. Due to the bilayered structure of the zona, resealing can be observed at times when drilling with acid tyrodes. This is noticed when the blastomere cannot be aspirated into the pipette with ease. The only way to proceed is to apply some more acid tyrodes at the site of resealing.

The second blastomere is not in the same plane as the hole. This is common, as the blastomeres of the embryo can be repositioned when biopsying the first blastomere. An extension to the present hole should be made rather than making another hole to access the second blastomere.

Deciding whether to take out one or two blastomeres: This decision is to be made by the embryologist in conjunction with the geneticist. For some diagnoses, due to the risk of mosaicism it is advisable to have two blastomeres for independent analysis. If there is blastomere lysis during the biopsy, then the embryologist might have to biopsy another blastomere. However, it is important to remember that the aim is to obtain an intact blastomere as well as an embryo with most of its blastomeres present to give it a fair chance of further development.

Laser drilling

1. A noncontact 1.48-μm diode laser can be used for zona drilling instead of acid Tyrodes.
2. Use the lowest power (around 10 m/s exposure) of the laser during drilling because lasers release heat at the site of application, and higher exposure times have shown to be detrimental to further embryonic development. Hence, 2–3 shots may be needed before an adequate-sized hole is made.

It must be remembered that the hole drilled is more in the shape of a groove than spherical and can lead to constriction of the cell during aspiration it if the opening is not big enough.

RECORD OF EMBRYO BIOPSY FOR PGD

Name: _____________________________   Med No: _____________________________

Screening For: _________________   Date: _____________________________

Biopsy Practitioner: _________________

Embryo ref.					
No of cells prior to biopsy					
Grade of embryo before biopsy					
Preincubation time in Ca^{2+}/Mg^{2+}-free medium					
No. of cells biopsied					
No. of cells intact					
No. of cells in embryo postbiopsy					
Grade of embryo postbiopsy					
Time taken to biopsy					
Video reference					
Comments					

No. of cells (day 4)					
Grade of embryo on day 4					
Diagnosis of embryo					
Fate of embryo					

Embryos to be transferred: ________ ________ _______

Signed: ______________ Witness: ______________ Date: ________

References

1. Harper, J. C., and Delhanty, J. D. A. (2000). *Curr. Opin. Obstet. Gynaecol.*, **12**, 67.
2. ESHRE Preimplantation Genetic Diagnosis (PGD) Consortium (1999). *Hum. Reprod.*, **14**, 3138.
3. ESHRE Preimplantation Genetic Diagnosis (PGD) Consortium (2000). *Hum. Reprod.*, **15**, 2673.
4. Verlinsky, Y., Ginsberg, N., Lifchez, A., Valle, J., Moise, J., and Strom, C. M. (1990). *Hum. Reprod.*, **5**, 826.
5. Muggleton-Harris, A. L., Glazier, A. M., and Pickering, S. J. (1993). *Hum. Reprod.*, **8**, 2197.
6. Veiga, A., Sandalinas, M., Benkhalifa, M., Boada, M., Carrera, M., Santalo, J., Barri, P. N., and Menezo, Y. (1997). *Zygote*, **5**, 351.
7. Van de Velde, H., De Vos, A., Sermon, K., Staessen, C., De Rycke, M., Van Assche, E., Lissens, W., Vandervorst, M., Van Ranst H., Liebaers, I., and Van Steirteghem, A. (2000). *Prenat. Diagn.*, **20**, 1030.
8. Wells, D., and Sherlock, J. K. (1998) *Prenat. Diagn.*, **18**, 1389.
9. Griffin, D. K., Handyside, A. H., Harper, J. C., Wilton, L., Atkinson, G., Soussis, I., Wells, D., Kontogianni, E., Tarin, J., and Geber, S. (1994). *J. Assist. Reprod. Genet.*, **11**, 132.
10. Staessen, C., Van Assche, E., Joris, H., Bonduelle, M., Vandervorst, M., Liebaers, I., and Van Steirteghem, A. (1999). *Mol. Hum. Reprod.*, **5**, 382.
11. Conn, C. M., Harper, J. C., Winston, R. M., and Delhanty, J. D. (1998). *Hum. Genet.*, **102**, 117.
12. Van Assche, E., Staessen, C., Vegetti, W., Bonduelle, M., Vandervorst, M., Van Steirteghem, A., and Liebaers, I. (1999). *Mol. Hum. Reprod.*, **5**, 682.
13. Iwarsson, E., Malmgren, H., Inzunza, J., Ährlund-Richter, L., Sjöblom, P., Rosenlund, B., Fridström, M., Hovatta, O., Nordenskjöld, M., and Blennow, E. (2000). *Prenat. Diagn.*, **20**, 1038.
14. Munné, S., Magli, C., Cohen, J., Morton, P., Sadowy, S., Gianaroli, L., Tucker, M., Marquez, C., Sable, D., Ferraretti, A., Massey, J., and Scott, R. (1999). *Hum. Reprod.*, **14**, 2191.
15. Handyside, A. H., Kontogianni, E. H., Hardy, K., and Winston, R. M. (1990). *Nature*, **344**, 768.

16. Piyamongkol, W., Harper, J. C., Sherlock, J. K., Doshi, A., Serhal, P. F., Delhanty, J. D. A., and Wells, D. (2001). *Prenat. Diagn.*, **21**, 753.

17. Harper, J. C., and Wells, D. (1999). *Prenat. Diagn.*, **19**, 1193.

18. Delhanty, J. D. A., Harper, J. C., Ao, A., Handyside, A. H., and Winston, R. M. L. (1997). *Hum. Genet.*, **99**, 755.

19. Conn, C. M., Cozzi, J., Harper, J. C., Winston, R. M., and Delhanty, J. D. (1999). *J. Med. Genet.*, **36**, 45.

20. Iwarsson, E., Ahrlund-Richter, L., Inzunza, J., Rosenlund, B., Fridstrom, M., Hillensjo, T., Sjoblom, I., Nordenskjold, M., and Blennow, E. (1998). *Mol. Hum. Reprod.*, **4**, 719.

21. Reubinoff, B. E., Abeliovich, D., Werner, M., Schenker, J. G., Safran, A., and Lewin, A., (1998). *Hum. Reprod.*, **13**, 1887.

22. Verlinsky, Y., Cieslak, J., Ivakhnenko, V., Evsikov, S., Wolf, G., White, M., et al. (1998). *J. Assist. Reprod. Genet.*, **15**, 285.

23. Pellicer, A., Rubio, C., Vidal, F., Minguez, Y., Giminez, C., Egozcue, J., et al. (1998). *Fertil. Steril.*, **71**, 1033.

24. Harper, J. C., Coonen, E., Handyside, A. H., Winston, R. M. L., Hopman, A. H. N., and Delhanty, J. D. A. (1995). *Prenat. Diagn.*, **15**, 41.

25. Munné, S., Grifo, J., Cohen, J., and Weier, H. U. G. (1994). *Am. J. Hum. Genet.*, **55**, 150.

26. Evsikov, S., and Verlinsky, Y. (1998). *Hum. Reprod.*, **13**, 3151.

27. Ruangvutilert, P., Delhanty, J. D. A., Serhal, P., Rodeck, C., and Harper, J. C. (2000). *Prenat. Diagn.*, **20**, 552.

28. Wells, D., Sherlock, J. K., Handyside, A. H., and Delhanty, J. D. A. (1999). *Nucl. Acids Res.*, **27**, 1214.

29. Wells, D., and Delhanty, J. D. A. (2000). *Mol. Hum. Reprod.*, **6**, 1055.

30. Vouliare, L., Slater, H., Williamson, R., and Wilton, L. (2000). *Hum. Genet.*, **105**, 210.

31. Viville, S., and Pergament, D. (1998) *Prenat. Diagn.*, **18**, 1374.

32. First, N. L. (1991). In: *Preimplantation Genetics*, Y. Verlinsky and A. M. Kuliev, eds., p. 49. Plenum Press, New York.

33. Kaplan, B., Wolf, G., Kovalinskaya, L., and Verlinsky, Y. (1995). *J. Assist. Reprod. Genet.*, **12**, 747.

34. Munné, S., Dailey, T., Sultan, K. M., Grifo, J., and Cohen, J. (1995). *Mol. Hum. Reprod.*, **1**, 1014.

35. Verlinsky, Y., Cieslak, J., Freidine, M., Ivakhnenko, V., Wolf, G., Kovalinskaya, L., White, M., Lifchez, A., Kaplan, B., Moise, J., Valle, J. Ginsberg, N., Strom, C., and Kuliev, A. (1996). *J. Assist. Reprod. Genet.*, **13**, 157.

36. Verlinsky, V., Cieslak, J., Ivakhenko, V., Evsikov, S., Wolf, G., et al. (1999). *J. Assist. Reprod. Genet.*, **16**, 165.

37. Munné, S., Scott, R., Sable, D., and Cohen, J. (1998). *Fertil. Steril.*, **69**, 675.

38. Verlinsky, Y., and Kuliev, A. (1994). *Bailliere Clin. Obstet. Gynaecol.*, **8**, 177.

39. Verlinsky, Y., Rechitsky, S., Cieslak, J., Ivakhnenko, V., Wolf, G., Lifchez, A., Kaplan, B., Moise, J., Walle, J., White, M., Ginsberg, N., Strom, C., and Kuliev, A. (1997). *Biochem. Mol. Med.*, **62**, 182.

40. Strom, C. M., Ginsberg, N., Rechitsky, S., Cieslak, J., et al. (1998). *Am. J. Obstet. Gynaecol.*, **178**, 1298.

41. Malter, H. E., and Cohen, J. (1989). *Fertil. Steril.*, **51**, 139.

42. Montag, M, van der Ven, K., Delacretaz, G., Rink, K., and van der Ven, H. (1998). *Fertil. Steril.*, **69**, 539.

43. Navidi, W., and Arnheim, N. (1992). *Hum. Reprod.*, **7**, 288.

44. Tarin, J. J., Conaghan, J., Winston, R. M. L., and Handyside, A.H. (1992). *Fertil. Steril.*, **58**, 970.

45. Tarin, J. J., and Handyside, A. H. (1993). *Fertil. Steril.*, **59**, 943.

46. Nijs, M., and Van Steirteghem, A. C. (1987). *Hum. Reprod.*, **2**, 421.

47. Krzyminska, U. B., Lutjen, J., and O'Neill, C. (1990). *Hum. Reprod.*, **5**, 203.

48. Hardy, K., Martin, K. L., Leese, H. J., Winston, R. M. L., and Handyside, A. H. (1990). *Hum. Reprod.*, **5**, 708.

49. Yang, X., and Foot, R. H. (1987). *Biol. Reprod.*, **37**, 1007.

50. Monk, M., Muggleton-Harris, A. L., Rawlings, E., and Whittingham, D. G. (1988). *Hum. Reprod.*, **3**, 377.

51. Wilton, L. J., Shaw, J. M., and Trounson, A. O. (1989). *Fertil. Steril.*, **51**, 513.

52. Gordon, J. W., and Talansky, B. E. (1986). *J. Exp. Zool.*, **239**, 347.
53. Depypere, H. T., and Laybaert, L. (1994). *Fertil. Steril.*, **61**, 319.
54. Garrisi, G. J., Sapira, V., Talansky, B. E., Navot, D., Grunfeld, L., and Gordon, J. W. (1990). *Fertil. Steril.*, **54**, 671.
55. Payne, D., McLaughlin, K. J., Depypere, H. T., Kirby, C. A., Warnes, G. M., and Mathews, C. D. (1991). *Hum. Reprod.*, **6**, 423.
56. Santalo, J., Veiga, A., Calafell, J. M., Calderon, G., Vidal, F., Barri, P.N., Gimenez, C., and Egozcue, J. (1995). *Fertil. Steril.*, **64**, 44.
57. Cui, K. H., Pannall, P., Cates, G., and Matthews, C. D. (1993). *Hum. Reprod.*, **8**, 1906.
58. Cui, K. H., Barua, R., and Matthews, C. D. (1994). *Hum. Reprod.*, **9**, 1146.
59. Viville, S., Messaddeq, N., Flori, E., and Gerlinger, P. (1998). *Hum. Reprod.*, **13**, 1022.
60. Palanker, D., Ohad, F., Lewis, A., Simon, A., Schenkar, J., et al. (1991). *Lasers Surg. Med.*, **11**, 580.
61. Germond, M., Nocera, D., Senn, A., Rink, K., Delacretaz, G., and Fakan, S. (1995). *Fertil. Steril.*, **64**, 604.
62. Boada, M., Carrera, M., De La Iglesia, C., Sandalinas, M., Barri, P.N, and Veiga, A. J. (1998). *Assist. Reprod. Genet.*, **15**, 302.
63. Feichtinger, W., Strohmer, H., and Radner, K. M. (1992). *Lancet*, **340**, 115.
64. Feichtinger, W., Strohmer, H., Fuhrberg, P., Radivojevic, K., Antinori, S., et al. (1992). *Lancet*, **339**, 811.
65. Blanchet, G. B., Russel, J. B., Fincher, C. R., and Portmann, M. (1992). *Fertil. Steril.*, **57**, 1337.
66. Neev, J., Gonzales, A., Licciardi, F., Alikani, M., Tadir, Y., Berns, M., and Cohen, J. (1993). *Hum. Reprod.*, **8**, 939.
67. El Danasouri, I., Westphal, L. M., Neev, Y., Gebhardt, J., Louie, D., and Berns, M. W. (1993). *Hum. Reprod.*, **8**, 464.
68. Tadir, Y., Wright, W. H., Vafa, O., Liaw, L. H., Asch, R., and Burns, M. W. (1991). *Hum. Reprod.*, **6**, 1011.
69. Rasmussen, R. E., Hammer-Wilson, M., and Berns, M. W. (1989). *Photochem. Photobiol.*, **49**, 413.
70. Antinori, S., Versaci, C., Fuhrberg, P., Panci, C., Caffa, B., and Gholami, G. H. (1994). *Hum. Reprod.*, **9**, 1891.
71. Rink, K., Delacretaz, G., Salathe, R. P., Senn, A., Nocera, D., Germond, M., De Grandi, P., and Fakan, S. (1996). *Lasers Surg. Med.*, **18**, 52.
72. Dokras, A., Sargent, I. L., Ross, C., Gardner, R. L., and Barlow, D. H. (1990). *Hum. Reprod.*, **5(7)**, 821.
73. Carson, S. (1991). In: *Preimplantation Genetics*, Y. Verlinsky and A. Kuliev, eds., p. 85. Plenum Press, New York.
74. Joris, H., Van den Abbeel, E., De Vos, A., and Van Steirteghem, A. (1999). *Hum. Reprod.*, **14**, 2833.
75. Cohen, J., Elsner, C., Kort, H., Malter, H., Massey, J., Mayer, M. P., and Weimar, K. (1990). *Hum. Reprod.*, **5**, 7.
76. Cohen, J., and Feldberg, D. (1991). *Mol Reprod Dev.*, **30**, 70.
77. Thompson, L. A., Srikantharajah, A., Hamilton, M. R., and Templeton, A. (1995). *Hum. Reprod.*, **10**, 659.
78. Roudebush, W. E., Kim, J. G., Minhas, B. S., and Dodson, M. G. (1990). *Am. J. Obstet. Gynecol.*, **162**, 1084.
79. Pierce, K. E., Michalopoulos, J., Kiessling, A. A., Seibel, M. M., and Zilberstein, M. (1997). *Hum. Reprod.*, **12**, 351.
80. Pierce, K. E., Fitzgerald, L. M., Seibel, M. M., and Zilberstein, M. (1998). *Mol. Hum. Reprod.*, **4**, 167.
81. Gordon, J. W., and Gang, I. (1990). *Biol. Reprod.*, **42**, 869.
82. Levinson, G., Field, R. A, Harton, G. L, Palmer, F. T, Maddalena, A., Fugger, E. F., and Schulman, J. D. (1992). *Hum. Reprod.*, **7**, 1304.
83. Muggleton-Harris, A. L., Glazier, A. M., Pickering, S. J., and Wall, M. (1995). *Hum. Reprod.*, **10**, 183.
84. Inzunza, J., Iwarsson, M., Fridstrom, B., Rosenlund, P. et al. (1998). *Prenat. Diagn.*, **18**, 1381.

85. Chen, S.-U., Chao, K.-H., Wu, M.-Y., Chen, C.-D., Ho, H.-N., and Yang, Y.-S. (1998). *Fertil. Steril.*, **69**, 569.
86. Bavister, B. D. (1995). *Hum. Reprod. Update*, **1**, 91.
87. Hardy, K., Warner, N., Winston, R. M. L., and Becker, D. L. (1996). *Mol. Hum. Reprod.*, **2**, 621.
88. Dumoulin, J. C., Bras, M., Coonen, E., Dreesen, J., Geraedts, J. P., and Ever, J. L. (1998). *Hum. Reprod.*, **13**, 2880.
89. Santalo, J., Grossman, M., and Egozcue, J. (1996). *Hum. Reprod. Update*, **2**, 257.
90. Van Blerk, M., Nijs, M., and Van Steirteghem, A. C. (1991). *Hum. Reprod.*, **6**, 1298.
91. Liu, J., Lissens, W., Devroey, P., Liebaers, I., and Van Steirteghem, A. (1993). *Fertil. Steril.*, **59**, 815.
92. Liu, J., Tsai, Y-L., Zheng, X-Z., Yazigi, R. A., Baramki, T. A., Compton, G., and Katz, E. (1998). *Mol. Hum. Reprod.*, **4**, 972.
93. Liu, J., Van den Abbeel, E., and Van Steirteghem, A. (1993). *Hum. Reprod.*, **8**, 1481.
94. Kuo, H.-C. , Mackie Ogilvie, C., and Handyside, A. H. (1998). *J. Assist. Reprod. Genet.*, **15**, 276.
95. Magli, M. C., Gianaroli, L., Fortini, D., Ferraretti, A. P., and Munné, S. (1999). *Hum. Reprod.*, **14**, **3**, 770.
96. Lee, M., and Munne, S. (2000). *Fertil. Steril.*, **73**, 645.
97. Freedman, M., Farber, M., Farmer, L., Leibo, S. P., Heyner, S., and Rall, W. F. (1988). *Obstet. Gynaecol.*, **72**, 502.
98. Monk, M., and Handyside, A. H. (1988). *J. Reprod. Fertil.*, **82**, 365.
99. Summers, P. M., Campbell, J. M., and Miller, M. W. (1988). *Hum. Reprod.*, **3**, 389.
100. Han, Y. M., Yoo, O. J., and Lee, K. K. (1993). *J. Assist. Reprod. Genet.*, **10**, 151.
101. Dokras, A., Sargent, I. L., Gardner, R. L., and Barlow, D. H. (1991). *Hum. Reprod.*, **6**, 1453.
102. Holding, C., Bently, D., Roberts, R., Bobrow, M., and Mathew, C. (1993). *J. Med. Genet.*, **30**, 903.
103. Jones, G. M., Trounson, A. O., Lolatgis, N., and Wood, C. (1998). *Fertil. Steril.*, **70**, 1022.
104. Vandervorst, M., Liebaers, I., Sermon, K, Staessen, C., De Vos, A., et al. (1998). *Hum. Reprod.*, **13**, 3169.
105. James, R. M., and West, D. (1994). *Hum. Genet.*, **93**, 603.
106. Magli, M. C., Jones, G. M., Gras, L., Gianaroli, L., Korman, I., and Trounson, A. O. (2000). *Hum. Reprod.*, **15**, 1781.
107. Janny, L., and Menezo, Y. (1996). *Mol. Reprod. Dev.*, **45**, 31.
108. Gardner, D. K., and Lane, M. (1997). *Hum. Reprod. Update*, **3**, 362.
109. Veiga, A., Gil, Y., Boada, M., Carrera, M., Vidal, F., Boiso, I., Menezo, Y., and Barri, P. N. (1999). *Prenat. Diagn.*, **19**, 1242.
110. Geber, S., Winston, R. M. L., and Handyside, A.H. (1995). *Hum. Reprod.*, **10**, 1492.
111. Fridstrom, M., Ahrlund-Richter, L., Iwarsson, E., Malmgren, H., Inzunza, J., Rosenlund, B., Sjoblom, P., Nordenskjold, M., Blennow, E., and Hovatta, O. (2001). *Prenat. Diagn.*, **21**, 781.

7

MICHELLE LANE

JAY M. BALTZ

DAVID K. GARDNER

Analysis of Intracellular Ions in Embryos: pH and Calcium

1. ROLE OF pH AND CALCIUM IN PREIMPLANTATION EMBRYO DEVELOPMENT

To understand cell physiology and the regulation of development, it is essential to understand the role and control of intracellular ions. Ions such as calcium and protons (pH) are universal regulators of cell function. Temporal and spatial changes in the levels of these ions regulate and signal many cellular functions such as division, differentiation, metabolism, and secretion in response to stimuli such as growth factors. In reproduction, changes in intracellular calcium and pH are key factors in fertilization and in initiation of development of sea urchin and *Xenopus* eggs. Much of the work on these species was performed in the 1960s and 1970s using electron probes that could be inserted into the oocytes or cells of the embryo to take measurements. This was possible due to the large volume of the oocytes and embryos and further facilitated by the ability of the oocytes and embryos from these species to develop external to the body. Similar studies on mammalian oocytes and embryos had not been possible due to the small size of the cells and also due to an inability to maintain developmental competence external to the body for long periods of time.

Since the 1990s, there have been major advances in the development of fluorescent microscopy technology coupled with specific ratiometric dyes that make it possible to accurately measure levels of ions in single living cells. Coupled with advances in embryo culture technology (see Chapter 3), it is now possible to maintain mammalian embryos in culture for a sustained period of time with minimal loss in viability. Therefore, much can be learned about the control of intracellular ions and how they change with and regulate development of the mammalian embryo. This chapter outlines some of the advances in the field and presents techniques to measure intracellular ions and examples of measurement of transport systems in mammalian embryos.

2. PRINCIPLES OF RATIOMETRIC ANALYSIS

The ability to study cells as they grow and develop is a powerful method to establish how cell functions are regulated. With the development of quantitative fluorescence microscopy and ratiometric ion-sensitive fluorochromes, it is possible to study cellular and subcellular events in individual living cells. These fluorochromes shift excitation or emission wavelengths and quantum efficiency upon ligand binding (Figure 7.1) (1–9). Therefore alternation between wavelengths can be used to distinguish between dye that is not bound to ligand (controls for the level of dye that enters the cell) and dye bound to the ligand. By taking a ratio of these measurements, it is possible to accurately quantitate the level of the ion within the cellular compartment being studied without the level of dye entering the cell interfering with the analysis.

Further advantages of ratiometric measurement over the use of single-wavelength dyes is that the effects of cell thickness, dye content, or instrumental efficiency that interfere with the interpretation of measurements at single wavelengths are largely eliminated (1). This enables the signal obtained to be accurately calibrated and also enables comparisons to be made between samples and within different areas of the same sample and between different days of experimentation. Combining this powerful ratiometric technique with ultrasensitive low-light–level video cameras or photometers and digital image processing enables the study of both spatial and temporal distribution of ions within single cells (10).

3. TYPES OF EQUIPMENT

3.1 Fluorescence Microscope

All the ion-sensitive fluorophores discussed here rely on the excitation of fluorescence and quantitative measurement of fluorescence emission. For cells that are available in large quantities, quantitative fluorescence measurements can be made with large populations of cells in suspension or on coverslips, using macroscopic detection systems (e.g., a spectrofluorometer). However, mammalian oocytes and embryos are obtainable only in small numbers, and therefore quantitative fluorescence microscopy must be used because it allows the fluorescent signal to be measured in single or small groups of embryos. This requires a fluorescence microscope fitted with a detection system that permits the quantitative measurement of fluorescence emission. In practice, any high-quality fluorescence microscope, such as those manufactured by Zeiss, Nikon, and others, will suffice. The most convenient configuration for oocytes and embryos is an inverted microscope, which allows the imaging of the bottom of a culture dish or chamber filled with culture medium. Particular attention must be paid to the stability of the fluorescence source, as fluctuations in excitation intensity during measurements will introduce substantial errors. Because epifluorescence excitation is standard, the objectives chosen must be optimal for the transmission of both excitation and emission wavelengths. Thus, if UV-excited fluorophores are to be used, the objectives must have high transmission in the near UV.

The system used to measure fluorescence must be sensitive enough to detect the low-level fluorescence emitted by intracellular fluorophores and should also have an output that is linearly proportional to the light signal. These properties, sensitivity and linearity, are crucial determinants of the usefulness of any given quantitative fluorescence system (4). Currently, a number of systems are commercially available that are specifically designed for quantitative, ratiometric fluorescence measurements. These systems are based on either of two different detectors: photomultipliers or ultrasensitive video cameras. Photomultipliers are exceedingly sensitive and give a linear output over a large range of input intensities, making them ideal for low-light–level applications. Micros-

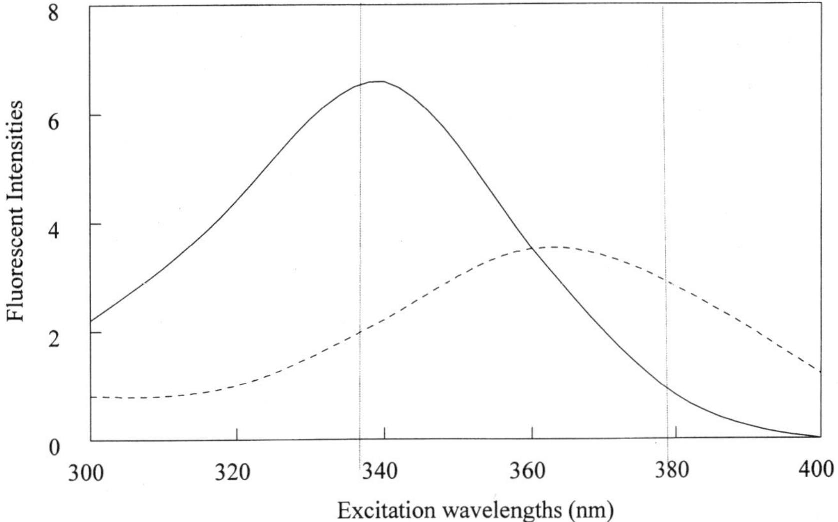

Figure 7.1. Changes in fluorescence of fura-2 when bound to calcium. All data are collected at a wavelength of 510 nm. For analysis the levels of fluorescence are ratioed at the excitation wave lengths of 340 nm, which has high levels of fluorescence when bound to the ligand and is therefore indicative of free calcium and 380 nm, which demonstrates low levels of fluorescence when bound to the ligand and is therefore representative of the levels of fluorochrome within the cells. Solid line indicates levels of fluorescence in the presence of saturating calcium. Dotted line indicates levels of fluorescence in the absence of calcium. Adapted from Grynkiewicz et al. (1).

copy systems using photomultipliers usually employ a pinhole aperture, which is used to select the portion of the microscope field for measurement; thus measurements can be obtained from only one oocyte or embryo, which is a significant limitation. Cameras, in contrast, permit the entire field of view to be captured at once, so that a number of oocytes or embryos can be assessed simultaneously. The newer cooled charge-coupled device (CCD) cameras rival photomultipliers in sensitivity and linearity and are the usual choice for quantitative fluorescence applications. This method also has the advantage that spatial information can be preserved from within a cell or multicellular structure so that both temporal and spatial data can be obtained simultaneously.

A basic requirement for any quantitative fluorescence microscope is the ability to select the appropriate excitation wavelengths. Because most useful ion-selective probes are ratiometric, the system must be able to rapidly switch between two different excitation or emission wavelengths. For excitation wavelengths this is most commonly done using a filter wheel with two or more openings for holding optical filters that pass a narrow range of wavelengths of light. By selecting two appropriate optical filters, the desired excitation wavelengths for ratiometric fluorescence excitation can be supplied. Such filter wheels are generally coupled with a shutter that blocks the excitation light except during actual measurements, which is crucial for limiting photobleaching and phototoxicity.

The emission wavelength must also be selected. For applications, such as with fluorophores where an excitation ratio is obtained, the emission wavelength does not need to be changed during an experiment. Thus, a single optical filter is placed between the specimen and the camera (or other detector) to determine the emission wavelength. An added constraint on the filter is that it must not introduce any optical distortion because, unlike the filters used to determine the excitation wavelengths, the image itself must pass through the filter. Some fluorophores, such as the calcium-sensitive Indo-1 or the pH-sensitive SNARF-1, are ratiometric using the emission wavelengths rather than excitation wavelengths and thus require rapid switching between emission wavelengths.

In theory, a filter wheel similar to those used in selecting excitation wavelengths could be introduced into the optical path just before the camera. However, in practice, this introduces far too much mechanical vibration. Thus, to obtain emission ratios, non-mechanical means of selecting wavelengths are required. One solution to this problem is the use of dual cameras or dual photomultipliers, each having a different emission wavelength selected by a fixed optical filter, with the image split before passing through the filters. Another solution is the use of tunable optical filters, such as liquid crystal monochrometers, which allow rapid switching between wavelengths by controlling the optical properties of the liquid crystal. Either of these options is practical; however, because of this constraint, the cost of equipment for emission ratiometric analysis is higher than for excitation ratiometric analysis.

A final requirement for the system is the ability to store and quantitatively analyze images. This almost universally involves the use of a computer-based imaging system with appropriate software, permitting the storage of images as well as their analysis. Minimum requirements, besides sufficient image-storage capacity, is the ability to obtain and subtract background from images and the ability to quickly compute ratios (i.e., to generate ratiometric images by dividing the image obtained at one wavelength by that obtained at the other on a pixel-by-pixel basis, or to divide the mean intensity in a defined area of the image at one wavelength by that of the same area in the paired image at the other wavelength). The ratios obtained, either displayed graphically or numerically, are the crucial measurement when measuring intracellular ion concentrations with ratiometric fluorophores. A number of systems are commercially available that integrate the microscope and detection system with a computer and software that both capture and analyze the images, as well as control wavelength selection, opening of the shutter, the camera or photomultiplier, and other functions.

3.2 Incubation Chamber

The oocytes or embryos must be maintained in an environment that sustains their viability during fluorescence measurements. A temperature-controlled chamber is required for any but the most short-term measurements. A number of chambers are available that control the temperature in a central well to within a narrow range, usually by means of a heating element in the walls coupled with a temperature sensor within the fluid in the chamber providing feedback to the controller. This is desirable to avoid undesired heating and cooling cycles associated with normal thermostats. The base of the chamber is optically clear and thin, allowing microscopic observation of cells on the bottom using an inverted microscope. Often, this base is a cover slip clamped into the chamber before it is filled. The chamber should be covered to minimize evaporation. Holding the temperature in the chamber relatively constant is a crucial factor, as both the fluorescent properties of ion-sensitive fluorophores, and, for example, the pH of cytoplasm and media, are highly temperature dependent. This can be accomplished by most commercial temperature-controlled chambers, which maintain temperature within a range of less than ±1°C.

One example of a chamber that fulfils these requirements is the PDMi-2 open perfusion microchamber/TC-202 bipolar temperature controller (Medical Systems, Greenvale, NY; Figure 7.2). It is recommended that a precision thermometer or temperature probe independent of the controller be used to periodically confirm that the temperature settings are accurate and to ensure that temperature does not vary excessively over time or across the chamber. A problem that can occur with temperature-controlled chambers is that they respond to solution changes, which tend to introduce somewhat cooler medium, by overcorrecting the temperature and thus overshooting. This can be quite detrimental to the embryos. Therefore, it is imperative to follow the temperature in the chamber while the solution is changed and detect any transient overshoot. Overshoots

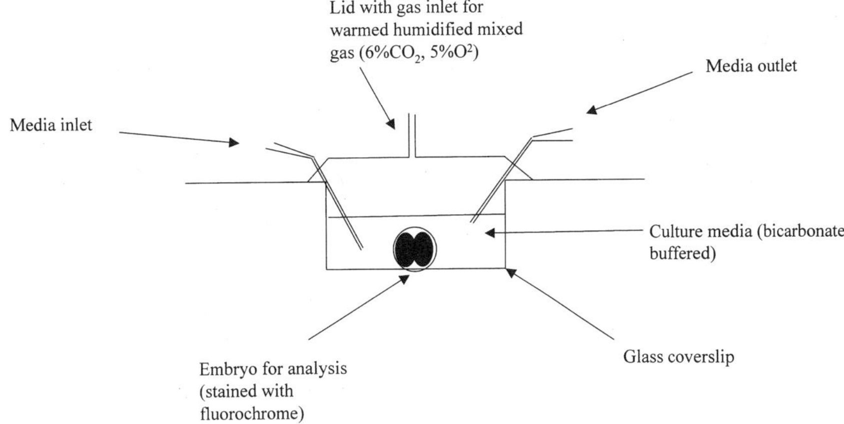

Figure 7.2. Diagram of the incubation chamber for the measurement of intracellular ions. This chamber regulates not only temperature but can also be connected to an external gas source to maintain a specific gas phase. Additionally, the levels of solutions can be easily changed by the use of a syringe pump connected to the inlet and outlet ports.

can be largely eliminated by prewarming the medium so that it enters the chamber at just slightly below the desired temperature.

In addition to temperature control, it is useful to control the gas phase as well, as this allows the use of HCO_3^-/CO_2-buffered media during fluorescence measurements. We have successfully used a simple scheme in which an inverted plastic culture dish serves as a cover for the chamber, into which a constant flow of gas (e.g., 5–6% CO_2) is introduced at a low rate (about 30 ml/min). This maintains the atmosphere above the medium in the chamber, while not cooling it significantly. Care must be taken that the tubing through which the gas passes is thin (to allow rapid transit) and not too long because loss of CO_2 across the walls of the tubing can be significant. For long-term observations, evaporation of medium from the chamber can be a significant problem. In this case, the gas must first be passed through warm water before entering the chamber. Bubbling through distilled water at 60°C raises the humidity to near 100%. A short passage through thin tubing after humidifying ensures that the gas cools sufficiently before entering the chamber. The tubing must be configured to prevent the water that condenses in the tubing from running into the chamber.

For many applications, the solutions in the chamber need to be changed during the experiment. This can be accomplished using a home-made apparatus using either gravity feed or a syringe driver to provide prewarmed solution at a relatively low rate of flow. As the solution is introduced, a small piece of tubing (e.g., stainless steel the size of an 18-G needle) on the opposite side of the chamber can be used to remove liquid as it rises near the top of the chamber, by means of a suction pump or aspirator. Because of vibration, this is used only to change solutions. It is also possible to constantly superfuse solutions over oocytes or embryos during measurements using configurations commonly employed to change solutions during electrophysiological measurements. However, great care must be exercised to minimize mechanical movement of the cells. To prevent loss of oocytes or embryos during solution changes, they must be made to adhere to the chamber bottom. Polylysine coating of the coverslip has been used to render them sticky to the zona pellicuda, causing adhesion. However, in some cases the zona of embryos will adhere strongly to a clean glass surface in the absence of protein or other macromolecular components of the medium, and this is sufficient to allow solution changes (more details regarding methods of immobilizing oocytes and embryos can be found in reference 11). The oocytes or embryos must not move even slightly during the acquisition of images used to calculate a ratio because any misregistering of such paired images will introduce a large error in calculated ratios.

3.3 Calibrating Equipment

The ability to calculate ion concentrations from ratiometric fluorescence measurements is predicated on the assumption that the system responds linearly to input intensity (i.e., the emitted fluorescence intensity). In other words, the same ratio must be obtained for a given intracellular ion concentration regardless of other conditions such as the concentration of the dye or excitation intensity. Detailed procedures for exhaustively validating an imaging system are described elsewhere (12, 13). These require some effort, but it is well worth confirming that a newly set up imaging system is indeed functioning correctly. It is recommended that any imaging system to be used for ratiometric fluorescence measurements be thoroughly validated before experiments are begun with oocytes or embryos.

The most basic set of tests to ensure that a system is functioning adequately involve demonstrating that the ratios obtained are independent of changes in parameters such as excitation and emission intensity, fluorophore concentration, and so on. A simple test uses the fluorophore of interest in solution rather than fluorophore loaded into cells. Therefore, the unmodified form of the fluorophore must be used, rather than the cell-permeant form (usually with hydrophobic groups, such as acetoxymethyl groups, bound to the dye via ester bonds), which is cleaved intracellularly to yield the ion-sensitive, fluorescent form. Solutions of the unmodified form of the fluorophore are prepared in buffer at several different concentrations of the relevant ion (e.g., Ca^{++} or pH). A simple microcuvette can be fabricated on a microscope slide by cutting a small circular or square hole in a piece of black electrical tape (with a razor blade or scalpel) and firmly attaching it to the slide. The solution is placed in the well formed by the tape, filling the well so that the drop protrudes slightly over the top of the tape. The well is then carefully sealed with a coverslip, which will squeeze some liquid out and provide a relatively bubble-free sample. This yields a deep, uniform sample of the dye solution to use on the microscope.

First, the uniformity of fluorescence obtained across the microscope field should be assessed. The dye sample is placed on the microscope stage, the focus is adjusted to be within the fluorescent solution, away from the glass interfaces, and an image at one excitation wavelength is obtained. Ideally, the image should be uniformly bright across the entire image, with no major bright or dark spots. If this is not the case, then the alignment and focus of the fluorescence excitation source should be adjusted to give the most uniform fluorescence possible while still maximizing overall brightness. When the fluorescence image is as uniform as possible, a ratiometric image should be obtained. The value of the ratio should not vary significantly across the image. Again, if there is excessive variation, the excitation illumination can be adjusted to minimize it. If necessary, portions of the field of view where the ratio deviates too greatly can be marked and avoided during experiments. This procedure should be repeated periodically and should be carried out each time the excitation bulb is replaced or readjusted.

Once the illumination is adequately uniform, the repeatability of ratios obtained with the system should be checked. Ratiometric measurements should be obtained for several different concentrations of the relevant ion (e.g., Ca^{2+} or pH); these should yield significantly different values for the ratio. Measurements should be repeated several times for each sample to ensure that the values are the same over several replicates. Because the ratio should ideally depend only on the ion concentration, it should be determined at several concentrations of the dye. Care should be taken not to use an excessively high dye concentration, or a phenomenon known as "self-screening" will occur in which the fluorophore begins to absorb its own fluorescence, distorting the signal. Essentially the same ratio should be obtained at a given ion concentration despite variations in the concentration of the fluorophore. Note, however, that when the fluorescence signal gets too low, the signal-to-noise ratio will become too small, and errors in the ratio will appear. Conversely, when fluorescence is very high, the image will begin to saturate, causing the ratio to deviate. However, away from the extremes, the ratio should be constant for a given ion concentration.

An additional test that can be performed to determine whether the ratios remain constant is to vary the intensity of the excitation light. This can be done by adjusting an aperture in the excitation light path, where available, or via the use of neutral density filters (available from the same sources as the optical filters used to select excitation and emission wavelengths) placed in the excitation path. Again, the ratio should remain essentially constant despite excitation intensity changes, except where the signal-to-noise becomes too small or the image begins to saturate.

Another assurance that the system is working properly will be obtained routinely during the course of experiments. Conversions of the ratios or fluorescence intensities obtained to intracellular ion concentrations are accomplished primarily by methods in which the intracellular ion concentration is clamped to a known set of concentrations (see below). This serves as an internal control for system function. First, the ratio obtained should be essentially identical in each oocyte or embryo in the field of view when they are clamped to the same ion concentration. Second, the calibration curve obtained should closely follow the expected dependence (often linear with ion concentration over the range of interest). Finally, the calibration curves obtained should not deviate significantly from experiment to experiment, except when a change has been made in the system, such as the bulb being replaced in the excitation light source.

It is vital to ensure that an imaging system is working correctly before performing actual measurements or experiments on embryos. Controls should also be built into any experiments wherever possible that are designed to exclude artifactual changes in fluorescence intensity that can be mistaken for changes in intracellular ion concentrations.

4. CHOICE OF A FLUOROCHROME FOR MEASUREMENT OF INTRACELLULAR ION LEVELS

There are many dyes for analyzing ion levels in cells. It is essential that the correct dye be chosen for each experiment. You may be limited in your choices of dye depending on whether your equipment can support dual excitation and/or dual emission (see above). Table 7.1 lists commonly used dyes for the measurement of calcium, pH, magnesium, sodium, and potassium.

There are several factors to take into account when deciding on a dye: binding affinity, cytotoxicity, photobleaching, specificity (interference from ionic strength), quickness of response, and autofluoroscence. All of these will have a profound affect on the outcome of the experiment. Most of these parameters such as binding affinity (K_d) and specificity can be easily located in the catalogues of the major manufacturers of these probes and also from the original papers that describe their development and function (1, 9). Additionally, photobleaching can be reduced by limiting the duration of exposure with a shutter mechanism or optical switch to prevent illumination during intervals between recordings (13; see section 3.1).

A major concern when assessing mammalian embryos is the potential cytotoxicity of the probes. Mammalian oocytes and embryos are notoriously sensitive to their environment. Some probes have been shown to be more toxic to embryos than others; the pH probe BCECF is significantly more toxic than is SNARF-1 to mouse two-cell embryos (14). The toxicity of the probe on the oocytes or embryos that are to be measured should be assessed before intracellular ion levels are assessed. We suggest loading embryos with a range of concentrations of the probe and length of loading times that enable a reliable signal to be detected. The oocytes or embryos loaded with the probe

Table 7.1. Commonly used Fluorochromes for Assessment of Ion Levels in Living Cells

Ion for Measurement	Flurochrome	Ratiometric Analysis	References for Use in Mammalian Embryos
Calcium	Fura-2	Dual excitation	(26–29)
Calcium	Indo-1	Dual emission	(30, 31)
Calcium	Quin-2	Dual excitation	—
pH	BCECF	Dual excitation	(17–20, 25, 32, 33)
pH	SNARF-1	Dual emission	(21, 23, 34–36)
Magnesium	Mag-Fura-2	Dual excitation	(28)
Magnesium	Mag-Indo-1	Dual emission	(31)
Sodium	SBFI	Dual excitation	—
Potassium	PBFI	Dual excitation	—

should then be exposed to the wavelengths of excitation for the length of time of the proposed experiment. The embryos are then returned to the incubator and development is assessed compared to non treated controls. Choose loading conditions (concentration and time) that result in either no effect or minimal effect on embryo development to ensure that the measurements taken are not artifactual. This is especially important for experiments that require longer term measurements.

Autofluoroscence is an additional concern for embryos because they have comparatively high levels of autofluoroscence compared to some cell types. Therefore, whenever possible, avoid dyes that have excitation wavelengths or collection wavelengths close to 340 nm such as Fura-2 (excitation wavelengths are 340 and 380 nm). However, if using such dyes is unavoidable, appropriate controls need to be included to control for the levels of autofluorescence. Autofluorescence is usually determined by measuring the fluorescence of unloaded cells, and this fluorescence can be subtracted from the readings obtained. This can be achieved easily with most software packages.

Compartmentalization of a dye into organelles can also be a concern. This can be detected by a change in the fluorescence of the non–ligand-bound dye or by hot spots in the free dye excitation due to preferential accumulation of the dye. Some dyes such as Fura-2 are prone to compartmentalization into organelles, and this should be considered when choosing a dye (15, 16).

5. LOADING OOCYTES/EMBRYOS WITH FLUOROCHROME

The probes used for the intracellular ion measurement contain carboxylate groups that make them impermeable to the cell membrane. However, it is possible to purchase cell-permeant forms of the dyes that contain hydrophobic groups such as acetoxymethyl esters (AM) bound to the dye via ester bonds that mask the carboxylate groups. The addition of these AM derivatives makes the dye permeable to the cell membrane and enables the dye to diffuse into the cell. Once inside, the cell's endogenous esterases hydrolyze the dye and revert the dye back to a hydrophilic compound that is impermeable to the cell membrane, thereby trapping the dye within the cell. The dye then binds to its ligand within the cell. These cell-permeant dyes have been used successfully for oocytes and embryos from several species (17–23; Protocol 7.1). Determination of the timing of loading and the concentration of the dye to use can be obtained from the literature for most dyes. However, it is advisable to use a time and concentration that enables consistent levels of loading with the cells to be measured. This is usually achieved by loading dyes for increasing levels of time and determining when the level of dye entering the cell has reached a plateau. This will ensure that consistent levels of dye are routinely obtained. Once this has been achieved, the cytotoxicity of the dye level should be assessed as described above.

If a particular cell is resistant to the loading of the AM esters of the dye, loading can be facilitated by the use of the nonionic surfactant Pluronic F-127 at a low concentration such as 0.03%. This has been used for the loading of pig oocytes, but its use should be avoided if possible (24).

6. CALIBRATION OF FLUOROCHROMES

The fluorescence of the dye in both absorption and emission spectra vary in intracellular environments compared to that measured in buffers. This is due to a variety of reasons such as intracellular light scattering, changes in polarity, and viscosity, and therefore calibrations should occur in vivo whenever possible.

In situ calibration of ion levels within cells is achieved by incubating the cells in the presence of a specific ionophore which permealizes the cell to the ion being measured. This enables the internal environment of the cell to equilibrate with the external environment, thereby matching the intracellular levels of the ion to the known extracellular level.

There are two general methods used for the calibration of intracellular ion levels. For most ions the levels of intracellular ions are calibrated using a standard curve; however, for ions present in cells at very low levels such as calcium, an alternative calibration procedure is used (1).

7. CALIBRATION OF RATIOMETRIC SIGNALS

7.1 Calcium

Intracellular calcium levels are quantitated using a two-point in situ calibration (Table 7.2). For calibration of intracellular calcium levels, nonfluorescent ionophores such as ionomycin or 4-bromo-A-23187 are used. Ionophore A-23187 gives a fluorescent signal at some wavelengths and should not be used. Cells are permeablized using the ionophore and incubated in a buffer that contains a saturating concentration of calcium to determine the maximal fluorescent signal. This is then followed by a buffer that contains no calcium (it contains the calcium chelator EGTA) that enables the minimal levels of fluorescence to be determined (see Protocol 7.2). The levels of calcium are then calibrated using the following equation.

Table 7.2. Composition of Solutions for Calibration
of Calcium Levels

Component	mM
Solution A (Ca^{2+}-free solution)	
KCl	100
MOPS	10
K_2H_2-EGTA	10
Solution B (Ca^{2+}-saturating solution)	
KCl	100
K-MOPS	10
K_2H_2-EGTA	10
$CaCl_2$	1

Titrate the solution to pH 7.2 using a 2 M KOH solution and filter
through a 0.2-μm filter. Store at 4°C. Solutions A and B can be purchased as kits from Molecular Probes (Eugene, OR).

Calcium levels are related to the measured fluorescence by

$$[\text{Ca}^{2+}] = \beta \times K_d \times (R - R_{min})/(R_{max} - R),$$

where R is the ratio of the fluorescence intensities, R_{min} is the ratio in the absence of calcium, R_{max} is the ratio of Ca^{2+} saturated dye, β is the ratio of the fluorescence intensities at the wavelength chosen as the denominator of R (e.g., 480 nm emission for Indo-1) in zero and saturating $[\text{Ca}^{2+}]$, and K_d is the dissociation constant of the indicator for Ca^{2+} (1).

The K_d of the dye used can be affected by the experimental conditions. Therefore, to determine an accurate value of Ca^{2+} levels in the cell, the K_d should be estimated under the experimental conditions used. K_d is influenced by temperature, pH, viscosity, and ionic strength, and therefore the K_d estimated in one cell type may differ slightly in a different cell type. Details on how to calculate a K_d for a particular cell type can be found in detail in Grynkiewisc et al. (1). Cells are incubated in a series of solutions where the concentrations of calcium are known. The fluorescent values are the different excitation (e.g., Fura-2) or emission (e.g., Indo-1) wavelengths that are determined. These values can then be used to determine the K_d of the cells:

$$[\text{free Ca}^{2+}] = K_d \times (\text{Ca-EGTA complex})/ (\text{free EGTA}).$$

A convenient alternative to calculating the K_d is provided on the Molecular Probes website (www.probes.com) by using the dissociation constant calculator. The concentrations of calcium used in the external samples are entered along with the fluorescent values obtained at both wavelengths. The program then calculates the ratio and plots the $\log([\text{Ca}^{2+}]\text{free})$ versus the $\log(\text{bound/free Ca}^{2+})$ and then calculates the K_d. Calibration kits can also be purchased from Molecular Probes for determination of K_d.

7.2 pH

Intracellular pH levels are quantitated using a situ standard curve calibration (Protocol 7.3). Intracellular pH is matched to known extracellular values using the ionophores nigercin and valiomycin in buffers with suitable transmembrane K^+ and Na^+ gradients (Table 7.3). Cells are then incubated in ionophores in solutions of increasing pH. It is preferable to generate a standard calibration curve containing at least four points. This curve can then be plotted to ensure that a linear relationship has been obtained between the pH levels and the ratiometric values using regression analysis. Typically, standard curves should have an r value >.99. This curve can then be used to determine intracellular pH levels in the samples.

Table 7.3. Composition of Calibration Solutions for pH

Component	mM	g/l
KCl	100	7.455
NaCl	17.1	1.46
Sucrose	75.0	25.68
Hepes	21.0	5.004
Penicillin	100 IU	0.06

The 1-l solution is divided into 200-ml aliquots and each solution adjusted to the required pH with NaOH. Nigericin is stored as a 10 mg/ml stock in ethanol at −20°C. For use, 20 µl of this stock is added to 20 ml of calibration solution. The final concentration of nigericin in the calibration solution is 10 µg/ml. Valinomycin is stored as a 10 mg/ml stock in DMSO and stored at 4°C. For use, 10 µl of this stock is added to 20 ml of calibration solution. The final concentration of valinomycin in the calibration solution is 5 µg/ml.

8. MEASUREMENT OF TRANSPORT SYSTEMS FOR PH IN MAMMALIAN EMBRYOS

Using the procedures outlined above, it is possible to not only measure absolute concentrations of intracellular ions but also to assess the activity and kinetics of specific transport systems. For example, the activity of the Na^+/H^+ antiporter can be detected by determining the Na^+-dependent recovery from an acid load (Figure 7.3). This is simply achieved using a chamber where the solutions can be changed quickly and measurements taken in response to the solution changes. This has been used to determine the presence and kinetics of transport systems in mammalian embryos (19, 25).

9. CONCLUSIONS

The ability to quantiate intracellular levels of ions throughout preimplantation embryo development is an exciting and growing field of research. Little is currently known about the regulation of homeostasis of the mammalian embryo. With the use of ratiometric fluorescent analysis as detailed in this chapter, developmentally important questions about the regulation of physiology and biochemistry of the mammalian embryo can now be answered.

Protocol 7.1. Loading oocytes/embryos with fluorochrome

Purchase the fluorochromes in small aliquots to increase the shelf-life. Most fluorochromes can be purchased in lots of $20 \times 50\ \mu g$.

1. Dilute fluorochromes in DMSO (purchased in ampules rather than in a large bottle) and make stocks that are 1000 times the final concentration. This stock solution can then be pipetted into $10\ \mu l$ aliquots into sterile eppendorf tubes and stored at $-20°C$.
2. Immediately before staining the embryos, remove an aliquot of the fluorochrome from the freezer and add 1 ml of media to the tube. This media is usually protein-

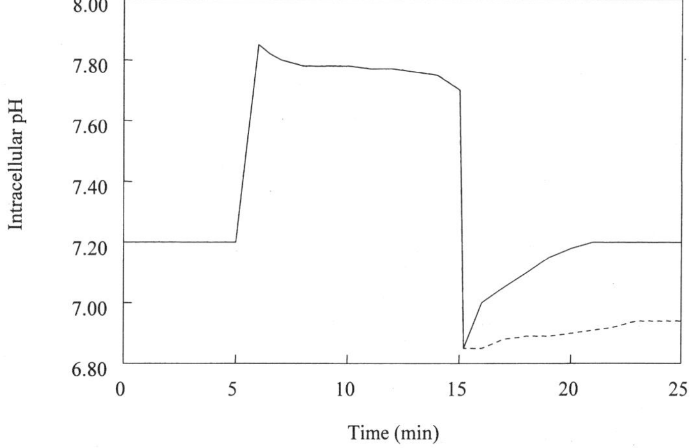

Figure 7.3. Measurement of the Na^+/H^+ antiporter in mammalian embryos. An intracellular acid load is induced by a 10-min pulse of 25 mM NH_4^+. Upon removal of the NH_4^+, the intracellular pH is reduced, and Na^+-dependent recovery is indicative of the presence of the Na^+/H^+ antiporter. The solid line indicates recovery in the presence of Na^+. The dotted line indicates recovery in the absence of Na^+. The kinetics of the antiporter can be determined by the rate of recovery from an acid load in the presence of different concentrations of Na^+.

free to assist later attachment to the microscope slide for measurement. Place the mixed fluorochrome into an organ-well dish.

3. Add the embryos to the dish containing the fluorochrome for the staining period. The length of time that embryos are stained needs to be determined for each different fluorochrome (see section 5 of text). If the medium used for staining is a bicarbonate-buffered medium, the embryos should be placed in the gassed incubator for the staining period. For HEPES/MOPS-buffered media, the embryos can be stained on a warm stage and kept in the dark by covering with foil. The media for staining the embryos should be the same as the media that the embryos are to be measured in.

4. At the end of the staining procedure, wash the embryos twice in the media that the embryos are to be measured in. They can then be placed into the incubation chamber for measurement.

Protocol 7.2. Calibration of intracellular calcium levels

1. Load cells with calcium probe and place embryos in chamber where intracellular measurements are to be taken in base medium. Mimic the measurement conditions wherever possible (e.g., temperature, volumes of fluids to be used).
2. Completely flush chamber with solution A containing a calcium ionophore (ionomycin or 4-bromo-A-23187).
3. Let the embryos sit for 10 min to equilibrate the intra- and extracellular fluids.
4. Record the fluorescent readings at both wavelengths.
5. Flush the chamber completely with solution B containing a calcium ionophore (ionomycin or 4-bromo-A-23187).
6. Let the embryos sit for 10 min to equilibrate the intra- and extracellular fluids.
7. These values can then be placed into equation 1 (see text) for the calibration of all experimental values.
8. Wash chamber, all tubing, and so on, very well, as any traces of the ionophores will affect subsequent readings.

Protocol 7.3. Calibration of intracellular pH levels

1. Load cells with pH probe and place embryos in chamber where intracellular measurements are to be taken in base medium. Mimic the measurement conditions wherever possible (e.g., temperature, volumes of fluids to be used, gas phase).
2. Completely flush chamber with the first calibration solution containing the ionophores nigercin and valinomycin. Start with the solution at the lowest pH. Ensure that the range of pH values used for the calibration curve cover the range of pH values expected within the experiment.
3. Let the embryos sit for 10 min to equilibrate the intra- and extracellular fluids.
4. Record the ratiometric readings.
5. Flush the chamber completely with the next calibration solution also containing the ionophores nigercin and valinomycin.
6. Let the embryos sit for 10 min to equilibrate the intra- and extracellular fluids.
7. Record the ratiometric readings.
8. Repeat steps 5–7 for all calibration solutions. Calibration curves should ideally have a minimum of 4 points.
9. Wash chamber, all tubing, and so on, very well, as any traces of the ionophores will affect subsequent readings.
10. Calculate the linear regression of the ratiometric values obtained with the known pH values of the solutions and establish that regression values are within acceptable range ($r > .99$).
11. This calibration curve can then be used to calculate intracellular pH (Figure p7.1).

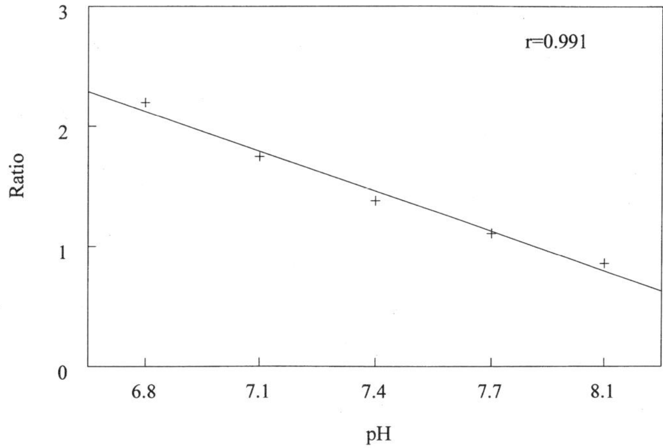

Figure p7.1. Calibration curve for the calculation of intracellular pH levels.

References

1. Grynkiewicz, G., Poenie, M., and Tsien, R.Y. (1985). *J. Biol. Chem.*, **260**, 3440.
2. Diarra, A., Sheldon, C., and Church, J. (2001). *Am. J. Physiol.*, **280**, C1623.
3. Henke, W., Cetinsoy, C., Jung, K., and Loening, S. (1996). *Cell Calcium*, **20**, 287.
4. Silver, R. A., Whitaker, M., and Bolsover, S. R. (1992). *Pflugers Arch.*, **420**, 595.
5. Tsien, R. Y. (1983). *Annu. Rev. Biophys. Bioeng.*, **12**, 91.
6. Tsien, R. Y. (1989). *Annu. Rev. Neurosci.*, **12**, 227.
7. Tsien, R. Y. (1989). *Methods Cell. Biol.*, **30**, 127.
8. Tsien, R. Y. (1988). *Trends Neurosci.*, **11**, 419.
9. Tsien, R. Y. (1986). *Soc. Gen. Physiol. Ser.*, **40**, 327.
10. Bouchelouche, P. N. (1993). *Scand. J. Clin. Lab. Invest. Suppl.*, **214**, 27.
11. Baltz, J. M., and Phillips, K. P. (1999). In: *A Comparative Methods Approach for the Study of Oocytes and Embryos*, J. D. Richter, ed., p. 39. Oxford University Press, New York.
12. Bright, G. R., Fisher, G. W., Rogowska, J., and Taylor, D. L. (1987). *J. Cell. Biol.*, **104**, 1019.
13. Bright, G. R., Fisher, G. W., Rogowska, J., and Taylor, D. L. (1989). *Methods Cell. Biol.*, **30**, 157.
14. Phillips, K. P., Zhou, W. L., and Baltz, J. M. (1998). *Zygote*, **6**, 113.
15. Poenie, M., and Tsien, R. (1986). *Prog. Clin. Biol. Res.*, **210**, 53.
16. Almers, W., and Neher, E. (1985). *FEBS Lett.*, **192**, 13.
17. Baltz, J. M., Biggers, J. D., and Lechene, C. (1990). *Dev. Biol.*, **138**, 421.
18. Baltz, J. M., Biggers, J. D., and Lechene, C. (1991). *J. Biol. Chem.*, **266**, 6052.
19. Lane, M., Baltz, J. M., and Bavister, B. D. (1999). *Biol. Reprod.*, **61**, 452.
20. Lane, M., Baltz, J. M., and Bavister, B. D. (1999). *Dev. Biol.*, **208**, 244.
21. Phillips, K. P., Leveille, M. C., Claman, P., and Baltz, J. M. (2000). *Hum. Reprod.*, **15**, 896.
22. Phillips, K. P., and Baltz, J. M. (1999). *Dev. Biol.*, **208**, 392.
23. Zhao, Y., and Baltz, J. M. (1996). *Am. J. Physiol.*, **271**, C1512.
24. Ruddock, N. T., Machaty, Z., and Prather, R. S. (2000). *Reprod. Fertil. Dev.*, **12**, 201.
25. Lane, M., Baltz, J. M., and Bavister, B. D. (1998). *Biol. Reprod.*, **59**, 1483.
26. Cheung, A., Swann, K., and Carroll, J. (2000). *Hum. Reprod.*, **15**, 1389.
27. Lane, M., and Bavister, B. D. (1998). *Biol. Reprod.*, **59**, 1000.
28. Lane, M., Boatman, D. E., Albrecht, R. M., and Bavister, B. D. (1998). *Mol. Reprod. Dev.*, **50**, 443.
29. Logan, J. E., and Roudebush, W. E. (2000). *Early Pregnancy*, **4**, 30.
30. Carroll, J., and Swann, K. (1992). *J. Biol. Chem.*, **267**, 11196.

31. Lane, M., and Gardner, D. K. (2001). *Mol. Reprod. Dev.*, **60**, 233.
32. Baltz, J. M., Biggers, J. D., and Lechene, C. (1993). *Development*, **118**, 1353.
33. Baltz, J. M. (1993). *Bioessays*, **15**, 523.
34. Phillips, K. P., and Baltz, J. M. (1999). *Dev. Biol.*, **208**, 392.
35. Zhao, Y., Doroshenko, P. A., Alper, S. L., and Baltz, J. M. (1997). *Dev. Biol.*, **189**, 148.
36. Zhao, Y., Chauvet, P. J., Alper, S. L., and Baltz, J. M. (1995). *J. Biol. Chem.*, **270**, 24428.

8

DAVID K. GARDNER
MICHELLE LANE

Assessment of Nutrient Uptake, Metabolite Production, and Enzyme Activity

1. MICROFLUOROMETRIC TECHNIQUES

The ability to measure the uptake and production of substrates by single embryos is an important tool for understanding how embryos control metabolism and embryo production. Little is known about the control of metabolism in the mammalian embryo compared to somatic cells or embryos from marine species or *Xenopus*. One of the main reasons for this is that the small amounts of material in the early embryo make measurement technically difficult. In recent years microfluorometric techniques have been developed that are capable of quantitatively analyzing the nutrient uptake and metabolite release from single cells. These techniques are based on miniaturizations of conventional methods of enzymatic analysis and can therefore be adapted to measure a great variety of metabolites (1–3) (Figure 8.1). An advantage of these procedures over the radiolabel procedures described in Chapter 9 is that they are totally noninvasive. Thus, microfluorometric techniques have the potential to be used as viability assays for mammalian embryos (2).

In addition, microfluorometry can be used to measure the enzyme activity in a single cell. It is possible to examine the activity and kinetics of a specific enzyme from a single oocyte or embryo using these procedures. In this chapter we provide details about how to use ultra-microfluriometry to measure nutrient uptakes and enzyme activities in mammalian embryos. Additionally, information is provided on how to calculate the kinetic properties of the enzymes and how to identify the isoforms present using electrophoresis.

2. ULTRAMICROFLUORIMETRY

Microfluorometric techniques developed in the 1980s are capable of quantitatively analyzing the nutrient uptake and metabolite release from single cells (4). These techniques are miniaturizations of conventional fluorometric methods of enzymatic analysis. Rather than occurring in cuvettes, the assays are scaled down with the use of micropipettes to measure picomole/femtomole levels of substrates.

139

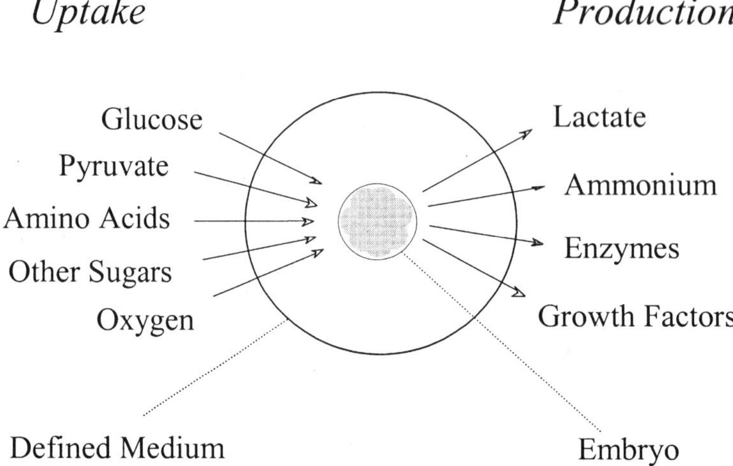

Figure 8.1. Diagram of how substrate uptake and production by embryos can be measured using ultramicrofluorimetry. Individual embryos are placed in known volumes of defined medium. The uptake of a substrate can be determined by measuring its rate of disappearance from the medium, whereas the production of a metabolite can be measured by its rate of appearance in the medium.

Fluorometric assays are based on the generation or consumption of the reduced pyridine nucleotides, NADH and NADPH, in coupled enzymatic reactions. These nucleotides fluoresce when excited with light at 340 nm, whereas the oxidized forms, NAD^+ and $NADP^+$, do not. Thus, the reaction:

$$\textit{Lactate dehydrogenase}$$
$$Pyruvate + NADH + H^+ \rightarrow Lactate + NAD^+$$

may be followed by monitoring a drop in fluorescence, or optical density at 340 nm, as NADH is converted to NAD^+. Under specific conditions the levels of change in fluorescence are proportional to the amount of substrate consumed in the reaction, and therefore the amounts can be calibrated using standard curves. Reactions are conventionally carried out in cuvettes and the fluorescence or absorbance measured by a fluorimeter or spectrophotometer. However, to study single-cell biochemistry, reactions are scaled down to occur in nanoliter volumes so that substrate levels in the picomole and femtomole range can be measured.

It is possible to measure any substrate that can be linked to the reaction that contains pyridine nucleotides or another fluorescent tag (Figure 8.1).

3. EQUIPMENT REQUIREMENTS

Assays for substrate levels and enzyme activity rely on quantiating the fluorescence of the reduced forms of the pyridine nucleotides (NADH, NADPH) under UV light. Therefore, the essentials of the equipment for this type of analysis are excitation of a sample with a mercury lamp so that excitation wavelengths are in the UV range (340 nm), a diaphragm to limit the excitation and emission to the drops to be analyzed, and quantification of the low emission levels from the sample. This is best achieved by the use of a photometer attachment with software that can convert the levels of emission light into a numerical value. The photometer scale is linear and is expressed in arbitrary units that can be calibrated using standard curves. Additionally, for the measurement of enzyme activity the use of an optical switch (as opposed to the use of shutters) is preferable because it enables real-time continuous kinetic measurement of enzyme activity.

Additional equipment to the fluorimetry measurement system that are required are a micromanipulator and stereo microscope for manipulation of micropipettes, a warm stage for embryos, and a warm stage on fluorimetry equipment for measurement of enzyme activities.

4. MICROPIPETTES

For analysis of substrate uptakes, metabolite production, or enzyme activity by embryos, the conventional assays are scaled down to occur in submicroliter volumes. The submicroliter volumes are manipulated by specially constructed constriction micropipettes. These pipettes are made from borosilicate glass capillaries (o.d. 1.0 mm, i.d. 0.8 mm) pulled over a flame to produce an inner diameter of 50–100 μm. The tubing is snapped in half and a small hook made on each end. Using a microforge, a constriction is made in the glass by placing the heated filament close to the glass capillary. A small weight (paperclip) is then placed onto the end of the hook of the capillary, and the tip is made by again heating the glass with the microforge filament. As the glass heats up, the weight will result in the glass pulling to a tip. The tip can then be broken using a pair of watchmaker's forceps. The size of the tip and constriction will control the speed and accuracy at which the pipette can be filled. The glass pipettes are mounted in 16-gauge stainless-steel tubing and sealed using sealing wax. Filling and expelling fluids from the pipettes is achieved using an air-filled syringe attached to the pipette via thin tubing. Before use, the micropipettes are siliconized to make manipulation of fluids easier (Figure 8.2). The volume of the pipettes between the tip and the constriction is calibrated using tritiated water. The pipettes are then held in a micromanipulator and the volumes manipulated under a microscope. Using this procedure, it is possible to accurately pipette volumes in the nano- and picoliter range (4).

5. NUTRIENT ASSAYS

The assays are housed in submicroliter droplets on siliconized microscope slides under heavy mineral oil (heavy white grade, Sigma Chemical Co., St. Louis, MO). For each assay, reagent cocktail solutions are prepared that contain a buffer and all of the cofactors and enzymes needed for the reaction (see Protocols 8.1–8.5). Typically, 10- to 20-nl drops (depending on the substrate to be analyzed) of this reagent cocktail are placed onto a siliconized slide under the mineral oil. The fluorescence of each droplet is measured in turn using a 20× objective by exposing the pyridine nucleotides in the cocktail to the UV light source. Drops are routinely exposed for up to 0.25 s because there is no detectable photo-oxidation of NADH or NADPH during this time.

Following this initial determination of fluorescence, a 1- to 5-nl sample (depending on the substrate to be analyzed) is added to the reagent cocktail drop. The addition of

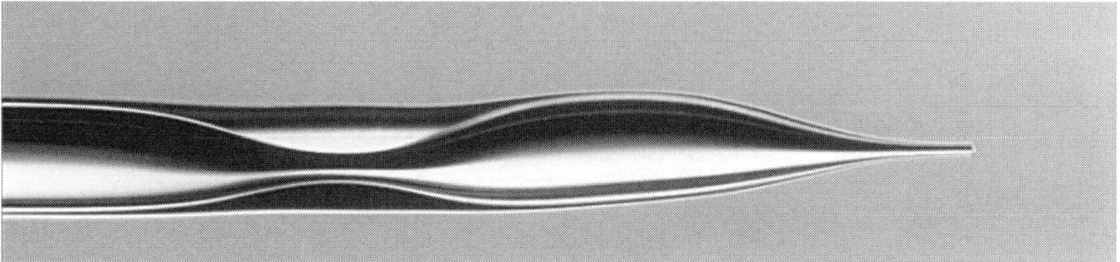

Figure 8.2. Micrograph of a 1nl microconstriction pipette.

the substrate initiates the reaction. The drops on the slide are then left until the reaction has gone to completion. This time can vary between 3 min and 1 h depending on the substrate to be analyzed. (This must be determined for each assay for a substrate by taking readings over time to determine when the reaction has reached completion.) The fluorescence of the drops is again determined. The change in fluorescence between the reagent cocktail drop before and after addition of the sample should be linear within the concentrations to be assessed. This is determined with a set of standards run on each day of experiment. Unlike spectrophotometric assays, the Beer-Lambert law does not hold for fluorescence microscopy. Therefore it is necessary to run a new set of standards with each day of experiment and to redetermine that the fluorescence range is linear within the concentration of the substrate. An acceptable linear regression value is typically $R > .99$ (Figure 8.3). Once a linear standard curve has been achieved, the concentration of substrates from samples can be calculated from this curve. The reactions and assay conditions for assays for pyruvate, lactate, glucose, ammonium, and glutamine are shown in Protocols 8.1–8.5.

6. MEASUREMENT OF NUTRIENT UPTAKES BY INDIVIDUAL EMBRYOS

Embryos are incubated in a medium containing the substrates for measurement at levels typically between 0.5 and 1.0 mM. This medium can either be bicarbonate buffered, in which case the incubation must occur in the CO_2 incubator, or HEPES/MOPS buffered, in which case the incubation can occur on a calibrated heating stage. Embryos are typically incubated in this medium for 1–4 h depending on the volume of the drop and the stage of embryo to be analyzed. For all uptake experiments, linear rates of uptakes should be determined whenever possible. This measurement will ensure that there are not alterations in substrate uptake or metabolite release as a result of the culture conditions. It has been demonstrated that the culture environment can significantly alter glucose metabolism in just 3 h (5).

For incubation of the embryos, drops of metabolic incubation medium are placed under oil in a dish. Use dishes with low sides, or use the lid of a 35-mm Petri dish. For mouse embryos it is typical to incubate embryos in drops of medium in the 20–50 nl range, while for domestic animal embryos or human embryos volumes of 200 nl to 1 µl are used. Wash embryos well in the metabolic incubation medium to ensure there is no carryover of substrates to the small incubation drops. The embryos are then picked up in a very small amount of volume using a pulled pipette just larger than the embryos.

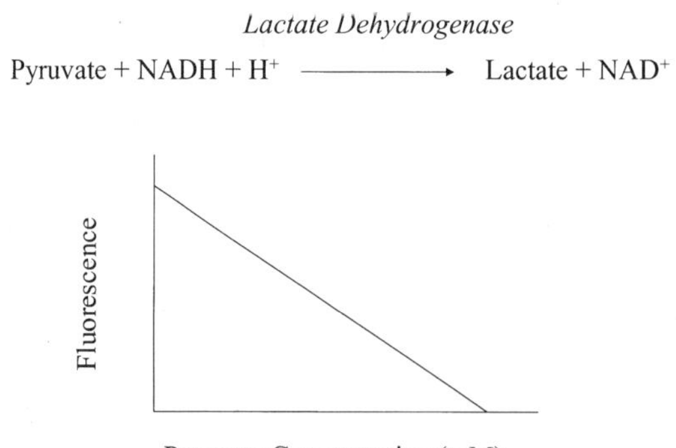

Figure 8.3. Standard curve for pyruvate. Levels of pyruvate in a sample can be assessed by a linear decrease in fluorescence with an increase in pyruvate concentration.

This will ensure that there is a minimal carryover of extra volume into the drops (6). The amount of volume that is carried over into the drops can be determined using tritiated water.

For determination of linear rates of uptakes or production, serial samples are taken at 20–45 min intervals. A minimum of three readings is required to determine linear rates. For endpoint determinations, an extra three or four drops of medium alone should be included in the incubation to determine the exact amounts of nutrients that were present in the medium and to control for any breakdown that may have occurred during the incubation period. Ensure that the concentration of substrate to be measured does not fall below the K_m of the transport system for the substrate. If the samples are not to be analyzed immediately, the embryos can be removed from the drops at the end of the incubation and the media taken up in 1–5 µl capillary tubes surrounded by oil on each end to avoid evaporation. These can then be stored in plastic insemination straws and stored at –80°C. There is negligible breakdown of most substrates during storage for several days under these conditions.

7. ENZYME ANALYSIS

Energy metabolism is controlled by precise regulation of a large number of enzymes. Cells can control the function of enzymes by regulating the availability of cofactors and substrates and by feedback inhibition. Such dynamic changes in metabolic regulation rarely occur by regulating mRNA expression levels or by protein synthesis. Cells can also regulate enzyme activity by changing the isoform of an enzyme. As different isoforms have different kinetic properties, changes in the abundance of a particular isoform can also regulate the flux of a substrate through an enzyme and therefore control pathway activity. Therefore the analysis of enzymes and their kinetics is fundamental to understanding how cells regulate their metabolism and development. Compared to somatic cells, our knowledge of the regulation of enzymes in embryos and how they control metabolism is quite limited. Most of the studies that have been performed have only established the presence of the enzymes and their maximal activity in the presence of nonlimiting conditions (zero-order kinetics). Most often this analysis has been a single reading at the end of an extended incubation and has not used continuous kinetic analysis. The reason for this is the small amount of material present in embryos, which has necessitated the use of complicated procedures such as enzyme cycling to detect enzyme activity (7–9). However, in more recent years, with the further development of microfluorimetry procedures and the increased sensitivity of fluorescence equipment, particularly those with optical switches, it is now possible to establish the kinetics (K_m, V_{max}) of both cytoplasmic and mitochondrial enzymes in single cells. The procedures described below outline how to determine the kinetic activity of a single enzyme from either a single cell or embryo.

8. ENZYME KINETICS

Enzyme kinetics can be complicated by the various control mechanisms in a cell such as feedback inhibition. The measurement of enzyme activity and establishment of kinetic data can be simplified by making the measurement conditions such that these control mechanisms are eliminated in the measurement conditions. For the assays and procedures listed below, continuous kinetic analyses are used in zero-order kinetics to establish enzyme activity. Using this type of analysis, a single factor (such as the concentration of the substrate or the level of an effector) can be altered in the assay conditions, and the direct effect on enzyme activity can be determined.

Two parameters, K_m and V_{max}, describe the kinetic properties of an enzyme. The K_m of an enzyme is the concentration of the substrate that enables the enzyme to function at half-maximal activity and is therefore a measure of the specificity of a substrate for the enzyme. For example, the K_m of lactate dehydrogenase for pyruvate in mouse embryos is 0.23 mM (10). The lower the K_m for a substrate, the higher the specificity of the substrate for the enzyme. Different isoforms of an enzyme have differing values of K_m for a given substrate. Therefore, a change in K_m can indicate a change in the relative abundance of different isoforms that can alter pathway activity. For example, the glycolytic enzyme hexokinase IV (glucokinase), which is the isoform predominantly found in the liver, has a K_m about 250 times higher (~ 10 mM) than the isoforms I–III. Isoform IV is able to control metabolism in the presence of very high concentrations of glucose that may arise in the liver.

The K_m for hexokinase in mouse embryos changes from the zygote to the blastocyst stage, indicating a change in the distribution of different isoforms in the embryo as development proceeds (11). This change in enzyme isoform may enable the embryo to alter the activity of the glycolytic pathway. The second parameter that characterizes the kinetics of an enzyme is V_{max} or maximal velocity. Both K_m and V_{max} are determined by establishing the activity of the enzyme at different substrate concentrations using continuous rate measurements (see below). The significance of determining enzyme kinetics is that K_m and V_{max} can establish the specificity of an enzyme for its substrates and determine the type of reaction that the enzyme controls between steady-state and equilibrium control. This information thereby enables one to determine the contribution of the particular enzyme to the control of a metabolic pathway.

Continuous rate measurement and establishment of kinetics has rarely been performed for enzymes in the mammalian embryo. Instead, velocity has been reported at a single concentration of substrate after several minutes (9, 12–19). Although this measurement ascertains that there is some activity of the enzyme, it tells little about the actual activity of the enzyme and tells nothing about its regulation or importance in the regulation of overall pathway activity. For most enzymes in the mammalian embryo, this highly valuable information is yet to be collected. Measurement of the actual kinetics of an enzyme will provide information about the action of an enzyme within a pathway. Further, it can establish how the enzyme's activity changes in response to changes in substrate availability and the presence or absence of specific effectors (10). This information is essential for understanding how the embryo controls the changes in metabolic activity from a carboxylic acid-based metabolism at the zygote stage to a glucose-based metabolism at the blastocyst stage throughout development.

9. ANALYSIS OF ENZYME ACTIVITY IN EMBRYOS

Activity and kinetics of enzymes in individual cells/embryos can be assessed using ultramicrofluorometric analysis. Similar to the nutrient assays described above, the analysis of enzymes are based on the consumption or production of the pyridine nucleotides NAD(P)H/NAD(P)$^+$ in coupled reactions. The equipment required for the measurement of enzyme activity is similar to that listed in section 3. However, for enzymes it is desirable for the equipment to have an optical switch that can control the levels of exposure of the samples to UV light to nanoseconds and also therefore able to measure real time enzyme activity.

This type of analysis has been used to determine the kinetics of lactate dehydrogenase (LDH) in the preimplantation mouse embryo (10) and will be used as an example in this chapter. However, numerous enzymes can be measured in the same fashion by coupling the reactions to the pyridine nucleotides. Fluorescent assays for many enzymes can be found in Bergmeyer and Gawehn (1). Similar to the assays for assessing substrate concentrations, these reactions can then be scaled down to measure the activity in single cells using specially constructed micropipettes.

9.1 Extraction of Enzymes

Enzymes are extracted from individual or groups of embryos using the same procedure. The enzymes are extracted by freezing and thawing the embryos in an extraction buffer (19). A detailed description of the extraction procedure used in our laboratory is shown in Protocol 8.6. This extraction procedure, coupled with the ultra-miroflurometric assay, results in dilution of the cell lysate, which makes endogenous regulators ineffectual. Thus the enzyme/factor of interest can be measured without the need to do extensive purifications that would require a large amount of material.

9.2 Analysis of Enzyme Activity

For each enzyme to be analyzed, a reagent cocktail is prepared in a buffer with the optimal pH and concentrations of substrates and cofactors to ensure zero-state kinetics (i.e., all factors are in excess compared to activity of the enzyme to be measured). The composition and preparation of the reagent cocktail for LDH is described in Protocol 8.7. In addition, a reagent cocktail without substrate is prepared as a control to establish baseline autofluroscence and changes in fluorescence.

For kinetic analysis, the reagent cocktail is prepared with four to six different substrate concentrations (see Protocol 8.7). In addition, a standard must be prepared for each day of experimentation to calibrate and calculate the consumption or production of the pyridine nucleotides.

For all enzymes to be measured, the activity of the assay should be validated by determining the activity of an enzyme standard. Standards can be purchased from several suppliers with actual activities of the lot of enzyme provided (e.g., Roche, Sigma). Using the procedure outlined below, a dilution of the purified enzyme (for LDH 1:2000 dilution) can be substituted for the embryo sample and activity measured. The specific activity can then be calculated and compared with the activity provided by the supplier to validate the assay.

9.3 Measurement of LDH Activity

The K_m and V_{max} for pyruvate of LDH is measured using the following reaction. LDH activity can be followed by a decrease in fluorescence over time as the fluorescent NADH is oxidized to NAD^+.

$$LDH$$
$$Pyruvate + NADH + H^+ \rightarrow Lactate + NAD^+.$$

Enzyme extracts are thawed and expelled onto a siliconized glass slide under mineral oil. The extract should be kept on ice until all assays are completed. The assays are housed in droplets under mineral oil on siliconized glass slides. All enzyme activity should be measured at 37°C. This reaction drop is incubated at 37°C for at least 2 min before the initiation of the reaction by the addition of preequilibrated enzyme (sample) or substrate. Upon initiation of the reaction, the change in fluorescence is monitored for several minutes until the reaction has reached a plateau.

A standard curve of NAD(P)H is run before enzyme samples to enable calibration. For LDH a standard curve of NADH concentrations of between 0 and 0.1 mM is used. Standard curves should have a minimum of four concentrations to determine the linear regression with the fluorescence obtained. A linear regression value of $R > .995$ is advisable to ensure that the calibration will result in accurate enzyme activities.

The change in fluorescence can then be calculated in units of pyridine nucleotide from the standard curve, taking into account the rate of any autofluorescence that may have resulted from the exposure to UV light. This rate of NAD(P)H consumption or production is then plotted against time.

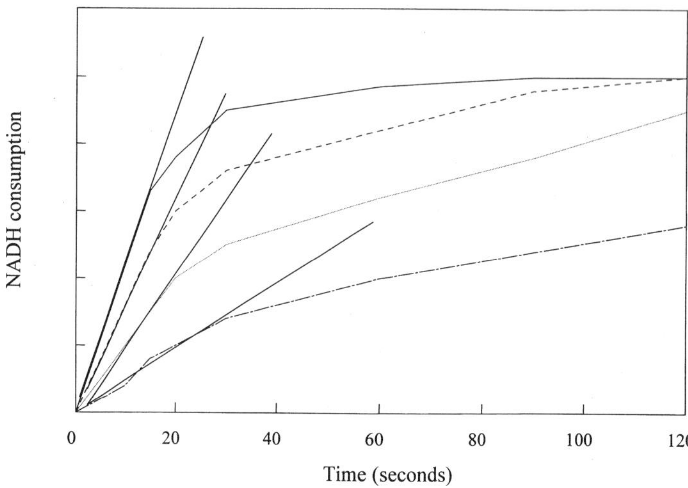

Figure 8.4. Calculation of activity of lactate dehydrogenase. Rates of NADH consumption are determined over a range of pyruvate concentrations, 0.6 mM, 0.3 mM, 0.15 mM, and 0.075 mM. The velocity of lactate dehydrogenase activity is determined by plotting a tangent to the initial velocity and is measured as picomoles NADH consumed/hour.

To assess the kinetics of the enzyme, this is repeated on each sample at different substrate concentrations. For determining the K_m of an enzyme, use 4–6 concentrations to obtain accurate calculations.

From each embryo sample, it is possible to obtain six different enzyme activity curves for the six different concentrations used (Figure 8.4). These data can then be used to determine a V_i for each concentration. This is the initial velocity(V_i) of the enzyme and can be determined by drawing a tangent to the initial linear portion of the curve. It can be seen from the example in Figure 8.4 that this type of analysis yields vastly different results from endpoint calculations taken after several minutes. Using this approach a V_i can be obtained for each of the substrate concentrations used. This can then be plotted as V_i against substrate concentration. This graph will produce a Michaelis-Menton plot where the velocity of the reaction is dependent on the substrate concentration.

$$V_i = \frac{V_{max}[S]}{K_m + [S]},$$

This confirms that the zero-order kinetics was correct and that there was no carryover of endogenous regulators. However, it is not possible to determine K_m or V_{max} from these plots. Instead, to determine the kinetic parameters of the enzyme, the equations must be transformed to give either a Lineweaver-Burk or Eadie-Hofstee equations that result in convenient graphical representation of the data that enables easy calculation of V_{max} and K_m.

The Lineweaver-Burk plot is the most straightforward of these equations. This results in a double-reciprocal graph that plots $1/V_i$ against $1/[S]$. The slope of this graph is K_m/V_{max}, and the x-intercept gives $-1/K_m$ and the y-intercept gives $1/V_{max}$ (see Figure 8.5). This procedure relies on few assumptions and therefore is beneficial over the least square means analysis, which makes assumptions that are difficult to verify with the small amount of enzyme available from a single embryo.

Using this analysis it is possible to determine a K_m and V_{max} for an enzyme for each individual embryo. The parameters of K_m and V_{max} are important because they provide information such as

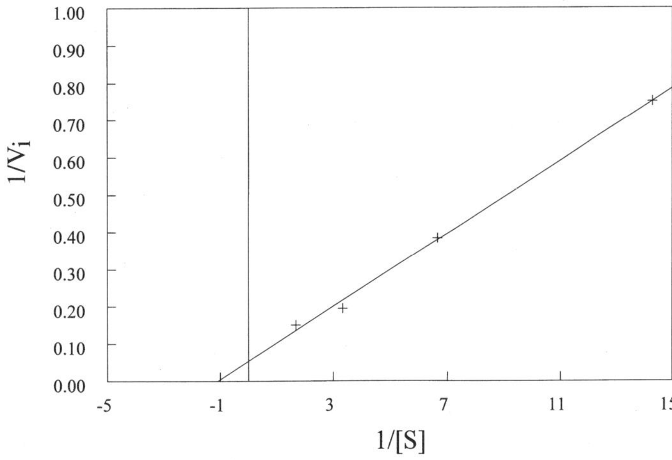

Figure 8.5. Lineweaver-Burk plot for determination of K_m. Rates of lactate dehydrogenase activity at increasing levels of pyruvate are plotted on a double reciprocal plot. The linear regression of this line is $r = .998$. The equation of the line is $y = 0.049x + 0.53$. The y-intercept of graph the $1/V_{max}$, and the x-intercept is $-1/K_m$.

1. Determining if there are changes in enzyme isoforms
2. Characterizing the specificity of an enzyme for a substrate (ratio of V_{max}/K_m is highest for the substrate with the highest specificity)
3. Deciding between steady-state and equilibrium mechanisms
4. Assessing the type of inhibition by negative effectors
5. Indicating the role of an enzyme in metabolism (by relating the K_m to the concentration of substrate present in the cell, it is possible to determine the relative activity of the enzyme present).

10. DETERMINATION OF ENZYME ISOFORMS

Many enzymes have different forms, or isoforms. Isoforms arise from genetically determined differences in amino acid sequences. Different isoforms of the enzyme have different kinetic properties, and therefore a change in the isoform present can change the flux through an enzyme and therefore pathway activity. Therefore, the determination of which isoform of an enzyme that is present in a cell is essential for the interpreting and understanding the control of pathway activity. For embryos, there is little information about the isoform of an enzyme that is present throughout development. The most notable exception is for LDH, which was first shown to change from isoform I (heart type) during the preimplantation period in the mouse to isoform V (muscle form) during the peri-implantation period more than 30 years ago (20, 21).

Different isoforms of an enzyme can be separated by electrophoresis. Isoforms are named based on the extent of migration in an electric field, starting with the species with the greatest mobility toward the anode. In this manner it is possible to distinguish between isoforms in different tissues/embryos by examining their migration by electrophoresis.

For embryos, we have found that the best procedure for separation of isoforms is electrophoresis in a 6% acrylamide gel (Protocol 8.8), followed by a staining protocol customized for the individual enzyme using either tetrazolium or diazonium salts. Two extensive books have been published that contain staining protocols for a wide variety of enzymes that can be used for staining isoforms of embryos (22, 23). This method of staining is based on the detection of the activity of the enzymes themselves. Therefore, it is not necessary to purify the sample because only the reaction specific to the enzyme of interest will stain on the gel. This approach enables the detection of isoform patterns of embryos. An example of the staining procedure for LDH is shown in Protocol 8.9. We have successfully used these protocols for staining isoforms for many different enzymes and for distinguishing between cytosolic and mitochondrial enzymes.

Protocol 8.1. Preparation of pyruvate assay reagents and standards

$$LDH$$
$$\text{Pyruvate} + \text{NADH} + \text{H}^+ \rightarrow \text{Lactate} + \text{NAD}^+$$

Buffer

   2.52 g EPPS
   10 mg penicillin
   10 mg streptomycin

   Make up to 200 ml and adjust pH to 8.0 with 1 M NaOH.

Reagent cocktail (can be frozen in aliquots)

   14 ml Buffer
   0.3 ml 5 mM NADH
   0.4 ml lactate dehydrogenase

   Make 5mM NADH by adding 17.7 mg in 5 ml water.

Standard curve preparation

Prepare a 1 mM pyruvate solution by dissolving 0.0110 g pyruvate in 100 ml water. Serially dilute the 1 mM pyruvate solution to give final concentrations of 0, 0.0625, 0.125, 0.25, and 0.5 mM. Prepare these solutions daily and use to generate a standard curve.

Protocol 8.2. Preparation of lactate assay reagents and standards

$$LDH$$
$$\text{Lactate} + \text{NAD}^+ \rightarrow \text{Pyruvate} + \text{NADH} + \text{H}^+$$

Buffer

   7.5 g glycine
   5.2 g hydrazine
   0.2 g EDTA

   Dissolve in 49 ml of water; add 51 ml 2 M NaOH.

Reagent cocktail (prepared day of assay)

   0.45 ml Buffer
   0.40 ml Water
   25 µl lactate dehydrogenase
   75 µl NAD$^+$ solution

   Make the NAD$^+$ solution 40 mg/ml and store as 75 µl aliquots in freezer.

Standard curve preparation

Prepare a 1 mM lactate solution by dissolving 0.0112 g lactate in 100 ml water. Serially dilute the 1 mM lactate solution to give final concentrations of 0, 0.0625, 0.125, 0.25, and 0.5 mM. Prepare these solutions daily and use to generate a standard curve.

Protocol 8.3. Preparation of glucose assay reagents and standards

Hexokinase
Glucose + ATP → glucose-6-phosphate + ADP

Glucose-6-phosphate dehydrogenase
Glucose-6-phosphate + $NADP^+$ → 6-phosphogluconate + NADPH + H^+

Buffer

2.52 g EPPS
10 mg Penicillin
10 mg Streptomycin

Make up to 200 ml and adjust pH to 8.0 with 1 M NaOH.

Reagent cocktail (can be frozen in aliquots)

15 ml Buffer
2 ml 5 mM DTT
2 ml 37 mM $MgSO_4$
1 ml 10 mM ATP
3 ml 10 mM NADP
1 ml hexokinase/glucose-6-phosphate dehydrogenase

Compound	Dilution
5 mM DTT	7.72 mg in 10 ml water
37 mM $MgSO_4$	91.2 mg in 10 ml water
10 mM ATP	30.3 mg in 5 ml water
10 mM $NADP^+$	39.4 mg in 5 ml water

Standard curve preparation

Prepare a 1 mM glucose solution by dissolving 0.0180 g glucose in 100 ml water. Serially dilute the 1 mM glucose solution to give final concentrations of 0, 0.0625, 0.125, 0.25, and 0.5 mM. Prepare these solutions daily and use to generate a standard curve.

Protocol 8.4. Preparation of ammonium assay reagents and standards

Glutamate dehydrogenase
α-ketoglutarate + NADH + H^+ → Glutamate + H_2O + NAD^+

Buffer

9.3 g triethanolamine (TEA)-HCL
95 mg ADP
670 mg α-ketoglutarate

Dissolve in 70 ml water; adjust pH to 8.0 with 5 M NaOH. Make up to 100 ml.

NADH solution

10 mg NADH
20 mg $NaHCO_3$

Dissolve in 2 ml water.

Reagent cocktail (prepared day of assay)

250 μl Buffer
500 μl Water

25 µl glutamate dehydrogenase
25 µl NADH solution

Standard curve preparation

Prepare a 10 mM ammonium chloride solution by dissolving 0.053 g ammonium chloride in 100 ml water. Dilute by adding 1 ml of the 10 mM solution to 9 ml of water. Serially dilute the 1 mM solution to give final concentrations of 0, 0.0625, 0.125, 0.25, and 0.5 mM. Prepare these solutions daily and use to generate a standard curve.

Protocol 8.5. Preparation of glutamine assay reagents and standards

Step 1

Glutaminase

$$Glutamine + H_2O \rightarrow Glutamate + NH_3$$

Buffer

6.8 g sodium acetate in 100 ml H_2O
2.9 ml acetic acid in 97.1 ml H_2O 2.9

Mix 68 ml of sodium acetate with 32 ml of acetic acid. Check that the pH is 5.0.

Reagent cocktail

40 µl Buffer
20 µl Water
20 µl Glutaminase

Dissolve 0.5 mg glutaminase into 1 ml of acetate buffer and freeze in aliquots at −20°C.

Step 2

Glutamate dehydrogenase

$$Glutamate + H_2O + NAD^+ \rightarrow \alpha\text{-ketoglutarate} + NADH + NH_4^+$$

Buffer

7.5 g glycine
5.2 g hydrazine
0.2 g EDTA

Dissolve in 49 ml of water, add 51 ml 2 M NaOH.
For 33.5 mM ADP, add 14.3 mg ADP in 1 ml water. For 27 mM NAD^+, 18.0 mg in 1 ml water.

Reagent cocktail

200 µl Buffer
20 µl NAD^+
10 µl ADP
50 µl glutamate dehydrogenase

Standard curve preparation

Prepare a 1 mM glutamine solution by dissolving 0.0147 g glutamine in 100 ml water. Serially dilute the 1 mM glutamine solution to give final concentrations of 0, 0.0625, 0.125, 0.25, and 0.5 mM. Prepare these solutions daily and use to generate a standard curve.

Protocol 8.6. Enzyme extraction from embryos

Extraction buffer

100 mM K_2HPO_4
30 mM KF
1 mM EDTA
5 mM β-mercaptoethanol
2 g/l BSA

Adjusted to pH 7.5.

Phenylmethylsulfonyl fluoride (PMSF) stock

Dissolve 10 g/l PMSF in ethanol and store at 4°C. Immediately before use, add 50 µl of the PMSF stock to 950 µl of the extraction buffer.

1. Wash individual or groups of embryos in saline supplemented with 4 mg/ml BSA.
2. Transfer embryos in a minimal amount of volume (<1 µl) into a 500 µl drop of the extraction buffer.
3. Pick up embryos in a 200-nl to 1-µl drop of the extraction buffer and place the drop containing the embryos under oil.
4. Using a washed glass capillary (1–5 µl, Drummond) take up the drop containing the embryo surrounded on either end with oil, which will prevent evaporation.
5. Seal the glass capillary within a plastic insemination straw and kept frozen at –80°C until analysis.

Note: For some enzymes such as phosphofructokinase the enzyme is not stable stored at –80°C and therefore should be immediately thawed and analyzed.

Protocol 8.7. Preparation of lactate dehydrogenase reagents and standards

Buffer

K_2HPO_4 34.0 mM
KH_2PO_4 7.3 mM

For kinetic analysis, prepare 5 solutions containing pyruvate:

1. 100 ml buffer containing 1.25 mM pyruvate
2. 100 ml buffer containing 0.63 mM pyruvate
3. 100 ml buffer containing 0.31 mM pyruvate
4. 100 ml buffer containing 0.16 mM pyruvate
5. 100 ml buffer containing 0.0.8 mM pyruvate

NADH solution

Prepare the 11.3 mM NADH solution fresh daily. Add 9.3 mg NADH and 10 mg $NaHCO_3$ to 1 ml of water and dissolve.

Assay cocktail

For each buffer solution, a cocktail solution is prepared consisting of 1 ml buffer and 5 µl NADH solution.

Standard curve preparation

Prepare a 10 mM NADH solution by dissolving 7.1 mg NADH into 1 ml water. Dilute by adding 10 µl of 10 mM solution to 990 µl of water to give a 100 µM NADH solution. Serially dilute the 100 μM NADH solution to give final solutions of 0, 3.125 μM, 6.25 μM, 12.5 μM, 25.0 μM, 50 μM, and 100 μM. Prepare these solutions daily and use to generate a standard curve for calibration of lactate dehydrogenase activity.

Protocol 8.8. Preparation of a 6% acrylamide gel

Component	10 ml	20 ml	30 ml	50 ml
H_2O	5.3	10.6	15.9	26.5
30% Acrylamide mix	2.0	4.0	6.0	10.0
1.5M Tris base (pH 8.8)	2.5	5.0	7.5	12.5
10% Ammonium persulfate (w/v)	0.1	0.2	0.3	0.5
TEMED	8 μl	16 μl	24 μl	40 μl

Mix all the solutions except for the TEMED into a glass flask. Add the TEMED immediately before the gel is poured.

Running buffer

Dissolve 4 g glycine and 1.2 g Tris base in 2 l water and adjust pH to 8.3. Run gel at 40 amps (around 250 V) for 1.5 h. Keep the gel cool by either running in the cold room or by cooling with cold water.

Protocol 8.9. Staining solution for lactate dehydrogenase isoforms

Staining of the gel relies on the activity of the enzyme. The gel is stained using the dye combinations of tetrazolium salt nitro-BT.

Phenazine methosulfate (PMS) acts by accepting a hydrogen from the NADH and passing it to the tetrazolium salt nitro-BT. When the nitro-BT is reduced, the dye is blue-black, and therefore wherever there is LDH activity it will appear as a blue-black band.

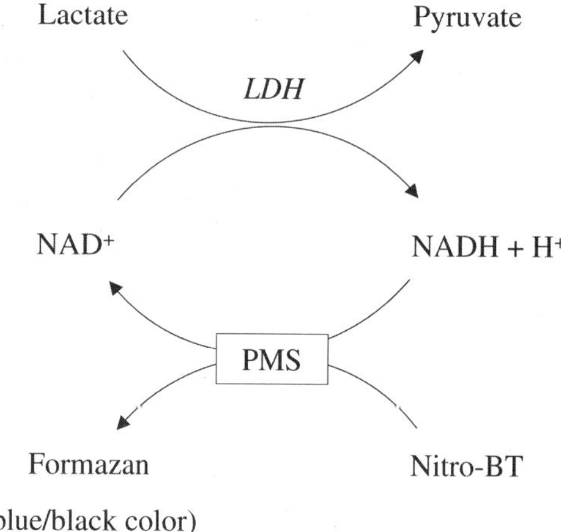

Staining solution

    0.2 M Tris-HCl buffer, pH 8.0
    0.5 M lactate
    10 mg/ml NAD+
    1 mg/ml nitro-BT
    2.5 mg/ml PMS

Incubate gel in staining solution for about 1 h in the dark at 37°C or until bands appear.

References

1. Bergmeyer, H. U., and Gawehn, K. (1974). *Methods of Enzymatic Analysis*. Academic Press, New York.
2. Gardner, D. K., and Leese, H. J. (1999). In: *Handbook of in Vitro Fertilization*, 2nd ed., A. O. Trounson and D. K. Gardner, eds., p. 347. CRC Press, Boca Raton, FL.
3. Gardner, D. K., and Leese, H. J. (1993). In: *Handbook of in Vitro Fertilization*, A. Trounson and D. K. Gardner, eds., p. 195. CRC Press, Boca Raton, FL.
4. Mroz, E. A., and Lechene, C. (1980). *Anal. Biochem.*, **102**, 90.
5. Lane, M., and Gardner, D. K. (1998). *Hum. Reprod.*, **13**, 991.
6. Gardner, D. K., and Leese, H. J. (1986). *Hum. Reprod.*, **1**, 25.
7. Barbehenn, E. K., Wales, R. G., and Lowry, O. H. (1974). *Proc. Natl. Acad. Sci. USA*, **71**, 1056.
8. Barbehenn, E. K., Wales, R. G., and Lowry, O. H. (1978). *J Embryol. Exp. Morphol.*, **43**, 29.
9. Chi, M. M., Manchester, J. K., Yang, V. C., Curato, A. D., Strickler, R. C., and Lowry, O. H. (1988). *Biol. Reprod.*, **39**, 295.
10. Lane, M., and Gardner, D. K. (2000). *Biol. Reprod.*, **62**, 16.
11. Gardner, D. K., Pool, T. B., and Lane, M. (2000). *Semin. Reprod. Med.*, **18**, 205.
12. Brinster, R. L. (1965). *Biochim. Biophys. Acta*, **110**, 439.
13. Brinster, R. L. (1966). *Biochem. J.*, **101**, 161.
14. Brinster, R. L. (1966). *Exp. Cell Res.*, **43**, 131.
15. Brinster, R. L. (1967). *Biochim. Biophys. Acta*, **148**, 298.
16. Brinster, R. L. (1968). *Enzymologia*, **34**, 304.
17. Brinster, R. L. (1971). *Wilhelm Roux Arch. Entwicklungsmech. Org.*, **166**, 300.
18. Brinster, R. L. (1968). *J. Reprod. Fertil.*, **17**, 139.
19. Martin, K. L., Hardy, K., Winston, R. M., and Leese, H. J. (1993). *J. Reprod. Fertil.*, **99**, 259.
20. Auerbach, S., and Brinster, R. L. (1967). *Exp. Cell Res.*, **46**, 89.
21. Auerbach, S., and Brinster, R. L. (1968). *Exp. Cell Res.*, **53**, 313.
22. Rothe, G. M. (1994). *Electrophoresis of Enzymes*. Springer Verlag, Berlin.
23. Manchenko, G. M. (1994). *Handbook of Detection of Enzymes on Electrophoretic Gels*. CRC Press, Boca Raton, FL.

9

DON RIEGER

Metabolic Pathway Activity

1. SIGNIFICANCE OF METABOLISM IN EMBRYO DEVELOPMENT

The early development of the mammalian embryo is directed by its inherent genetic programming and affected by its environment. Both the genetic program and the environment act as stimuli to the early embryo to induce developmental responses, initially in its intracellular biochemical activity, and ultimately in cell division, differentiation, and function. Techniques for manipulating gene expression and development, such as gene transfer, cloning, IVF, ICSI, and culture, similarly act as stimuli to induce the desired biochemical and developmental responses.

Whatever the source of the stimulus, genetic or environmental, natural or artificial, one or more intracellular processes are required to produce a developmental response. As illustrated in Figure 9.1, the pathway between the stimulus to a cell and its response can include any or all of transcription, translation, post-transitional modification, and export, activation, and function of an enzyme or other protein. Directly or indirectly, all of these processes rely on energy metabolism to provide cellular energy in the form of ATP, reducing equivalents in the form of NADH and NADPH, and precursors for the synthesis of macromolecules.

Consider, for example, the transcription of a 1200-nucleotide strand of mRNA. This would require the anaerobic metabolism of 5250 molecules of glucose for the production of approximately 10,500 ATP molecules for nucleotide synthesis and a further 1200 molecules of glucose to provide the ribose moieties. A single translation of that mRNA strand into a 400-amino-acid polypeptide would require 1200 ATP molecules, representing the anaerobic metabolism of 600 molecules of glucose. Large amounts of ATP are similarly required for DNA and lipid synthesis during cell division and for the activity of the sodium–potassium ATPase pump during formation of the blastocoele.

The energy metabolism of the early mammalian embryo can be studied indirectly by culture in varying concentrations and combinations of energy substrates. A more direct approach is to measure the disappearance of substrates from, and the release of metabolic products into, the medium. The breakdown or incorporation of radiolabeled substrates can be used to measure the activity of specific metabolic pathways, and the activity

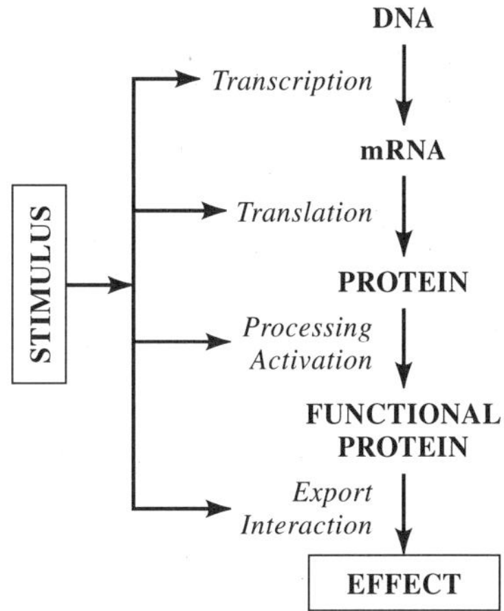

Figure 9.1. The mechanisms involved in the response of a cell to a stimulus.

of specific enzymes can be measured in broken-cell preparations (see Rieger [1]). The technique described here is used to measure the production of $^{14}CO_2$ and ^3H_2O from ^{14}C- and ^3H-labeled energy substrates by intact embryos, and the discussion will be limited to studies that have used that approach. More comprehensive reviews of embryo metabolism can be found elsewhere (2–7).

2. LABELED SUBSTRATES USED TO EVALUATE ENERGY METABOLISM PATHWAYS

Figure 9.2 shows the relationships among the carbon atoms of glucose, pyruvate, and lactate. Glucose can be metabolized through the pentose-phosphate pathway (PPP) to produce ribose, with a concomitant release of carbon-1 (C-1) as CO_2. When metabolized through the Embden-Meyerhof pathway (EMP) of anaerobic glycolysis, each molecule of glucose produces two molecules of pyruvate. Glucose C-1 and C-6 become pyruvate C-3, glucose C-2 and C-5 become pyruvate C-2, and glucose C-3 and C-4 become pyruvate C-1. Pyruvate and lactate are interconvertible, and the carbons retain the same numbers. Pyruvate C-1 (from glucose C-3 and C-4) is released as CO_2 in the conversion of pyruvate to acetyl-CoA by pyruvate dehydrogenase. The two carbons of acetyl-CoA (from glucose C-1, C-2, C-5, and C-6, or pyruvate C-2 and C-3) are released as CO_2 after two or more cycles through the mitochondrial Krebs cycle of oxidative metabolism. The hydrogen on glucose C-5 is released as H_2O in the conversion of 2-phosphoglycerate to phospho-enolpyruvate, the second to last step in the EMP. Glutamine is converted to glutamate and then to α-ketoglutarate, and its carbon and hydrogen atoms are released as CO_2 and H_2O in the Krebs cycle and electron transport chain, respectively.

Given these relationships, it is possible to evaluate the activity of the PPP, the EMP, and of mitochondrial oxidative metabolism by measuring the production of $^{14}CO_2$ and ^3H_2O from ^{14}C- and ^3H-labeled energy substrates. The production of ^3H_2O from [5-^3H]glucose is a simple measure of the activity of the EMP and can be confirmed using iodoacetate, an inhibitor of aldolase.

The production of $^{14}CO_2$ or ^3H_2O from ^{14}C- or ^3H-labeled glutamine is a measure of the activity of the Krebs cycle and can be confirmed using 2,4–dinitrophenol, an indirect stimulator of the cycle, malonate, an inhibitor of the cycle; or inhibitors of the electron transport chain such as antimycin-A, rotenone, sodium azide, and cyanide. The activity of the Krebs cycle can also be measured using [6-^{14}C]glucose, [2-^{14}C]pyruvate

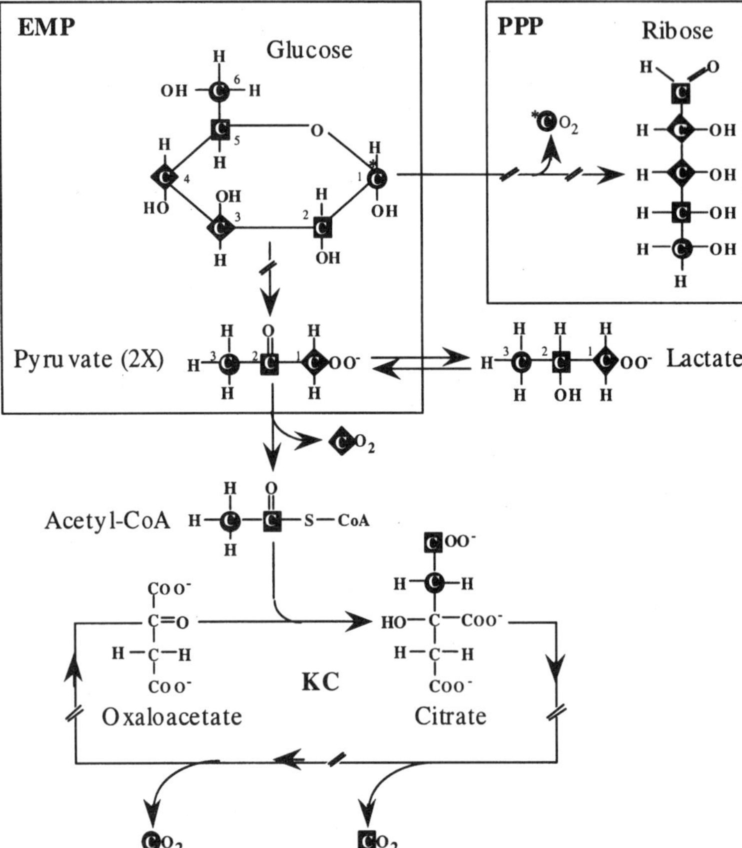

Figure 9.2. The fate of the carbon atoms of glucose, pyruvate, and lactate metabolized through the Embden-Meyerhof pathway (EMP), the pentose-phosphate pathway (PPP), and the Krebs cycle (KC). Broken lines indicate two or more reactions.

or, [2-^{14}C]lactate, and [1-^{14}C]pyruvate or [1-^{14}C]lactate can be used to evaluate the activity of pyruvate dehydrogenase.

The evaluation of the PPP is more difficult. Because glucose C-1 is released as CO_2 in both the PPP and Krebs cycle, whereas C-6 is released as CO_2 only in the Krebs cycle, the ratio of $^{14}CO_2$ produced from [1-^{14}C]glucose to that produced from [6-^{14}C]glucose (the C-1/C-6 ratio) has been used as a measure of the relative activity of the PPP. Due to the possibility of recycling of the products of the PPP to fructose and rearrangement of the carbon atoms, this approach is highly questionable (8, 9). However, if the ratio is very high (10:1 or more), it is reasonable to assume that production of $^{14}CO_2$ from [1-^{14}C]glucose is a reliable measure of PPP activity. This can be confirmed using brilliant cresyl blue or phenazine ethosulfate, stimulators of the PPP.

The production of $^{14}CO_2$ from [U-^{14}C]glucose may reflect the activity of any or all of the PPP, pyruvate dehydrogenase, and the Krebs cycle and has been used as a measure of the total metabolism of glucose. The production of $^{14}CO_2$ from [U-^{14}C]lactate similarly reflects the activity of both pyruvate dehydrogenase and the Krebs cycle.

To my knowledge, the first report of the metabolism of radiolabeled substrates by the mammalian embryo was that of Fridhandler (10), who adapted the manometric Cartesian diver to measure the production of $^{14}CO_2$ from [1-^{14}C]glucose or [6-^{14}C]glucose by rabbit embryos. The C-1/C-6 ratio was high before the blastocyst stage (as much as 20:1), suggesting that glucose metabolism was mainly by the PPP. After formation of the blastocyst, the ratio decreased to approximately 1:1, suggesting that glucose was metabolized through the Embden-Meyerhof pathway and the Krebs cycle. In the mouse embryo, the production of $^{14}CO_2$ from [U-^{14}C]glucose increased almost 5-fold between the unfertilized and fertilized ovum and 100-fold from the unfertilized ovum to the blastocyst stage (11). The C-1/C-6 ratio was significantly >1 at all stages, but the maximum was 2.33, at the two-cell stage. A similar study of the rabbit embryo (12) showed that the production of $^{14}CO_2$ from [U-^{14}C]glucose was not different between the unfertilized and fertilized ovum but increased 8000-fold by the day-6 blastocyst stage. The C-1/C-6 ratio was high until the morula stage and then dropped to approximately 1.5 thereafter. In addition to defining the basic pattern of glucose metabolism during the early development of mammalian embryos, these studies demonstrated that glucose metabolism differs significantly between embryos of different species.

Menke and McLaren (13) compared the production of $^{14}CO_2$ from [U-^{14}C]glucose by mouse blastocysts that had been cultured from the eight-cell stage or recovered from the uterus. They found that glucose metabolism by cultured blastocysts was generally greater that that by the uterine blastocysts and was increased by including fetal calf serum in the culture medium. These results demonstrated that the prior environment of the embryo can affect glucose metabolism.

The uptake of [U-^{14}C]malate, a Krebs cycle intermediate, and its metabolism to $^{14}CO_2$, was significantly greater by eight-cell mouse embryos than by two-cell embryos, suggesting that the cell membrane becomes permeable to malate between the first and third cleavage divisions (14). Conversely, the two-cell mouse embryo could produce $^{14}CO_2$ from [1-^{14}C]pyruvate, [2-^{14}C]pyruvate, [1-^{14}C]lactate, and [2-^{14}C]lactate, and the amount of $^{14}CO_2$ produced from C-1 was two- to threefold greater than that produced from C-2 (15). These results demonstrated that the Krebs cycle was active in the two-cell mouse embryo but that a significant portion of the amount of pyruvate that was converted to acetyl-CoA did not enter the Krebs cycle. It is important to note that the observation that little glucose carbon passes through to the Krebs cycle in the two-cell mouse embryo (11) is not inconsistent with the observation that $^{14}CO_2$ is produced from [2-^{14}C]pyruvate, because the intracellular concentration of pyruvate being taken up from the medium is likely much greater than that arising from the metabolism of glucose. Further reports of the energy metabolism of mouse and rabbit embryos continued sporadically throughout the 1970s and are covered elsewhere (2–7).

Interest in the energy metabolism of early embryos was stimulated by the advent of commercial embryo transfer in cattle and by the observation that the uptake of glucose by day-10 and day-11 cattle embryos was directly related to their development to term after transfer to recipients (16). Unfortunately, the method of measuring glucose uptake was not sensitive enough to be used for day-7 embryos, when cattle embryos are commonly transferred. However, it was possible to measure the production of $^{14}CO_2$ from ^{14}C-labeled glucose and lactate by individual in-vivo–produced, day-7 cattle embryos, and this offered a possible approach to evaluating the viability of cattle embryos before transfer (4).

O'Fallon and Wright (17) developed a technique in which individual mouse embryos were suspended in a droplet of medium above a NaOH trap to collect $^{14}CO_2$ produced from ^{14}C-labeled substrates and ^3H_2O from [5-^{3}H]glucose. Rieger and Guay (18) initially developed a technique in which an individual embryo was contained in a droplet of medium in a small well within a larger well containing 25 mM $NaHCO_3$. At the end of the incubation, NaOH was added to the bicarbonate to convert it to carbonate for counting. Rieger et al. (19) subsequently showed that the metabolism of one ^{14}C-labeled and one ^{3}H-labeled substrate by an embryo could be measured simultaneously. The hanging drop technique of O'Fallon and Wright (17) was then modified to use a 25 mM $NaHCO_3$ exchange reservoir in place of the NaOH trap (20), as described below. If an NaOH trap is used, the culture environment is completely free of CO_2 and bicarbonate during the incubation, which is disadvantageous because CO_2/bicarbonate favors embryo development by mechanisms unrelated to pH (7). Conversely, the CO_2/bicarbonate environment is maintained by the bicarbonate exchange reservoir, which satisfies the suggestion that "meaningful measurements are made only under conditions that support embryo development" (7).

In addition to determining the general pattern of energy metabolism during early development of the cattle embryo (20), my colleagues and I have shown that glucose metabolism is related to the sex of the early cattle embryo (21) and viability following cryopreservation of cattle blastocysts (22) and is increased by early exposure to glucose in culture (23). We have also studied the patterns of energy metabolism in cattle oocytes during in vitro maturation (24, 25), in horse (26) and hamster (27) embryos, and in bird sperm (28).

3. OVERVIEW OF THE METABOLIC MEASUREMENT ASSAY

The metabolism assay apparatus is show in Figure 9.3. Individual embryos (or oocytes) are taken up in 2 μl of culture medium, with or without metabolic stimulators, inhibitors, or other test substances, and placed in the cap of a sterile 2.0-ml screw-cap microvial (Sarstedt Inc., Montreal, Canada). To this is added 2 μl of culture medium containing one ^{14}C-labeled substrate or one ^{3}H-labeled substrate, or a mixture of one ^{14}C-labeled substrate and one ^{3}H-labeled substrate. The vial contains 1.5 ml of 25 mM $NaHCO_3$ that has been equilibrated with the gas mixture, to act as an exchange reservoir for the $^{14}CO_2$ and $^{3}H_2O$ produced during the incubation period. The vial is flushed with the gas mixture (usually 5% CO_2, 5% O_2, 90% N_2) just before the caps are fitted. At the end of the incubation period, the embryos are recovered and returned to culture. The $NaHCO_3$ is mixed with NaOH and scintillation fluid, and counted in a scintillation counter to determine the content of $^{14}CO_2$ and/or $^{3}H_2O$.

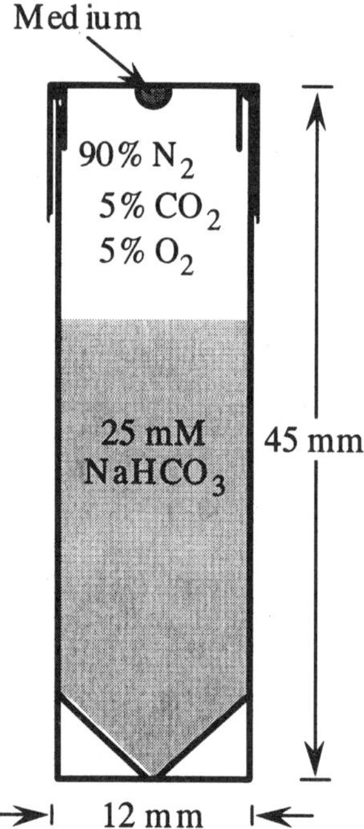

Figure 9.3. Apparatus used for the measurement of the metabolism of individual embryos. The embryo is contained in the 4-μl suspended droplet of culture medium. (Redrawn from Rieger et al. [20], with permission.)

4. CULTURE MEDIUM

As noted above, the technique is designed to use a bicarbonate-based culture medium. The basic medium is usually that used for routine embryo culture in the laboratory, supplemented with 10–20 mM HEPES to maintain the pH of the incubation droplets while they are being prepared. The pH in the microdroplets can be monitored by including phenol red in the medium.

5. PREPARATION OF THE RADIOLABELED SUBSTRATES

The labeled substrates can be purchased from the major suppliers (e.g., Dupont, Amersham). In addition to the nuclide and the position of the label, the preparations differ in their specific activity, concentration, and solvent. In general, crystalline preparations of high specific activity are the most convenient, but they are not available for all substrates. They are often supplied in water or water:ethanol at relatively low concentrations and must be dried and then resuspended in the basic culture medium. A simple approach to drying is to insert two syringe needles into the vial and pass dry N_2 into the vial through one needle, leaving the other needle open for the gas to escape. The end of the needles should be at least 1 cm above the surface of the liquid, and the gas flow rate should be only sufficient to create a dimple in the surface of the liquid. The vial can be placed into a hot water bath to hasten drying. Alternatively, the labeled substrates can be dried using a vacuum evaporator.

The concentration of the labeled substrate in the culture medium depends on the requirements of the experiment, the rate of metabolism of the substrate by the embryo, and the specific activity of the labeled substrate. Optimally, the concentration of labeled substrate should be just enough to result in the production an amount of $^{14}CO_2$ or ^3H_2O that can be reliably measured, while the total amount of labeled and unlabeled substrate should not be significantly greater than that normally used in the culture medium. In 4 µl of medium, 0.5 µCi of [5-^{3}H]glucose with a specific activity of 15 Ci/mmol would add only 0.01 mM to the total glucose concentration, a negligible amount. Conversely, 0.5 µCi of [1-^{14}C-]glucose with a specific activity of 50 mCi/mmol would yield 2.5 mM. If the basic medium contained 5.55 mM glucose, then the total concentration would be 8.05 mM. This might be too great a glucose concentration to be acceptable, and less [1-^{14}C]glucose would have to be used, or the concentration of unlabeled glucose in the basic medium would have to be reduced. At the extreme, 0.5 µCi of [2-^{14}C]pyruvate with a specific activity of 10 mCi/mmol would yield 12.5 mM, a far greater concentration than that normally used in embryo culture media. Fortunately, the metabolism of [2-^{14}C]pyruvate is relatively high in early embryos, and 0.05 µCi or less is usually sufficient to produced measurable amounts of $^{14}CO_2$.

As an example, suppose that the metabolism of [1-^{14}C]glucose (specific activity = 50 mCi/mmol) is to be measured in a 4-µl droplet, that the base medium contains 1.0 mM glucose, and that the final total concentration (labeled plus unlabeled) of glucose is to be 1.5 mM. A 4-µl droplet of the base medium would contain 4000 pmol of unlabeled glucose, and the total required glucose would be 6000 pmol. The difference of 2000 pmol would require 0.1 µCi (2.22×10^5 dpm) of [1-^{14}C]glucose. The radiolabeled substrate solution would therefore have to be made to contain 2.22×10^5 dpm/2 µl.

6. PREPARATION OF THE EMBRYOS

Depending on the experimental protocol, the embryos may be exposed to test procedures or conditions before the metabolic measurement. For example, an experiment might involve comparing the metabolic activity of embryos that have been cultured under different culture media or conditions or comparing activity between fresh and frozen-thawed embryos. Aside from this, the embryos require no special treatment except that they are normally passed through several washes of culture medium and sorted into experimental groups as appropriate. Each group is then transferred into a final wash just before being placed, individually, into the caps of the metabolic measurement vials. The final wash may also contain metabolic stimulators or inhibitors, which must be at twice the desired final concentration.

7. PREPARATION OF THE METABOLIC MEASUREMENT VIALS

To avoid drying and loss of CO_2 from the droplets, the vials should be prepared in batches of a maximum of 10. A single embryo is taken up in 2 μl of the final wash and placed in the cap of the incubation vial. To this is added 2 μl of the mixture of radiolabeled substrates, to produce a total volume of 4 μl. The vials are then loaded with 25 mM $NaHCO_3$ using the apparatus shown in Figure 9.4. The delivery tubing is cut to length to have a volume of 1.5 ml. The needle at the end of the delivery tubing is placed in the incubation vial, and the 50-ml culture tube is inverted to fill the tubing with $NaHCO_3$. When the bicarbonate solution reaches the end of the delivery tubing, the culture tube is turned upright, and the gas forces the $NaHCO_3$ solution out of the tubing into the vial. The needle at the end of the delivery tube is left in the vial and the cap held against the needle for a few seconds to flush the vial with the gas mixture. The needle is then withdrawn and the cap screwed onto the vial.

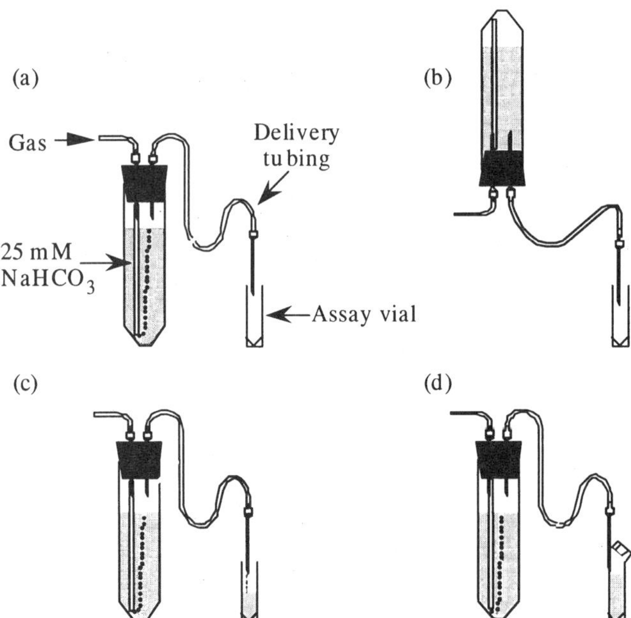

Figure 9.4. The apparatus and procedure used to load the metabolism assay vials with 25 mM NaHCO3. (a) The delivery tubing is cut to length to have a volume of 1.5 ml. The needle at the end of the delivery tubing is placed in the incubation vial. (b) The 50 ml culture tube is inverted to fill the tubing with $NaHCO_3$. (c) When the bicarbonate solution reaches the end of the delivery tubing, the culture tube is turned upright, and the gas forces the $NaHCO_3$ solution out of the tubing, into the vial. (d) The needle at the end of the delivery tube is left in the vial and the cap held against the needle for a few seconds to flush the vial with the gas mixture. The needle is then withdrawn and the cap screwed onto the vial.

It is possible to use snap-cap Eppendorf or other small vials rather than the screw-top Sarstedt vials and to simply pipette the gassed bicarbonate into the vials (29), rather than using the bicarbonate delivery apparatus shown in figure 9.4. However, the Sarstedt vial caps have the advantage of having an inner well for the droplets, and securing the screw-caps is less likely to disturb the droplets than is securing a snap-cap. Moreover, Sarstedt sells a rack that holds the vials from turning so that the caps can be screwed on with one hand. The bicarbonate delivery apparatus is very cheap and easy to construct and is convenient to use.

In addition to the vials containing the embryos, each assay must include a minimum of three total count vials and a minimum of three sham vials for each treatment group. Total count vials are prepared by placing 2 μl of the solution of radiolabeled substrates into the cap of a vial containing 1.5 ml of 25 mM $NaHCO_3$. The cap is screwed onto the vial, and the vial is inverted to mix the radiolabels with the bicarbonate. Sham vials are used to correct for counting background, chemiluminesence, spontaneous degradation of the labeled substrates, microbial contamination, and any other source of nonspecific counts. They are prepared exactly as for the vials containing the embryos, except that the droplets contain 2 μl of the final wash and 2 μl of the solution of radiolabeled substrates, but no embryo.

8. INCUBATION

Once capped, the vials should be held at the appropriate temperature, but there is no need for a humidified or gassed incubator. However, in practice, it is more convenient to culture them in the same incubator used for routine culture. We routinely culture the vials for 3 h, which is sufficient to produce measurable quantities of $^{14}CO_2$ or ^3H_2O from labeled glucose, pyruvate, and glutamine and to ensure equilibration of the labeled products between the culture drop and the bicarbonate reservoir (see Figure 9.5).

9. TERMINATING THE ASSAY

At the end of the incubation period, the bicarbonate is transferred to scintillation vials for counting, and the embryos are recovered for return to culture or further analysis. Working in batches of a maximum of 10 vials, the caps are removed and the bicarbonate decanted into 20-ml scintillation vials containing 200 μl of 0.1 N NaOH to convert the dissolved CO_2 and bicarbonate to carbonate.

To recover the embryos, 20–100 μl of fresh culture medium is added to the droplet in each cap. The embryos are picked up in 2 μl of medium and washed and returned to

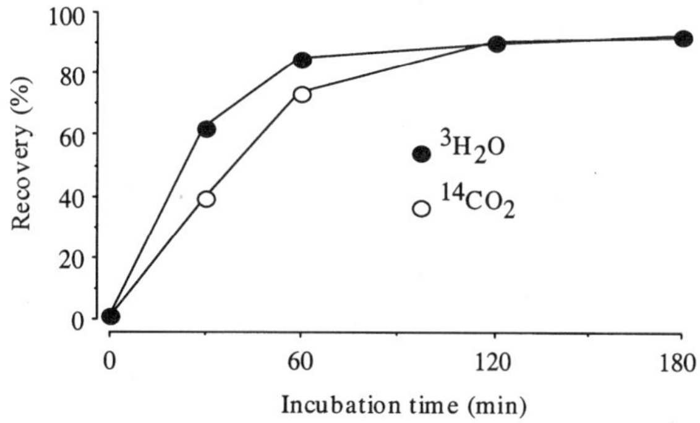

Figure 9.5. The recovery of ^3H_2O and $^{14}CO_2$ from the incubation droplet into the bicarbonate exchange reservoir of the metabolism assay vial. (Redrawn from Rieger et al. [20], with permission.)

culture, fixed, or analyzed by other procedures. To assure that the assay is not having any severe deleterious effects on the embryos, embryos that have been subject to the metabolic measurement assay should be left in culture for several days and their development compared to embryos that have not be subjected to the assay. When all the incubation vials have been decanted, 15 ml of a scintillation fluid for aqueous samples is added to each scintillation vial.

10. SCINTILLATION COUNTING

The scintillation vials are counted in a liquid scintillation beta counter for a minimum of 5 min each, and the raw counts per minute are corrected for quenching and converted to disintegrations per minute. If two labeled substrates (one ^{3}H-labeled and one ^{14}C-labeled) are used, then the sample must be counted in two counting windows, and the counts corrected to determine the disintegrations per minute for each label. These determinations are done automatically by most modern scintillation counters, and this requires only that the counting program be properly defined.

11. DETERMINATION OF PRODUCT RECOVERY

To determine the proportions of $^{14}CO_2$ or ^3H_2O produced that are recovered in the bicarbonate, known amounts of $NaH^{14}CO_3$ and ^3H_2O in 4-μl droplets of culture medium are cultured in the incubation vials, exactly as for the embryos. Three to five vials are removed from culture after 0, 30, 60, 120, and 180 min and the bicarbonate counted (Figure 9.5). The recovery correction factor is calculated as the reciprocal of the integrated proportion under the curve for each of $^{14}CO_2$ and ^3H_2O. Recovery is greater than 70% for both. Note that this determination need only be done once.

12. CALCULATIONS

The amount of each substrate (in picomoles) metabolized by an embryo is calculated by first subtracting the mean sham disintegrations per minute from the disintegrations per minute for the embryo to correct for nonspecific counts. The difference is divided by the mean total disintegrations per minute to give the proportion of labeled substrate metabolized by the embryo, which is, by definition, equal to the proportion of the total amount of substrate (labeled plus unlabeled) metabolized. This proportion is multiplied by the total amount of substrate and the recovery correction factor to give the amount of substrate metabolized. The formula is:

$$S_m = \frac{(D_e - D_s) \times R \times (S_l + S_u)}{D_t},$$

where S_m = the amount of substrate metabolized, D_e = the number of disintegrations per minute of $^{14}CO_2$ or ^3H_2O produced by the embryo, D_s = the mean number of disintegrations per minute for the group sham preparations, D_t = the *mean* number of disintegrations per minute of ^{14}C- or ^{3}H-labeled substrate in the droplet (as counted in the total count preparations). R = the recovery correction factor, S_l = the amount of labeled substrate in the droplet, S_u = the amount of unlabeled substrate in the droplet. The amount of labeled substrate in the droplet (S_l) in picomoles is determined from the mean total counts (D_t) and the specific activity (Sp. Act.) in milliCuries per millimole:

$$S_l = \text{Sp. Act.} \times (D_t/2.22).$$

The amount of unlabeled substrate in the droplet (S_u) in picomoles is determined from its concentration (C) in the medium (millimolar), and the volume of the droplet (V) in microliters:

$$S_u = C \times V \times 1000.$$

We handle these calculations with a spreadsheet program that includes a test that the disintegrations per minute for an embryo equal or exceed the sensitivity of the assay (defined as the mean sham disintegrations per minute plus 2 standard errors), and any value below that is set to zero. The spreadsheet also calculates the embryo treatment group means and variances, which can be cut and pasted to other programs for data storage, statistical analysis, and graphing.

The variance of the measurements of metabolism of embryos of a single cleavage stage is usually sufficiently homogeneous that the data can be statistically analyzed without transformation. For experiments involving blastocysts or comparisons across stages, it is often necessary to log-transform the results to provide homogeneity of variance.

13. COMMON PROBLEMS

The procedures have been developed to make this assay as simple as possible and relatively trouble free. There are only two aspects that are of significant concern and need to be monitored closely: the nonspecific counts measured in the sham preparations and the viability of the embryos after being subjected to the procedure.

Although the sham values are included in the calculations to correct for nonspecific counts, they should be reduced to the minimum to increase the signal-to-noise ratio. Machine background can be checked by counting scintillation vials containing only scintillation fluid. If these produce more than 50 cpm, then the counter should be checked, cleaned, and adjusted by a qualified technician. Chemiluminesence can be checked by preparing counting scintillation vials containing 4 μl of culture medium, 200 μl of 0.1 N NaOH, and 1.5 ml of the bicarbonate solution together with 15 ml of scintillation fluid, but no radioactive material. Background noise can be produced by NaOH, but this is unlikely to be a significant problem because of the small amounts used. It is possible that unusual components of the culture medium could be luminescent and might have to be eliminated or reduced in concentration. Microbial contamination must be eliminated by proper sterile procedures. Chemical breakdown of the radiolabeled substrates can be a significant source of nonspecific counts. This is especially true for [5-^{3}H]glucose, where the label is somewhat labile and can be spontaneously transferred to water. We also see relatively high sham counts with [2-^{14}C]pyruvate, possibly due to conversion to pyropyruvate. However, this has never been a serious problem because the metabolism of [2-^{14}C]pyruvate is also relatively high.

As noted above, the embryos must be capable of normal development after being subjected to the assay in order to be certain that the measurements reflect normal metabolic function. It is important to note that apparent morphological normality immediately after the metabolic measurement is not a reliable indicator of viability. We have seen a significant loss of viability on only two occasions. In the first case, the problem was traced to the N_2 used to dry the radiolabeled substrates, and in the second, to the batch of Percoll used to prepare the sperm for in vitro fertilization. As these two examples demonstrate, it can be difficult to determine the exact cause of a loss of viability, and it may be necessary to test all parts of the culture system and the assay.

14. CONCLUSIONS

The survival and development of the early mammalian embryo depends on its ability to metabolize energy substrates, and any manipulation that affects energy metabolism may disturb or inhibit development. Evaluating the effects of molecular and mechanical manipulations on the energy metabolism of the early embryo may lead to improvements in techniques such as cryopreservation, gene transfer, cloning, and the production of stem cells. A notable example is the incidence of large lambs and calves from embryos produced by cloning (30), which may well be due to very early metabolic defects.

Acknowledgments   The development and application of this technique were made possible by the efforts and kindness of my collaborators at the Universities of Montréal, Guelph, Wisconsin, Milan, Louvain-la-Neuve, Melbourne, and Dublin, the INRA research station, Nouzilly, France, and AgResearch, Hamilton, New Zealand.

References

1. Rieger, D. (1992). *Theriogenology*, **37**, 75.
2. Biggers, J. D., and Stern, S. (1973). *Adv. Reprod. Physiol.*, **6**, 1.
3. Wales, R. G. (1975). *Biol. Reprod.*, **12**, 66.
4. Rieger, D. (1984). *Theriogenology*, **21**, 138.
5. Kaye, P. L. (1986). In: *Experimental Approaches to Mammalian Embryonic Development*, J. Rossant, and R. A. Pederson, eds., p. 267. Cambridge University Press, Cambridge.
6. Leese, H. J. (1991). In: *Oxford Reviews of Reproductive Biology*, S. R. Milligan, ed., Vol. 13, p. 35. Oxford University Press, Oxford.
7. Barnett, D. K., and Bavister, B. D. (1996). *Mol. Reprod. Dev.*, **43**, 105.
8. Larrabee, M. (1989). *J. Biol. Chem.*, **264**, 15875.
9. Larrabee, M. G. (1990). *Biochem. J.*, **272**, 127.
10. Fridhandler, L. (1961). *Exp. Cell Res.*, **22**, 303.
11. Brinster, R. L. (1967). *Exp. Cell Res.*, **47**, 271.
12. Brinster, R. L. (1968). *Exp. Cell Res.*, **51**, 330.
13. Menke, T. M., and McLaren, A. (1970). *J. Reprod. Fertil.*, **23**, 117.
14. Wales, R. G., and Biggers, J. D. (1968). *J. Reprod. Fertil.*, **15**, 103.
15. Wales, R. G., and Whittingham, D. G. (1973). *J. Reprod. Fertil.*, **33**, 207.
16. Renard, J. P., Philippon, A., and Ménézo, Y. (1980). *J. Reprod. Fertil.*, **58**, 161.
17. O'Fallon, J. V., and Wright, R. W. Jr. (1986). *Biol. Reprod.*, **34**, 58.
18. Rieger, D., and Guay, P. (1988). *J. Reprod. Fertil.*, **83**, 585.
19. Rieger, D., Palmer, E., Lagneau-Petit, D., and Tiffin, G. (1989). *Theriogenology*, **31**, 249.
20. Rieger, D., Loskutoff, N. M., and Betteridge, K. J. (1992). *J. Reprod. Fertil.*, **95**, 585.
21. Tiffin, G., Rieger, D., Betteridge, K. J., Yadav, B. R., and King, W. A. (1991). *J. Reprod. Fertil.*, **93**, 125.
22. Rieger, D., Pollard, J. W., and Leibo, S. P. (1993). *Cryobiology*, **30**, 631.
23. Rieger, D., Grisart, B., Semple, E., Van Langendonckt, A., Betteridge, K. J., and Dessy, F. (1995). *J. Reprod. Fertil.*, **105**, 91.
24. Rieger, D., and Loskutoff, N. M. (1994). *J. Reprod. Fertil.*, **100**, 257.
25. Rieger, D., Luciano, A. M., Modina, S., Pocar, P., Lauria, A., and Gandolfi, F. (1998). *J. Reprod. Fertil.*, **112**, 123.
26. Rieger, D., Bruyas, J.-F., Lagneau, D., Bézard, J., and Palmer, E. (1991). *J. Reprod. Fertil. Suppl.*, **44**, 411.
27. Barnett, D. K., Rieger, D., and Bavister, B. D. (1993). *Theriogenology*, **39**, 185.
28. Rose, K. A., and Rieger, D. (1998). In: *Gametes: Development and Function*, A. Lauria, F. Gandolfi, G. Enne, and L. Gianaroli, eds., p. 287. Serono Symposia, Rome.
29. Spindler, R. E., Renfree, M. B., Shaw, G., and Gardner, D. K. (1998). *Biol. Reprod.*, **58**, 1425.
30. Kruip, T. A. M., and den Daas, J.H.G. (1997). *Theriogenology*, **47**, 43.

10

CALVIN SIMERLY
RICARDO MORENO
JOÃO RAMALHO-SANTOS
LAURA HEWITSON
GERALD SCHATTEN

Confocal Imaging of Structural Molecules in Mammalian Gametes

1. ADVANCES IN REPRODUCTIVE BIOLOGY

The fields of cellular, developmental, molecular, and reproductive biology and genetics emerged from studies on eggs at fertilization (1). These early studies relied on systems like frogs and sea urchins, in which fertilization in vitro and embryo culture occurred using simple solutions like seawater or pond water at room temperature. The recent design of methods for routinely and reliably obtaining excellent in vitro fertilization of many mammals now permits detailed experimentation on molecular and structural features of their development. These investigations have led to many important and unexpected basic discoveries including genomic imprinting (2–4); gametic recognition involving unique receptors and galactosyltransferases (5–7); atypical maternal inheritance pattern of the centrosome in mice (8, 9); both paternal and maternal inheritance of mitochondria (10); and unexpected signal transduction pathways for fertilization and cell cycle regulation (11–13), to name but a few.

The move into investigating molecular and cellular events in mammals is swiftly advancing. There is a superb, accurate, and growing literature, and for fertilization researchers, the malleability of many oocytes is becoming more routine. Rodents, such as the mouse, are easily maintained in the lab, and development of zygotes occurs with good synchrony. In addition, mice have a large and expanding number of genetic strains for investigational purposes. In domestic species like the cow and pig, technical improvements in the maturation, fertilization, and development for the production of offspring are now available (14). Primates such as the rhesus monkey are also unique resources for avoiding the complexities of working with fertilized human zygotes while garnering vital information that cannot be extrapolated directly from rodents (1). Their hormonal regulation closely parallels that of humans, and the cytoskeletal arrangements observed during meiotic maturation and fertilization mirror events observed in humans (15, 16). Notwithstanding these advances, mammalian oocytes and embryos provide unusual challenges for the researcher interested in detecting cytoplasmic or nuclear structural features: they are large cells (80–150 μm),

165

some oocytes are nearly opaque (e.g., bovine, porcine, canine), and the extracellular zona pellucida and cumulus cells can cause problems in introducing the imaging probes (e.g., antibodies, vital dyes). In addition, oocyte numbers can be limiting and their manipulation labor intensive and costly.

The methods used in the detection of cytoskeletal and nuclear architectural structures in mammalian oocytes during fertilization and in spermatogenic cells have now been described (17–22). Immunocytochemical investigations on microtubule organization and DNA configurations during spermatogenesis as well as after both in vitro fertilization and ICSI have been investigated in a number of mammals using laser-scanning confocal microscopy (15, 18, 23–25). Questions addressed in these studies demonstrated that all mammals except rodents follow a paternal centrosome inheritance pattern (9). Microtubules assembled from the introduced sperm centrosome were found to be essential for pronuclear apposition during fertilization in nonrodents, a key crucial step necessary to accurately conclude the fertilization process in mammals. The mechanisms underlying cell cycle progression and differentiation during spermatogenesis are also tightly regulated with the microtubule network, especially with regard to spermatid organelle movement and nuclear-shaping events (18–20). Finally, the results from ICSI fertilization events in the rhesus monkey demonstrated the utility of studying this model for predicting ICSI events in the human, especially in relation to the steps necessary for sperm activation after injection of sperm into the cytoplasm (23, 25, 26).

Analyzing cellular structure, physiology, and function by immunofluorescence has benefited tremendously from advances in light microscopy. Among the many technical achievements, the advent of laser scanning confocal microscopy (LSCM) for immunofluorescent applications has yielded enormous benefits (27–31). Confocal microscopy permits information to be collected from defined optical sections rather than from the entire specimen, thus reducing out-of-focus fluorescence and increasing contrast, clarity, and detection sensitivity of images. Because optical sectioning is noninvasive, living cells as well as fixed static images can be observed by confocal microscopy. Furthermore, specimens can be optically sectioned in various planes (i.e., either perpendicular or parallel to the optical axis of the microscope) so that variations throughout specimen height can be investigated (28). Because stacks of optical sections taken at successive focal planes (the z-series) can be reconstructed within the computer, it is possible to generate three-dimensional views in static samples or four-dimensional imaging (three-dimensional over time) in living cells, giving the investigator a powerful means to address biological problems with increased clarity. Therefore, confocal microscopy is an important tool for bridging the information provided by either conventional light or electron microscopy.

The goals of this chapter are to consider the methods used in the detection of cytoskeletal and nuclear architectural structures in mammalian gametes using confocal microscopy. We describe the methods applied in our laboratory for static immunocytochemical localization of microtubules and DNA in fixed human oocytes. We also examine microtubule-based cytoskeletal configurations in mammalian spermatids and discuss how this technique provides mechanistic information on vesicular trafficking and organelle movement during spermatogenesis. Finally, we describe how confocal microscopy has provided clinically relevant information with regard to abnormal sperm decondensation following ICSI, a concern given the widespread application of this assisted reproductive technology and the reported increase in sex chromosome anomalies observed in ICSI embryos (25, 26, 32). This chapter is principally aimed at prospective light or confocal microscopists who want to apply imaging technology to mammalian gametes. Additional information on the theories, principles, and types of confocal microscopy or application in systems other than mammals can be found in a number of excellent reviews (27–29, 30, 31, 33–35).

2. THE PRINCIPLES AND APPLICATIONS OF IMMUNOFLUORESCENCE ON FIXED MAMMALIAN GAMETES

A number of excellent reviews are recommended to acquaint readers with the basic principles of immunocytochemical (ICC) techniques (36–38). ICC coupled with modern imaging microscopes provides invaluable clues for detecting specific intracellular proteins and their subcellular localization within oocytes. The application of ICC can even be extended to functional analysis of intracellular proteins by first microinjecting primary antibodies into oocytes and then processing them for antigen detection, providing a powerful tool for investigating the role of proteins in oocyte and embryonic development (39). ICC is routinely performed on fixed, static specimens, but it also can be adapted to investigate the localization and function of proteins or ions in living gametes. For instance, the detection of the meiotic spindle in living oocytes can be approached using fluorescently tagged proteins such as rhodamine tubulin (26, 40, 41) and their dynamics explored by examining fluorescent recovery after photobleaching (FRAP) or UV photoactivation of caged fluorescein derivatives (40). Ion imaging using both conventional and confocal microscopy can be artfully used to explore the role of cations such as calcium during oocyte activation (42). Likewise, the sensitivity of fluorescent techniques in living spermatozoa combined with flow cytometry is now sufficient to provide quantitative data, permitting assessments of sperm viability, organelle integrity and function, as well as fertilization potential (43, 44).

The sequential steps for implementing static ICC in mammalian gametes include gamete preparation, fixation, antibody labeling, detection of the antibody–antigen complex, and imaging. For oocytes, removal of cumulus cells and the zona pellucida before fixation facilitates better fixation, antibody penetration, and removal during immunostaining, as well as microscopic imaging during analysis. In addition, gametes are not attached to substrates for easy handling, making it cumbersome to process them by standard ICC protocols. To facilitate easier ICC processing, oocytes or spermatogenic cells can first be attached to glass surfaces. Compounds such as charged polyamino acids (poly-L-lysine; 45), protamine sulfate (46), phytohemagglutinin (47), or a fibrin clot attached to a microscope slide (48) have all been successfully used to immobilize oocytes onto glass for ICC processing.

A critical first step in performing indirect ICC is the selection of the proper fixation method to preserve intracellular proteins and structure. Significant amounts of soluble proteins and cytoplasmic inclusions in mammalian gametes can increase nonspecific antibody binding and degrade image quality (39). The decision of whether to use an organic solvent (i.e., methanol, ethanol) or cross-linking reagent (e.g., formaldehyde, glutaraldehyde) for fixation must take into account the nature of the protein(s) being targeted for detection, the primary antibody characteristics, and the type of detection method used (38, 39). The choice of which fixative to use is somewhat empirical and best decided by experimentation with both fixatives, especially in cases where antibody characteristics have not been previously defined in other systems such as tissue culture cells (38). For mammalian oocytes, it is often useful to first extract gametes in detergent-based stabilization buffers (39). The use of these stabilization buffers helps increase antibody penetration, lower nonspecific antibody binding, reduce monomer concentration for polymer detection, and allow greater sensitivity for the detection of small cytoplasmic structures such as centrosomes or kinetochores. However, it is important to remember that detergent-based permeabilization buffers preserve only the detergent-insoluble proteins. In some instances, soluble proteins or proteins with transient or weak associations with intracellular structures may not be maintained in the specimens. In these cases, it may be advisable to fix gametes directly in methanol or apply cross-linking fixatives before detergent treatment. Fixation before permeablization will invariably have increased background staining, necessitating the use of blocking reagents before primary antibody application to limit degradation of image quality (38).

The application of a primary antibody will depend on the type used (monoclonal, polyclonal, or ascites fluid). Polyclonal antibodies contain many different types of antibodies directed at many antigens, creating problems in binding to only one targeted epitope within gametes. Monoclonal antibodies, however, are homogenous and extremely epitope specific, but they are costly and time-consuming to prepare and have much greater variability in practical ICC application (38). Whatever the type of antibody, it may be necessary to experiment with the concentrations of the antibodies to obtain a detectable signal while keeping nonspecific background fluorescence to a minimum. The number of proteins detectable after immunostaining will depend on the number of species of primary antibodies used and the number of fluorochromes that can be imaged by the microscopic system. Best results are achieved if each antibody is added sequentially, but it is possible to mix primary antibodies together for a single-step application. For some intracellular proteins that are present in extremely low copies, it may be advisable to amplify the detectable signal by labeling with the biotin/streptavidin system. Details of this protocol can be found in other publications (49).

Because fluorochromes are susceptible to extreme fading under fluorescent irradiation, stained oocytes are mounted in an antifade reagent to retard the rapid loss of fluorescence during microscopic examination. There are several varieties of aqueous or nonaqueous buffers and antifade reagents that work with fluorescent samples (38, 39) and that can be purchased commercially. Antifade reagents which are designed to work well with confocal microscopic imaging (e.g., SlowFade, Molecular Probes, Eugene, OR) and which do not quench the applied flurochrome are ideal candidates for mounting immunostained coverslips. For most indirect immunofluorescent techniques using commonly used flurochromes, a nonpermanent mounting medium is sufficient to retard photobleaching and preserve samples for several weeks.

Immunostained specimens should be viewed and imaged first by conventional epifluorescence using quality microscopes with high numerical aperture objectives before confocal laser scanning microscopy (50). Best results are obtained if imaging is performed as soon as possible after mounting, securing a permanent record of the data before fading or any redistribution of the fluorochromes occurs. If film is to be used, we recommend fast-speed black-and-white professional films (e.g., Kodak Tri-X 400), which can be push-processed in high contrast developer to obtain publication-quality photographs. This will help minimize potential bleaching problems by reducing exposure of immunostained samples to UV excitation. Color films of ASA 160 or greater are also useful to obtain presentation-quality slides. Digital data can be recorded using any number of chilled charge-coupled device (CCD) scientific cameras interfaced to the microscope and a computer system. The highest resolution is usually obtained using black-and-white CCD chip cameras. Most important, digital data needs to be archived using erasable magneto optical, Zip, or Jazz discs or recordable CDs. Digital data can then be downloaded to film recorders or printers for the preparation of slides or plates for publication.

After conventional microscopy has been applied, specimens can be subjected to confocal microscopy for imaging the three-dimensional volume of gametes. There are different types of confocal microscopes that can be useful for imaging fixed biological samples (31). A slow-scan CLSM equipped with 60× Plan-Apo, 1.4 numerical aperture oil-immersion objective and a krypton-argon ion laser works well for imaging fluorescein- and rhodamine-immunostained gametes, but other lasers are needed for detecting stained DNA. Specific setting of the confocal microscope will vary with the application, but whatever the CLSM, try to limit laser power for sample exposure and use neutral density filters to attenuate the beam further to minimize sample deterioration.

3. PROTOCOLS FOR INDIRECT IMMUNOFLUORESCENCE AND CONFOCAL IMAGING ON FIXED MAMMALIAN GAMETES

We consider two fixation methods used for the detection of microtubules and DNA in mammalian gametes, although adoption of other probes or antibodies can be easily substituted. Examples of the data collected by either fixation method are presented (in Protocols 10.1–10.4).

4. APPLICATIONS OF CONFOCAL FLUORESCENT MICROSCOPY IN MAMMALIAN GAMETES

Motility and cytoskeletal rearrangements are essential for successful completion of mammalian fertilization (9). Our understanding of cytoskeletal organization during spermatogenesis, oocyte maturation, fertilization, and development have benefitted tremendously from applying static immunocytochemistry technology and confocal microscopy. These investigations led to important discoveries about the basic features that prime and position the parental genomes for merging during fertilization. Among the discoveries is the observation that the restoration of the centrosome, the cell's microtubule organizing center, is accomplished at fertilization by the attraction of a number of maternal centrosomal components to the sperm centriole introduced at insemination. The resultant zygotic centrosome organizes the sperm aster structure at the base of the sperm head that ultimately directs the apposition of the male and female pronuclei within the activated oocytes cytoplasm. The sperm centriole is highly favored for the completion of fertilization in most mammals, as the maternal centrosomes found at the meiotic spindle poles do not appear to reproduce within the oocyte's cytoplasm (52, 53). With the exception of rodents, which do not inherit an active sperm centriole during fertilization (21, 23, 54), every mammalian species thus far examined, including primates, domestic species, and even evolutionary primitive marsupials, inherit their centrosomes from their fathers (9).

The pattern of centrosomal inheritance during human fertilization is presented in Figure 10.1 following antitubulin immunostaining and confocal microscopy (15). Microtubules are only present in the metaphase-arrested second meiotic spindle in the unfertilized oocyte (Figure 10.1A). Within 6 h after insemination, a small microtubule sperm aster emanates from the base of the sperm head (Figure 10.1B, C). The activated oocyte extrudes the second polar body, shown in Figure 10.1 attached to the developing female pronucleus by the midbody structure (Figure 10.1B, C). Sperm astral microtubules continue to elongate throughout the cytoplasm during early development (Figure 10.1D, E). Some of the sperm aster microtubules make contact with the decondensing maternal pronucleus, initiating pronuclear migration (Figure 10.1E, F). The zygotic centrosome duplicates and splits apart during late interphase, as microtubules emanate from between the eccentrically positioned, juxtaposed male and female pronuclei (Figure 10.1F). Mitotic prophase commences with the separate chromosomal condensation of the male and female pronuclei, as the zygotic centrosomes nucleates the microtubules of the bipolar mitotic spindle apparatus (Figure 10.1G). By late prometaphase, the chromosomes intermix and align along the equator of the bipolar, anastral mitotic spindle (Figure 10.1H), completing the fertilization process in humans.

These results have been further supported by observations of polyspermy, parthenogenesis, and naturally occurring defects in the sperm centrosome of some men, which lead to fertilization failures (15, 55–57). In the latter case, confocal microscopy was instrumental in demonstrating how defective sperm centrosomes sometimes start microtubule nucleation, but subsequently fail to continue microtubule growth. Other examples observed by confocal microscopy included sperm centrosomes prematurely

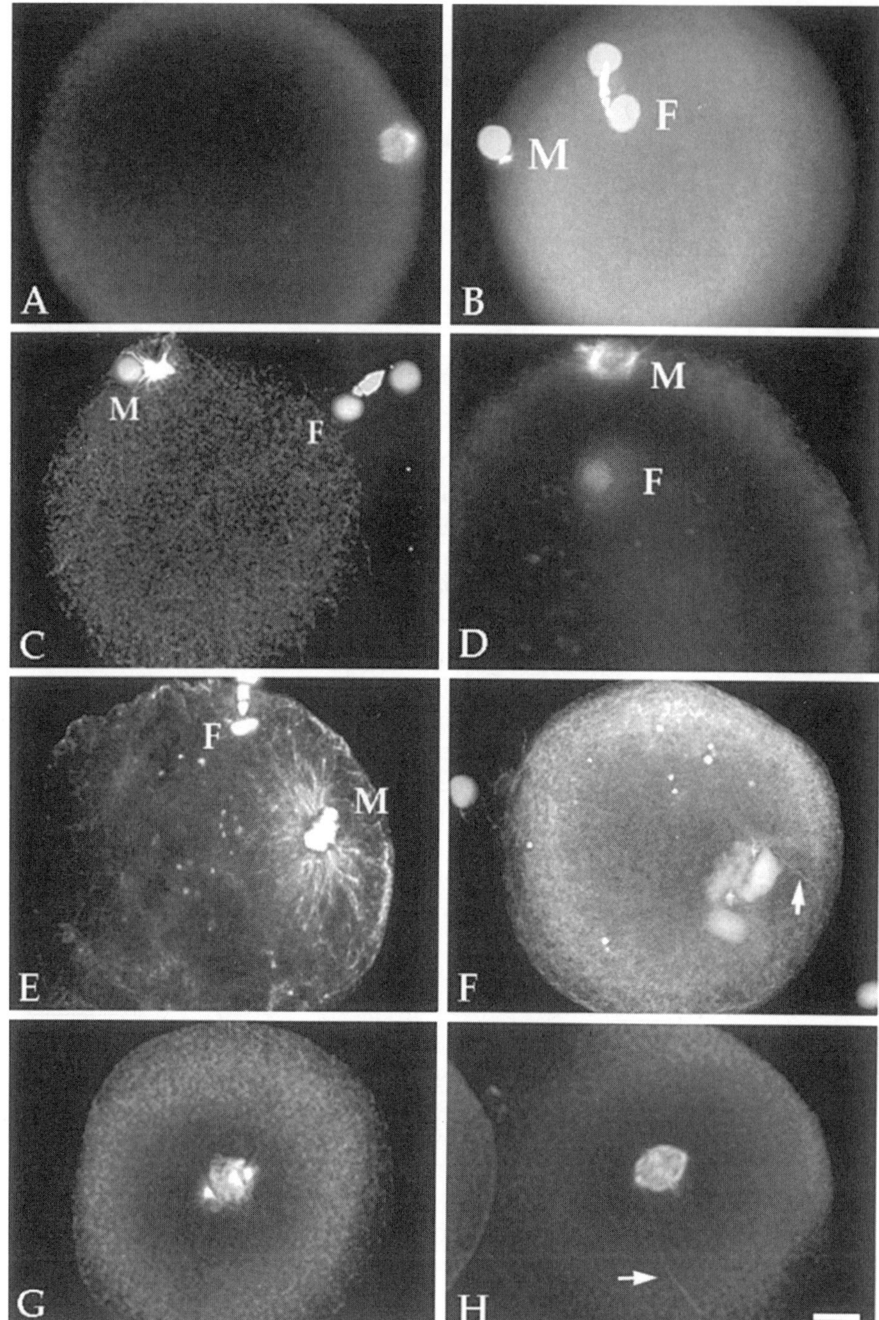

Figure 10.1. Microtubule organization and DNA patterns in normal inseminated human oocytes. (A) The meiotic spindle in mature, unfertilized human oocytes is anastral, oriented radially to the cell surface, and asymmetric, with a focused pole abutting the cortex and a broader pole facing the cytoplasm. No other microtubules are detected in the cytoplasm of the unfertilized human oocyte. (B–D) Shortly after sperm incorporation (3–6.5 h postinsemination), sperm astral microtubules assemble around the base of the sperm head, as the inseminated oocytes complete second meiosis and extrude the second polar body (M = male pronucleus; F = female pronucleus). The close association of the meiotic midbody identifies the female pronucleus. Short, sparse disarrayed cytoplasmic microtubules can also be observed in the cytoplasm upon confocal microscopic observations of these early activated oocytes (C). (E) As the male pronucleus continues to decondense in the cytoplasm (M), the microtubules of the sperm aster enlarge, circumscribing the male pronucleus. (F) By 15 h postinsemination, the centrosome splits and organizes a bipolar microtubule array that emanates from the tightly apposed pronuclei. The sperm tail is associated with an aster (arrow). (G) At first mitotic prophase (16.5 h postinsemination), the male and female chromosomes condense separately as a bipolar array of microtubules marks the developing first mitotic spindle poles. (H) By prometaphase, when the chromosomes begin to intermix on the metaphase equator, a barrel-shaped, anastral spindle forms in the cytoplasm. The sperm axoneme remains associated with a small aster found at one of the spindle poles (arrow). Bars = 10 μm. Modified with permission from Simerly et al. (15).

detaching from the sperm head or organizing a large, disorganized sperm aster defective in initiating pronuclear migration. Collectively, these results suggest that fertilization arrest may occur at several points in the cell cycle as a result of improper centrosomal functioning. These types of defects indicate male infertility factors because excess oocytes from some of the same patients were successfully fertilized with donor sperm, though not with spousal sperm.

Small cells like spermatocytes and spermatids can present significant problems for the clear resolution of intracellular structures despite the marvelous detection probes available and improvements in conventional imaging techniques. Confocal microscopic analysis has facilitated characterization of the microtubule and DNA patterns in mammalian spermatids and their implications for nuclear shaping and organelle movement. Unlike images of tissue slices derived from cryopreserved specimens (58–60), confocal microscopy has the ability to produce high resolution images and informative three-dimensional renderings of stacks of optical sections for the detection of the entire microtubule arrays (18). As shown in Figures 10.2 and 10.3, bovine spermatids contain short microtubules running at the postacrosomal domain near the nuclear surface (arrows). These microtubules are not nucleated from a traditional microtubule organizing center (MTOC) present in the spermatid cytoplasm (21) but appear to originate from the cortical region, as observed for many other types of polarized cell types (61, 62). Confocal microscopy has also identified post-translationally modified forms of microtubules during spermatogenesis (18, 58, 63, 64). Spermatocytes and spermatids both contain acetylated α-tubulin at the cell cortex as well as in the axial fibers destined to form the mature sperm tail (18) (Figure 10.4). These microtubules have been correlated with microtubule stability and are not observed in the manchette or other microtubule-based structures at other differentiation stages. The array of acetylated α-tubulin is best observed after confocal microscopic three-dimensional reconstruction of spermatids at steps 3–6 (Figure 10.5), where the developing axial filament is seen engulfing the nucleus around a centrosomelike area. Conversely, glutamylated tubulin is first observed in step 7 spermatids bundled around the nucleus opposite of the acrosome but is then found restricted to the manchette microtubules and sperm tail in later stages of spemiogenesis (58, 64). These major microtubule reordering processes may be important for changes observed in organelle translocation and nuclear-shape changes observed in the latter stages of spermiogenesis (19, 20, 65, 66). For example, the Golgi apparatus has been shown to move away from the acrosome toward the opposite pole of the cell, perhaps by a mechanism involving microtubules and the motor proteins kinesin and dynein (67–69). Eventually, this structure is shed in the cytoplasmic droplet along with most other spermatid organelles (70).

Fundamental research in relevant animal models is now permitting investigations of the cellular and molecular events underlying commonly used assisted reproductive techniques such as ICSI. Despite the clinical successes of ICSI, discoveries in animal research models suggest that this method of fertilization differs from classical fertilization and from in vitro fertilization techniques (1). For instance, certain sperm proteins, removed at the egg surface during insemination, are retained after ICSI, and sperm DNA decondensation occurs asynchronously (25, 26). Sperm components like the acrosome, perinuclear theca, and SNAREs (soluble *N*-ethylmalameide-sensitive factor attachment protein receptors) are all introduced into the egg after ICSI, and it is unclear whether the retention of these proteins can have a detrimental role in transformation of the sperm nucleus into a functional male pronucleus (24–26, 71–73). As shown in Figure 10.6, confocal imaging of an antibody that recognizes vesicle-associated membrane protein (VAMP), a type of vesicular carrier SNARE (v-SNARE) found in sperm (25) is retained as a collar in the sperm equatorial segment after ICSI (Figure 10.6A). This VAMP-containing region separates the decondensing DNA at the posterior end of the sperm head from the condensed anterior (acrosomal) portion (Figure 10.6B–E). Although this asynchronous decondensation of the male DNA may eventually be overcome (Figure 10.6E, F), it has been suggested that the higher rate of sex chromosome disorders in embryos and fetuses conceived using

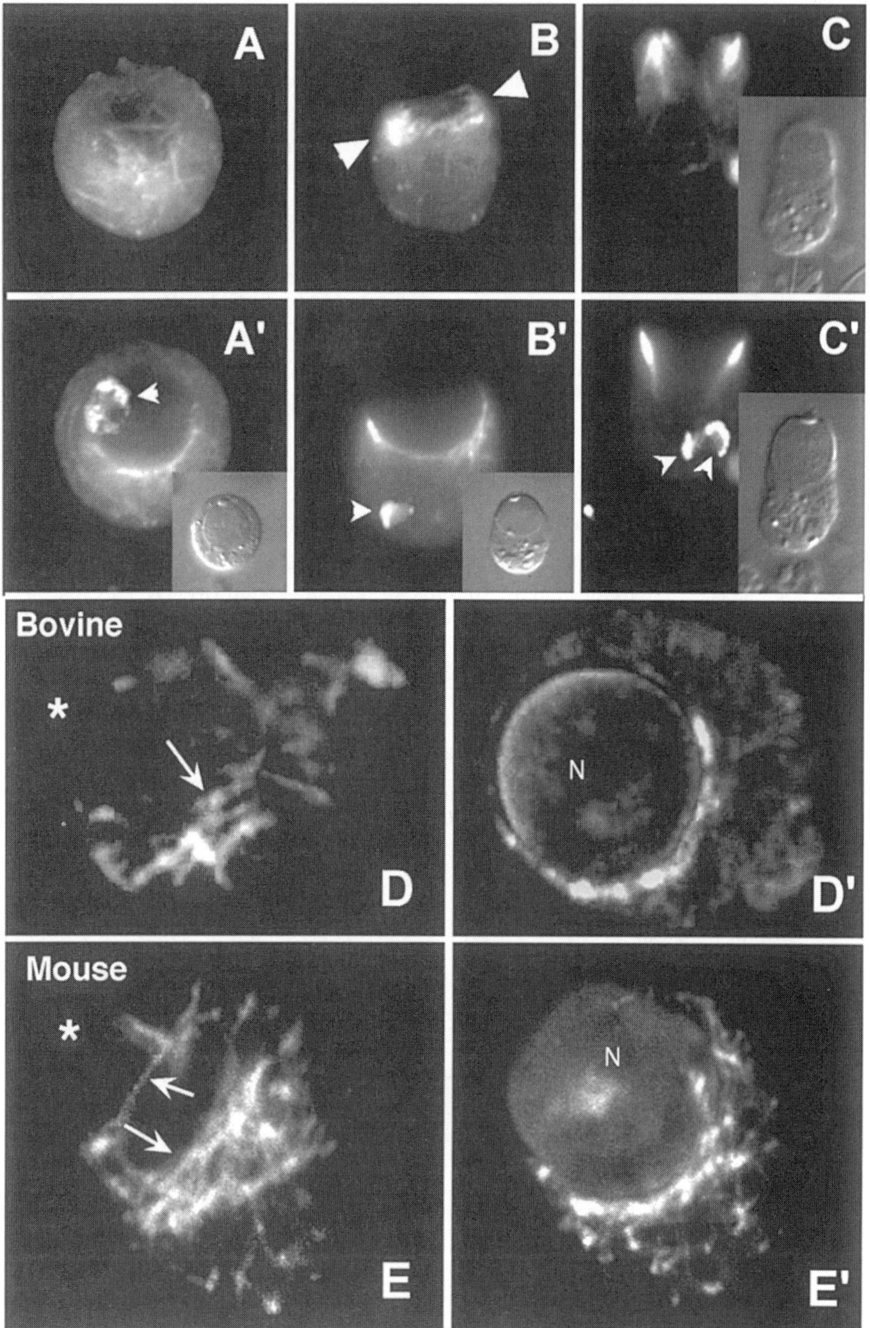

Figure 10.2. Microtubule dynamics and manchette formation in elongating spermatids. (A) Early elongating spermatids in stage 7 have a cortical microtubule network without any apparent focal point, as described for the earlier stage. (A') In a deeper focal plane it is possible to observe an increase of tubulin on the nuclear surface starting at the edges of the acrosomal vesicle (arrow). The Golgi apparatus has moved from the acrosomal area. (B) At stage 8, a bundle of microtubules oriented perpendicular to the main axis of the spermatids runs around the postacrosomal region of the nucleus (arrowheads). (B') The same cell in another focal plane shows that the nucleus still has a strong label on the postacrosomal region all the way in between the rims of the acrosome. The Golgi apparatus sits in the cytoplasmic lobe without contact in any microtubular structure (arrowheads). (C, C') At later stages of elongation, the clear manchette appears surrounding the postacrosomal region of the spermatid. The microtubules run parallel to the main axis of the cell and there are a few microtubules in the cytoplasmic lobe (arrowheads). (D, D') Laser scanning confocal microscopy of bovine and mouse (E, E') spermatids at stages 7 of differentiation. (D, E) Spermatids display a bundle of cortical microtubules running around the postacrosomal domain (arrows). The asterisks indicate the position of the acrosome in each cell. They do not show any obvious microtubule organizer center either at the cortex or inside the cell (D', E'). A number of microtubules run into the cytoplasm surrounding the postacrosomal domain of the nucleus of bovine (D') and mouse (E'). The insets show the differential interference contrast picture of each cell. Modified with permission from Moreno and Schatten (18).

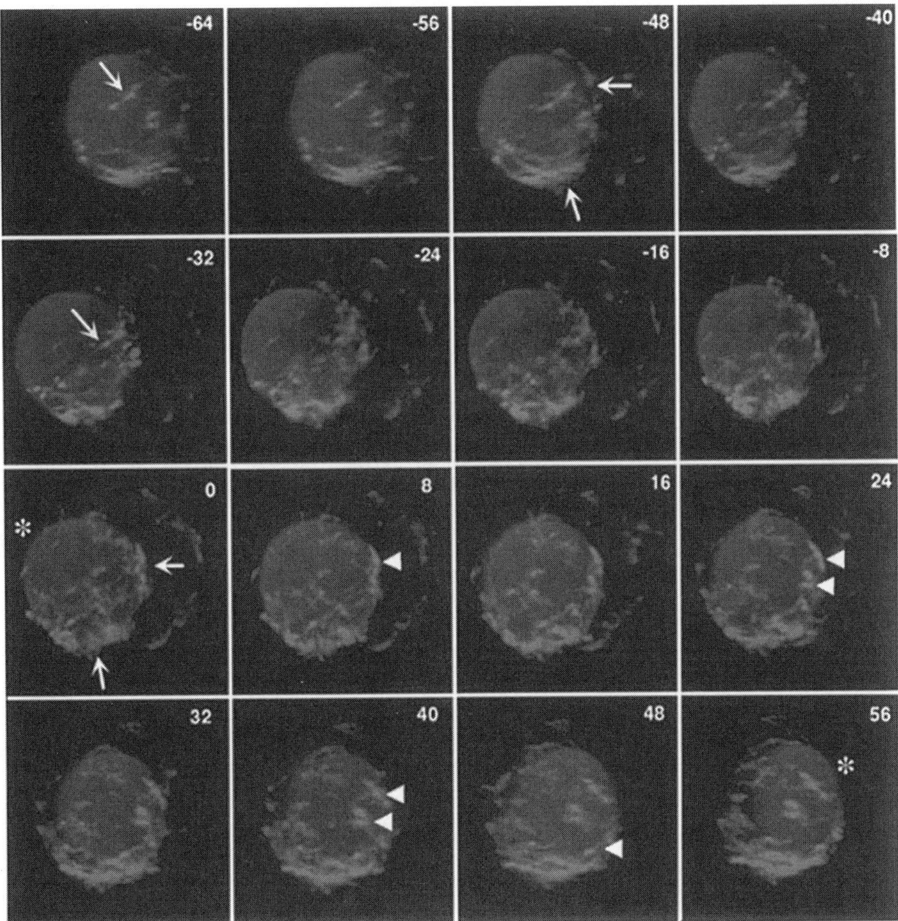

Figure 10.3. Three-dimensional reconstruction of microtubule configuration of the round spermatid. Shown here is a three-dimensional reconstruction of α-tubulin of a bovine round spermatic in stage 6. These microtubules cover the entire postacrosomal domain of the nucleus (arrows). Short microtubules run around the nucleus perpendicular to the main axis of the cell (arrow). These short microtubules are in close contact with the postacrosomal domain of the nuclear envelope (arrowheads). The acrosomal domain (asterisks) contains a few microtubules around the acrosome. The number indicates the angle of rotation left to right. Modified with permission from Moreno and Schatten (18).

ICSI may be related to the abnormal sperm remodeling during early fertilization (26). Support for this notion comes from studies demonstrating that the X and Y chromosomes reside principally in the apical portion of the sperm head (74, 75). Furthermore, the nuclear importation of cytoplasmic proteins such as nuclear mitotic apparatus protein is atypical, being initially excluded from the still-condensed regions of paternal chromatin, and DNA synthesis onset is delayed in sperm that undergo asynchronous decondensation (26, 75). Variations in the dynamics of fertilization after ICSI might well lead to a diminished ability of the oocyte to express or be exposed to important paternal genes or gene products, thus leading to improper embryo formation.

5. TROUBLESHOOTING AND COMMON PROBLEMS ASSOCIATED WITH IMAGING

In this section we consider some of the more prevalent problems associated with applying ICC to mammalian gametes.

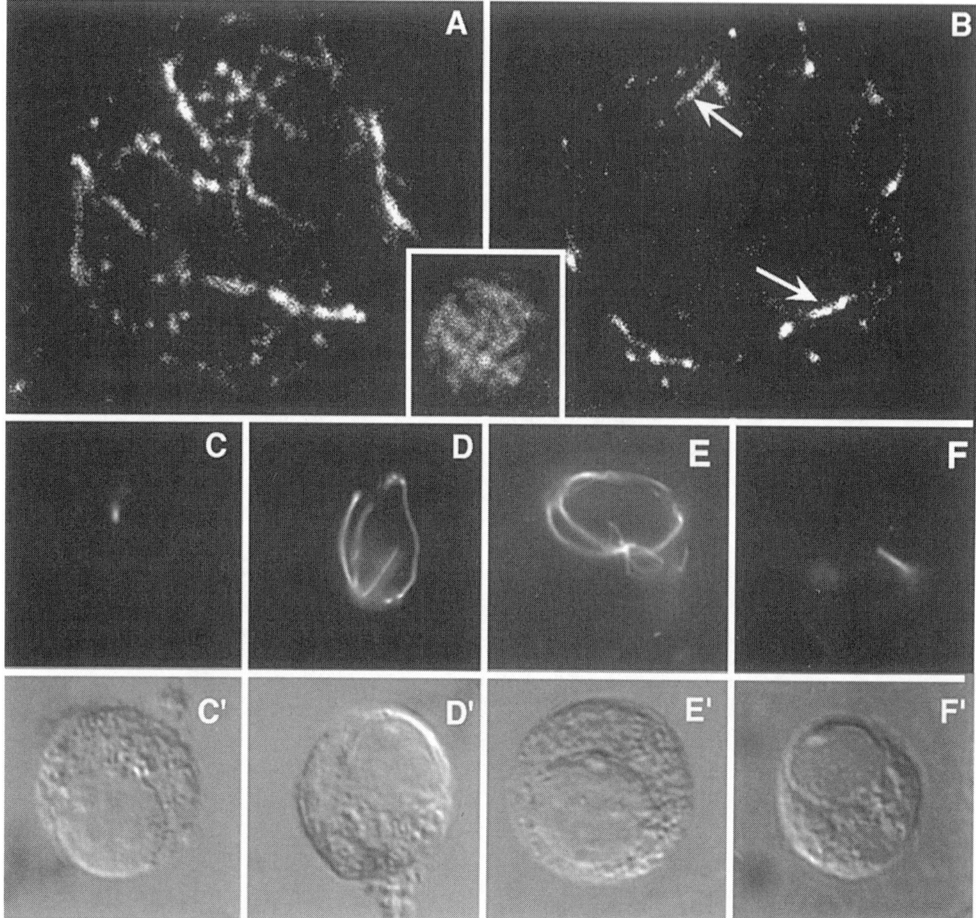

Figure 10.4. Acetylated microtubule dynamics during spermiogenesis. (A) Pachytene spermatocytes contain a subset of acetylated microtubules at the cell cortex, as evaluated by laser scanning confocal microscopy. (B) In a deeper focal plane it is possible to see that some of the microtubules penetrate into the cytoplasm (arrows). The inset shows the chromatin structure of the cell. (C) In stage 1–3 round spermatids, acetylated tubulin appears as a spot near the acrosomal vesicle (C'). (D, F) In later stages the acetylated configuration is like a wire wrapping around the nucleus (D', F'). This wirelike structure may correspond to the axial filament because at later stages it is found attached to the spermatid nucleus (D, D'). Later on, in spermatids at stage 6 (E), single microtubules form an asterlike configuration. These microtubules disappear at stage 7, and only the sperm tail contains acetylated tubulin (F, F'). Modified with permission from Moreno and Schatten (18).

5.1 Attachment of Oocytes to Poly-L-lysine-coated Coverslips

A technical problem often encountered in processing mammalian gametes for ICC is adhering oocytes or spermatogenic cells to substrates. For oocytes, both the cumulus and zona must be completely removed to get good adherence to polylysine-coated coverslips. If using hyaluronidase to remove the cumulus, the treatment should be kept short (5–7 min) to prevent potential parthenogenetic activation of metaphase-arrested oocytes. Incubation times with enzymes or acidified culture medium to remove the zona should be kept to a minimum to reduce intracellular damage. Recovery periods of 30 min or longer in normal culture medium should be permitted after zona removal and before fixation. Simultaneous removal of adhering cumulus cells and the zona pellucida is not recommended to prevent the accessory cells from sticking to the oocyte surface and interfering with imaging. The use of high molecular-weight polylysine (mw > 300,000)

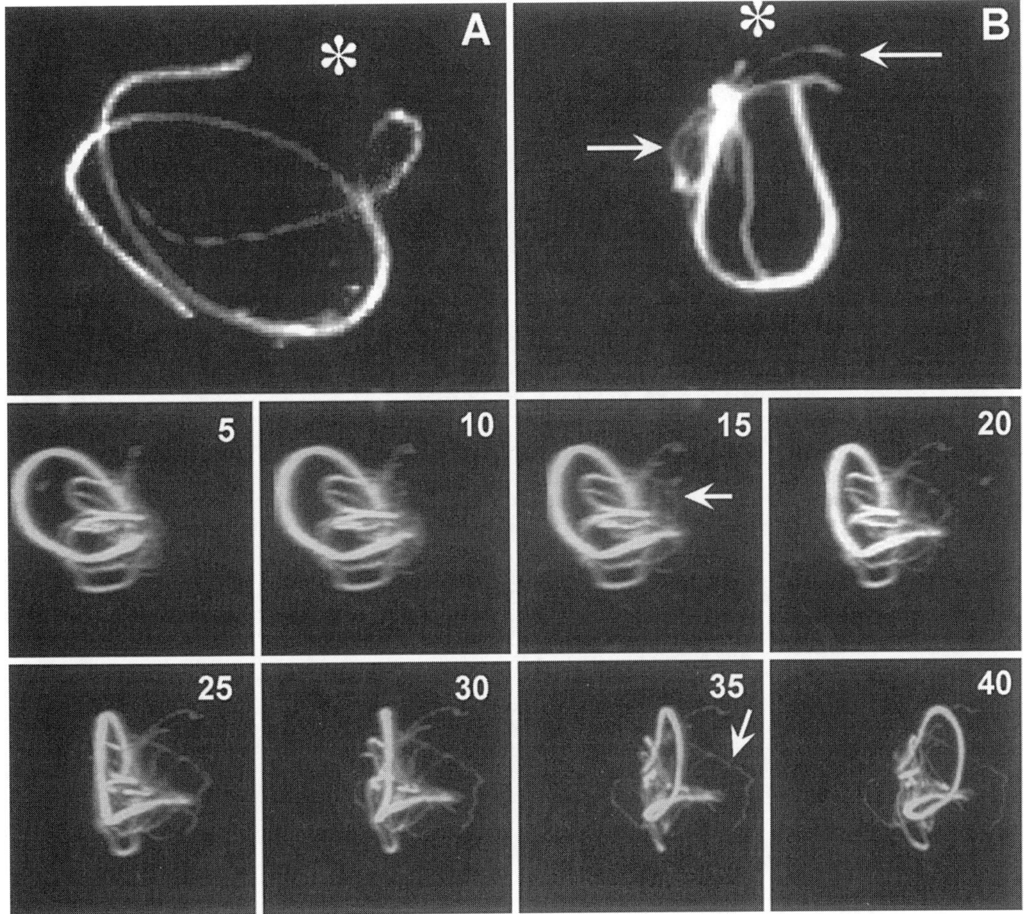

Figure 10.5. Acetylated α-tubulin configurations in round spermatids. (A) A laser scanning confocal microscopy observation of acetylated tubulin in stage 3–4 shows that only the axial filament contains this tubulin isoform. (B) At stage 6 there are several acetylated microtubules converging to a focal point near one end of the axial filament. The asterisks indicate the position of the acrosome in each cell. The figure shows a three-dimensional reconstruction of acetylated microtubules of a round spermatid in stage 5–6. The main labeled structure is the axial filament (arrow). The rotation of the structure allows identification of microtubules (arrowhead) concentrated around a centrosomelike area. The number indicates the angle of rotation in between each frame. Modified with permission from Moreno and Schatten (18).

and clean coverslips is crucial to facilitating good polylysine coating to the glass surface and gamete adherence. If necessary, coverslips can be cleaned in 95% alcohol and the surfaces wiped clean with a lint-free cloth before applying the polylysine solution. Also, good gamete attachment will only occur if the culture medium or PBS solution does not contain proteins such as BSA or serum. First rinse gametes in protein-free culture medium in an agar-coated dish (to prevent zona-free oocytes from sticking to the plastic surface) immediately before transferring them to the polylysine-coated coverslip. Agar dishes can be prepared by adding 2.5 ml of a 1% bacto-agar solution dissolved in distilled water to a small 10 × 35 mm Petri dish. Plates can be prepared in advance and stored at 4°C for up to 1 week.

5.2 Fixation Protocols

Fixation techniques should avoid steps that can alter the distribution of the proteins in vivo (76). For instance, many soluble membrane proteins will not be retained or can

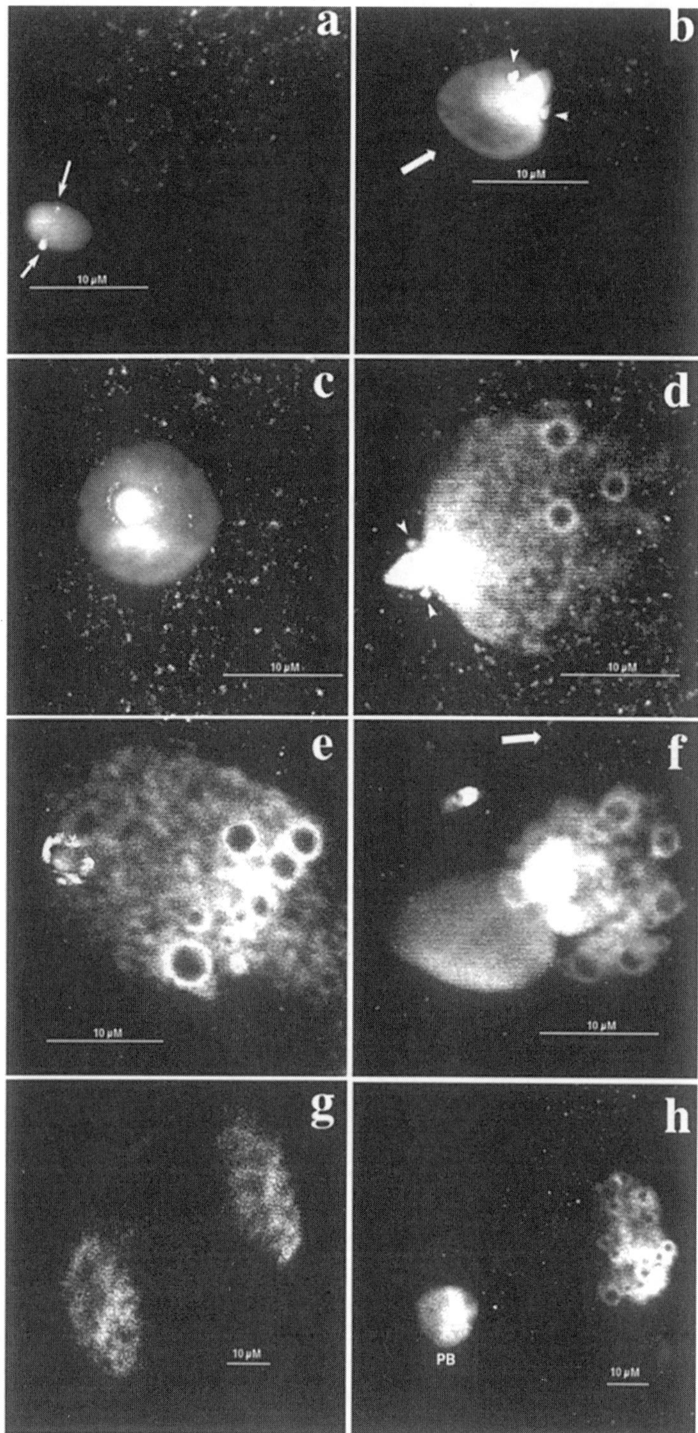

Figure 10.6. Confocal imaging of the persistence of vesicle-associated membrane protein (VAMP) on the sperm head after ICSI in the rhesus monkey. (a) Sperm with a typical VAMP (arrows) pattern (30 min postinjection). (b–d) DNA decondensation at the anterior portion of the sperm head seems to be prevented or retarded by a VAMP two-piece "collar" 4 h postinjection (b and d, side views; c, top view; arrowheads). Eventually the male DNA decondenses completely, forming a pronucleus with VAMP remnants in the vicinity (e, 8 h postinjection), sometimes persisting up to the stage of pronuclear apposition (f, 20 h postinjection). VAMP became undetectable by first mitosis (g, first mitotic anaphase). Arrows denote the presence of the sperm tail as observed by phase-contrast microscopy and Mitotracker labeling. Nucleoli are detected as dark regions within the labeled male chromatin. VAMP is not associated with the female pronuclei nor with the polar body (h, PB, polar body). Bars = 10 μm. Reprinted with permission from Ramalho-Santos et al. (25).

be redistributed intracellularly if detergents are used before fixation or if organic solvents are used for cell preservation. Likewise, glutaraldehyde often deleteriously affects the intracellular antigenicity of proteins and/or increases nonspecific fluorescent background in the wavelengths commonly used for ICC labeling, giving rise to false-positive or false-negative staining profiles. Several observations are worth noting when using buffer M detergent extraction before fixation. Permeabilization will not be complete unless the zona is removed from the oocytes. Also, low concentrations of detergent (< 0.1%) in buffer M can lead to insufficient extraction and cell lysis, perhaps as a result of a differential osmolarity gradient between the external buffer and the egg cytoplasm. It may be desirable to have low extracellular calcium present during the permeabilization step to retard calcium-induced alterations in the cytoskeleton, especially in the preservation of microtubules. Attachment of oocytes to polylysine-coated coverslips before detergent permeabilization is useful for oocyte recovery and to minimize structural damage to the extracted cells. For reattaching detergent-permeabilized oocytes to new polylysine-coated coverslips after buffer M treatment, take the following steps to help ensure that oocytes remain attached: (1) prewash the new coverslip in PBS–Triton X-100 to assist the process of oocyte reattachment (this probably removes electrostatic repelling characteristics of the glass coverslip and the glycerol-based buffer); (2) keep the volume of the buffer transferred to a minimum so oocytes do not float away from the coverslip surface; and (3) do not expel the oocytes and the fluid all at once at the center of the coverslip, or the cells may not stick. Gently sweep the pipette across the surface of the coverslip to get the best reattachment. When using cold organic solvents like methanol to fix the cells, add slowly, or detachment and/or damage to the extracted eggs can occur. It is helpful to perform this step using a dissecting microscope set at low magnification. Finally, do not allow detergent-extracted cells to become exposed to the air–liquid interface after attachment to polylsine and all subsequent steps.

Other cross-linking reagents like dimethyl 3,3'dithiopropionimidate dihydrochloride (DTBP) and ethylene glycol bis (succinimidyl succinate) (EGS) can be used in situations where antigenicity is compromised by organic solvents or by aldehyde fixation. Be aware that nonionic detergents such as Triton X-100 can reverse the cross-linking property of formaldehydes, resulting in poorly preserved oocytes (38).

5.3 Immunostaining Protocols

All antibodies should be diluted into an appropriate aqueous buffer solution (i.e., PBS or TBS) containing a protein carrier like BSA (3 mg/ml) or an immunoglobulin (10%) derived from the same species as the detection reagent (e.g., normal goat, human, or rabbit serum). It is also advisable to briefly spin each diluted antibody solution in a microcentrifuge for 2 min to remove insoluble particulates.

The length of antibody staining can be varied. For instance, it is possible to apply the diluted primary antibody overnight at 4°C. However, do not shorten the PBS wash steps between application of the primary and secondary antibodies to ensure lower background staining. The use of a humidified chamber for immunostaining will prevent antibody evaporation.

5.4 Mounting Immunostained Cells for Fluorescent Observation

Specimens are placed in mounting media to prevent dehydration and fluorescent signal fading during storage and imaging. If three-dimensional data are to be collected by confocal microscopy, it may be useful to mount coverslips using spacers between the coverslip and slide before mounting. Spacers can be either broken strips of a coverslip (type 1) or a small amount of a gel substance (e.g., silicone grease or petroleum jelly) added to corners of the coverslips containing the specimen. After mounting and sealing, store slides in the dark at 4°C in a light-tight box. Specimens mounted and sealed properly will be stable for several months.

5.5 Controls

The localization of intracellular antigens by any immunofluorescence method must be confirmed by careful scrutiny against a number of controls (36). Staining patterns observed after replacing the primary antibody with preimmune sera, after preabsorption of primary antibody with excess antigen, or after staining with secondary antibody alone should be routinely performed. Similar dilutions of each control solution should be used for comparison with the staining patterns obtained with the primary antibody. Also, image exposure and acquisition times should be similar for controls and antibody-labeled specimens when photographing during microscopy.

6. CONCLUSION

In the past, the study of cellular structure and function of mammalian gametes has benefited tremendously from light and electron microscopy. Although each of these microscopy techniques has its advantages, there are limitations to deriving complete information. Transmission electron microscopy produces superior resolution of ultrastructural components, but it is generally limited to two-dimensional, static imaging on fixed specimens. Conventional light microscopy, especially in combination with fluorescent probes, offers a superb means to examine living specimens dynamically and with a global view of the cellular organization. However, image quality by conventional microscopy can be limited by out-of-focus fluorescent information, especially on thick specimens such as oocytes. Recent advances in image processing and computer deconvolution programs (77, 78) have greatly improved conventional microscopy. In particular, confocal microscopy has proven to be a powerful method for exploring cellular structure and function in mammalian gametes, and it nicely complements light and electron microscopy. The application of confocal microscopy is sufficiently advanced to permit the analysis of both living and fixed specimens labeled with multiple probes. Furthermore, the ability to do rapid optical sectioning limits artifacts introduced by physical sectioning while practically eliminating out-of-focus information, resulting in improved contrast and effective resolution of cellular structures. Finally, the ability to take stacks of optical sections derived from images collected at different focal planes and to determine the three-dimensional organization of intracellular components has yielded enormous information in structure and organization. With the establishment of analysis software, powerful computers, and fantastic probes for detecting intracellular structures, the capability of doing three-dimensional imaging over time (i.e., four-dimensional imaging) is now possible. Emerging data from confocal studies of mammalian gametes are providing exciting and fundamental discoveries in cell, developmental, and reproductive biology, permitting the modern microscopist to explore subtle relationships between cellular structure and function.

Acknowledgments   We are indebted to our many wonderful colleagues and collaborators, including R. Balczon, J. Holy, E. Jacoby, C. Martinovich, M. Miller, K. Mueller, C. Payne, P. Sutovsky, D. Takahashi, and S. Wright. We thank Michelle Emme for editorial assistance. This research was supported by NICHD/NIH through a cooperative agreement (U54 18185) as part of the Specialized Cooperative Centers Program in Reproduction Research. Other support by the National Institutes of Heath and the Mellon Foundation is grateful acknowledged. J.R.-S. is a recipient of a Praxis XXI postdoctoral fellowship from Fundação para a Ciência e Tecnolgia (Portugal) and received additional support from Fundação Luso-Americana para o Desenvolvimento (Portugal). Studies on clinically discarded human oocytes were supported by nonfederal sources. We thank Ares Serono, Inc. (Randolph, Massachusetts) and Organon, Inc. (West Orange, New Jersey) for donating hormones used in monkey stimulations. All animal procedures were approved by the ORPRC, Institutional Animal Care and Use Committee, and the use of clinically discarded human oocytes was approved by Human Subjects Institution Review Board.

Protocol 10.1. Attaching oocytes or spermatids to polylysine-coated coverslips

1. Prepare 2 mg/ml poly-L-lysine (PL) hydrobromide (mw > 300,000; Sigma Chemical Co., St. Louis, MO) in distilled water, aliquot, and stored frozen at –20°C.
2. Add 50 μl of the PL stock solution to a 18–22 mm^2 coverslip for 5 min. The PL solution should not form a bead on the coverslip but should smoothly coat the entire surface.
3. After 5 min, rinse the coverslip in dH$_2$O to remove excess PL and then allow it to air dry.
4. Place PL-coated coverslips in a 6-well, flat-bottom assay plate (Falcon #3046).
5. Add 5 ml of calcium-free and protein-free culture medium at 37°C and keep warm on a slide warmer.
6. Carefully pipette washed zona pellucida-free eggs or isolated spermatocytes onto PL-coated coverslips in the assay plate; incubate without moving the coverslips at 37°C for 2–3 min to permit strong adherence.

Protocol 10.2. Fixation techniques for mammalian gametes

Oocyte fixation in methanol following permeabilization in a microtubule stabilization buffer

1. Remove most of the culture medium from the 6-well, flat-bottom assay plate containing the PL-attached oocytes but do not expose the attached eggs to the liquid–air interface.
2. Carefully pipette into plate approximately 6 ml buffer M containing 1% Triton X-100, 1 mM 2-mercaptoethanol, and 0.2 mM PMSF (pH 6.8, 37°C). Incubate 10 min at 37°C.
3. Prepare a new PL-coated coverslip; place in a 4-well quad-style petri dish (Falcon #1009) and add PBS containing 0.1% Triton X-100 (PBS-TX) for 1 min. Aspirate the coverslips washed in PBS-TX until completely dry.
4. Using a flame-drawn pipette slightly larger than diameter of the extracted eggs, transfer the permeabilized oocytes to the new PL-coated coverslip; transfer minimum amount of extraction buffer with the oocytes. Continually move pipette across glass surface as oocytes are gently expelled, but do not introduce air bubbles. Ideally, this step should be performed with a dissecting scope equipped with darkfield optics and set at low zoom power.
5. Immediately add 3 ml warm buffer M without detergent or methanol additives. Do not allow the attached, permeabilized oocytes to air dry.
6. Slowly add –10°C absolute methanol into buffer M until saturation. Incubate 10 min at room temperature.
7. Remove methanol by rinsing in PBS-TX three times.
8. Permeabilized, fixed oocytes can be stored overnight in PBS-TX before starting immunostaining procedure.

Oocyte or spermatogenic cell fixation using formaldehyde

1. Attach zona-denuded oocytes or spermatogenic cells to PL-coated coverslips in a 6-well, flat-bottom assay plate and incubate at 37°C for 2–3 min.
2. Remove most of the culture medium from plate but do not expose attached gametes to the liquid–air interface.
3. Carefully pipette into the assay plate 6 ml of fresh 2% formaldehyde (EM grade; methanol free; Polysciences, Inc., Warrington, Pennsylvania) prepared in protein-free culture medium (see below) or PBS at 37°C for 1–24 h to fix sample.
4. After fixation, rinse in PBS-TX three times, 5 min between exchanges.
5. Permeabilize/extract fixed specimens in PBS containing 1.0% Triton X-100 for 40 min.

6. Incubate specimens in PBS containing 150 mM glycine and 3 mg/ml BSA (blocking solution) for 30 min.
7. Rinse in PBS-TX three times, 5 min between exchanges.
8. Samples can be stored for 24 h before immunostaining as described below.

Buffer M constituents

50 mM imidazole·HCl, pH 6.8
50 mM KCl
0.5 mM $MgCl_2$
0.1 mM EDTA
1 mM EGTA
25% (v/v) glycerol

Before use, add 1mM ß-mercaptoethanol, 0.2 mM phenylmethylsulfonyl fluoride (PMSF; prepared as a 0.1 M stock in absolute methanol and stored at –20°C) and 1–3% nonionic detergent (e.g., Triton X-100 or Nonidet P-40).

PBS with detergent

137 mM NaCl
2.7 mM KCl
10.1 mM Na_2HPO_4
1.8 mM KH_2PO_4

Dissolve salt components in distilled water and adjust pH to 7.2. Add 0.1–1% Triton X-100 from a 10% stock solution of detergent in distilled water. Filter-sterilize and store at 4°C.

Calcium-free/protein-free TL-HEPES culture medium

Component[a]	Concentration
Penicillin G	10,000 U/100 ml
phenol red	1 mg/100 ml
HEPES (Na salt)	5.0 mM
NaCl	127.0 mM
KCl	3.16 mM
$MgCl_2 \cdot 6H_2O$	0.5 mM
Na lactate[b]	[0.185 ml per 100 mls]
$NaH_2PO4 \cdot H_2O$[c]	0.35
Glucose	5.0
$NaHCO_3$	2.0

Modified from Boatman (80).

[a]Final osmotic pressure should be between 280–290 mOsmol. Sterilize by filtration and store for maximum of 1 week. Warm to 37°C before use.
[b]Sodium lactate is 60% syrup.
[c]Sodium dihydrogen phosphate should be added slowly to prevent formation of precipitates in the medium.

Protocol 10.3. Immunostaining

1. Blot excess PBS-TX off coverslip by touching the edge of the coverslip with a Kimwipe.
2. Add 90 μl of the antitubulin antibody (i.e., E7 culture supernatant; 1:5 in PBS containing 5% normal goat serum; Developmental Studies Hybridoma Bank, Iowa City, IA) to the surface of the coverslip placed on the quad-plate divider of the 4-well petri dish. Add a small amount of PBS to each petri well to humidify the chamber and incubate at 37°C for 60 min.

3. Gently decant off the primary antibody. Place the coverslip in the bottom well of the quad dish and add about 6 mls of PBS-TX. Incubate at room temp for 10 min, then exchange twice more with fresh PBS-TX, each for 10 min (30 min total).
4. Blot off excess rinse solution; add 90 μl of fluorescein-conjugated goat anti-mouse IgG secondary antibody. Incubate at 37°C for 60 min in the quad plate humidified chamber.
5. Rinse in PBS-TX solution as above.
6. Remove the rinse solution; add 90 μl of 2.5 μg/ ml Hoechst 33342 or DAPI to detect DNA (5 min, room temperature).
7. Blot off excess DNA stain; repeat the rinse step in PBS-TX.

Protocol 10.4. Mounting immunostained coverslips for fluorescent microscopy

1. Rinse immunostained coverslips once in dH$_2$O to remove salt crystallization on surface of coverslip.
2. Using fine forceps, blot excess water off the coverslip by touching a Kimwipe to the edge of the glass, invert, and carefully lay the coverslip over 30 μl (for a 22-mm^2 coverslip) of Vectashield antifade (Vector Laboratory, Inc., Burlingame, CA) added to a glass slide. Avoid introducing air bubbles in this step by touching the edge of the coverslip to the antifade drop first and slowly lowering the inverted coverslip until the solution is dispersed over the entire coverslip evenly.
3. Wick away any excess water/antifade solution with a Kimwipe, but do not bump the coverslip or try to remove the coverslip from the antifade drop after application, or the gametes may detach or be crushed in the process.
4. Place slides on a 37°C slide warmer for 5 min to dry the water from the exposed surface of the coverslip before sealing with nail polish. Avoid getting any nail polish on top of the coverslip. After sealing, replace on slide warmer for 3–5 min to completely dry the nail polish.
5. Store samples at 4°C in a light-tight slide box until confocal microscopy.

References

1. Schatten, G. (1999). In: *Molecular Biology in Reproductive Medicine*, B. C. J. M. Fauser, A. J. Rutherford, J. F. Strauss III, A. Van Steirteghem, eds. Parthenon Publishing, New York.
2. Surani, M. A., Kothary, R., Allen, N. D., Singh, P. B., Fundele, R., Ferguson-Smith, A. C., and Barton, S. C. (1990). *Dev Suppl.*, 89.
3. Rossant, J. (1985). *Phil. Trans. R. Soc. Lond. B*, **312**, 91.
4. Solter, D. (1988). *Annu. Rev. Genet.*, **22**, 127.
5. Wassarman, P. M. (1995). *Curr. Opin. Cell Biol.*, **7**, 658.
6. Shur, B. D., and Neely, C. A. (1988). *J. Biol. Chem.*, **263**, 17706.
7. Snell, W. J., and White, J. M. (1996). *Cell*, **85**, 629.
8. Schatten, G., Simerly, C., and Schatten, H. (1991). *Proc. Natl. Acad. Sci. USA*, **88**, 6785.
9. Schatten, G. (1994). *Dev. Biol.*, **165**, 299.
10. Gyllensten, U., Wharton, D., Josefsson, A., and Wilson, A. C. (1991). *Nature*, **352**, 255.
11. Jaffe, L. A. (1990). *J. Reprod. Fertil. Suppl* **42**, 107.
12. Tombes, R. M., Simerly, C., Borisy, G. G., and Schatten, G. (1992). *J. Cell Biol.*, **117**, 799.
13. Clark, E. A., and Brugge, J. S. (1995). *Science*, **268**, 233.
14. Sirard, M. A., Leibfried-Rutledge, M. L., Parrish, J. J., Ware, C. M., and First, N. L. (1988). *Biol. Reprod.*, **39**, 546.
15. Simerly, C., Wu, G. J., Zoran, S., Ord, T., Rawlins, R., Jones, J., Navara, C., Gerrity, M., Rinehart, J., and Binor, Z. (1995). *Nat. Med.*, **1**, 47.
16. Wu, G. J., Simerly, C., Zoran, S. S., Funte, L. R., and Schatten, G. (1996). *Biol Reprod.*, **55**, 260.
17. Simerly, C. R., Hecht, N. B., Goldberg, E., and Schatten, G. (1993). *Dev. Biol.*, **158**, 536.
18. Moreno, R., and Schatten, G. (2000). *Cell Motil. Cytoskeleton*, **46**, 235.

19. Moreno, R. D., Ramalho-Santos, J., Chan, E. K., Wessel, G. M., and Schatten, G. (2000). *Dev. Biol.*, **219**, 334.
20. Moreno, R. D., Ramalho-Santos, J., Sutovsky, P., Chan, E. K., and Schatten, G. (2000). *Biol Reprod.*, **63**, 89.
21. Manandhar, G., Sutovsky, P., Joshi, H. C., Stearns, T., and Schatten, G. (1998). *Dev. Biol.*, **203**, 424.
22. Manandhar, G., Simerly, C., and Schatten, G. (2000). *Hum. Reprod.*, **15**, 256.
23. Hewitson, L. C., Simerly, C. R., Tengowski, M. W., Sutovsky, P., Navara, C. S., Haavisto, A. J., and Schatten, G. (1996). *Biol. Reprod.*, **55**, 271.
24. Sutovsky, P., Hewitson, L., Simerly, C. R., Tengowski, M. W., Navara, C. S., Haavisto, A., and Schatten, G. (1996). *Hum. Reprod.*, **11**, 1703.
25. Ramalho-Santos, J., Sutovsky, P., Simerly, C., Oko, R., Wessel, G. M., Hewitson, L., and Schatten, G. (2000). *Hum. Reprod.*, **15**, 2610.
26. Hewitson, L., Dominko, T., Takahashi, D., Martinovich, C., Ramalho-Santos, J., Sutovsky, P., Fanton, J., Jacob, D., Monteith, D., Neuringer, M., Battaglia, D., Simerly, C., and Schatten, G. (1999). *Nat Med.*, **5**, 431.
27. Inoue S. (1989). In: *The Handbook of Biological Confocal Microscopy*, J. Pawley, ed., p. 92. IMR Press, Madison, WI.
28. Wilson, T. (1990). *Confocal Microscopy*. Academic Press, New York.
29. Holy, J., Simerly, C., Paddock, S., and Schatten, G. (1991). *J. Electron Microsc. Tech.*, **17**, 384.
30. Gard, D. L., and Kropf, D. L. (1993). *Methods Cell Biol.*, **37**, 147.
31. Wright, S. J, Centonze, V., Stricker, S., DeVries, P., Paddock, S., and G. Schatten. (1993). *Methods Cell Biol.*, **38**, 1.
32. In't Veld, P., Brandenburg, H., Verhoeff, A., Dhont, M., and Los, F. (1995). *Lancet*, **346**, 773.
33. Wright, S. J., Walker, J. S., Schatten, H., Simerly, C., McCarthy, J. J., and Schatten, G. (1989). *J. Cell Sci.*, **94**, 617.
34. Wright, S. J., Schatten, H., Simerly, C., and Schatten, G. (1990). In: *Optical Microscopy for Biology*, B. Herman and K. Jacobson, eds., p. 29. Alan R. Liss, New York.
35. Vos, I. W., Valster, A. H., and Hepler, P. K. (1999). *Methods Cell Biol.*, **61**, 413.
36. Brinkley, B. R., Fistel, S. H., Marcum, J. M., and Pardue, R. L. (1979). *Int. Rev. Cytol.*, **63**, 59.
37. Osborn, M., and Weber, K. (1982). *Methods Cell. Biol.*, **24**, 97.
38. Harlow, E., and Lane, D. (1988) In: *Antibodies: A Laboratory Manual*, E. Harlow and D. Lane, eds., Cold Spring Harbor Laboratory, Cold Spring Harbor, NY.
39. Simerly, C., and Schatten, G. (1993). *Methods Enzymol.*, **225**, 516.
40. Gorbsky, G. J., Simerly, C., Schatten, G., and Borisy, G. G. (1990). *Proc. Natl. Acad. Sci. USA*, **87**, 6049.
41. Dominko, T., Mitalipova, M., Haley, B., Beyhan, Z., Memili, E., McKusick, B., and First, N. L. (1999). *Biol Reprod.* **60**, 1496.
42. Stricker, S. A. (1999). *Dev. Biol.*, **211**, 157.
43. Garner, D. L., Thomas, C. A., Joerg, H. W., DeJarnette, M., and Marshall, C. E. (1997). *Biol. Reprod.*, **57**, 1401.
44. Graham, J. K., Kunze, E., and Hammerstedt, R. H. (1990). *Biol. Reprod.*, **43**, 55.
45. Mazia, D. (1978). In: *Future Research on Mitosis*, p. 196.
46. Stricker, S. A., Centonze, V. E., Paddock, S. W., and Schatten, G. (1992). *Dev. Biol.*, **149**, 370.
47. Pratt, H. P. M. (1987). In: *Mammalian Development: A Practical Approach*, M. Monk, ed., p. 13. IRL Press, Oxford.
48. LeMaire-Adkins, R., Radke, K., and Hunt, P. A. (1997). *J. Cell Biol.*, **139**, 1611.
49. Schatten, G., Simerly, C., Asai, D. J., Szoke, E., Cooke, P., and Schatten, H. (1988). *Dev. Biol.*, **130**, 74.
50. Wright, S. J., and Schatten, G. (1991). *J. Electron Microsc. Tech.*, **18**, 2.
51. Mazia, D., Schatten, G., and Sale, W. (1975). *J. Cell Biol.*, **66**, 198.
52. Sluder, G., Miller, F. J., and Lewis, K. (1993). *Dev. Biol.*, **155**, 58.
53. Sluder, G., Miller, F. J., Lewis, K., Davison, E. D., and Rieder, C. L. (1989). *Dev. Biol.*, **131**, 567.
54. Manandhar, G., Simerly, C., Salisbury, J. L., and Schatten, G. (1999). *Cell Motil. Cytoskel.*, **43**, 137.

55. Asch, R., Simerly, C., Ord, T., Ord, V. A., and Schatten, G. (1995). *Hum. Reprod.*, **10**, 1897.

56. Rawe, V. Y., Olmedo, S. B., Nodar, F. N., Doncel, G. D., Acosta, A. A., and Vitullo, A. D. (2000). *Mol. Hum. Reprod.*, **6**, 510.

57. Van Blerkom, J., Davis, P. W., and Merriam, J. (1994). *Hum. Reprod.*, **9**, 2381.

58. Fouquet, J. P., Edde, B., Kann, M. L., Wolff, A., Desbruyeres, E., and Denoulet, P. (1994). *Cell Motil. Cytoskeleton*, **27**, 49.

59. Fouquet, J. P., Kann, M. L., Combeau, C., and Melki, R. (1998). *Mol. Hum. Reprod.*, **4**, 1122.

60. Connors, S. A., Kanatsu-Shinohara, M., Schultz, R. M., and Kopf, G. S. (1998). *Dev. Biol.*, **200**, 103.

61. Bre, M. H., Peppekok, R., Hill, A. M., Levilliers, N., Ansorge, W., Stelzer, E. H., and Karsenti, E. (1990). *J. Cell Biol.*, **111**, 3013.

62. Mays, R. W., Beck, K. A., and Nelson, W. J. (1994). *Curr. Opin. Cell Biol.*, **6**, 16.

63. Hermo, L., Rambourg, A., and Clermont, Y. (1980). *Am. J. Anat.*, **157**, 357.

64. Kann, M. L., Prigent, Y., Levilliers, N., Bre, M. H., and Fouquet, J. P. (1998). *Cell Motil. Cytoskel.*, **41**, 341.

65. Susi, F. R., Leblond, C. P., and Clermont, Y. (1971). *Am. J. Anat.*, **130**, 251.

66. Hermo, L, Oko, R., and Hecht, N. B. (1991). *Anat. Rec.*, **229**, 31.

67. Hall, E. S., Eveleth, J., Jiang, C., Redenbach, D. M., and Boekelheide, K. (1992). *Biol. Reprod.*, **46**, 817.

68. Lippincott-Schwartz, J., Cole, N. B., Marotta, A., Conrad, P. A., and Bloom, G. S. (1995). *J. Cell Biol.*, **128**, 293.

69. Vaisberg, E. A., Grissom, P. M., and McIntosh, J. R. (1996). *J. Cell Biol.*, **133**, 831.

70. Sakai, Y., and Yamashina, S. (1989). *Anat. Rec.*, **223**, 43.

71. Bourgain, C., Nagy, Z. P., De Zutter, H., Van Ranst, H., Nogueira, D., and Van Steirteghem, A. C. (1998). *Hum. Reprod.*, **13 Suppl. 1**, 107.

72. Yanagimachi, R. (1998). *Hum. Reprod.*, **13 Suppl. 1**, 87.

73. Sutovsky, P., Tengowski, M. W., Navara, C. S., Zoran, S. S., and Schatten, G. (1997). *Mol. Reprod. Dev.*, **47**, 79.

74. Luetjens, M., Payne, C., and Schatten, G. (1999). *Lancet*, **353**, 1240.

75. Terada, Y., Luetjens, C. M., Sutovsky, P., and Schatten, G. (2000). *Fertil. Steril.*, **74**, 454.

76. Melan, M. A., and Sluder, G. (1992). *J. Cell Sci.*, **101**, 731.

77. Gustafsson, M. G., Agard, D. A., and Sedat, J. W. (1999). *J. Microsc.*, **195**, 10.

78. Kam, Z., Hanser, B., Gustafsson, M. G., Agard, D. A., and Sedat, J. W. (2001). *Proc. Natl. Acad. Sci. USA*, **98**, 3790.

79. Bershadsky, A. D., Gelfand, V. I., Svitkina, T. M., and Tint, I. S. (1978). *Cell Biol. Intl. Rep.*, **2**, 425.

80. Boatman, D. E. (1987). In: *The Mammalian Preimplantation Embryo*, B. D. Bavister, ed., Plenum Press, New York.

11

LEEANDA WILTON

Fluorescent in Situ Hybridization for Detection of Aneuploidy in Single Human Blastomeres for Preimplantation Genetic Diagnosis

1. DETECTION OF CHROMOSOME ERRORS USING FISH

The impact of numerical chromosome errors, or aneuploidies, on fetal development is well documented. Approximately 60% of spontaneously aborted fetuses are aneuploid (1, 2), demonstrating the lethality of the condition. Aneuploidies of the sex chromosomes can be viable, such as X0 (Turner's syndrome) and XXY (Klinefelter's syndrome) and some, like XXX or XYY, have an essentially normal phenotype. However, autosomal aneuploidies are almost always lethal; the notable exception to this is trisomy 21 (Down's syndrome).

Most of our knowledge of the effect of aneuploidies comes from analysis of spontaneously aborted conceptuses, but for many years there has been great interest in studying the chromosomal complement of early embryos before implantation. Initial studies using traditional karyotyping techniques were problematic because of the limited number of cells available and the difficulties in obtaining nuclei in metaphase (3, 4), which is essential for karyotypic analysis.

The advent of molecular karyotyping techniques such as FISH has made it possible to enumerate some chromosomes in interphase nuclei. FISH uses sequences of DNA (probes), labeled with different-colored fluorochromes, which are complementary to specific sequences on human chromosomes. FISH exploits the double-stranded nature of DNA. When it is denatured at high temperature, the two strands of DNA separate. Upon cooling the strands reanneal, providing the opportunity for the fluorescently labeled probe to hybridize with its complementary DNA sequence in the target. Detection of this fluorescent probe using a sensitive fluorescence microscope and a computerized imaging system enables enumeration of that chromosome (Figure 11.1). The accuracy of the hybridization of the probe and its target DNA depends on the stringency conditions of the reaction. If the stringency is low, then nonspecific hybridization will occur, and if the stringency is too high, it may not be possible to visualize the fluorescent signal.

The first report of the application of FISH to human blastomeres was by Griffin et al. (5), who determined the gender of blastomeres using probes to the X and Y chromosomes. This led to the first use of FISH for preimplantation genetic diagnosis (PGD), where prediction of the sex of embryos from couples where the female was a carrier of

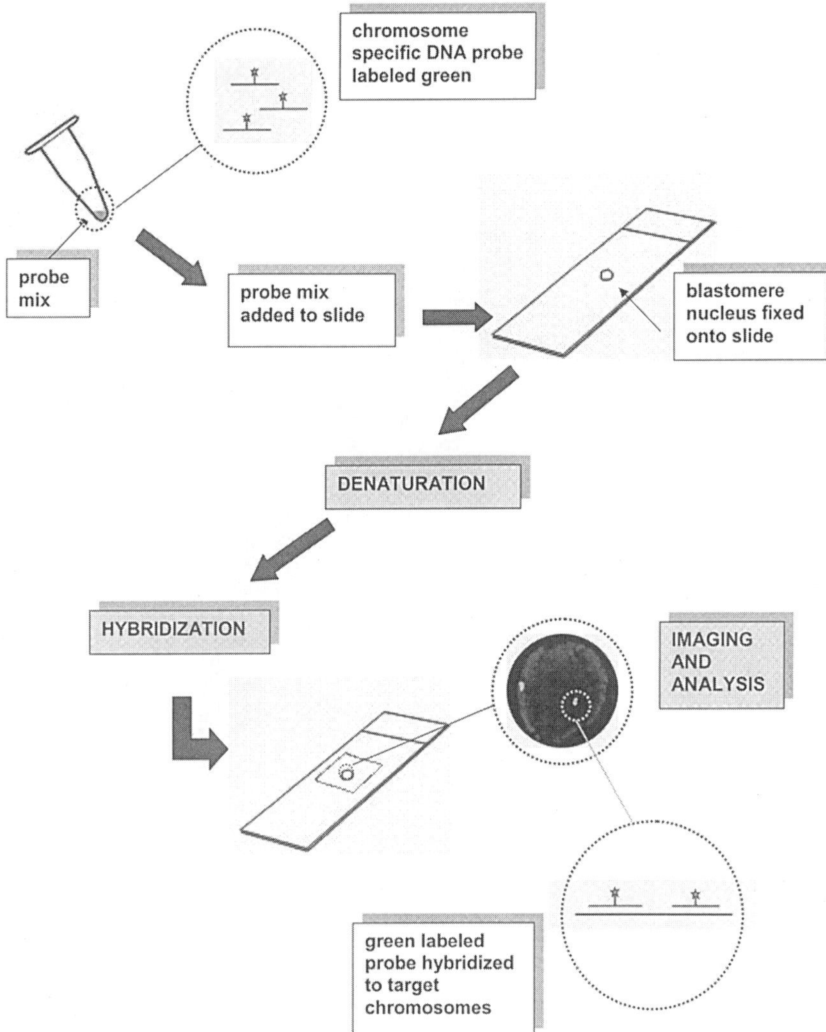

Figure 11.1. Flow diagram showing the important steps and principles of FISH.

an X-linked recessive disorder was achieved using specific probes for the X and Y chromosomes on a single blastomere biopsied from embryos (6).

Since this time FISH has been used extensively to examine chromosomes in a variety of single cells including oocytes, sperm, and blastomeres. Up to five probes can be used simultaneously, and rehybridization using other probes now means that about nine different chromosomes can be detected in a single cell. The wealth of literature describing the use of FISH to detect aneuploidies in human embryos has recently been reviewed (7). This chapter will describe the techniques required to use FISH to enumerate whole chromosomes in single interphase blastomeres for PGD.

2. TYPES OF FISH PROBES

2.1 Centromeric

Centromeric probes generally hybridize to the centromere of the chromosomes and exploit the fact that there are highly repeated sequences of α-satellite DNA specific to the centromeres of most human chromosomes. Because the target sequences are present in many copies, the fluorescent signal is usually very bright and easy to detect, and only short hybridization times are required. Specific centromeric probes are not available for all chromosomes. In particular, chromosomes 13 and 21 and also 14 and 22 have high sequence homology in the centromeric region, resulting in cross-hybridization of the probes.

2.2 Locus Specific

These probes are homologous to specific loci or genes outside the centromeric region. The target may be a small sequence of DNA, and so longer hybridization times are required to obtain detectable fluorescent signals. Locus-specific probes are available for the detection of chromosomes 13, 14, 21, and 22, circumventing the problems of identifying these chromosomes using centromeric probes.

2.3 Chromosome Paints

Chromosome paints consist of cocktails of many probes specific to many different regions of one chromosome. This results in the chromosome appearing to be "painted" along its length by the fluorescent probe. Chromosome paints are very useful for analysis of metaphase chromosomes, but in interphase cells they often form a large signal that could represent one or more chromosomes, making confident determination of ploidy very difficult.

3. AVAILABILITY OF PROBES

FISH probes can be produced in house, but this can be technically demanding, requiring artificial chromosome libraries from bacteria or yeast. Once probes are isolated, they need to be labeled with fluorophores, usually by nick translation, and thoroughly tested for hybridization accuracy and efficiency. A number of commercial sources of centromeric and locus-specific probes are now available from several companies.

4. PREPARATION OF BLASTOMERES FOR FISH

There are two methods of preparing blastomeres for FISH, and both are described in detail in Protocols 11.1 and 11.2. The Tarkowski (8) method involves fixing the cells with a solution of methanol and acetic acid and is based on the longstanding techniques for preparing metaphase chromosome spreads for routine cytogenetic analysis. The second method, first published by Coonen et al. (9), involves lysing the cell in a solution of detergent and hydrochloric acid and allowing the nucleus to adhere onto the microscope slide without actual fixation.

Each method has advantages and disadvantages. The Tarkowski (8) method results in a well-spread nucleus, but there is not a great deal of control over the extent of spread. An overspread nucleus can mean very diffuse signals that are difficult to read, particularly if many probes are used simultaneously. Also, the nuclei cannot be readily visualized after fixing. The Coonen (9) method results in more compact nuclei that give well-defined signals, although the number of probes that can be used simultaneously may be limited because of the smaller size of the nucleus and the risk of overlapping signals. The nuclei can easily be seen after spreading using phase-contrast microscopy.

5. PRECISE LOCATION OF NUCLEI BEFORE FISH

Some FISH protocols use DAPI as the DNA counterstain. This bright purple/blue color can be readily seen using a 10× or 20× fluorescence objective, and the nuclei are easy to find to read the signals after FISH is complete. Other FISH probe cocktails, including the PGT probe for chromosomes X, Y, 13, 18, and 21 (Vysis Inc., Downers Grove, IL) and the PB probe for chromosomes 13, 16, 18, 21, and 22 (Vysis) use an aqua DNA counterstain that is extremely difficult to see except under a 100× objective. Unless the nucleus has been accurately located before the FISH a great deal of time can be spent trying to find each nucleus to read the signals.

Precise location can be achieved using an England finder and a graticule marked with cross-hairs fitted to the eyepiece of the microscope. After FISH is complete, the microscope stage is set in the exact predetermined position, and the nucleus is visualized by moving to the correct focal plane for the slide. This is consistent for each microscope and requires one or two revolutions of the fine-focus control.

6. FISH

The protocol described (Protocol 11.3) is similar to that first described by Harper et al. (10). It should be noted that each FISH probe may have different optimal conditions for obtaining the brightest signals, and, if commercial probes are used, it is worthwhile to follow the manufacturer's instructions particularly for the denaturation, hybridization, and posthybridization washing steps. If a cocktail of probes is used, it may be difficult to determine conditions that suit every probe. A good starting point is to use the protocol that is optimal for the weakest probe in the set.

7. SUBSEQUENT ROUNDS OF HYBRIDIZATION

A subsequent round of hybridization using probes for different chromosomes than those detected in the first round enables enumeration of more chromosomes. For the second round of FISH, it is not necessary to wash the slides in pepsin or paraformaldehyde again. The protocol can be started by removing the coverslips and washing the slides in PBS and water. The denaturation step follows, and the protocol is the same as the initial round of hybridization from that point. It is important to note that successive rehybridizations result in some degradation of the DNA and loss of efficiency of hybridization (11).

8. VISUALIZATION AND SCORING OF FISH SIGNALS

FISH signals need to be visualized at high magnification using a fluorescence microscope fitted with filters appropriate for the fluorochromes that the probes are labeled with. Multiband pass filters enable several different fluorochromes to be observed at once, eliminating the need to change filters for each color signal. It is imperative to have the correct filters for the fluorochromes being used, otherwise signals will not be visible.

Figures 11.2, 11.3, and 11.4 show some examples of FISH on human blastomeres. These images are obtained using a CCD camera and computerized analysis system that captures an image of each color plane and then overlays them to produce a composite image of all color planes. Scoring of signals is critical to obtaining the correct results. Confident signal interpretation is possible when there is no background fluorescence and the signals are of similar size and brightness. Single signals that have split could be

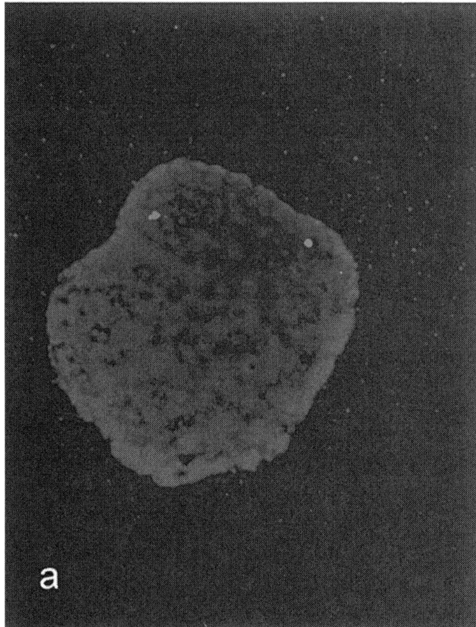
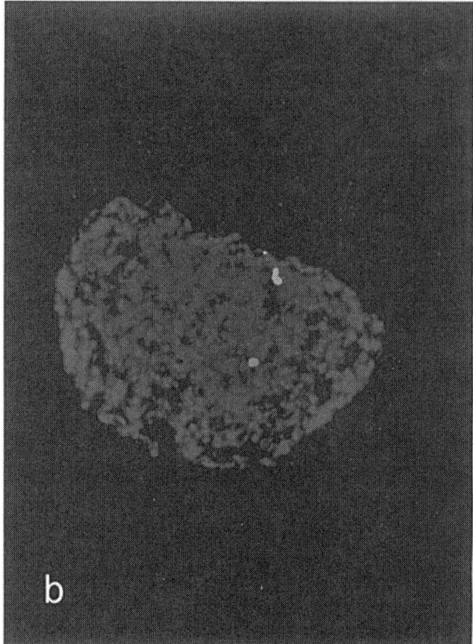

Figure 11.2. FISH image of blastomere nuclei counterstained with DAPI which shows the DNA in purple/blue. X chromosomes are labeled with red fluorochrome and Y chromosomes with green fluorochrome. (a) A nucleus with two red signals from a female embryo; (b) a nucleus with one red and one green signal from a male embryo. See color insert.

misread as two signals. Most investigators use the criteria of Hopman et al. (12), which suggest that discrete signals should be at least one signal width apart.

Overlapping signals can also result in misinterpretation of FISH results. Many investigators increase the number of FISH probes used in each analysis by using combination labeling, where some chromosomes will be detected by probes labeled with a single fluorochrome and others with a combination of two fluorochromes. For example, a 3:2 proportion of aqua and red fluorochromes is visualized as magenta. The problem with this approach is that, using this example, it would be impossible to tell if the magenta signal is hybridized to one locus, indicating one chromosome, or if it is actually an overlap of the two probes (and hence the two chromosomes) that were labeled in aqua and red. An example of this is shown in figure 11.2. Using only monocolor probes avoids this problem (13).

9. TROUBLESHOOTING

The efficiency of FISH depends on many factors, including the type and source of the probes. In optimal conditions an efficiency of approximately 95% per probe can be expected. The two most common problems that occur with FISH are weak/failed signals or excessive background. Weak or failed signals may be caused by (1) inadequate denaturation of the probe, target DNA, or both; (2) hybridization time too short; (3) excessive or insufficient humidity in the hybridization chamber; and (4) stringency of posthybridization wash too high. Excessive background may be caused by (1) improper cleaning of microscope slides; (2) poor preparation of the nuclei with too much cytoplasmic debris present; (3) insufficient pretreatment of the nucleus with pepsin; and (4) stringency of posthybridization wash too low.

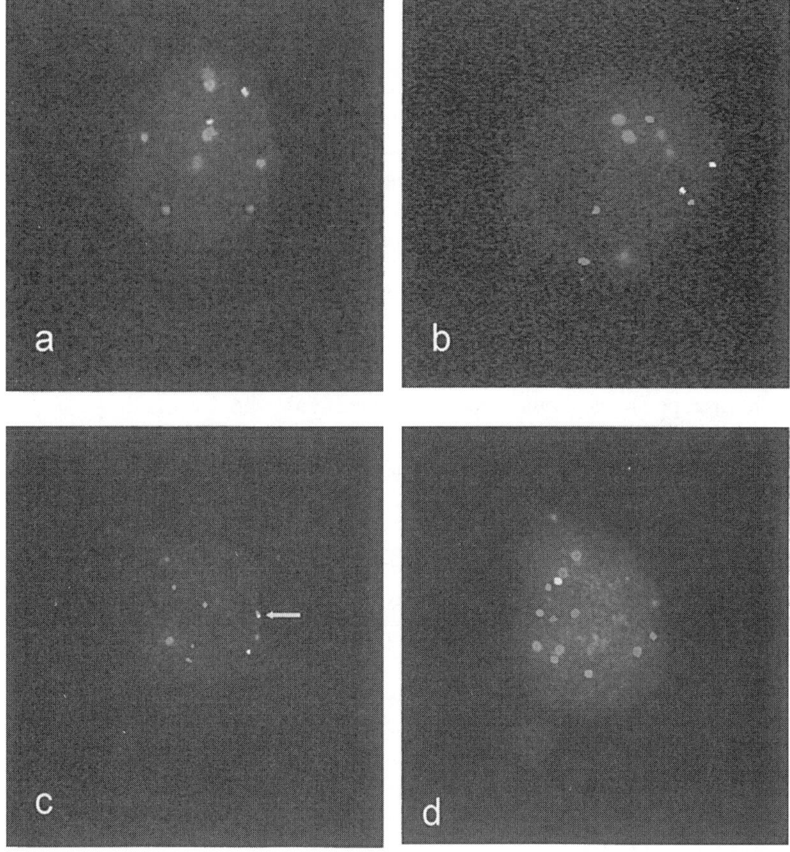

Figure 11.3. FISH images of blastomere nuclei counterstained in aqua. The probes used are for chromosomes 13 (red), 16 (aqua), 18 (blue), 21 (green) and 22 (yellow). (a) A normal nucleus with two signals of each color, (b) nucleus with an extra blue signal diagnosed as trisomy 18, (c) nucleus with an extra red signal diagnosed as trisomy 13. Note that one green and one yellow signal (arrow) are closely adjacent. (d) This nucleus approximates tetrasomy, as it has four signals of each color, with the exception of chromosome 22 (yellow), for which there is only one signal. See color insert.

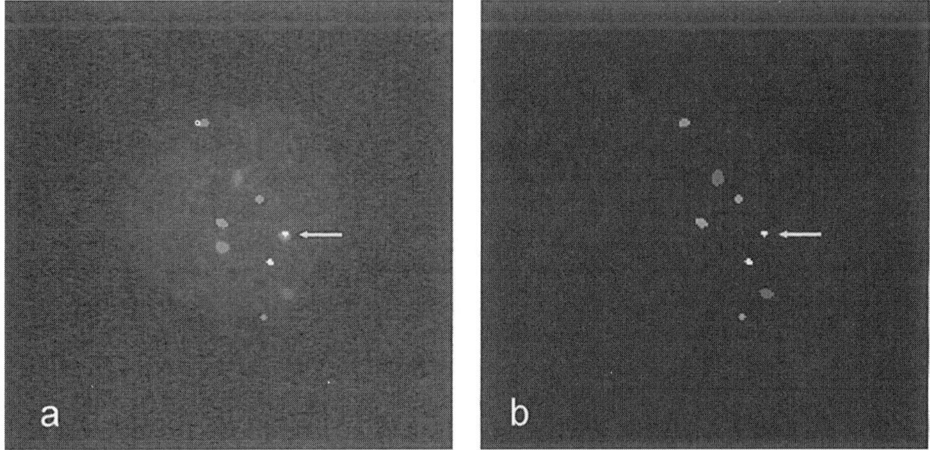

Figure 11.4. The advantages of using only monochromatically labeled probes and an imaging system. (a) This nucleus could easily be diagnosed as monosomy 22 with only one yellow signal. The arrow shows that one aqua signal may be overlying a yellow signal. (b) The same nucleus as in panel a, but the imaging system has been used to remove the aqua color plane allowing unequivocal confirmation that a yellow signal is underneath the aqua. See color insert.

10. CONCLUSION

FISH is an effective technique for detecting aneuploidy in single cells. Its application to blastomeres has provided a wealth of information on chromosome errors in early embryos. The technique is relatively straightforward. It is important to closely monitor the conditions of denaturation, hybridization, and posthybridization washing. Good results can be expected if protocols are followed and one has a basic understanding of the principles of the technique.

Protocol 11.1. Methanol:acetic acid fixation of blastomeres

1. Microscope slides should be clean and preferably acid washed.
2. Using a diamond pencil, etch a small circle, approximately 2–3 mm diameter, on the underside of each microscope slide.
3. Prepare a hypotonic solution of 0.6% trisodium citrate.
4. Prepare a solution of 3:1 anhydrous methanol:glacial acetic acid fixative. This must be prepared fresh and should be discarded after about 4–6 h.
5. Incubate individual blastomeres in hypotonic solution for approximately 1 min.
6. Place the blastomere on the microscope slide over the etched circle in a minimal volume of hypotonic solution.
7. Observe the blastomere down the microscope and just at the moment before the cell dries out, use a finely drawn Pasteur pipette to apply a drop of fixative on top of the cell. The cell should be followed as it moves and a second drop of fixative added just as the first drop dries. A third drop of fixative should be applied and the cell allowed to air dry. The approximate location of cell in relation to the etched circle should be noted.
8. Usually it is not possible to accurately locate the nucleus, except if it is stained using a 10% Giemsa solution.

Protocol 11.2. Tween 20:HCl spreading of blastomeres

1. Precoat clean, acid-washed microscope slides by incubating them in a 1/10 dilution of poly-L-lysine in water for 5 min. Air dry the slides and store them at 4°C, but discard after about 3 months.
2. Make a 1% stock solution of Tween 20 by adding 1 ml of Tween 20 to 99 ml of water. This can be stored at 4°C.
3. Make the final spreading solution of 0.1% Tween 20 in 0.1 N HCl just before use by adding 1 ml of Tween 20 stock solution and 0.1 ml of 1 N HCl to 8.9 ml of water.
4. Using a diamond pencil, etch a small circle, approximately 2–3 mm diameter, on the underside of each microscope slide.
5. Briefly wash the blastomere in PBS.
6. Place a small drop of spreading solution on the microscope slide over the etched circle.
7. Using a finely drawn pipette and in a minimal volume of PBS, move the blastomere into the drop of spreading solution.
8. Immediately transfer the slide to the stage of an inverted microscope and observe the blastomere under a total magnification of 100–200×.
9. After a few seconds up to about 1 min the cell membrane will lyse, and the cytoplasmic contents will begin to disperse. The nucleus should become clearly visible. The cytoplasm will gradually clear and the nucleus will move slowly around the slide. Follow it until it dries and record its final location with respect to the circle.
10. As soon as possible, wash the slide in PBS for 5 min and dehydrate through a graded series of 70%, 90%, and 100% ethanol for 1 min each. Air dry.

Protocol 11.3. FISH

The protocol described here is similar to that first described by Harper et al. (10). Note that each FISH probe may have different optimal conditions for obtaining the brightest signals, and, if commercial probes are used, it is worthwhile to follow the manufacturer's instructions, particularly for the denaturation, hybridization, and posthybridization washing steps. If a cocktail of probes is used, it may be difficult to determine conditions which suit every probe. A good starting point is to use the protocol that is optimal for the weakest probe in the set.

1. Incubate slides in 100 μg/ml pepsin in 0.1 N HCl at 37°C for 20 min to remove cytoplasmic debris. A pepsin stock can be made up of 250 mg pepsin dissolved in 25 ml of 1 N HCl and 0.5 ml aliquots stored at –20°C. One aliquot is added to 49.5 ml of water for use.
2. Wash slides briefly in water and then in PBS.
3. Incubate slides in 1% paraformaldehyde in PBS at 4°C for 10 min.
4. Wash slides briefly once in PBS and twice in water.
5. Dehydrate slides through a graded series of 70%, 90%, and 100% ethanol for 1 min each. Air dry.
6. The next step is to denature both probe and target DNA. It can be useful to work in dim light from this point to minimize quenching of fluorescence.

Co-denaturation

For some probes denaturation can be achieved simultaneously by adding the appropriate volume of probe (e.g., 2.5 μl if using a 12-mm diameter round coverslip or 7 μl if using an 18-mm square coverslip) to the microscope slide over the target DNA, adding the coverslip and heating the slide to 68–72°C for 2–8 min, depending on the probe.

Separate denaturation

The slide needs to be denatured in 70% formamide in 2 × SSC at 72–74°C for approximately 5 min. At the same time the probe should be denatured at the same temperature for approximately 5 min. The probe and the target DNA should be maintained between 50° and 60°C to minimize renaturation while the appropriate volume of probe is added to the target area of the slide.

1. Hybridize the probe to the target by incubating the slides in the dark at 37°C for 1 h to overnight, depending on the probe used. If longer hybridization times are required, it is necessary to maintain humidity. This can be achieved placing a small piece of damp sponge in the chamber. It may also be necessary to seal the hybridization area by applying rubber cement around the coverslips.
2. After hybridization remove the coverslips and wash the slides in a solution of, for example, 0.7 × SSC/0.3% Nonidet P-40 (NP40) detergent for 2–5 min at 72–74°C. The exact concentrations of SSC and detergent, the temperature and time depend on the probe that has been used, and if it is a commercial probe, then the manufacturer's instructions should be followed.
3. Briefly wash the slides in 2 × SSC/0.1% NP40 at room temperature.
4. Dehydrate slides through 70%, 90%, and 100% ethanol for 1 min each and air dry.
6. Apply the appropriate counterstain in antifade solution to the hybridization area and apply a coverslip to each slide.

Note: SSC solutions are diluted from a 20 × SSC stock, which consists of a solution of 3 M sodium chloride and 0.3 M sodium citrate.

References

1. Hassold, T., Chen, J., Funkhouser, J., Jooss, T., Manuel, B., Matsuura, J., Matsuyama, A., Wilson, C., Yamane, J. A., and Jacobs, P. A. (1980). *Ann. Hum. Genet.*, **44**, 151.
2. Jacobs, P. A. (1992). *Oxford Rev. Reprod. Biol.*, **14**, 47.
3. Wilton, L. (1993). In: *Gamete and Embryo Micromanipulation in Human Reproduction*, S. Fishel and M. Symonds, eds., p. 187. Edward Arnold, London.
4. Jamieson, M. E., Coutts, J. R. T., and Connor, J. M. (1994). *Hum. Reprod.*, **9**, 709.
5. Griffin, D. K., Wilton, L. J., Handyside, A. H., Winston, R. M. K., and Delhanty, J. D. A. (1992). *Hum. Genet.*, **89**, 18.
6. Griffin, D., Wilton, L., Handyside, A., Winston, R., and Delhanty, J. (1993). *Br. Med. J.*, **306**, 1382.
7. Wilton, L. (2002) *Prenat. Diagn.*, **22**, 1.
8. Tarkowski, A. K., and Wroblenska, J. (1967). *J. Embryol. Exp. Morphol.*, **18**, 155.
9. Coonen, E., Dumoulin, J. C. M., Ramaekers, F. C. S., and Hopman, A. H. N. (1994). *Hum. Reprod.*, **9**, 533.
10. Harper, J., Coonen, E., Ramaekers, F. C. S., Delhanty, J. D. A., Handyside, A. H., Winston, R. M. L., and Hopma, A. H. N. (1994). *Hum. Reprod.*, **9**, 721.
11. Liu, J., Tsai, Y-L., Zheng, X-Z., Yazigi, R. A., Baramki, T. A., Compton, G., and Katz, E. (1998). *Mol. Hum. Reprod.*, **4**, 972.
12. Hopman, A. H. N., Raemakers, F. C. S., Raap, A. K., Beck, J. L. M., Devilee, P., van der Ploeg, M., and Vooijs, G. P. (1998). *Histochemistry*, **89**, 307.
13. Bahçe, M., Escudero, T., Sandalinas, M., Morrison, L., Legator, M., and Munné, S. 2000. *Mol. Hum. Reprod.*, **6**, 849.

12

W. A. KING

D. H. BETTS

W. BUREAU

P. K. BASRUR

Assessment of Ploidy, Telomere Length, and Telomerase Activity in Oocytes and Embryos

1. CYTOGENETICS PERSPECTIVES

Early cytogenetic investigation on plants and animals based on sections and squash preparations of proliferating tissues such as root tips, bone marrow, and the testis indicated that chromosomal imbalance was often the cause of semisterility and/or embryo loss. The advent of hypotonic pretreatment (1), postfixation air drying (2), and lymphocyte culture procedures (3, 4) provided the technical means for confirmation of these observations. With the use of these techniques, the number and morphology of chromosomes could be precisely delineated, which, in turn, led to the linking of chromosome abnormalities with specific postnatal malformations in humans (5) and with reproductive failure in cattle (6). These techniques were successfully applied to testicular material, yielding details on the process of meiosis in the male. However, the impact of abnormal chromosome contributions of the oocyte on subsequent embryo development could not be assessed until the 1960s and 1970s. Techniques for the preparation of chromosomes from oocytes and embryos, based on procedures for visualizing chromosomes of somatic cells, were described for the mouse (7), pig (8), human (9), and cattle (10). The groundbreaking study on human spontaneous abortions by Carr (11) confirmed that chromosome abnormalities are a significant cause of first trimester pregnancy failures.

Application of the techniques for delineation of oocytes and embryos prepared under a variety of in vivo and in vitro conditions continues to enhance our understanding of the impact of chromosome defects on the final steps of oocyte maturation and the early events in embryo development. The emergence of molecular biology, which has vastly expanded our knowledge of the specific function of individual genes and gene products, also holds promise for identifying the effects of deletions and duplications of noncoding genetic sequences. In this chapter we outline methods for chromosome preparation and analysis (Protocols 12.1–12.6) and describe molecular techniques for assessing telomeric DNA integrity (Protocols 12.7–12.8) during embryo development. Changes in DNA integrity have been linked to success or failure of meiosis and mitosis and are thought to have profound effects on the success of stem-cell culture and animal cloning.

2. CHROMOSOMAL COMPLEMENT AND DEVELOPMENTAL POTENTIAL

Technical advancements in chromosome preparations have allowed assessment of the impact of chromosome variations on specific stages in mammalian development, including those present in the germ line before meiosis, those that arise during meiosis and fertilization, and those that occur subsequent to fertilization. Chromosome changes constituting a deviation from the species-specific karyotype are associated with growth arrest and/or developmental failure, malformations, and aging (12). The most frequently observed deviations from the normal karyotype in oocytes are aneuploidy, with near-haploid chromosome number resulting from the loss (or an excess) of individual chromosomes, and diploidy, with double the haploid number at the second meiotic metaphase resulting from the failure of meiosis I (first division of meiosis) (13–15). Abnormal karyotypes of the zygote and the early cleavage stages reflect the processes that were affected either at meiosis or at fertilization. They include variations such as aneuploidy, polyploidy (multiples of the haploid set of chromosomes, in excess of two), and mixoploidy (cell types with two or more different chromosomal complements within the same embryo). Rates of development, as well as the survival of the early embryos, are affected by the embryo's chromosome complement. Among bovine embryos produced in vitro, those with a diploid chromosome complement consistently display a significantly higher cell number and a faster rate of development to the blastocysts stage than do polyploid or mixoploid embryos (16).

A systematic study of developing embryos showed that the prevalence and types of chromosome variations change as development progresses (13, 16). These variations often lead to stage-dependent death of the embryos or to the elimination of individual cells through apoptosis, depending on the stage of embryogenesis at which the defects occurred, the type of chromosome involved, and the severity of the defect. Thus, a loss of an autosome during early postfertilization divisions might lead to the loss of the embryo, whereas a similar change at a later stage may retard the growth rate due to the elimination of cells that sustained the loss. In cattle, the majority of chromosomally abnormal embryos are eliminated during the first 2 weeks of development (17, 18). It has been hypothesized that there is a threshold level of chromosomally abnormal cells that are tolerated as the embryos grow and still survive. Embryo survival, therefore, may be thought of as a balance between the proportion of abnormal cells and the ability of the conceptus to eliminate such cells by selective cell death (apoptosis) or to sequester them into the trophectoderm lineage (19, 20).

3. TELOMERES

Telomeres are the physical ends of linear chromosomes. Early cytological and genetic studies demonstrated that these terminal regions are essential for maintaining chromosome integrity and that chromosomes with truncated ends are unstable and fuse with other chromosomes or are lost during cell division (20, 21). As demonstrated in yeast, removal of telomeres even from a nonessential chromosome can lead to the dramatic loss of that chromosome during cell division (22). Association of telomeres with each other and with the nuclear membrane during early stages of meiosis and mitosis are interactions that are probably of fundamental importance for homologous chromosome pairing (23), recombination (24), and the positioning of chromosomes within the nucleus (25). Telomeres play a crucial role in the correct segregation of chromosomes (26, 27) and buffer the loss of terminal DNA that occurs during DNA replication (reviewed by Blackburn [28]). Hence, damage to, or deficiencies in, the telomere structure can have profound effects on chromosome maintenance and function.

In most eukaryotic species, the telomere consists of a repeat sequence of a small number of deoxynucleotides and specific binding proteins that cap both ends of each chromosome. Mammalian telomeric DNA contains the hexameric repeat, TTAGGG, which may be tandemly reiterated up to 15 kilobases in humans (29, 30). The telomere length varies widely among species and also among the cells of an individual (reviewed by Preston [31]). A G-rich, single-stranded overhang extends beyond the double-stranded telomere section, which is folded back into the DNA helix by means of specific telomere-binding proteins, forming a T-loop structure that disguises the single-stranded region (32). The inherent shortening of telomere length in continuously dividing cells is considered to be primarily due to the end replication problem (33, 34). Because of the uni-directional synthesis of DNA polymerases and the requirement of a 5' primer, DNA replication of the distal 3' end of each lagging strand is incomplete, leading to telomere loss of approximately 50–200 bp per cell division in vitro and in vivo (35–38). This shortening of telomere length has been suggested to be a mitotic clock regulating the proliferative capacity of cells by signaling senescence (an irreversible cell-cycle arrest), when dividing cells reach a critically short telomere length and the G-rich, single-stranded overhang is unmasked (32, 39).

4. TELOMERASE

The vulnerability of the progressive shortening of the telomere is overcome by the ribonucleoprotein telomerase, which is expressed in the germ line (40–42) and ensures transmission of full-length chromosomes to the progeny. Telomerase is a multi-subunit reverse transcriptase that uses its RNA component to anchor to the chromosome end and to synthesize telomeric DNA by way of the complementary repeat sequence (43). Telomerase activity has also been detected in other regenerative tissues such as the endometrium, stem cells, blood cells, epidermis, and liver (44–47). In addition, telomerase reactivation is thought to be essential for stabilizing telomere length and acquisition of cellular immortality in most human cancer cells (reviewed by Shay and Bacchetti [48]). Generally, the presence of telomerase activity and maintenance of telomere length are associated with cellular immortality, and the lack of telomerase activity and shortening of telomere length are attributed to cellular aging and senescence.

5. TELOMERE LENGTH DURING MAMMALIAN DEVELOPMENT

The transmission of full-length telomeres to the subsequent generations is accomplished by the expression of telomerase in the germ line. However, telomeres shorten in dividing cells upon suppression of telomerase activity during cellular differentiation (49). Although there are significant telomere-length differences between individuals, there is telomere-length synchrony among human fetal tissues, which is eventually lost during postnatal life (50). Telomere length is similar in different tissues of newborn mice but differs between tissues of each adult animal (51). The telomere-length variations among and within newborns probably reflect the different proliferative rates of cells (50) and the rates of cell differentiation (52) in various tissues. However, in contrast to human tissues, adult mouse tissues exhibit telomerase activity and telomeres that are much longer than those of human chromosomes (51).

The telomere hypothesis of aging can be tested in vivo using nuclear transfer technology because it allows the production of cloned animals from adult and cultured somatic cells without the involvement of the germ line. Nuclear transfer of aged and culture-propagated bovine somatic cells into enucleated bovine oocytes allows restoration of telomere length in cloned bovine fetuses and offspring (53–55). Although it is not clear when telomere restoration occurs during cloned bovine embryo development, the rebuilding of telomere length from the shorter telomeres of the donor cells could be due to the nuclear reprogramming of telomerase activity, detected at the blastocyst stage of cloned embryo development (55). These results differ from studies in which telomere erosion did not appear to be repaired after nuclear transfer in sheep (56, 57) and where telomere length of cloned calves was extended beyond those of newborn and age-matched control animals using near-senescent cells as nuclear donors (58). Differences between donor cells in telomere-binding proteins (59), the extent of nuclear reprogramming of telomerase activity and its ability to extend telomeres (60), the regenerative ability of donor cells of different species, cell types, and the age of donors may account for these differences. It is not known whether the longevity of these animals will be reflected in their telomere lengths. Despite having shorter mean terminal restriction fragment (TRF) lengths, cloned sheep are healthy, fertile, and display physiological and morphological characteristics typical of sheep of their breed and age (57). However, cells derived from cloned bovine fetuses display a 50% longer proliferative capacity than those obtained from control fetuses (58). Nevertheless, these results suggest that bovine oocytes and normal and reconstructed embryos do contain a protective mechanism(s) to ensure telomere-length maintenance, repair, and regeneration, providing chromosomal stability during meiosis and embryogenesis. Further research using somatic cell nuclear transfer will allow a better understanding of the relationship between physiological aging, telomere length, telomere maintenance/lengthening mechanisms, and the replicative capacity of cells, both in vivo and in vitro.

Loss or acquisition of a whole chromosome or a chromosomal segment generally has a profound and often deleterious effect on embryo survival. Developmentally regulated changes such as late replication of the inactive X chromosome, errors incurred during postfertilization mitoses, and telomere shortening occurring as the embryo cleaves and begins to differentiate lead to more subtle restrictions on the developmental potential of the embryo through their specific impact on individual cells and their descendants. Understanding and overcoming these restrictions pose challenges for the successful reprogramming of cells in conjunction with cloning and stem cell culture. The techniques described in the remainder of this chapter (Protocols 12.1–12.8) provide vital tools for the advancement of research in this area.

6. CHROMOSOME PREPARATIONS FROM OOCYTES, EMBRYOS, AND FETAL TISSUES

Successful chromosome analysis of oocytes, embryos, and fetal tissue requires fixed, well-spread chromosomes devoid of cytoplasm and cytoplasmic and other cell debris (Protocols 12.1, 12.2). Oocytes progress from germinal vesicle stage to metaphase II (MII) in a species-dependent manner when removed from follicles and placed in culture in vitro. Thus the particular stages of meiosis can be examined by fixing oocytes at defined times after the initiation of in vitro maturation (Protocol 12.1). After fertilization, only the first two or three cleavage divisions are synchronous, and mitosis only takes 2–3 h out of a 12- to 18-h cell cycle. It is therefore advantageous to synchronize the mitotic process in blastomeres (Protocol 12.1). The most common approach is to use mitotic spindle inhibitors such as colcemide or nocadazol. While microtubule inhibitors arrest the nuclei in metaphase, the individual chromatin threads continue to contract. Prolonged exposure to microtubule inhibitors results in highly contracted chromosomes that are difficult to analyze. It is therefore important to optimize the concentration and duration of exposure to the specific spindle inhibitor used for each species and culture condition.

To obtain well-spread nuclei devoid of cytoplasmic debris, it is necessary to pretreat cells with a hypotonic solution to allow the cells to swell and to ensure dispersal of cytoplasm during fixation. Briefly incubating cells in a hypotonic saline solution will induce swelling of the cytoplasm and aid in separation of the chromosomes. This is a critical step because remnants of cytoplasm and other cellular debris will obscure the visibility of chromosomes and reduce the accessibility of the hybridization probes to the specific target, thereby leading to high levels of background staining. As with microtubule inhibitors, each method must be optimized for the individual system, species, and culture conditions used.

The fixative generally used for high-quality chromosome preparation is a precipitating fixative consisting of a mixture of methanol and acetic acid or ethanol and acetic acid (by volume), usually in the proportions of 3:1. Embryos are rapidly fixed directly on clean, lint- and grease-free slides. There is species variation in the response of the embryos to fixation. For example, species such as cattle require a higher proportion of acetic acid to dissolve the zona pellucida and let the chromosome adhere to the slide. Occasionally, chromosomes and nuclei from embryos do not adhere to slides and can be lost during staining or in situ hybridization. To avoid such loss, slides may be pre-treated with gelatin-chromic alum (61) or poly-L-lysine (62). Alternatively, the slides pre-coated with positively charged groups (Superfrost Plus slides, Kindler, Freiburg, Germany) may be used. For in situ hybridization (ISH), prolonged fixation should be avoided to ensure that the target sequences are maintained. Examples of preparations obtained from metaphase and interphase nuclei from bovine oocytes and embryos are provided in Figure 12.1.

7. FLUORESCENT IN SITU HYBRIDIZATION

The in situ hybridization (ISH) technique was first described in the late 1960s by two independent groups (63, 64). The method (Protocols 12.3–12.6) is based on the visualization of specific sequences of DNA or RNA in intact nuclei (or on chromosomes) by their complementary annealing through hydrogen bonds formed between bases attached to the sugar-phosphate backbone. For a successful outcome, ISH requires skills in molecular biology and cytology, availability of suitable probes for specific hybridization to the chromosome or to the gene of interest, and accessibility of the target nucleic acid for hybridization with the probe.

Depending on the nature of the material to be hybridized, the target nucleic acid is most often cross-linked and embedded in a complex matrix. As a consequence, the accessibility of the probe decreases. Initially, only radioisotopes and autoradiography were used for visualization of specific hybridization. In more recent years, fluorescent labels have become increasingly popular. The development of these sensitive, safe, and rapid labeling systems has been a crucial step toward expanding the use of this technique in a number of fields of biomedical research including cell and developmental biology, genetics, and pathology.

The use of FISH (Table 12.1) for localizing genes on chromosomes of embryos is presented in this section. FISH is now more frequently used for chromosome analysis of human embryos than conventional karyotype analysis. It has allowed characterization of numerous chromosomal abnormalities in preimplantation embryos (65). The increasing popularity of FISH compared to karyotyping is mainly due to the problems encountered in culturing embryos and obtaining metaphase spreads from a representative number of blastomeres within an embryo (66, 67). In addition, when metaphases are obtained, they are often either poorly or overly spread. Moreover, the identification of individual chromosomes is usually more difficult in embryos in which chromosomes are highly contracted due to the long exposure to the microtubule inhibitor. Therefore, limited information is gathered from embryo karyotyping due to the low number of analyzable cells (68). This problem is circumvented by the use of FISH because it enables the use of chromosome-specific probes for the analyses of interphase nuclei. Various types of nucleic acid probes for ISH have been described. These can be divided into three different categories:

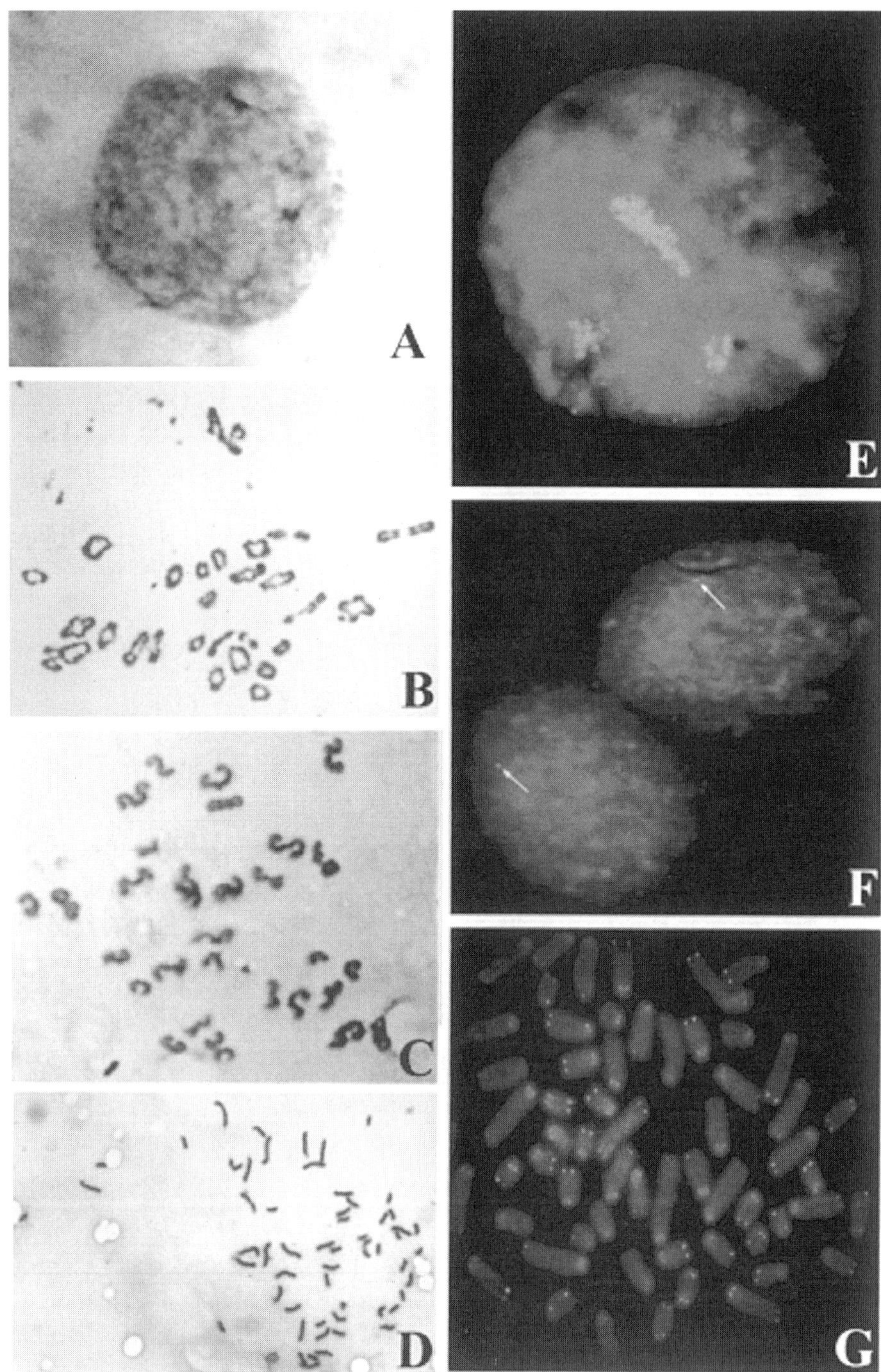

Figure 12.1. Methanol-acetic acid fixed preparations from bovine oocytes and two-cell and blastocyst stage embryos. Nuclei (interphase and metaphase) were stained with 4% Giemsa (A–D) or subjected to FISH using fluoresceinisothiocynate-labeled detection systems (green) and counterstained with propidium iodide (red). (A) Giemsa-stained germinal vesicle stage oocyte; (B) Giemsa-stained metaphase I; (C) Giemsa-stained metaphase II; (D) two Giemsa-stained metaphase spreads from a two-cell-stage embryo; (E) a blastomere nucleus after FISH using a repetitive DNA sequences specific for chromosomes 14, 20, and 25; (F) two blastomere nuclei from a haploid parthogenetically activated oocyte after FISH using a unique DNA sequence specific for chromosome 1 (signal is indicated by the arrow); (G) a metaphase spread from a blastocyst after FISH using a PNA telomere repeat (TTAGGG) probe. See color insert.

Table 12.1. Reagents Required for FISH

Reagents	Supplier	Catalogue No.
BioNick labeling system	Gibco BRL (Gaithersburg, MD)	18247-015
Herring sperm	Sigma (St. Louis, MO)	D-7290
DNA calf thymus	Sigma	D-8661
Sephadex GX50	Pharmacia (Uppsala, Sweden)	17-057301
Formamide	Fisher (Houston, TX)	BP228-100
Dextran sulfate	Sigma	D-8906
Antibiotin antibody	Applied Biosystems (Foster City, CA)	Sp.300
FITC-ANTI-IgG	Tebu (be Perray-en-Yvelines, France)	009RAGF
Propidium iodide	Sigma	P4170
Vectashield	Vector (Burlingame, CA)	H-1000
BSA	Sigma	A-9205

1. Chromosome-painting probes derived from individual chromosome libraries isolated by flow cytometry. These probes are used mainly to detect structural chromosome abnormalities.
2. Probes derived from unique or low copy sequences (figure 12.1e). These probes enable the physical localization of genes on specific chromosomes and detection of gene deletions and transgene insertions. However, they are the most difficult of all ISH probes.
3. Probes derived from repeated sequences. These probes are powerful tools for detecting numerical abnormalities of individual chromosomes. In ISH they display distinct bright signals in metaphase as well as interphase nuclei (figure 12.1f). Such probes are already isolated and cloned for several individual chromosomes in humans. This is not the case for other species, including cattle, for which only a few repetitive DNA probes are available (69, 70).

In addition, peptide nucleic acid probes (PNA) have been recently developed for detecting single-copy loci as well as repetitive sequences such as telomeres (figure 12.1g). The main advantage of these probes is that they are neutral, and consequently they access their target better than DNA probes. However, these probes are expensive, and their use is limited to very short sequences.

8. DETECTION OF TELOMERASE ACTIVITY

The development of the telomeric repeat amplification protocol (TRAP; Table 12.2) has allowed large-scale screening of telomerase activity in samples of small cell numbers and limited tissue (40). TRAP is a PCR-based assay that involves the addition of telomeric repeats $(TTAGGG)^n$ to the 3' end of a radiolabeled oligonucleotide substrate (^{32}P-TS primer) that telomerase recognizes and binds to (Protocol 12.7). The extended TRAP products are amplified by the PCR using the substrate oligonucleotide and reverse primers, generating a ladder of products with 6-base increments starting at 50 nucleotides (e.g., 50, 56, 62, 68). TRAP reaction products are resolved by gel electrophoresis and then analyzed by densitometry (Figure 12.2).

9. MEASUREMENT OF TELOMERE LENGTH BY TERMINAL RESTRICTION FRAGMENT ANALYSIS

A variety of methods have been described to detect and measure telomere length (Protocol 12.8). FISH using a telomere DNA probe $(CCCTAA)^3$ can localize telomeric DNA to

Table 12.2. Materials and Reagents Required for the TRAP Assay

Material/Reagent	Supplier	Catalogue No.
2-propanol (isopropanol)	Sigma	I-9516
40% Polyacrylamide/bisacrylamide stock (19:1)	Bio-Rad (Hercules, CA)	161-0144EDU
Ammonium persulfate (APS)	Bio-Rad	161-0700EDU
Bio-Rad Protein Assay Kit	Bio-Rad	500-0002
Boric acid	Sigma	B-0252
Bromophenol blue	Sigma	B-6131
EDTA	Sigma	E-5134
EGTA	Sigma	E-3839
Ethanol	Fisher	—
Filter paper	Fisher	—
Fuji Medical X-ray Film (Super RX)	Fisher	—
Glacial Acetic Acid	Fisher	—
Glycerol	Sigma	G-2025
Heat block	Fisher	—
Magnesium chloride (MgCl$_2\cdot$6H$_2$O)	Sigma	M-2670
Nonidet P-40 (NP40)	Sigma	N-6507
PAGE vertical gel apparatus	BioRad	—
PhosphorImager	Molecular Dynamics	—
PMSF	Sigma	P-7626
Potassium chloride	Sigma	P-4504
Potassium phosphate (KH$_2$PO$_4$)	Sigma	P-5655
Power supply	BioRad	—
Pronase	Calbiochem (LaJolla, CA)	537088
Ribonuclease (RNase) inhibitor	Gibco BRL	15518-012
Sodium acetate	Sigma	S-8625
Sodium chloride (NaCl)	Sigma	S-3014
Sodium deoxycholate	Sigma	D-6750
Sodium phosphate (Na$_2$HPO$_4\cdot$7H$_2$O)	Sigma	S-9390
Sterile H$_2$O	—	—
SYBR Green	Molecular Probes (Eugene, OR)	S-7563
T4 polynucleotide kinase	Gibco BRL	18004-010
Taq polymerase	Gibco BRL	10342020
TEMED	Bio-Rad	161-0800EDU
Thermocycler (PCR)	—	—
TRAPeze Telomerase Detection Kit	Intergen Company (Norcross, GA)	S7700
Tris base	Sigma	T-1503
Xylene cyanol	Sigma	X-4126
b-mercaptoethanol	Sigma	M-7522
^{32}P-ATP	Perkin-Elmer (Wellesley, MA)	BLU502H

metaphase chromosome spreads (42). Fluorescent quantification of hybridization signals
(Q-FISH), alongside standard curves of fluorescent intensity of telomeric DNA of known
length, can determine telomere lengths of individual chromosomes from metaphase nuclei
(75). To measure the average length of telomere repeats in individual cells, a flow cytometry
method using FISH (flow FISH) has been developed (76). However, the traditional technique of measuring telomere length employs Southern blot analysis of TRF obtained by
restriction endonuclease digestion of genomic DNA. The absence of restriction recognition sequences within the telomere allows telomere length to be determined. The chromosomal DNA is cut into small fragments except for the telomeres and subtelomeric regions
(DNA adjacent to the telomere), which together comprise the TRF (Figure 12.4). After digestion, the DNA fragments are separated by gel electrophoresis, blotted, and TRF profiles
are visualized directly or indirectly by hybridization with a labeled telomeric DNA oligonucleotide that is complementary to the telomeric repeat sequence. Finally, the size distribution of the TRFs can be compared to a DNA-length standard (35). A few commercial
kits have been developed recently that use nonradioactive chemiluminescent probe to determine telomere length. TRF analysis can be carried out using either the TeloQuant Te-

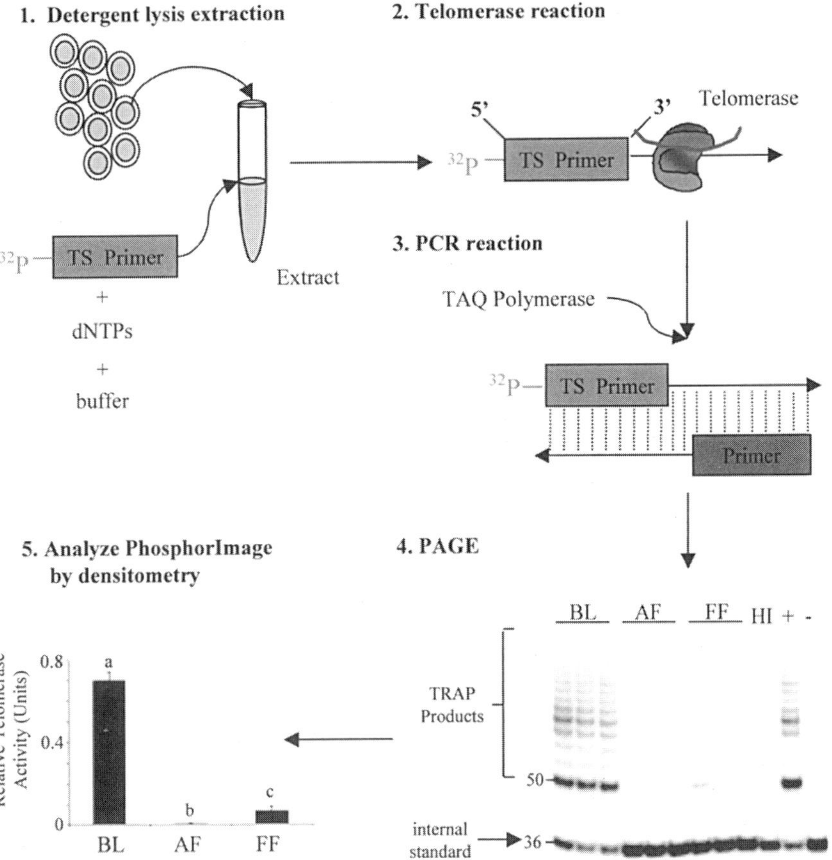

Figure 12.2. Detection of telomerase activity using the telomeric repeat amplification protocol (TRAP) assay. 1. Extracts are prepared by detergent lysis of cells, tissues or embryos. 2. Telomerase in the extract elongates 32P-labeled TS-primers producing DNA products heterogeneous in length. 3. The telomerase products are amplified by the polymerase chain reaction (PCR). 4. The amplified products are run in a 10% polyacrylamide gel and then exposed to a PhosphorImagerTM screen. TRAP products appear as a ladder of DNA products with 6 base increments starting at 50 nucleotides. A 36-bp internal standard is produced as a PCR amplification control and equal loading control between samples. In this example the lanes are: Bovine lymphoma (BL); adult fibroblasts (AF); fetal fibroblasts (FF); heat-inactivated control (HI); positive control (+); negative lysis control (–). 5. The relative telomerase activity (RTA) between samples is determined by densitometric analysis of the phosphorImageTM.

lomere Length Assay Kit (BD PharMingen, San Diego, California) or *Telo TAGGG* Telomere Length Assay (Roche Molecular Biochemicals; see Table 12.3).

10. CONCLUSIONS

The techniques described in this chapter have evolved since the mid-1960s and reflect the progressive refinements of methods used for studying early embryo development. Before these innovations, our understanding of the chromosome makeup of preimplantation embryos was based on projections of observations made during gametogenesis and fetal development. The application of techniques such as the ones described here has filled in many of the gaps in our knowledge of the developing embryo and has provided new insights into the consequences of chromosome variations. These advances, which permit the shift from examination of whole chromosomes to the identification of

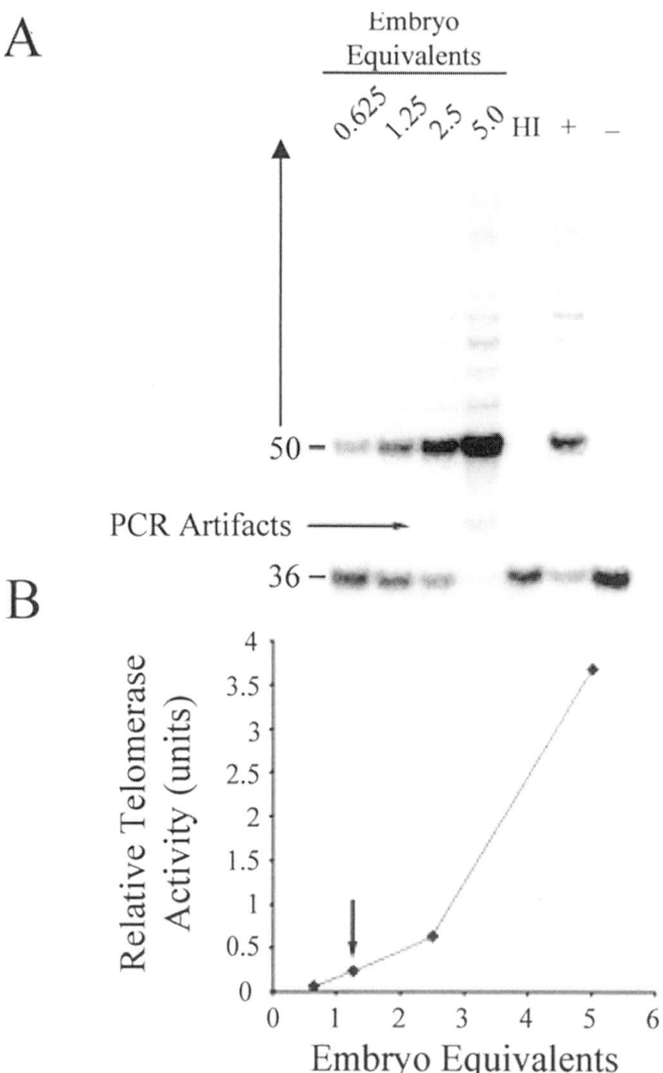

Figure 12.3. Embryo dilution series for telomerase detection by the TRAP assay. To determine the optimal embryo equivalents to use in the TRAP assay, serial dilutions of cell extracts are analyzed. (A) PhosphorImageTM of a PAGE displaying a ladder of telomeric repeat amplification protocol products, with a 36 bp internal standard to control for equal sample loading and aberrant PCR amplification. Telomerase activity is highest in samples of larger embryo equivalents (5.0) and decreases as embryo equivalents are reduced. However, at high protein concentrations, PCR artifacts are apparent due to oligonucleotide primer dimers and *Taq* inhibitors that also reduce the 36 bp internal control signal. Heat inactivated (HI) control, positive 293 cell extract control (+) and negative lysis buffer control (−) are used in each TRAP assay. (B) Densitometric analysis of TRAP reaction products reveals the relative telomerase activity between samples of different embryo quantities. A linear distribution is evident with the chosen optimal TRAP embryo content (1.25 embryo equivalents) for subsequent experiments indicated by the arrow.

chromosomal regions and individual gene sequences, are the result of advances in reproductive technologies and molecular biology. In this regard, advances in in vitro embryo production procedures have created a source of material for the development of techniques and experimental systems for studying the causes and effects of chromosome variations. Recent studies have shown the effects of the in vitro culture environment on chromosome constitution in general and on telomeres in particular. The

Table 12.3. Materials and Reagents Required for the TRF Assay

Material/Reagent	Supplier	Catalogue No.
Telomere Length Assay Kit	BD PharMingen (San Diego, CA)	559838
TeloTAGGG Telomere Length Assay Kit	Roche Molecular Biochemicals (Basel, Switzerland)	2 209 136
Hinf I	Gibco BRL	15223-019
Hybond-N+ Nylon Transfer Membrane	Amersham Pharmarcia Biotch (Piscataway, NJ)	RPN 303B
Rsa I	Gibco BRL	15424-013
Fuji Medical X-ray Film (Super RX)	Fisher	03G050
Agarose	Gibco BRL	15510027
NaOH pellets	Sigma	S-5881
Hydrochloric acid, 50% v/v (1+1)	Fisher	LC15130-1
Lauryl sulfate (SDS)	Sigma	L-4509

long-term effects of subtle variations in choromosomes that arise during in vitro culture and manipulation of embryos and oocytes have yet to be determined.

Acknowledgments   The financial support of the Natural Sciences and Engineering Research Council of Canada, the Canadian Institutes of Health Research, the Food Systems Biotechnology Centre, and the Ontario Ministry of Agriculture, Food and Rural Affairs, and the Ontario Research and Development Challenge Fund is gratefully acknowledged. We thank Ed R. Reyes and Liz St. John for technical support and for assistance with the preparation of this chapter and Dr. Lena King for reviewing the manuscript and for helpful discussions.

Protocol 12.1. Air-dried chromosome spreads from zona pellucida-enclosed oocytes and embryos

1. Introduce embryos to equilibrated, prewarmed culture medium containing a microtubule inhibitor. (Colcemide at a final concentration of 0.05 μg/ml medium is one of the most popularly used.)
2. Incubate at 39°C in 5% CO_2 for up to 5 h (minimum of 30 min).
3. Transfer the embryo into a hypotonic solution (0.88% sodium citrate) and keep at room temperature for 3–5 min (requires optimization).
4. Place the embryo with a small drop of hypotonic solution onto a precleaned slide. The hypotonic solution should not be allowed to evaporate, as this will dry the embryo or (oocyte) and prevent spreading of the chromosomes.
5. Drop 1–2 drops (~ 50 μl) of freshly prepared fixative onto the embryo (1:1 methanol:acetic acid for bovine embryos and 3:1 for mouse embryos).
6. Mark a circle on the back of the slide to indicate the location of the embryo.
7. Air-dry by blowing on the slide.
8. Place the slide into fresh fixative (3:1 methanol:acetic acid) for 30 min (if bovine embryos).
9. Remove the slide from the fixative and air-dry at room temperature in a fume hood or well-ventilated area.
10. Stain in 4% Giemsa (ddH_2O) for 4 min.
11. Mount a # 1 coverglass with DPX mounting medium.

Note: Oocytes do not require incubation with spindle inhibitors. Begin at step 3. If the chromosomes do not spread or are overspread, adjust the duration of the hypotonic treatment (step 3).

Protocol 12.2. Extraction of embryonic interphase nuclei for FISH

Method

1. Transfer embryo to a petri dish containing lysing buffer (0.01 N HCl, 0.1% Tween 20).
2. Place the embryo with 3–5 µl lysing buffer on a microscope slide.
3. The zona pellucida and the blastomere cytoplasm dissolve gradually. Observe the embryos constantly under an inverted phase-contrast microscope (100×). It may be necessary to blow gently on the sample in order to remove the debris and disperse the nuclei. Use needle in difficult cases. Spreading usually takes 15 min at room temperature.
4. Immediately before the nuclei dry out, add fixative dropwise (3:1 methanol: glacial acetic acid).
5. The specimens are then fixed in 3:1 methanol: glacial acetic acid at 4°C for 24 h.
6. Air dry the specimens.
7. To harden nuclei, incubate at 60°C overnight or at 65°C for 2 h.
8. Store at –20°C or at –80°C until use.

Note: Amount of lysing buffer varies with embryo size and ambient temperature. Dead or dying embryos are difficult to spread.

Protocol 12.3. Air-dried chromosome preparations from hatched embryos and fetal tissue

1. Cut embryos or fetal biopsies into 1-mm³ pieces.
2. Incubate in equilibrated, prewarmed culture medium containing colcemide (0.05 µg/ml) at 37°C for 2–4 h.
3. Transfer the pieces individually into 1.0 to 2.0 ml hypotonic solution (0.88% sodium citrate) in a petri dish or multiwell plate and incubate at room temperature for 10 min.
4. Transfer the pieces individually to 1.0 ml of freshly prepared fixative methanol-acetic acid (3:1) in a tube with a round or conical bottom. Let stand at room temperature for 10 min.
5. Remove all of the fixative, leaving the tissue attached, but not dried, to the bottom of the tube.
6. Add one or two drops (50 µl) 50% acetic acid (in distilled water) to cover the pieces. This step requires optimization to accommodate the size and cell density of the embryos and tissues.
7. Gently agitate the acetic acid solution to facilitate dissociation of the cells.
8. Aspirate the acetic acid solution containing the cells into a pipette and repeatedly express and reaspirate onto the surface of a precleaned glass microscope slide warmed to 40°C. The acetic acid solution will form a bead on the slide. As it is aspirated back into the pipette, it will leave a trail of cells attached to the slide.
9. Stain in 4% Giemsa (in ddH₂O) for 4 min.

Note: It is essential to ensure that the fixative is totally removed (step 5). If fixative remains in the tube, the cells will not dissociate in the acetic acid solution. If this happens, the amount of acetic acid added can be adjusted. However, the suspension of cells will become very diluted, and the metaphase spreads will be difficult to locate.

Protocol 12.4. FISH analysis

The technique described here is adapted for mammalian embryo FISH using repetitive DNA sequence probes. The theory and practice of each of the FISH step (probe labeling, denaturation-hybridization, and detection of the ISH sites) are discussed below. A list of reagents required for FISH, along with suppliers and catalog numbers, is presented in Table 12.1.

Hybridization pretreatment

Pretreatment of material before hybridization is a standard step in most ISH protocols and helps increase probe penetration and accessibility to the target and reduce nonspecific hybridization. RNAse to digest single-stranded RNA and enzymes including proteinase K, Pronase, or pepsin (in HCl) to digest proteins are the pretreatment reagents most widely used. However, this step is not essential and can be omitted when performing FISH with repetitive probes, especially when nuclei are free from cytoplasm and other cellular material.

After fixation and before hybridization, the material is usually dehydrated in an ascending series of ethanol baths (70%–90%–100%) for 5 min each to ensure that the probe will not be diluted.

Probe labeling overview

For numerical chromosome analysis, probes derived from repeated satellite sequences are the best for the identification of individual chromosomes in interphase nuclei. Painting probes can also be useful. However, in interphase nuclei the hybridization signals are large and sometimes hard to analyze when dealing with polyploidy or polysomy.

For FISH, probes are labeled by the incorporation of a modified nucleotide with a hapten group for which a specific antibody coupled to an enzyme is available. The nuclei are incubated with this specific antibody, and the use of a substrate for the enzyme produces the signal. The most commonly used haptens are biotin and digoxigenin, which can both be detected by immunohistochemistry.

Biotin can be incorporated into the nucleic acid in several forms, the most widely used being biotin-11-dUTP. Digoxigenin is a phytosteroid that can be incorporated into the nucleic acid in the form of digoxigenin-11-dUTP.

For double-stranded DNA cloned probes, there are two enzymatic labeling procedures: nick translation and random priming. The nick translation procedure for labeling the probe is described here, as random priming does not efficiently label circular DNA, and the labeling is not limited to the insert.

The nick translation reaction, first described by Rigby et al. (71), employs two enzymes: DNase I, which makes cuts in one strand (nicks) at random sites on the template DNA, and DNA polymerase, which incorporates labeled nucleotides. A mixture of both of these enzymes optimized to produce more than 50% of incorporation in 1 h is commercially available in kit form (BioNick labeling system).

Probe labeling method

The procedure uses the BioNick labeling (Gibco, Gaithersburg, MD) system, and the protocol is a modification of the manufacturer's instructions.

1. Add the following (in the order shown) to a 1.5 ml Eppendorf tube on ice:
   5 µl 10× dNTP mix
   500 ng DNA
   Distilled water to make up to a final volume of 45 µl
   5 µl 10× enzyme mix
2. Mix well, centrifuge at 15,000 G for 5 s).
3. Incubate at 16°C for 1 h.
4. Stop reaction by adding 5 µl stop buffer.
5. Remove unincorporated nucleotides by passing the labeled probe through a Sephadex GX50 column.

6. Measure the volume of the elute, then ethanol precipitate the purified, labeled probe by adding the following:

   0.1 mg Herring sperm DNA
   0.1 mg calf thymus DNA
   0.1 vol. 3 M sodium acetate, pH 5.6
   2–2.5 vol. ice-cold ethanol (100%)

   Mix well and place at –70°C for at least 1 h or at –20°C overnight.

7. Centrifuge in a microcentrifuge at 12,000 rpm for 15 min at 4°C. Pour off the supernatant, rinse with ice-cold ethanol (70%), and centrifuge as above. Pour off the supernatant and dry the pellet (air dry or in a vacuum dessicator); resuspend the pellet in 50 µl of hybridization mix (see below) to give a final concentration of 10 ng/µl.

8. Allow the DNA to dissolve overnight at 4°C.

Solutions

Hybridization mix (10 ml)

   7 ml (25 ml formamide + 10 ml SSCP, pH 7)
   1 g dextran sulfate
   3 ml H_2O

Incubate at 37°C for 20 min.

SSCP (1000 ml)

   70 g NaCl
   44 g $Na_3C_6H_5O_7{\cdot}2\ H_2O$
   500 ml H_2O
   500 ml A (phosphate buffer)

Solution A:

   300 ml $Na_2PO_4{\cdot}2\ H_2O$, 0.4 M
   288 ml $Na_2HPO_4{\cdot}12\ H_2O$, 0.4 M

Protocol 12.5. Denaturation and hybridization

Both the probe and the target sequences need to be denatured to make them single stranded for ISH. Usually denaturation is performed in heated formamide. Formamide reduces the melting temperature; T_m, the temperature at which 50% of the nucleic acids are dissociated into single strands (72) by disruption of hydrogen bonds between complementary bases.

The temperature of denaturation and the concentration of formamide should be optimized empirically for each material investigated. Denaturation of target DNA differs depending on the species, cell type, and the fixation procedure. Usually it is more difficult to deal with target DNA than with the probe because the target can be associated with proteins and sometimes can be coiled, especially in chromosomes, and thus can affect the efficiency of denaturation. A protocol of denaturation-hybridization specifically adapted for blastomeres derived from bovine embryos is presented below.

Once the probe and the target nucleic acids are single stranded, the probe is usually left to hybridize with the target overnight at 37°C. However, hybridization efficiency can be influenced by several factors, including probe concentration, duration of exposure, and temperature. Hybridization conditions are often optimized empirically. The duration of hybridization depends on the length, sequence complexity, and the concentration of the probe. Rapid hybridization usually occurs with short and simple sequence probes at high concentration levels.

Competitive in situ hybridization

An additional step can be done before hybridization for probes that are cloned in large DNA fragments, such as phage or cosmids, to block ubiquitous repetitive sequences. This is achieved by a short incubation before hybridization with unlabelled bovine competitive DNA such as calf thymus DNA. Prehybridization annealing is performed for 30–60 min.

Denaturation and hybridization

1. Denature the probe mixture by boiling for 10 min. Place tubes in crushed ice for 3 min. Place the probe mixture at 37°C for 1 h (for competitive ISH).
2. Just before hybridization, denature the nuclear DNA as follows:
   Incubate slides in denaturing solution at 72°C for 2 min.
   Wash slides in cold 2× SSC.
   Dehydrate using a cold ethanol series (70%–90%–100%).
   Air dry.
3. Place the probe mixture (25 μl) over the nuclei, cover with a plastic cover slip, and place the slides in a moist chamber at 37°C for overnight.
4. Remove the plastic coverslip and wash slides twice (2 min in each wash) in 50% formamide in 2× SSC, at 38°C and twice (2 min in each wash) in 2× SSC at 38°C.
5. Incubate slides in blocking solution for 10 min at room temperature.

Note: The posthybridization washing is usually optimized according to the homology between the probe and target sequences. It is advisable to start with a moderate washing and adjust the temperature and salt concentration according to the presence of background and intensity of signals observed. Both high temperature and low salt concentration can help eliminate nonspecific binding.

Protocol 12.6. Detection of ISH sites

For biotin-labeled probes, the visualization of labeled sites may be achieved by two detection systems: antibiotin antibody or biotin–avidin system. Antibody or avidin could directly be conjugated to the signal-generating system, which enables visualization of the sites of probe hybridization.

When using the biotin–avidin system, the signal may be amplified by using biotinylated antiavidin, followed by a further layer of avidin conjugated to the signal-generating system such as FITC. Nonspecific background labeling with the biotin-antibody system is less than that with biotin–avidin.

Fluorescent signals are observed in green under an epifluorescence microscope equipped with a 490-nm excitation filter and 550-nm emission filter. Nuclei are counterstained with propidium iodide (PI), which fluoresces red under green excitation and also at the same wavelengths used to excite FITC in ISH. Because fluorochromes are rapidly bleached by epifluorescence, the use of an antifade reagents to reduce the rate of bleaching is recommended.

1. Dilute 4 μl of stock antibody 1 (antibiotin antibody) in 1.0 ml blocking solution (see below). Use 60 μl for each slide, cover with plastic coverslip. Incubate in a moist chamber at 37°C, for 45 min.
2. Remove coverslip and wash three times in blocking solution at 38°C.
3. Dilute 60 μl of stock antibody 2 (FITC-anti-IgG) in 1.0 ml blocking solution, and proceed as above.
4. Dilute 1 μl of stock PI (1.0 mg/ml) in 1.0 ml of PBS. Use 60 μl for each slide and incubate for 10 min (for two-cell stage embryo), and 5 min (for blastocyst stage) in dark at room temperature.
5. Rinse slides in PBS.

6.   Air dry.
7.   Mount in 10 μl antifade (Vectashield).

Blocking solution (1000 ml)

1000 ml PBS
4 ml BSA
0.5 ml Tween 20

Protocol 12.7. Telomeric repeat amplification protocol

Measurement of telomerase activity in oocytes and embryos

Telomerase activity has been detected in bovine oocytes and in in vitro-produced embryos (42) and nuclear transfer embryos (55) using the TRAPeze telomerase detection kit (Intergen Co., Norcross, GA) with minor modifications (73) of the original TRAP assay (40). This protocol consists of three primary steps: (1) preparation of telomerase extracts from oocytes/embryos, (2) telomerase assay, and (3) polyacrylamide gel electrophoresis of TRAP products. A list of the materials and reagents required for the TRAP assay, along with suppliers and catalog numbers, is presented in table 12.1. Solutions and their composition are presented at the end of this protocol.

Preparation of telomerase extracts from oocytes/embryos

1.   Rinse pools of 10–25 oocytes/embryos three times in PBS and transfer them to a 0.5-ml microcentrifuge tube. Pulse-centrifuge to spin down cells and remove remaining PBS with a narrow-bore mouth pipette. Proceed to step 2 or snap-freeze sample in liquid nitrogen. Store at –80°C.
2.   Add NP40 lysis buffer (working solution made fresh) to a concentration of 1 oocyte/embryo/μl lysis buffer. Aspirate the lysis buffer repeatedly with a narrow-bore pipette to lyse the zona-intact oocytes/embryos. Freeze-thaw the sample three times using liquid nitrogen and let sit on ice for a minimum of 30 min.
3.   Alternatively, remove the zona pellucida surrounding live (not frozen) oocytes/embryos by a brief treatment (2–4 min) in 0.1% Pronase (dissolved in culture media) preequilibrated at 38.5°C in 5% CO_2 in air atmosphere. Wash 3 times in PBS, transfer to a centrifuge tube, and remove remaining PBS by way of a quick spin and mouth pipette removal. Add NP40 lysis buffer (1 oocyte/embryo per μl) and vortex for 30 s. Freeze-thaw the sample three times using liquid nitrogen and let sit on ice for a minimum of 30 min.
4.   Centrifuge samples at 12,000 *g* for 20 min at 4°C. Transfer the supernatant, leaving behind the cellular debris, into a new tube, snap-freeze in liquid nitrogen, and store at –80°C.

Note: The NP40 lysis-buffer is used as the cell extraction solution rather than 1× CHAPS supplied in the TRAPeze kit (Intergen) because it is a more potent extraction buffer. The sensitivity of the assay is therefore increased, allowing for the detection of telomerase activity from fewer oocytes and/or embryos.

Telomerase assay

The assay is performed in two steps: telomerase-mediated extension of an oligonucleotide primer (TS), which serves as a substrate for telomerase, and a PCR amplification of the resultant product using the TS and reverse primers.

Telomerase is a heat-sensitive enzyme. Therefore, as a negative control, sample extracts are tested for heat sensitivity by incubating 5 μl at 85°C for 10 min. A negative lysis control consisting of 2.0 μl of NP40 lysis buffer and a positive telomerase extract control of an immortal telomerase/human telomerase reverse transcriptase-positive 293 cell extract (0.5 µg) are used in each experiment. The optimal embryo/oocyte equivalents and protein content utilized for comparing relative telomerase activities between samples is determined by way of a standard curve (figure 12.3). Two microliters of a serial dilution of embryo/oocyte extracts (0.625–5.0 embryo equivalents) and control samples are each added to 48 µl of the "master mix."

Master mix step

1. Thaw reagents.
2. On ice, add master mix (48 µl per reaction):
   5 µl 10× TRAP buffer
   1 µl 50× dNTP mix
   2 µl ^{32}P-TS primer (1 µl if nonradioactive)
   1 µl TRAP primer mix
   0.4 µl Taq Poly (5 U/ul)
   38.6 µl dH$_2$O (39.6 µl if nonradioactive)
3. Make up heat-inactivated control samples: for each sample extract, incubate 4–5 µl at 85°C for 10 min.
4. In each tube pipette 48 µl of master mix, plus any one of the following:
   2 µl sample extract
   2 µl positive control (telomerase-positive cell line extract)
   2 µl negative control (NP40 lysis buffer).

^{32}P-TS primer (2 µl per reaction)

0.25 µl γ^{32}P-ATP
1 µl TS primer
   0.4 µl 5× exchange buffer
   0.05 µl T4 kinase (10 U/µl)
   0.3 µl dH$_2$O
Place TS primer to be labeled at 37°C for 20 min, 85°C for 5 min, ice (4°C).

Note: If a nonradioactive method of detection (i.e., SYBR Green stain) is used, this end-labeling step is omitted.

TS primer extension and PCR amplification

1. For telomerase extension of TS primer, place tubes at 30°C for 30 min, then transfer immediately to 94°C to inactivate telomerase enzyme.
2. For PCR amplification, 94°C, 4 min; 27 cycles of 94°C, 30 s and 60°C, 30 s; 4°C, soak. For nonradioactive TRAP, use 30 cycles of 94°C, 30 s and 60°C 30 s.

Note: These TRAP reaction conditions have been optimized for use with extracts of pooled in vitro-derived bovine oocytes or embryos. Nevertheless, before each series of experiments, a standard curve of relative telomerase activities from a serial dilution of each sample extract should be conducted to determine the optimal extract concentration for telomerase activity comparisons. The chosen concentration should be within the linear range of detection, excluding samples with excessive PCR artifacts and diminished 36-bp internal control products (Figure 12.3).

As a PCR amplification control, the TRAPeze primer mix contains internal control oligonucleotides K1 and TSK1 that, together with radiolabeled TS oligonucleotide, produce a 36-bp band in every lane (figures 12.2 and 12.3).

Polyacrylamide gel electrophoresis of TRAP products

To visualize and quantify telomerase reaction products, samples are loaded and electrophoresed on a 10% non-denaturing (ND) PAGE. The gel is then fixed and placed on a PhosphorImager (Molecular Dynamics, Amersham Biosciences, Piscataway, NJ) screen overnight. The PhosphorImage screen is scanned using the Molecular Analyst Software (Bio-Rad).

PAGE

1. Add 5 µl loading dye to each reaction.
2. Load 25 µl onto 10% ND-PAGE (no urea) in 0.6× TBE buffer.
3. Run 30 min at 100 V.
4. Run 3.0 h at 300 V until second dye is near the bottom.
5. Fix gel in TRAP gel fixative, 25 min.
6. Expose gel to PhosphorImager plate (or X-ray film) for 24 h.
7. Scan PhosphorImage or develop film.

Quantification of relative telomerase activity

The relative telomerase activity (RTA) is determined by the densitometric analysis of the ladder of TRAP reaction products for each sample (s), positive control extract (pc), background (b) signal, and the 36-basepair internal PCR control products for each sample (ic_s) and for the positive control (ic_{pc}). The internal PCR control product adjusts for sample-to-sample variation and equal gel loading, while the positive control accounts for any gel-to-gel differences in densitometric readings and provides the reference point for the relative telomerase activity for each test sample. Densitometry readings should cover the entire TRAP product profile (50, 56, 62, 72 bp, etc.) for each lane/sample, the densitometry area should remain the same from lane to lane for each analysis. Therefore, the densitometry area should be as large as the lane with the strongest TRAP ladder intensities and processivity (longest TRAP product). A smaller densitometry box will suffice for each 36-bp internal control signal. Quantification of the relative amount of telomerase activity is calculated using the following formula:

$$\text{RTA (units)} = [(s - b)/ic_s]/[(pc - b)/ic_{pc}].$$

Optimization of the TRAP assay and its applications

The method described above has been optimized for extracts prepared from pools ($n = 10\text{–}25$) of bovine oocytes and embryos (42, 55). Quantifiable TRAP products have been obtained from sample extracts derived from as few as five pooled oocytes (data not published). The assay can indisputably compare RTAs from extracts equivalent to a concentration of less than one oocyte/embryo per reaction. However, there is opportunity within the protocol for improving the sensitivity of the TRAP assay. Increasing the incubation time for telomerase extension of the TS primer from 30 min to an hour or greater could increase the ability to detect and quantify telomerase activity from a single-cell sample. In addition, increasing the number of PCR cycles could increase the TRAP amplification products. With any modification to the assay, a new standard curve must be carried out with a serial dilution of sample extracts to determine the new linear range of telomerase detection (Figure 12.3).

It has been suggested that the alternative splicing variants of telomerase reverse transcriptase found in some human oocytes and embryos may be associated with the lack of telomerase activity detected in some human preimplantation embryos (74). Telomeric sequences in early mammalian embryos might act as stress sensors, signaling a permanent cell-cycle arrest (embryo senescence) upon telomeric DNA damage above a certain threshold. The presence of telomerase during embryo development may permit DNA and telomeric DNA repair for the resumption of development after a transient embryonic arrest. Therefore, it is possible that differing telomerase activities in individual oocytes and embryos may serve as a marker for developmental potential (42). Evaluation of whole embryos and/or blastomere biopsies for levels of telomerase activity may be a means to assess environmental factors that affect oocyte and embryo quality prior to embryo transfer.

Solutions

NP40 lysis buffer

Stock buffer (100 ml):
1 ml of 1.0 M stock 10 mM Tris (pH 7.5)
1 ml of 0.1 M stock 1 mM $MgCl_2$
1 ml of 0.1 M stock 1 mM EGTA
10 ml of 1.5 M stock 150 mM NaCl
10 ml 10% glycerol
87 ml sterile H_2O

Working solution:

4 µl 0.1 M PMSF (17.4 mg/ml in 2-propanol)
1.4 µl β-mercaptoethanol
0.84 µl 0.12 M sodium deoxycholate
10 µl NP40
40 µl RNase inhibitor (10 U/µl)

Bring to 4 ml with NP40 stock buffer.

PBS (10×)

80 g NaCl
2 g KCl
11.5 g $Na_2HPO_4 \cdot 7H_2O$
2 g KH_2PO_4

Adjust pH to 7.3; add to 1 l with sterile H_2O.

10% Polyacrylamide stock in 0.5× TBE (400 ml)

100 ml 40% polyacrlamide solution (19:1)
40 ml 5X TBE buffer
260 ml deionized H_2O

10% Polyacrylamide gel (50 ml)

49.5 ml 10% polyacrylamide stock (19:1)
0.5 ml 10% ammonium persulfate
0.05 ml TEMED

TAE buffer (50×)

242 g TRIS base
57.1 ml glacial acetic acid
100 ml 0.5 M EDTA (pH 8.0)

Add sterile H_2O to 1 l, autoclave.

TBE buffer (5×)

54 g Tris base
27.5 g boric acid
20 ml 0.5 M EDTA

Add deionized water to 1 l.

TRAP gel fixative (1 l)

29.2 g NaCl
3.28 g Na acetate
800 ml 50% ethanol

Adjust pH to 4.2; add to 1 l with 50% ethanol.

Gel loading dye solution (5 ml)

2.5 ml glycerol
1.0 ml 1.25% bromophenol blue
1.0 ml 1.25% xylene cyanol
0.5 ml 0.5 M EDTA (pH 8.0)

Protocol 12.8. Terminal repeat fragment analysis

Because Q-FISH is complex and requires specialized equipment for the detection and measurement of fluorescent telomeric signals, the standard Southern blotting TRF method is the simplest approach to measure telomere length within cell and tissue samples. However, due to the requirement of a minimal quantity of genomic DNA (~2 µg), there are limitations with respect to the types of samples that can be analyzed. Therefore, early developmental stages cannot be analyzed for TRF length, but data have been obtained from 2-week-old in vivo bovine embryos that have been flushed out of reproductive tracts (D. H. Betts, unpublished data). Reliable data can be obtained from later stage embryos, embryo-derived cell lines (e.g., embryonic stem cells), and from fetal and newborn tissues. The protocol described below is a summary of the procedure described in the Telomere Length Assay Kit (BD PharMingen; table 12.2), emphasizing the key steps and listing helpful tips to generate consistent results.

1. Isolate genomic DNA using standard protocols. There are numerous commercial kits (e.g., Roche Molecular Biochemicals; Qiagen) now available to isolate genomic DNA from cells, blood, and tissues. *Note*: To accurately quantify the DNA concentration, RNase (table 12.2) treatment will remove any contaminating RNA from your samples. Precise DNA quantification will produce more consistent results because TRF length determination not only depends on the length of the fragments, but on the intensities of the TRF profiles. DNA overloading may hinder the comparison of relative TRF lengths between samples.
2. Digest genomic DNA using *RsaI/HinfI* enzyme mixture in a 1× final concentration of enzyme reaction buffer. Digest for 12–18 h at 37°C using 4 U the enzyme mix/µg DNA. At least 2 µg/sample should be used for cells with large telomere lengths.
3. Resolve TRFs on 0.6% agarose gel using 1× TAE buffer. *Note*: Use a horizontal gel apparatus that uses gel casts of a least 15 cm in length.
4. Load at least 2 µg/lane of digested genomic DNA for long telomeres, 7.5 µg/lane for very short telomeres (e.g., from tumor cells).
5. Load 12 marker lanes of biotinylated marker (*BstE*II and *Hind*III). Heat at 60°C for 3 min before loading.
6. Run gel at 1 V/cm overnight (24–28 h) to resolve TRF fragments.
7. Soak gel in 0.25 M HCl (table 12.2) for 15 min with gentle agitation. Repeat. *Note*: It is important to make fresh 0.25 M HCl from a concentrated stock solution just before use.
8. Rinse in dH$_2$O.
9. Soak gel in 0.4 N NaOH for 15 min with gentle agitation. Repeat.
10. Using 0.4 N NaOH as transfer buffer, prepare a Southern transfer (as depicted in Maniatis et al. [77]) to a positively charged nylon membrane (Hybond-N$^+$,

Amersham). Transfer at room temperature for 1 h with 2 inches of paper towels, then change paper towels and continue transfer for 1–2 h. *Note*: The efficiency of DNA transfer is optimal at transfer times of 3 h or more.

11. Rinse membrane in 2× SSC. Air-dry the filter for storage if desired. Store in desiccator at room temperature.

12. Denature biotinylated telomere probe by boiling (100°C) for 10 min, then chill on ice for 5 min before use.

13. Warm the hybridization buffer to bring components back into solution. Soak filter using 0.1 ml/cm^2 of membrane. Rotate in hybridization tube at 55°C, 30 min.

14. Add biotinylated telomere probe to hybridization buffer; rotate overnight at 55°C. The final concentration of the probe can range from 1 to 5 ng/ml (5 ng/ml most common).

15. Prewarm hybridization stringency wash buffer (2×) to bring components back into solution. Dilute buffer 1:1 with sterile H$_2$O. The resulting buffer contains 2X SSC/0.1% SDS.

16. Wash the membrane three times for 5 min each in 1X stringency wash buffer (0.2 ml/cm^2 of membrane) with gentle agitation at 55°C.

17. Pour off hybridization stringency wash buffer and add membrane-blocking buffer to cover membrane (0.25 ml/cm^2 of membrane). Incubate for 30 min at room temperature.

18. Pour off some block buffer into a separate tube and add stabilized streptavidin-horseradish peroxidase to make a 1:300 final dilution when added to the membrane. Add diluted streptavidin-horseradish peroxide to the membrane and incubate for 30 min at room temperature with gentle agitation.

19. Dilute wash buffer (4×) to 1× with sterile H$_2$O. Wash the membrane four times for 5 min per wash with gentle agitation.

20. In a clean wash tray, incubate membrane in the North2South Substrate Equilibrium Buffer (0.25 ml/cm^2 of membrane) for 5 min at room tempeerature with gentle agitation.

21. Prepare North2South substrate working solution (0.1 ml/cm^2 of membrane) by mixing equal volumes of the North2South Stable Peroxide Solution and North2South Luminol/Enhance Solution.

22. Transfer the membrane to the substrate working solution and incubate for 5–10 min at room temperature.

23. Pour off excess substrate from the membrane and place the filter between two pieces of plastic wrap. Remove any trapped air bubbles by rolling a pipette over the membrane with paper towels underneath and on top of it. Dry the outside of the plastic wrap with paper towel.

24. Place covered membrane in an X-ray film cassette and expose the blot to film for 10 s, 30 s, 2 min, or longer until a clear signal with low background is obtained on development.

25. Calculate the mean TRF length. Using Molecular Analyst Software (Bio-Rad), divide the scanned image into a grid consisting of 30 boxes per column. Position the grid over the entire vertical length of a lane so that many boxes overlay a signal telomere smear. For each sample, optical density (OD) and length (L) are computed for each grid box, where OD is the total signal intensity within a grid box, and L is the molecular weight at the midpoint of the grid box. A standard curve of molecular weight versus migration is determined from a densitometric analysis of the molecular weight markers on the scanned image. The mean TRF length for each sample is calculated using the formula:

$$L = \Sigma \, (Odi \times Li) \, / \, \Sigma \, (ODi)$$

where ODi and Li are the signal intensity and TRF length, respectively, at position i on the gel image, where i denotes the grid box number from 1 to 30 (Figure 12.4).

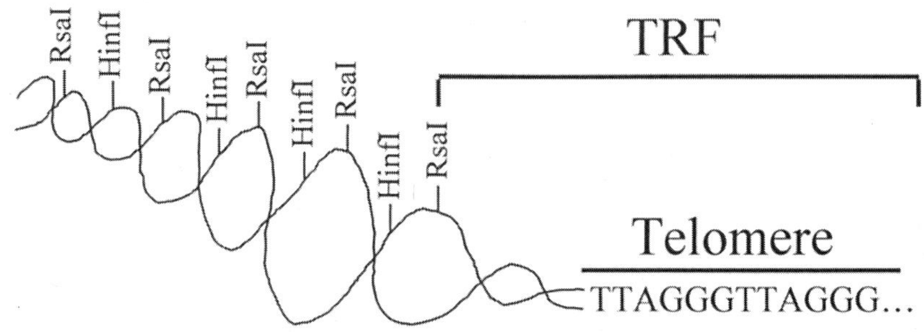

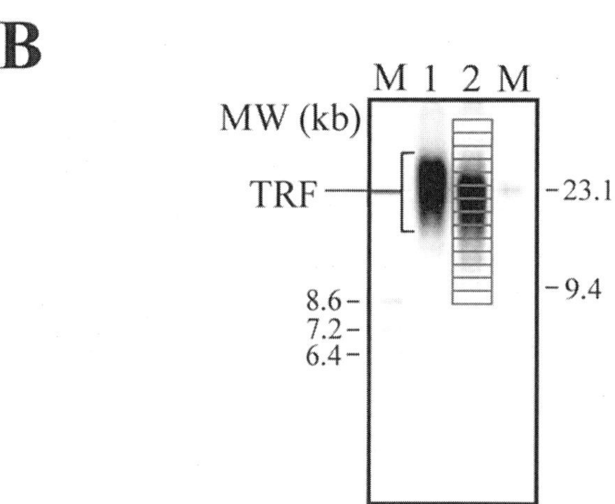

Figure 12.4. Measurement of telomere length by terminal restriction fragment (TRF) analysis. (A) The absence of restriction recognition sequences within the telomere allows the determination of telomere length. Double digestion of genomic DNA with RsaI and HinfI restriction enzymes cut the chromosomal DNA into small fragments. The distance from the last restriction site to the end of the telomere comprises the terminal restriction fragment (TRF). (B) The digested DNA (samples 1 and 2) is separated on a 0.6% agarose gel and transferred to positively-charged membrane for Southern blot analysis. The blot is probed with a biotinylated telomere-specific probe and the terminal restriction fragments are detected by a chemiluminescent detection system. The mean TRF length is calculated from the position of the detected signal (red boxes) relative to the position of known size standards (M).

Solutions

SSC (20×)

175.3 g NaCl
88.2 g Na citrate · 2 H_2O
800 ml Sterile H_2O

Adjust pH to 7.0 (add a few drops of 10 N NaOH); volume to 1 l with sterile H_2O, autoclave.

TAE buffer (50×)

242 g TRIS base
57.1 ml glacial acetic acid
100 ml 0.5 M EDTA (pH 8.0)

Add sterile H_2O to 1 l and autoclave.

References

1. Hsu, T. C., and Pomerat, C. M. (1953). *J. Hered.*, **44**, 23.
2. Rothfels, K. H., and Simnovitch, L. (1958). *Stain Technol.*, **33**, 73.
3. Moorhead, P. S., Nowell, P. C., Mellman, W. J., Batipps, D. M., and Hungerford, D. A. (1960). *Exp. Cell. Res.*, **20**, 413.
4. Basrur, P. K., and Gilman, J. P. W. (1964). *Nature*, **204**, 1335.
5. Lejeune, J., Gauthier, M., and Turpin, R. C. (1959). *C.R. Acad. Sci.*, **248**, 1721.
6. Gustavsson, I., and Rockborn, G. (1964). *Nature*, **203**, 990.
7. Tarkowski, A. K. (1966). *Cytogenetics*, **5**, 394.
8. McFeely, R. A. (1967). *J. Reprod. Fertil.*, **13**, 579.
9. Chandley, A. C. (1971). *J. Reprod. Fertil. Suppl.*, **14**, 1.
10. King, W. A., Linares, T., Gustavsson, I., and Bane, A. (1979). *Vet. Sci. Com.*, **3**, 51.
11. Carr, D. H. (1967). *J. Obstet. Gynecol.*, **97**, 283.
12. Hassold, T., Abruzzo, M., Adkins, K., Griffin, D., Merrill, M., Millie, E., Saker, D., Shen, J., and Zaragoza, M. (1996). *Environ. Mol. Mutagen.*, **28**, 167.
13. King, W. A. (1990). *Adv. Vet. Sci. Comp. Med.*, **34**, 229.
14. Placho, M. (1998). In *Fertility and Reproductive Medicine*, R. D. Kempers, J. Cohen, A. F. Haney, and J. B. Younger, eds., p. 711. Elsevier, New York.
15. Edirisinghe, W. R., Murch, A., Junk, S, and Yovich, J. (1997). *Hum. Reprod.*, **12**, 2784.
16. Kawarsky, S. J., Basrur, P. K., Stubbings, R. B., Hansen, P. J., and King, W. A. (1996). *Biol. Reprod.*, **54**, 53.
17. Dunne, L. D., Diskin, M. G., and Sreenan, J. M. (2000). A*nim. Reprod. Sci.*, **58**, 39.
18. Iwasaki, S., Hamano, S., Kuwayama., M., Yamashita., M., Ushijima, H., Nagaoka, S., and Nakahara, T. (1992). *J. Exp. Zool.*, **261**, 79.
19. Matwee, C., Betts, D. H., and King, W. A. (2000). *Zygote*, **8**, 57.
20. Muller, H. J. (1938). The collecting net. *Woods Hole*, **13**, 181.
21. McClintock, B. (1941). *Genetics*, **41**, 234.
22. Sandell, L. L., and Zakian, V. A. (1993). *Cell*, **75**, 729.
23. Ashley, T., and Cacheiro, N. L. A. (1993). *Chromosoma*, **102**, 112.
24. Scherthan, H., Bahler, J., and Kohli, J. (1994). *J. Cell Biol.*, **127**, 273.
25. Nimmo, E. R., Pidoux, A. L., Perry, P. E., and Allshire, R. C. (1998). *Nature*, **392**, 825.
26. Yu, G.-L., Bradley, J. D., Attardi, L. D., and Blackburn, E. H. (1990). *Nature*, **344**, 126.
27. Kirk, K. E., Harmon, B. P., Reichardt, I. K., Sedat, J. W., and Blackburn, E. H. (1997). *Science*, **275**, 1478.
28. Blackburn, E. H. (1991). *Nature*, **350**, 569.
29. Allshire, R. C., Dempster, M., and Hastie, N. D. (1989). *Nucleic Acids Res.*, **17**, 4611.
30. Moyzis, R. K., Buckingham, J. M., Cram, L. S., Dani, M., Deaven, L. L., Jones, M. D., Meyne, J., Ratliff, R. L., and Wu, J. R. (1988). *Proc. Natl. Acad. Sci. USA*, **85**, 6622.
31. Preston, R. J. (1997). *Radiat. Res.*, **147**, 529.
32. Griffith, J. D., Comeau, L., Rosenfield, S., Stansel, R. M., Bianchi, A., Moss, H., and de Lange, T. (1999). *Cell*, **97**, 503.
33. Olovikov, A. M. (1973). *J. Theor. Biol.*, **41**, 181.
34. Greider, C. W., and Blackburn, E. H. (1985). *Cell*, **43**, 405.
35. Harley, C. B., Futcher, A. B., and Greider, C. W. (1990). *Nature*, **345**, 458.
36. Allsopp, R. C., Chang, E., Kashefi-Aazam, M., Rogaev, E. I., Piatyszek, M. A., Shay, J. W., and Harley, C. B. (1995). *Exp. Cell Res.*, **220**, 194.
37. Lindsey, J., McGill, N. I., Lindsey, L. A., Green, D. K., and Cooke, H. J. (1991). *Mutat. Res.*, **25**, 45.
38. Hastie, N. D., Dempster, M., Dunlop, M. G., Thompson, A. M., Green, D. K., and Allshire, R. C. (1990). *Nature*, **346**, 866.
39. Harley, C. B., Vaziri, H., Counter, C. M., and Allsopp, R. C. (1992). *Exp. Gerontol.*, **27**, 375.
40. Kim, N. W., Piatyszek, M. A., Prowse, K. R., Harley, C. B., West, M. D., Ho, P. L., Coviello, G. M.,Wright, W. E., Weinrich, S. L., and Shay, J. W. (1994). *Science*, **266**, 2011.
41. Wright, W. E., Piatyszek, M. A., Rainey, W. E., Byrd, W., and Shay, J. W. (1996). *Dev. Genet.*, **18**, 173.
42. Betts, D. H., and King, W. A. (1999). *Dev. Genet.*, **25**, 397.
43. Collins, K., Kobayashy, R., and Greider, C. W. (1995). *Cell*, **81**, 677.

44. Kyo, S., Takakura, M., Kohama, T., and Inoue, M. (1997). *Cancer Res.* **57**, 610.
45. Broccoli, D., Young, J. W., and de Lange, T. (1995). *Proc. Natl. Acad. Sci. USA*, **92**, 9082.
46. Burger, A. M., Bibby, M. C., and Double, J. A. (1997). *Br. J. Cancer*, **75**, 516.
47. Harle-Bachor, C., and Boukamp, P. (1996). *Proc. Natl. Acad. Sci. USA*, **93**, 6476.
48. Shay, J. W., and Bacchetti, S. (1997). *Eur. J. Cancer*, **3**, 787.
49. Sharma, H. W., Sokoloski, J. A., Perez, J. R., Maltese, J. Y., Sartorelli, A. C., Stein, C. A., Nicols, G., Khaled, Z., Telonang, N. T., and Narayanan, R. (1995). *Proc. Natl. Acad. Sci. USA*, **92**, 12343.
50. Youngren, K., Jeanclos, E., Aviv, H., Kimura, M., Stock, J., Hanna, M., Skurnick, J., Bardeguez, A., and Aviv, A. (1998). *Hum. Genet.*, **102**, 640–643.
51. Prowse, K. R., and Greider, C. W. (1995). *Proc. Natl. Acad. Sci. USA*, **92**, 4818.
52. Ulaner, G. A., and Giudice, L. C. (1997). *Mol. Hum. Reprod.*, **3**, 769.
53. Kato, Y., Tani, T., and Tsunoda, Y. (2000). *J. Reprod. Fertil.*, **120**, 231.
54. Tian, X. C., Xu, J., and Yang, X. (2000). *Nat. Genet.*, **26**(3), 272.
55. Betts, D. H., Bordignon, V., Hill, J. R., Winger, Q., Westhusin, M. E., Smith, L. C., and King, W. A. (2001). *Proc. Natl. Acad. Sci. USA*, **98**, 1077.
56. Shiels, P. G., Kind, A. J., Campbell, K. H. S., Waddington, D., Wilmut, I., Colman, A., and Schnieke, A. E. (1999). *Nature*, **399**, 316.
57. Shiels, P. G., Kind, A. J., Campbell, K. H. S., Wilmut, I., Waddington, D., Colman, A., and Schnieke, A. E. (1999). *Cloning*, **1**, 119.
58. Lanza, R. P., Cibelli, J. B., Blackwell, C., Cristofalo, V. J., Francis, M. K., Baerlocher, G. M., Mak, J., Schertzer, M., Chavez, E. A., Sawyer, N., Lansdorp, P. M., and West, M. D. (2000). *Science*, **288**, 665.
59. Figueroa, R., Lindenmaier, H., Hergenhahn, M., Vang Neilson, K., and Boukamp, P. (2000). *Cancer Res.*, **60**, 2770.
60. Zhu, J., Wang, H., Bishop, J. M., and Blackburn, E. (1999). *Proc. Natl. Acad. Sci. USA*, **96**, 3723.
61. Gall, G., and Pardue, M. L. (1971). *Meth. Enzymol.*, **38**, 470.
62. Huang, W. M., Gibson, S. J., Facer, P., Gu, J., and Polak, J. M. (1983). *Histochemistry*, **77**, 275.
63. Gall, J. G., and Pardue, M. L. (1969). *Genetics*, **63**, 378.
64. John, H. A., Birnstiel, M. L., and Jones, K. W. (1969). *Nature*, **223**, 582.
65. Verlinsky, Y., Cieslak, J., Freidine, M., Luakhnentko, V., Wolf, G., Kovalinskaya, L., White, M., Lifchez, A., Kaplan, B., Moise, J., Valle, J., Ginsberg, N., Stom, C., and Kuliev, A. (1995). *Hum. Reprod.*, **10**, 1923.
66. Plachot, M., Junca, A. M., Mandelbaum, J., de Grouchy, J., Salat,-Barous, J. and Cohen, J. (1987). *Hum.. Reprod.*, **2**, 29.
67. Pellestor, F., Dufour, M. C., Arnal, F., and Humeau, C. (1994). *Hum. Reprod.*, **9**, 293.
68. Zenzes, M. T., and Casper, R. F. (1992). *Hum. Genet.*, **88**, 367.
69. Solinas-Toldo, S., Mezzelani, A., Hawkins, G. A., Bishop, M. D., Olsakeer, L., Mackinlay, A. Ferretti, L., and Fries, R. (1995). *Cytogenet. Cell. Genet.*, **69**, 1.
70. Slimane, W. Vaiman, D., Godard, S., Vaiman, A., Cribiu, E., and Renard, J. P. (2000). *Genet. Sel. Evol.*, **32**, 217.
71. Rigby, P., Dieckman, M., Rhodes, C., and Berg, P. (1977). *J. Mol. Biol.*, **113**, 237.
72. Thomas, C. A., and Dancis, B. M. (1973). *Mol. Biol.*, **77**, 44.
73. Holt, S. E., Aisner, D. L., Shay, J. W., and Wright, W. E. (1997). *Proc. Natl. Acad. Sci. USA*, **944**, 10687–10692.
74. Brenner, C. A., Wolny, Y. M., Adler, R. R., and Cohen, J. (1999). *Mol. Hum. Reprod.*, 5845.
75. Poon, S. S., Martens, U. M., Ward, R. K., and Lansdorp, P.M. (1999). *Cytometry*, **36**, 267.
76. Rufer, N., Dragowska, W., Thornbury, G., Roosnek, E., and Lansdorp, P. M. (1998). *Nat. Biotechnol.*, **16**, 743.
77. Maniatis, T., Fritsch, E. F., and Sambrook, J. (1982). *Molecular Cloning: A Laboratory Manual.* Cold Spring Harbor Laboratory Press, Cold Spring Harbor, NY.
78. Viuff, D., Greve, T., Avery, B., Hyttel, P., Brockhoff, P. B., and Thomsen, P. D. (2000). *Biol. Reprod.*, **63**, 443–1148.

13

C. WRENZYCKI
A. J. WATSON
M. CALDER

Microscale RNA Isolation from Mammalian Embryos

1. METHODS FOR ISOLATING RNA

Many different microscale techniques have been developed to study gene expression in samples of small numbers of cells or rare tissues. Gene expression analysis includes the study of RNA and protein levels. Alterations in mRNA levels are often associated with the corresponding protein level modification. Presence or absence of specific mRNAs could be associated with the presence or absence of the protein. Therefore, the use of RNA analysis techniques is directed at screening the entire RNA pool before an investigation of protein levels and synthesis.

A single mammalian cell contains a total of approximately 1×10^{-5} µg total RNA. Eighty-five percent is ribosomal RNA (rRNA), and 15–20% is composed of a variety of low molecular weight species, such as transfer RNA (tRNA). The remaining 1–5% is messenger RNA (mRNA) of both different size and sequence complements.

Several difficulties exist in the isolation of RNA from embryos or oocytes. The first is that the quantity of RNA is very limiting. The amount of RNA in a murine oocyte has been estimated as 0.35–0.47 ng (1, 2), in the bovine oocyte as 0.98–2.4 ng (2, 3), and it varies with stage from 0.7 to 5.3 ng in the bovine preimplantation embryo (3). Procedures such as Northern blot hybridization may require a minimum of 2–5 µg of total RNA for their application. Therefore, to overcome this limitation, RNA has been isolated from large pools (up to 300) of oocytes or embryos (4). Disadvantages with this approach include the expense and time to collect the large embryo pools required for one experiment and the fact that the use of large pools obscures the variability in mRNA expression between individual embryos or oocytes.

Polymerase chain reaction (PCR)-based techniques such as reverse transcription-polymerase chain reaction (RT-PCR), differential display RT-PCR (DD-RT-PCR), and suppressive subtractive hybridization-PCR (SSH-PCR) have been developed to incorporate RNA isolation from much smaller pools of embryos including individual embryos. Of these techniques, the most commonly used is RT-PCR, which allows the conversion of small quantities of single-stranded mRNA to double-stranded cDNA by reverse transcription and then amplification by PCR of a known sequence of interest following the design of appropriate primers. With these techniques, the ability to detect differences in mRNA expression even in small numbers or individual oocytes or embryos is greatly increased. However, care must be taken in optimizing PCR conditions and cycle numbers. DD-RT-PCR (see chapter 15) involves reverse transcription of a subset of mRNAs with a primer of poly-T anchored with one of the four bases, A, T, C, or G. PCR amplification is accomplished with 1 of 24 random decamers. Unknown products are cut from a gel, cloned, and sequenced. Similarly, SSH-PCR may identify both known and unknown products (see chapter 18). The SSH-PCR technique consists of using RNA from one treatment divided into two pools and amplified with one of two different ends and then combined with excess unlabeled cDNA from a distinct treatment group. Labeled and unlabeled cDNAs are allowed to hybridize to remove RNA sequences common to both treatments. Unpaired end-labeled cDNA is isolated and combined with cDNA with the opposite end label. Then PCR is performed with a system that exponentially amplifies only cDNA with different ends. The PCR products are ligated into plasmids, bacteria are transformed with these plasmids, and the products are cloned and sequenced.

2. RNA ISOLATION IN GENERAL

Successful RNA isolation depends on a number of parameters, including effective disruption of cells or tissue, denaturation of ribonuclease (RNase) activity, and removal of contaminating DNA and proteins. Immediate inactivation of RNases is extremely important, as these enzymes efficiently degrade RNA samples and are active under a wide variety of conditions (5). Unfortunately, RNases are extremely difficult to inactivate. In all procedures, because RNases are prevalent within tissues, it is important to inhibit endogenous RNases that could degrade RNA during the isolation procedure. In addition, because RNases are prevalent on skin surfaces, it is important not to accidentally introduce exogenous RNases into samples. Thus, gloves must always be worn to handle pipettors, tubes, and tips. Glassware can be sterilized by autoclaving and decontaminated by baking. Autoclaving alone is not sufficient to inactivate RNases. Use of disposable plasticware is advisable, as these items are generally RNase-free at manufacture. Subsequently, ungloved hands should not be inserted into bags or boxes of materials used for isolating RNA, and replacement pipette tips should be loaded into autoclavable boxes with gloves. A dedicated set of pipettors, which are not used for other purposes, should be used exclusively for isolating RNA. Diethyl pyrocarbonate (DEPC; 1 ml/L, 0.1%) should be used to treat most buffers and water used for RNA isolation. TRIS buffers cannot be treated directly with DEPC. Therefore, caution must be used to avoid RNase contamination when weighing out TRIS and DEPC-pretreated water, and treated glassware should be used when preparing TRIS buffers. Care must also be taken not to contaminate bottles after opening. Single-use buffer aliquots are favored when practical. RNase inhibitors (RNasins) can be added to buffers and water for resuspending RNA samples, but as RNasins are removed during extraction, they may have to be added at more than one stage during RNA isolation.

RNA can be isolated as either total RNA (Protocols 13.1–13.4) or poly(A)$^+$ RNA (Protocols 13.5–13.6) (6), depending on the downstream application for the isolated RNA. The use of poly(A)$^+$ RNA offers a variety of advantages, such as lower mRNA-independent synthesis during RT-PCR. When poly(A)$^+$ RNA is desired from limiting amounts of starting material, direct isolation may lead to better yields.

3. ISOLATION OF TOTAL RNA

Simultaneous disruption of cells and inactivation of RNases is achieved by lysing of cells in strong denaturing agents. Most commonly, tissue RNA is isolated by a modification of the Chomczynski and Sacchi (7) method using guanidium isothiocyanate (GITC), phenol, and chloroform/isoamyl alcohol (24:1); a procedure that may be performed on several samples in a few hours (Protocols 13.1, 13.2, 13.3). GITC is a chaotropic and deproteinizing agent, inhibiting RNases during tissue processing. Homogenization of the tissue shears DNA and increases the yield of RNA, as well as increasing the contact of GITC with cellular contents. Phenol and chloroform denature and extract proteins, thus separating RNases from contact with RNA. In addition, sarcosyl and β-mercaptoethanol, potent reducing agents, may denature tissue RNases. RNA is then precipitated with isopropanol or ethanol. Other methodologies use guanidine-HCl, dithiothrietol, and sodium lauryl sarcosinate (8) or GITC combined with a high-speed centrifugation overnight through a cesium chloride gradient (5). The latter technique, although effective, takes more time and requires an ultracentrifuge.

4. ISOLATION OF POLY(A)⁺ RNA

Edmonds et al. (10) and Aviv and Leder (6) originally used a solid support of d(T)-cellulose to extract mRNAs, and to remove rRNA and tRNA from their preparations. In addition, poly(U)-sepharose columns have also been used to separate mRNA (11). Such products are commercially available. RNA that has been previously isolated and precipitated is added to the column in TRIS, NaCl, EDTA buffer and allowed to bind. For d(T)-cellulose, mRNAs are later eluted with lowered ionic strength buffers; for poly(U)-sepharose columns, formamide and HEPES are also used (12). Often two column passes must be used. The diluted mRNA must be reprecipitated with ethanol and sodium acetate to concentrate.

5. COMMERCIALLY AVAILABLE KITS TO ISOLATE TOTAL OR POLY(A)⁺ RNA

Currently, several biotech companies market RNA isolation buffers or kits. Most are based on the Chomczynski and Sacchi (7) method. One such product, Trizol (Gibco, Burlington, ON), contains both GITC and phenol and can be used to isolate RNA, DNA, or proteins in one easy step, followed by alcohol precipitation steps. Other procedures, such as the Fast Trac 2.0 mRNA isolation kit (Invitrogen, Burlington, ON) use oligo(dT) cellulose to isolate poly(A)⁺ RNA (Protocol 13.5). Other products such as GlassMAX RNA microisolation Spin Cartridge system (Gibco) use GITC and sodium iodide in a silica column and remove contaminating DNA with DNAse I. RNeasy (Qiagen, Mississauga, OH) also uses GITC and silica spin column, avoiding phenol/chloroform extraction steps. Still other products use oligo d(T) bound to magnetic beads to extract poly(A)⁺ RNA (Protocol 13.6).

6. ISOLATION OF RNA FROM OOCYTES OR EMBRYOS

For microscale RNA isolation from single to small numbers of oocytes or embryos, several methods have been used, often modifications of the microscale method developed by Arcellana-Panlilio and Schultz (9) adapted from the phenol-chloroform extraction method originally described by Braude and Pelham (13) (Protocol 13.4). There are many protocols that are similar to those found in Temeles et al. (14). Often RNase inhibitors are used during RNA isolation and reverse transcription. Still, for RT-PCR applications with very limited amounts of RNA, it is necessary to carefully optimize PCR conditions. A technique for using reverse transcription on lysed cells without RNA extraction and purification was developed by Revel et al. (15) for construction of cDNA libraries from a small numbers of mouse oocytes. This method was later modified by Robert et al. (16) for use in DD-RT-PCR, with the proviso that fewer than 15 embryos must be used, otherwise proteins and genomic DNA may interfere with the efficiency of the reverse transcription.

7. COMMERCIALLY AVAILABLE MICROSCALE METHODS TO ISOLATE RNA

Commercially available products for RNA isolation such as Trizol (16), Ultraspec (Biotecx Laboratories, Houston, TX [18]) may be used in scaled down protocols to successfully isolate small quantities of RNA. Other technologies using commercially available kits employing oligo (dT) microcolumns or oligo (dT) magnetic beads (18, 19) have been developed in other laboratories. Some of the procedures differ in the type of carrier recommended; glycogen (16) and polyinosonic acid (17) have been used successfully.

There are several kits marketed by biotech companies to isolate RNA from small numbers of cells within 1–2 h. Oligo dT RNA isolation columns such as Gibco RNA microisolation Spin Cartridge system or the Stratagene Absolutely RNA Nanoprep Kit will effectively isolate mRNA from $<10^2$ cells. We have recently tried several of these kits (Stratagene Absolutely RNA Nanoprep Kit (La Jolla, CA), Qiagen RNeasy Mini kit, Sigma GenElut Mammalian Total RNA kit) in our laboratory. They differ in ease of use and the volume used to elute RNA. Some include a DNase treatment step to reduce genomic contamination. Samples are lysed in GITC containing buffer, with β-mercaptoethanol. Alcohol is then added and the sample placed on a silica gel column. The column is washed several times using a microfuge, then RNA is eluted in RNase-free water or buffer. We found that the Stratagene kit will effectively isolate mRNA from single embryos and can be used to amplify up to seven different RT-PCR amplicons from an individual embryo. Each manufacturer's recommendations should be followed to get optimal yields of mRNA, although multiple elution washes are recommended. The silica gel system is expected to remove some small rRNA and tRNAs thus enriching for mRNAs. In some cases, access to a vacuum desiccator may be required to reduce the elution volume before reverse transcription.

8. SUMMARY

Although the methods that have been developed for the efficient isolation of total RNA and polyA$^+$ mRNA from small pools or individual samples of oocytes and preimplantation stage embryos are diverse, they represent a critical starting point for most gene expression studies. Care must be taken in eliminating or reducing RNase activity. All of the procedures reported in this chapter are proven methods that will enable the transition to expression studies by the application of RT-PCR, DD-RT-PCR, or SSH-PCR methodologies.

Protocol 13.1. Chomczynski and Sacchi (7) method for tissue

Denaturing buffer (buffer D)

4 M GITC
25 mM sodium citrate, pH 7
0.5% sarcosyl
0.2 M β-mercaptoethanol, added just before use

Other materials

2 M sodium acetate
Saturated phenol
Chloroform:isoamyl alcohol (24:1)

1. Mince tissue using RNase-free tools and homogenize with denaturing buffer with a mortar and pestle or tissue homogenizer. For cultured cells or easily disrupted tissue, homogenization can be accomplished using a small-gauge needle and syringe. Large RNase-free tubes with resistance to chemicals should be used.
2. *After homogenization*, add per milliliter denaturing buffer 0.2 ml of 2 M sodium acetate, 0.2 ml phenol (saturated), 0.2 ml chloroform:isoamyl alcohol (49:1).
3. Vortex the mixture thoroughly and cool on ice 15 min. Centrifuge samples 20 min at 4°C at 10,000 g. Transfer the RNA (aqueous phase) to a new tube and precipitate with 1 volume isopropanol or 2 volumes absolute ethanol at –20°C for 1 h.
4. Centrifuge a second time at 10,000 g for 20 min at 4°C. A white pellet should be present in the bottom of the tube.
5. Resuspend the pellet in 0.3 ml of buffer D and move to a 1.5 ml Eppendorf tube. Add 1 volume isopropanol or 2 volumes absolute ethanol to the aqueous phase and place tubes at –20°C for at least 1 h, followed by centrifugation for 10 min at 10000 g at 4°C.
6. Wash the pellet in 75% ethanol, centrifuge, and resuspend the pellet in 50 µl 0.5% SDS at 65°C for 10 min, or DEPC water, or EDTA-containing buffer. Use immediately for other procedures or store at –70°C.

Protocol 13.2. Modified procedure for tissue (9)

GITC buffer

4 M guanidium thiocyanate
0.5% sarcosyl
25 mM sodium citrate, pH 7
0.1% antifoam A
5% β-mercaptoethanol

Other materials

TE-saturated phenol
TE-saturated chloroform
2 M sodium acetate, pH 4
isopropanol
DEPC-treated water
absolute ethanol (EtOH)
80% EtOH in DEPC-water
5 M NaCl

1. Homogenize tissue (0.5 g/5 ml) in 4 M GITC buffer in a polytron tissue homogenizer. Keep samples on ice and homogenize in short bursts so that the temperature does not increase. Clean the homogenizer tip in chloroform or ethanol between samples.
2. Add a 1/5 volume of 2 M Na acetate to homogenate and vortex. Add an equal volume of saturated phenol as the original volume of GITC buffer and vortex. Next add 1/5 volume of TE-saturated chloroform, vortex, and hold the mixture on ice for 15 min.
3. Centrifuge at room temperature at 8000 rpm. Transfer the upper aqueous layer to a new tube. To this, add an equal volume of isopropanol and hold the tubes at room temperature for 30 min.

4. Centrifuge the tube at 8000 rpm for 10 min to pellet the RNA. Remove the supernatant and wash the pellet with 80% EtOH (in DEPC), centrifuge again, remove supernatant, and air-dry the pellet. *Note*: The pellet should never become overdried because it becomes difficult to resuspend. Resuspend the pellet in 400 μl DEPC-water and transfer to a 1.5 ml Eppendorf tube.
5. Reprecipitate the RNA with 1/20 volume (20 μl) 5 M NaCl, 2.5 vol (1000 μl) absolute EtOH, and place at –20°C or –70°C for 30 min. Centrifuge for 20 min and carefully remove the supernatant by aspiration.
6. Wash the pellet in 80% EtOH in DEPC, centrifuge, and remove supernatant. Resuspend the pellet in 10–20 μl DEPC-water. Determine the concentration of RNA by spectrophotometry and adjust the RNA concentration to about 2 μg/μl. The RNA can be placed into smaller aliquots, depending on what it is to be used for, as repeated freeze-thaw cycles should be avoided.

Protocol 13.3. Trizol protocol

1. Homogenize 50–100 mg tissue in 1 ml Trizol and hold 5 min at room temperature.
2. Add 0.2 ml chloroform (per milliliter Trizol), shake the tubes vigorously, hold at room temperature for 2–3 min, and centrifuge at 12,000 rpm for 15 min at 4°C.
3. Aspirate the colorless upper aqueous phase and place into a fresh tube. Add 0.5 ml isopropanol (per ml Trizol) to this supernatant and hold at room temperature for 10 min, then centrifuge at 12,000 rpm for 10 min at 4°C.
4. Aspirate the supernatant from the pellet and wash the pellet with 1 ml 75% alcohol, mix, and centrifuge at 7500 rpm for 5 min at 4°C.
5. Remove the supernatant and dry the pellet (do not overdry) and suspend it in DEPC-water before spectrophotometry.

Protocol 13.4. Isolation of total RNA from embryos (9)

Extraction buffer

0.2 M NaCl
25 mM TRIS pH 7.4
1 mM EDTA in DEPC-treated water

Other materials

100 mM TRIS, pH 8
saturated phenol (or Gibco buffer saturated phenol)
chloroform:isoamyl alcohol (24:1)
Yeast transfer or ribosomal RNA (4 μg/μl)
95% EtOH
70% EtOH in DEPC-water

1. Freeze embryos in 0.5-ml Eppendorf tubes in a few microliters of culture media in liquid nitrogen and then transfer to –70°C for storage.
2. Before the RNA extraction, prepare and label one Eppendorf tube per sample containing 100 μl extraction buffer, 100 μl phenol, and 100 μl chloroform:IAA (isoamyl alcohol). Add 3 μl (12 μg) transfer or ribosomal RNA to the extraction phase and keep on ice. Prepare and label a second Eppendorf tube for each sample containing only 100 μl chloroform:IAA.
3. Directly at the freezer, before embryos thaw, add the bottom 100 μl phenol phase to one sample tube and shake while the embryos thaw. Then add the chloroform and extraction buffer phases, shake, and store on ice. Repeat for all tubes.

4. Vortex and shake the tubes for 3 × 10 s. Centrifuge at 13,000 rpm for 10 min at room temperature. Remove the upper aqueous phase and transfer it to the tube containing chloroform:IAA. Vortex briefly and centrifuge again at 13,000 rpm for 10 min.

5. Remove the supernatant and estimate the volume. Add about 2 volumes cold 95% ethanol (stored at –20°C) and precipitate tubes at –70°C for several hours or overnight.

6. Precipitate RNA by centrifuging at 10,000 rpm at 4°C for 20 min. Remove ethanol, wash with cold 70% ethanol, and centrifuge at 10,000 rpm at 4°C.

7. Remove ethanol and air-dry upside down on a clean surface at room temperature. Do not allow the pellets to overdry because they become difficult to resuspend. The RNA can be diluted and used for spectrophotometry. Although most of the A_{260} reading is from the carrier RNA, the reading can be used to estimate RNA recovery. If using the RNA for reverse transcription, the pellet can be resuspended directly in oligo(dT) or random hexamer mix.

Protocol 13.5. Isolation of poly(A)$^+$ RNA using messenger affinity paper (20)

GITC buffer

4 M guanidium thiocyanate
0.1 M TRIS, pH 7.4
1 M β-mercaptoethanol

Other materials

0.5 M NaCl
0.1 M TRIS, pH 8.3
0.5 M NaCl
70% EtOH

1. Cut a 2 × 2 mm^2 square of messenger affinity paper (mAP) and handle with RNase free, ethanol-flamed forceps and scissors and prewet with 0.5 M NaCl on sterile Whatman paper supported by parafilm. (Precut squares can be stored at –20°C in Eppendorf tube.)

2. Remove embryo samples of single or pools of embryos frozen in 10 µl GITC buffer from –70°C storage and thaw to room temperature.

3. If samples are to be quantified, add globin mRNA in a volume of 1 µl at 0.1 pg/ embryo equivalent to the embryo tube.

4. Blot the mAP square onto sterile Whatman paper supported by parafilm, then transfer with RNase-free tools into the embryo tube and leave at room temperature for 2–3 h at room temperature to allow poly(A)$^+$ RNA to bind to paper.

5. Transfer the mAP to fresh Whatman paper on parafilm and pipette the unabsorbed lysate onto the square.

6. Transfer the mAP square into a new tube containing 200 µl first-wash buffer (0.5 M NaCl, 0.1 M TRIS, pH 8.3) and gently invert several times. Remove the first-wash buffer and replace with fresh 200 µl buffer two more times, followed by three washes in 0.5 M NaCl alone and two washes in 70% ethanol. All buffers are DEPC treated.

7. Transfer the mAP square to a fresh tube and allow to dry on ice with the lid open for 5 min. The mAP can then be used directly for reverse transcription and is kept within the cDNA aliquot.

Note: The difficulty with the mAP paper isolation method is that mAP is no longer commercially available from the supplier, and an alternative supply has not been developed.

Protocol 13.6. Isolation of poly(A)⁺ RNA using Dynabeads mRNA Direct kit (18, 19)

1. Store pools of oocytes/embryos or single oocytes/embryos at –80°C in a minimum volume of washing medium (PBS plus 0.1% PVA) in siliconized tubes.

2. As a rapid lysis of oocytes/embryos is critical for obtaining intact mRNA, avoid thawing of the frozen material before the lysis step.

3. For semiquantitative RT-PCR, add 0.1 pg per oocyte/embryo equivalent (pools) or 1 pg (single oocyte/embryo) of globin RNA as an internal standard.

4. Lyse embryos or oocytes (20–50) in 150 µl (pools) or 30 µl (single embryos) lysis buffer (100 mM TRIS HCl, pH 8.0; 500 mM LiCl, 10 mM EDTA, 1% lithium dodecylsulfate, 5 mM dithiothreitol) and vortex for 10 s.

5. Centrifuge samples at 12,000 *g* for 15 s and incubate at room temperature for 10 min.

6. Add prewashed oligo dT Dynabeads (10 µl for pools or 5 µl for single embryos) to the sample. Incubate the sample by rotating on a mixer or roller for 10 min at room temperature to allow binding of poly(A)⁺ RNA to Dynabeads.

7. Place the tubes in the magnetic separator for 2 min.

8. After removing the supernatant, wash the beads once with buffer 1 (10 mM TRIS HCl, pH 8.0; 150 mM LiCl, 1 mM EDTA, 0.1% lithium dodecylsulfate) using 100 µl (pools) or 40 µl (single embryos) and three times with 100 µl buffer 2 (10 mM TRIS HCl, pH 8.0; 150 mM LiCl, 1 mM EDTA).

9. Elute the RNA from the beads with 11 µl sterile water, heat at 65°C for 2 min, and use directly for reverse transcription.

References

1. Piko, L., and Clegg, K. B. (1982). *Dev. Biol.*, **89**, 362.
2. Olszanska, B., and Borgul, A. (1993). *J. Exp. Zool.*, **265**, 317.
3. Bilodeau-Goeseels, S., and Schultz, G. A. (1997). *Biol. Reprod.*, **56**, 1323.
4. Bilodeau-Goeseels, S., and Schultz, G. A. (1997). *Mol. Reprod. Dev.*, **47**, 413.
5. Chirgwin, J. M., Przybyla, A. E., MacDonald, R. J., and Rutter, W. J. (1979). *Biochemistry*, **18**, 5294.
6. Aviv, H., and Leder, P. (1972). *Proc. Natl. Acad. Sci. USA*, **69**, 1408.
7. Chomczynski, P., and Sacchi, N. (1987). *Anal. Biochem.*, **162**, 156.
8. MacDonald, R. J., Swift, G. H., Przybyla, A. E., and Chirgwin, J. M. (1987). *Methods Enzymol.*, **152**, 221.
9. Arcellana-Panlilio, M. Y., and Schultz, G. A. (1993). *Methods Enzymol.*, **225**, 303.
10. Edmonds, M., Vaughn, M. H., and Nakazoto, H. (1971). *Proc. Natl. Acad. Sci. USA*, **68**, 1336.
11. Lindberg, U., and Persson, T. (1974) *Methods Enzymol.*, **34**, 496.
12. Jacobson, A. (1987). *Methods Enzymol.*, **152**, 254.
13. Braude, P. R., and Pelham, H. R. B. (1979). *J. Reprod. Fertil.*, **56**, 153.
14. Temeles, G. L., Ram, P. T., Rothstein, J. L., and Schultz, R. M. (1994). *Mol. Reprod. Dev.*, **37**, 121.
15. Revel, F., Renard, J. P., and Duranthon, V. (1995). *Zygote*, **3**, 241.
16. Robert, C., Barnes, F. L., Hue, I., and Sirard, M.-A. (2000). *Mol. Reprod. Dev.*, **57**, 167.
17. van Tol, H. T. A., van Eijk, M. J. T., Mummery, C. L., Van den Hurk, R., and Bevers, M. M. (1996). *Mol. Reprod. Dev.*, **45**, 218.
18. Wrenzycki, C., Herrmann, D., Carnwath, J. W., and Niemann, H. (1998). *J. Reprod. Fertil.*, **112**, 387.
19. Wrenzycki, C., Wells, D., Herrmann, D., Miller, A., Oliver, J., Tervit, R., and Niemann, H. (2001). *Biol. Reprod.* **65**, 323.
20. De Sousa, P., Westhusin, M. E., and Watson, A. J. (1998). *Mol. Reprod. Dev.*, **49**, 119.

14

P. A. DE SOUSA
A. J. WATSON

Relative mRNA Transcript Abundance in Early Embryos by Reverse Transcription-Polymerase Chain Reaction

1. QUANTIFICATION OF mRNA TRANSCRIPTS

Over the last decade methods for the microscale isolation of mRNA coupled with reverse transcription and cDNA amplification techniques have significantly enhanced our understanding of the genetic control of early mammalian development. In this regard, attempts to quantify changes in the relative abundance of specific gene transcripts have helped define their biological significance. Although there are many methods by which gene transcript abundance can be quantified, meaningful quantification requires an appreciation for the limitations and qualifications associated with these procedures. In this chapter our goal is to provide the context for understanding the principles and caveats of relative mRNA abundance calculations, together with a practical approach to meeting this objective by reverse transcription-polymerase chain reaction (RT-PCR).

1.1 Historical Perspective

Attempts to quantify the size and composition of RNA pools in early mammalian embryos began approximately 30 years ago using nucleic acid hybridization methods. The amount of tissue required for such methods necessitated working on species that could provide pools of hundreds to thousands of embryos as starting material or species whose early embryos were sufficiently large to bypass the need for large numbers. Thus, working on late blastocysts from rabbits, consisting of at least 50,000 cells, the first estimates of embryonic mRNA pool complexity were obtained by hybridizing genomic DNA to first total (primarily heterogeneous nuclear) and then cytoplasmic RNA (1, 2). Similar studies on large pools of embryos from mice, also known for their reproductive fecundity, provided much of the first estimates of the size and complexity of all subtypes of RNAs throughout embryogenesis (3–6).

With the emergence of gene-specific DNA probes for Northern analysis and Southern blotting of copy DNA came the first quantification of changes in the abundance of specific gene transcripts throughout early development. These began with examination of changes in histone and actin mRNAs, regarded as markers of nonadenylated and polyadenylated mRNA classes (7–9), followed throughout the 1980s by studies on a modest collection of genes including other members of the histone and actin families, ribosomal proteins, α-tubulin, E-cadherin, connexin 43, Na-K ATPases, hypoxanthine phosphoribosyltransferase (HPRT), β-glucoronidase, and IAP (intracisternal A particles)-1 and 2 (9–19).

In 1988 a new age for molecular embryology was ushered in with the first application of RT-PCR to preimplantation mouse embryos to profile the expression of cytokine mRNA transcripts (20). Since then the number of different gene transcripts examined during early mammalian development has increased by at least one order of magnitude. Between 1990 and 2000, at least 124 publications have reported the use of RT-PCR on preimplantation embryos from a broad range of species, as determined by a literature search of the National Center for Biotechnology Information PubMed database (www.ncbi.nlm.nih.gov/entrez). In contrast to earlier hybridization experiments, these studies were performed on single to tens of embryos and even on embryonic blastomeres (21–25).

Not long after the first reports of using RT-PCR to detect mRNAs in early embryos, different approaches to quantify transcript abundance were developed. Differences between methods have primarily related to the technique used to measure amplified mRNA products and the reference standard used for calculation of absolute or relative values. RT-PCR product abundance has been determined by densitometry after gel electrophoresis and ethidium bromide staining, capillary electrophoresis, incorporation of radiolabeled nucleoside triphosphates during amplification, hybridization of a complementary radiolabeled probe followed by RNAse protection, and most recently by hybridization and nuclease digestion of fluorogenic probes during real-time PCR detection (24–30). Absolute measurements of mRNA abundance can be made by referring amplified product abundance to standard curves defining the amplification of known dilutions of the same template (31). Although conceptually satisfying for its capacity to describe mRNA abundance in terms of transcript copy number, this approach does not account for variations between samples in RNA isolation, RT, and PCR. Such variations are inherently corrected for when mRNA abundance is described relative to exogenous or endogenous transcripts, with such measurements being sufficiently sensitive to detect twofold increases in transcript abundance (25, 26, 32–34). Relative measurements of RNA can thus be used to accurately assess the steady-state level of specific gene transcripts for comparison across developmental stages or in response to treatment conditions.

The first laboratory to pioneer relative mRNA determinations in early embryos was that of Richard Schultz at the University of Pennsylvania, with a method involving the co-amplification of exogenously supplied globin mRNA as a reference standard (26, 32). To control for differences between samples in RNA recovery, RT, and PCR, globin mRNA, not expressed by early embryos, was added to lysed samples before RNA isolation. In principle, the benefit of an exogenous RNA standard can also be achieved by other mRNAs not expressed in embryos, including recombinant versions of the template under study or synthetic RNAs flanked by gene-specific primer sequences (33, 34). Recombinant standards offer the advantage of being amplified by the same oligonucleotide primer sets used for the target template. However, this approach requires a greater investment in molecular biology to construct reference standards. Messenger RNA transcripts that are endogenous to an embryonic sample can also be used as reference standards, with the important qualification that these transcripts should not vary between samples as a result of development or treatment conditions. In this regard, 18 S RNA, classically used for Northern analysis, or glyceraldehyde-3-phosphate dehydrogenase represent preferred endogenous reference standards because their expression is normally constant within cells. Such standards also can control for differences between samples in cell number (30, 35).

In the succeeding sections we elaborate on the principles and caveats of relative mRNA quantifications specifically as they pertain to early mammalian embryos. We then describe a protocol exemplifying the use of exogenous globin mRNA as a reference standard (Protocol 14.1). Although the protocol described relies on densitometric quantification of amplified cDNA products by digital imaging-based gel documentation systems (Protocol 14.2), alternative approaches are discussed, including the most recent innovation of real-time fluorescence PCR.

2. PRINCIPLES AND CAVEATS OF RELATIVE mRNA ABUNDANCE MEASUREMENTS

The central tenet of mRNA quantification by RT-PCR is that the amplification of a gene transcript is related to its starting abundance and can be predicted if the number of cycles of amplification is known. This, of course, requires that the representation of mRNA in a sample be preserved at all times. In practical terms, this presumption is ambitious and is a challenge to maintain because RNA can be lost during its recovery, and its representation can be altered during both RT and PCR. Meaningful quantification of mRNA abundance therefore requires optimization of each of these component processes, which are summarized schematically in Figure 14.1.

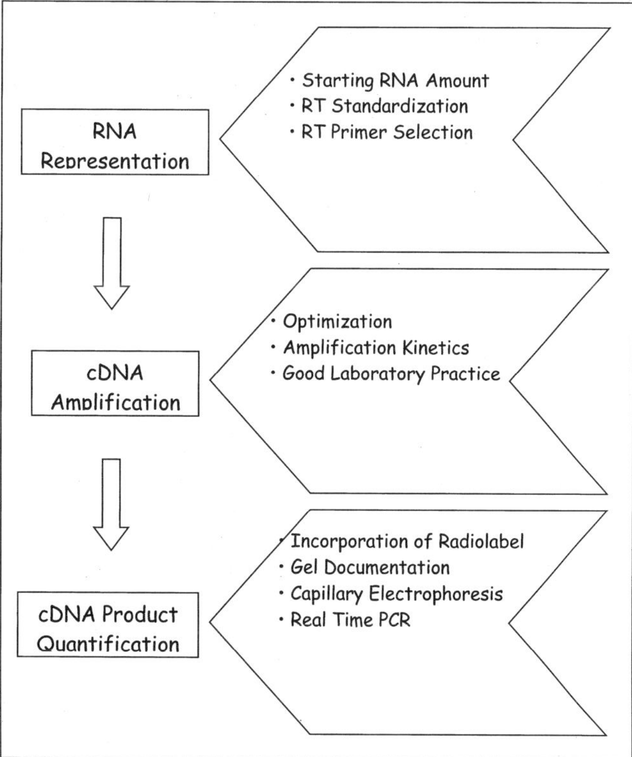

Figure 14.1. Schematic depicting the component steps of relative mRNA abundance analyses and associated considerations.

3. OPTIMIZING RNA REPRESENTATION

3.1 Amount of Starting RNA (Experiments on Single vs. Pooled Embryos)

The first question to resolve before engaging in a study of embryonic gene expression is whether the analysis can or should be performed on single versus small or large pools of embryos. Although in species other than the mouse, the lack of synchronous or homogenous specimens necessitates analyses on single or small pools of embryos (e.g., < 5), the ramifications of working on such limited sources of RNA should not be overshadowed by the technical capacity to detect targeted gene transcripts. Certainly, through optimization of PCR conditions and methodological tricks such as hot start priming or nested amplifications, it is possible to detect rare mRNAs present as fewer than 10 copies (21, 25, 31). However, reduction in the starting amount of a transcript results in increased variation in its amplification when replicate PCRs are performed on the same sample of reverse transcribed cDNA (25). The reason for this is not clear, but it presumably relates to uneven dispersal of reverse-transcribed cDNA despite thawing and rigorous mixing before use. Because relative abundance (RA) measurements are ratios of the amount of products amplified from targeted and reference transcripts, how much variation exists in the amplification of each transcript needs to be determined to define the precision of calculated values (25). It is only with this knowledge that differences in RA values between samples can be interpreted as biologically significant as opposed to technical artifact.

3.2 Reverse Transcription Reactions

Reverse transcription reactions on embryonic mRNA are normally performed, as for other tissues, according to the recommendation of the manufacturers of the reverse transcriptase used, and typically last 1–1.5 h. All conditions of these reactions should be standardized and conserved between reactions—most especially temperature and the selection of primer used to reverse transcribe. These parameters, along with the duration of an RT reaction, can alter the representation of mRNAs in a cDNA population. Increasing RT reaction temperature can serve to limit the population of mRNAs represented, as can primer selection. In addition, when conditions are not limiting, smaller mRNA will be more frequently represented in the resulting cDNA pool than will larger mRNAs (> 1–2 kb) that take longer to reverse transcribe. Although awareness of this point tends to be more critical in the design and construction of representative cDNA libraries (36), it may also affect the relation and quantification of a targeted gene transcript in relation to a standard when the size difference between these transcripts is too great.

3.3 RT Primer Selection

The types of oligonucleotide primers used in RT reactions have a direct bearing on the nature of subsequent quantitative strategies as well as on their interpretation. Unanchored oligo-dT primers are commonly used for RT reactions and consist of 21–30 bases of thymidine deoxyribonucleotide. Because this kind of primer anneals to poly-A tracts at the 3' end of mRNA, the representation of a mRNA is critically dependent on the extent to which it is polyadenylated. Changes in mRNA adenylation can therefore influence that apparent steady-state level of a transcript. Apparent increases in transcript abundance resulting from mRNA polyadenylation are suggested when they still can occur in the face of transcriptional inhibition (24, 37, 38). This can be confirmed using mobility shift assays (39) or a PCR-based assay for determining poly-A tail length (40, 41).

Alternative primers for reverse transcription that would be insensitive to changes in the adenylated status of message transcripts include anchored oligo-dT primers, gene-specific primers, or random hexamer sequences. Anchored oligo-dT primers contain an additional couple of bases of variable identity at their 5' end. Although the most common usage of anchored primers is in differential display analysis (42), there is no reason such a primer could not be used to selectively enhance mRNA representation for a transcript of interest as long as the reference transcript was also represented. Gene-specific primers can be used to enhance representation of specific mRNAs and are still suitable for relative mRNA measurements when a recombinant template is used as a reference standard (34). Alternatively, random hexamers would be the indicated choice when the reference RNA standard is nonadenylated (e.g., 18 S RNA) or is subject to changes in its adenylation (22, 35). Regardless of which primer is used for RT reactions, differences in primer efficacy and specificity can result in vastly different mRNA representations and quantitative results (43). Interpretation of experimental results should thus be appropriately qualified.

4. OPTIMIZING DNA AMPLIFICATION

4.1 Establishing PCR Conditions and Kinetics

As mentioned, the capacity of PCR primers to anneal and amplify specific gene transcripts is highly dependent on the amount of starting material from which RNA is isolated, its method of isolation, and the reverse transcription. Quantification of a gene-specific transcript further requires that it be represented by a single amplification product. This should be established during preliminary PCR experiments to optimize reaction conditions, most especially with respect to annealing temperature and $MgCl_2$ concentration (44). Failing to optimize reactions in this fashion may result in unpredictable competitive inhibitions of desired reaction products.

The most critical aspect of mRNA quantification by RT-PCR occurs during the amplification of a transcript-derived template. During PCR, an amplified template accumulates exponentially with each cycle until reaction components become limiting, at which time reactions are said to plateau or become saturated (Figure 14.2). It is only while reaction products accumulate exponentially that their abundance bears any relationship to the amount of starting material. Thus, the single most important consideration for mRNA quantification by RT-PCR is to establish the kinetics of PCR reactions for that transcript and to only rely on quantitative information acquired from exponentially accumulating product. In practical terms, exponential amplification of a targeted gene product by PCR is depicted by a linear relationship between the log of accumulated product and PCR cycle number.

4.2 Factors Affecting PCR Optimization

Representation of mRNA as a single PCR product can be influenced by whether multiple or single gene transcripts are amplified in the same reaction. Although theoretically the use of multiple primer pairs in a single reaction (referred to as multiplexing) can minimize errors in the sampling of a common cDNA sample, reactions can competitively inhibit each other as reagents become limiting. This competition can be minimized by adjusting relative primer and template concentration or by selectively controlling the timing of primer addition to a reaction (45, 46). The kinetics of a PCR amplification will also be affected by variations in the thermocycler used for an experiment, which differ in the amount of control, accuracy, and speed with which temperature cycling can be achieved. Thus, after optimizing amplification of a specific gene transcript with a given thermocycler, switching to an alternative cycler should be avoided, without rechecking optimized conditions.

Scarcity of target templates, as may occur when analyzing gene expression in single embryos, will alter PCR reaction kinetics by favoring mispriming of unrelated templates.

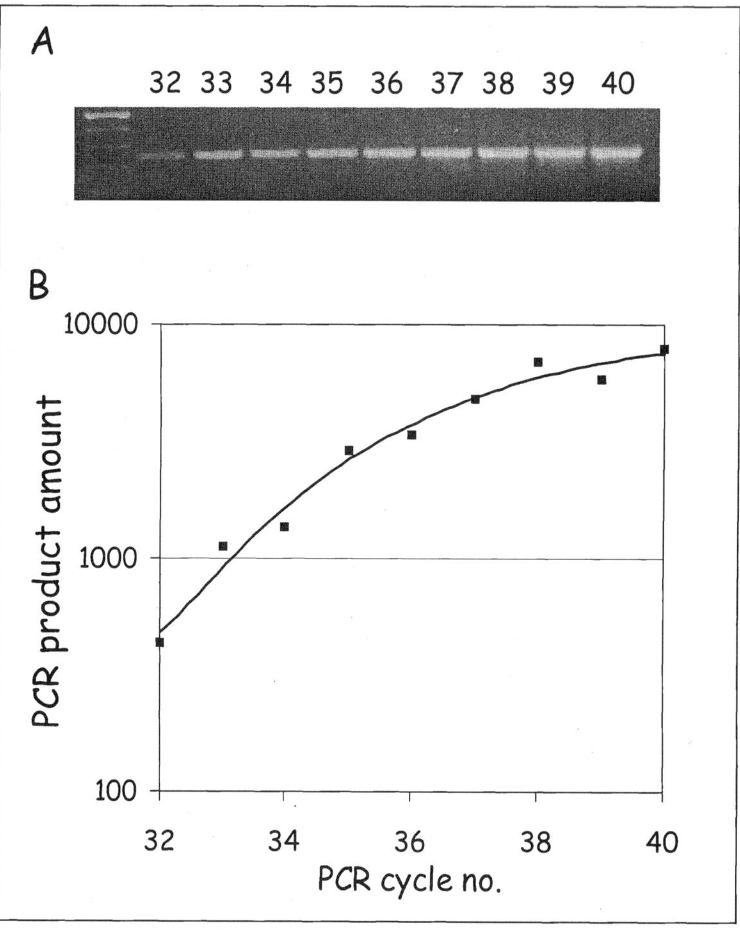

Figure 14.2. Kinetics of RT-PCR product amplification. (A) Ethidium bromide-stained gel profile of a 336-base product for bovine Na⁺/K⁺ ATPase α-1 mRNA amplified from single embryo cDNA equivalents after 32–40 cycles. (B) Corresponding graph depicting product abundance relative to PCR cycle number as determined using capillary electrophoresis.

This may be corrected by resorting to hot start PCR methodology (47, 48). Originally this method entailed the addition of Taq polymerase to a reaction mixture once it had reached the denaturation temperature, so as to minimize template elongation at sites where primers annealed nonspecifically. In recent years, recombinant Taq polymerase activated by high temperatures (e.g., Ampli-Taq Gold, PEC, Palo Alto, CA) have facilitated this approach by permitting all reagents to be mixed before temperature cycling, as done for conventional reactions (49). Hot start PCR with recombinant heat-activated Taq was key to our success in consistently quantifying mRNA in single bovine oocytes and embryos (25). However, the increased sensitivity this approach offered also increased our capacity to detect trace contaminants in negative control reactions presumably lacking template.

The potential for detecting contaminant cDNA is a real danger associated with PCR that cannot be taken lightly, especially when the methodology is optimized to work with limited amounts of tissue, such as single embryos. Measures to prevent the introduction of contaminants include (1) using separate pipettes for RNA isolation and RT versus PCR, (2) using either positive displacement pipettes or conventional pipettes with plugged (aerosol barrier) tips for all procedures, (3) preparing aliquots of reaction components as limited-use volumes, (4) using only ultra-pure water collected in single-use sterile plastic-ware in the preparation of all reagents, (5) avoiding any reagent or supply sterilized in the same autoclave used to decontaminate biological waste, and (6) performing RNA isolation and RTs in a separate work area from that designated for PCR setup.

4.3 Selecting a Method to Quantify PCR Product Abundance

4.3.1 Radiolabeled Nucleotides

The first studies to quantify relative mRNA abundance did so by spiking gene-specific PCRs with a radiolabled nucleotide (α-^{32}P-dCTP) (26, 32). Radiolabeled reaction products are separated on agarose gels that are then stained with ethidium bromide for product visualization. Products are then excised and measured using a scintiliation counter. Although the low cost and self-sufficiency of this method is ideally suited for small laboratories, the hazards of working with radioactivity make this approach less desirable than more modern approaches.

4.3.2 Gel Documentation Systems/Photography

In recent years gel documentation systems have become available as a departmental or divisional resource. In principle these systems enable densitometric calculations of amplification products visualized after ethidium bromide staining. Gel doc systems have been successfully applied in quantitative RT-PCR on early embryos (22, 23, 29). Although convenient in their capacity to provide greater control over image acquisition and analysis than conventional Polaroid photography, the capacity to alter image intensity and contrast can confound quantitative measurements. The greatest difficulty to be aware of is the capacity to create images whose intensity is saturated and no longer related to true signal strength. One way to monitor this is by running commercially available mass ladders on the same gels containing amplification products so that a standard curve relating mass to pixel intensity can be calculated. Post-electrophoretic staining of gels with ethidium bromide is preferable to including stain in the gel or run buffer because ethidium bromide migrates in the opposite direction to cDNA products during electrophoresis due to charge differences. Additional measures that ensure the accuracy and reproducibility of measurements of the amount of amplification product generated include subtracting pixel intensity determined from a neighboring region of gel not containing amplification product and standardizing all aspects of electrophoretic gel separation and gel staining.

4.3.3 Capillary Electrophoresis

Next to real-time fluorescence PCR (see below), capillary electrophoresis is perhaps one of the most accurate methods for quantifying PCR product abundance requiring minimal (1–5 μl) amount of reaction products. It entails electrophoretic separation of cDNA either straight from a PCR reaction mix or following desalting through a gel matrix preloaded in a glass capillary (50). Separation of products is on the basis of molecular weight, and run times of 20–30 min can resolve products differing by at least 10 nucleotides. Single nucleotide resolution can be attained by increasing both the run time and gel matrix composition, and product abundance is automatically calculated from the area under the peak corresponding to its size. Capillary electrophoresis has been used in the quantitation of gene transcripts from pools of mouse embryos and single bovine oocytes and embryos (25, 45). Despite the precision of this method, it has the disadvantage of only being able to run one sample at a time. Analysis of multiple samples representing either several gene products or developmental stages thus takes significantly longer than the other methods described.

4.3.4 Real-time Fluorescence PCR

In recent years, the integration of fluorogenic nuclease chemistry with proprietary technology for PCR and fluorescence detection from Perkin Elmer Applied Biosystems (now EG&G, Inc., Wellesley, MA) has made it possible to quantify cDNA template abundance noninvasively in real time (51). This technology is based on the annealing and cleavage of gene-specific oligonucleotide probes with fluorescent and quenching dyes attached during DNA amplification. In principle, probes anneal in a sequence-specific fashion downstream of one of the primer sites, when the target template is present. During primer extension, probes are then cleaved by the 5' nuclease activity of the Taq DNA polymerase. As a result, reporter and quencher dyes attached to the probe are uncoupled from proximity to one another, increasing the reporter dye signal, and the probe is removed from the target strand so that primer extension continues. By performing amplifications on an ABI PRISM 7700 sequence detector (EG&G, Inc., Wellesley, MA) or comparable facsimile, the fluorescent reporter signal is detected via laser-directed fiber optics and normalized to the fluorescence of an internal reference dye. Peak normalized reporter values are then averaged for each cycle and plotted versus cycle number to produce amplification plots comparable to those achieved by manual termination of reactions at the end of each cycle. Using real-time PCR, reactions are characterized by the point in time during cycling when amplification of a PCR product is first detected rather than the amount of PCR product accumulated after a fixed number of cycles using other approaches. This parameter is referred to as the threshold cycle, which subsequently can be used for either relative or absolute calculations of starting template abundance.

Several groups have used real-time PCR to analyze gene expression during early mammalian development. This includes studies on pooled bovine and single human and mouse oocytes and early embryos (30, 52, 53). Conceptually, the sensitivity of fluorescence-based detection of transcript amplification may make real-time PCR ideally suited for mRNA quantification in single embryos and cells. However, improved PCR sensitivity would still be subject to the same limitations on RNA representation imposed during RNA isolation and reverse transcription.

5. CONCLUSION

Relative mRNA measurements represent an indication of steady-state levels of a mRNA transcript and not transcription or degradation per se. Their accuracy is critically dependent on preserving mRNA representation during all aspects of RNA recovery, reverse transcription, and PCR. Establishing the kinetics of PCR amplifications to identify the exponential phase of either targeted or standard template accumulation is of paramount importance to the whole endeavor. Both exogenous and constantly expressed, nonregulated endogenous transcripts can serve as reference standards to control for variations in RNA representation between samples. Protocol 14.1 describes relative mRNA measurements using an exogenous RNA standard.

Protocol 14.1. Relative mRNA abundance determination using an exogenous globin standard

1. Lyse single or pooled embryo samples in a minimum volume (~1 µl) of their last wash by adding at least 10–100 µl of GITC lysis buffer (4 M guanidine isothiocyanate, 0.1 M Tris [pH 7.4], 1 M β-mercaptoethanol).
2. Lysed samples can be stored indefinitely at –80°C and transported if necessary in dry ice.
3. On day of RNA isolation, prepare working stock of globin mRNA from concentrated 50 ng/µl stock in sterile H_2O (50 ng/µl stock of globin mRNA [Gibco BRL, Gaithersburg, MD] keeps for up to 6 months at –80°C).

4. Thaw embryo samples and add to each 0.1 pg of globin RNA per embryo, in a volume not exceeding 1 μl. *Note*: Use diluted stocks immediately. Dilute globin mRNA is unstable and can deteriorate on ice within hours, or following freeze/thaw cycles.

5. Isolate RNA by microscale method of choice (see chapter 13 for alternatives).

6. Prepare oligo-dT master mix. For 1 reaction volume, mix 10 μl of DEPC-treated sterile H_2O (or if you are confident about your water, just sterile water) and 1 μl of oligo-dT (0.5 mg/ml; Gibco). For large volumes of master mix it is sometimes wise to prepare an extra volume to compensate for pipetting errors.

7. Prepare RT mix. For 1 reaction volume, mix 4 μl of Gibco BRL 5× First Strand buffer, 2 μl 0.1 M DTT, 1.5 μl 10 mM dNTPs, 1.5 μl of Superscript (300 U; 0.5 μg/μl stock; Gibco). *Note*: This amount of superscript may be in gross excess of what is needed. It is possible to cut back even further because 200 U are required for up to 5 μg of mRNA, and a single expanded mouse or bovine blastocyst has from 1–10 ng of total RNA. If you do choose to cut back, then increase the volume of H_2O in the oligo-dT mix accordingly.

8. On ice, add 11 μl of oligo-dT mix to each tube containing 1 μl of isolated RNA. *Note*: To accommodate increasing volumes of RNA template, the amount of H_2O in either the oligo-dT or RT master mixes can be reduced, so that the final volume of the reaction remains 20 μl.

9. Add 9 μl of RT mix to chilled RNA/oligo-dT mix and blend by repeated pipetting. If necessary, spin down tubes before beginning RT.

10. RT for 90 min at 43°C. Stop the reaction by treating for 5 min at 95°C and then chill on ice. *Note*: RTs can be performed on programmable thermocyclers instead of heating blocks. However, doing so runs the risk of RT reactions being contaminated by aerosols containing PCR cDNA products, which may be transferred between tubes during handling or via loose-fitting caps. Caps on plastic reaction tubes can fail if they are not high quality.

11. Store RTs at –20°C after spinning the tubes down.

12. PCRs can be performed on a 2- to 5-μl aliquot of the RT in a 25-μl reaction according to manufacturer's instructions.

13. From a 25-μl PCR reaction, a 10–20 μl volume is normally sufficient to visualize a product on an ethidium bromide-stained gel.

Protocol 14.2. Quantification of RT-PCR by gel documentation scanning

- For most typical RT-PCR product sizes (e.g., 200–800 bp) a 2% agarose gel is used (2 g in 100 ml 1× TAE) and contains 0.5 μg/ml of ethidium bromide. Gels are loaded with 20 μl of RT-PCR product per lane mixed with 4 μl of 6× loading dye, using the same set of pipettors for each gel to be quantified.

- In addition, 5 μl of a 100-bp marker (Fermentas, Hanover, MD) is run in a marker lane. The 500-bp band is the brightest, and we have used this marker for quantification.

- The gel is run for approximately 1 h at 115 V, or until the bromophenol blue dye marker is about 1 cm from the bottom of the gel.

- For PCR reactions to be quantified, the number of cycles must be optimized so that they fall within the linear amplification range for each primer set of interest.

- An image of the gel is taken with the gel doc system. We have used a Pharmacia Biotech system (Uppsala, Sweden). The UV lamp and the transmittance must be on and the black tray must be used. To achieve maximal image quality, the lens must be focused on the gel and the time, gain, and black level set and kept constant. The f-stop of the camera is usually set to about 2. Similar settings must be maintained for each gel. A hard copy may be printed with the thermal printer.

- For storage on the computer, the γ setting on the gel doc must be switched from 0.45 (pictures) to 1.00 (computer storage). This is important, because if it is not done, the Imagemaster program has difficulty discerning white bands from dark bands and will have to be reset manually.

- In the Imagemaster program (Pharmacia), in the experiment window set the function type to DNA analysis and enter ok. Then click the camera icon and click snap to take a picture. Again, the time, gain, and black level on the gel doc can be adjusted to get the best picture. This image should be saved as a tif file, and it can be analyzed immediately, or Imagemaster can be closed and the analysis done later.
- If the analysis is conducted later, open the Imagemaster program at that time and enter DNA analysis as before. On file menu, open the *.tif file. This will open to about a three-quarter screen.
- For quantification, start with the globin PCR, record the data for the 500-bp marker band and all globin bands. It is helpful if all other PCR gels are loaded in the same order. For other PCRs, record the data for the marker 500-bp band and other bands.

1. If the legend does not appear automatically, open this icon. First, go to the lanes menu and select add lanes, and a white, full-length lane will appear that can be moved on the screen with the mouse. Adjust the size (pixels) so the lane is a bit larger than the well and band size. Now mark all the lanes (each will be a different color), and when done click ok. The lanes can be moved individually to bracket the bands/wells with the mouse and or the size adjusted at any point. The length of the lanes can be reduced with the mouse at the top or bottom of lane to reduce the background.
2. On the lanes menu, go to the bands section. The sensitivity is usually on 5 and white bands are selected. Normally the automatic band detect is on.
3. Go to lane profile that gives a graphic representation of the fluorescence per lane. Select plot and select single lane. Here the brackets are adjusted on the fluorescent peaks, so that the band marker is just larger than the band in the photograph. Repeat for each marked band to be quantified.
4. Next, go to data to see data for all lanes horizontally. The leader should say absolute integrated optical density and the IOD box is checked. Make minor adjustments to brackets, lane position, and lane width while looking at the data until the best estimate is obtained.
5. Print the data and/or save to file. On data, go to file, data to file, and name the file. It will save as *.txt. Usually it is best to give the tif and txt files the same name so they can be easily matched.

- For quantification involving samples run on two or more gels simultaneously, we first determine the variation in ethidium bromide staining between gels by measuring the variation in intensity of the marker 500-bp band in each gel. This ratio of 500-bp marker lanes between gels generates an adjustment factor to standardize measurements of globin and target gene PCR product levels between gels. Intensities for globin and target gene products are measured in each lane, and they are adjusted by the marker 500-bp ratio factor. The globin-to-target gene ratio is then calculated for each band pair of interest. Duplicate and triplicate reactions should be prepared and measured for each sample to generate a mean and standard error of the mean for each experimental sample. Repeat for all lanes and all genes of interest.

References

1. Schultz, G. A., Manes, C., and Hahn, W. E. (1973). *Biochem. Genet.*, **9**, 247.
2. Manes, C., Byers, M. J., and Carver, A. S. (1981). In: *Cellular and Molecular Aspects of Implantation*, S. R. Glasser and D. W. Bullock, eds., p. 113. Plenum Press, New York.
3. Levy, I. L., Stull, G. B., and Brinster, R. L. (1978). *Dev. Biol.*, **64**, 140.
4. Piko, L., and Clegg, K. B. (1982). *Dev. Biol.*, **89**, 362.
5. Clegg, K. B., and Piko, L. (1983). *Dev. Biol.*, **95**, 331.

6. Schultz, G. A. (1986). In: *Experimental Approaches to Mammalian Embryonic Development*, J. Rossant and R. A. Pedersen, eds., p. 239. Cambridge University Press, New York.
7. Giebelhaus, D. H., Heikkila, J. J., and Schultz, G. A. (1983). *Dev. Biol.*, **98**, 148.
8. Giebelhaus, D. H., Weitlauf, H. M., and Schultz, G. A. (1985). *Dev. Biol.*, **107**, 407.
9. Graves, R. A., Marzluff, W. F., Giebelhaus, D. H., and Schultz, G. A. (1985). *Proc. Natl. Acad. Sci. USA*, **82**, 5685.
10. Paynton, V. B., Rempel, R., and Bachvarova, R. (1988). *Dev. Biol.*, **129**, 304.
11. Bevilacqua, A., Erickson, R. P., and Hieber, V. (1988). *Proc. Natl. Acad. Sci. USA*, **85**, 831.
12. Watson, A. J., Pape, C., Emanuel, J. R., Levenson, R., and Kidder, G. M. (1990). *Dev. Genet.*, **11**, 41.
13. Taylor, K. D., and Piko, L. (1990). *Mol. Reprod. Dev.*, **26**, 111.
14. Taylor, K. D., and Piko, L. (1991). *Mol. Reprod. Dev.*, **28**, 319.
15. Taylor, K. D., and Piko, L. (1992). *Mol. Reprod. Dev.*, **31**, 182.
16. Valdimarsson, G., De Sousa, P. A., Beyer, E. C., Paul, D. L., and Kidder, G. M. (1991). *Mol. Reprod. Dev.*, **30**, 18.
17. Nishi, M., Kumar, N. M., and Gilula, N. B. (1991). *Dev. Biol.*, **146**, 117.
18. Poznanski, A. A., and Calarco, P. G. (1991). *Dev. Biol.*, **143**, 271.
19. Kidder, G. M. (1992). *Dev. Genet.*, **13**, 319.
20. Rappolee, D. A., Brenner, C. A., Schultz, R., Mark, D., and Werb, Z. (1988). *Science*, **241**, 1823.
21. Collins, J., and Fleming, T. (1995). *Trends Genet.*, **11**, 5.
22. Liu, H.-C., He, Z.-Y., Tang, Y.-X, Mele, C. A., Veeck, L. L., Davis, O., and Rosenwaks, Z. (1997). *Fertil. Steril.*, **67**, 733.
23. Lequarre, A. S., Grisart, B., Moreau, B., Schuurbiers, N., Massip, A., and Dessy, F. (1997). *Mol. Reprod. Dev.* **48**, 216.
24. De Sousa, P. A., Watson, A. J., and Schultz, R. M. (1998). *Biol. Reprod.*, **59**, 969.
25. De Sousa, P. A., Westhusin, M. E., and Watson, A. J. (1998). *Mol. Reprod. Dev.*, **49**, 119.
26. Manejwala, F. M., Logan, C. Y., and Schultz, R. M. (1991). *Dev. Biol.*, **144**, 301.
27. Gaudette, M. F., and Crain, W. R. (1991). *Nucl. Acids Res.*, **19**, 1879.
28. Buzin, C. H., Mann, J. R., and Singer-Sam, J. (1994). *Development*, **120**, 3529.
29. Watson, A. J., De Sousa, P., Caveney, A., Barcroft, L. C., Natale, D. N., Urquhart, J., and Westhusin, M. E. (2000). *Biol. Reprod.* **62**, 355.
30. Prelle, K., Stojkovic, M., Boxhammer, K., Motlik, J., Ewald, D., Arnold, G. J., and Wolf, E. (2001). *Endocrinology*, **142**, 1309.
31. Pal, S. K., Crowell, R., Kiessling, A. A., and Cooper, G. M. (1993). *Mol. Reprod. Dev.*, **35**, 8.
32. Temeles, G. L., Ram, P. T., Rothstein, J. L., and Schultz, R. M. (1994). *Mol. Reprod. Dev.*, 37, 121.
33. Wang, A. M., Doyle, M. W., and Mark, D. F. (1989). *Proc. Natl. Acad. Sci. USA*, **86**, 9717.
34. Wang, A. M., and Mark, D. F. (1990). In: *PCR Protocols*, M. A. Innis, D. H. Gelfand, J. J. Sninsky, and T. J. White, eds., p. 60. Academic Press, New York.
35. Young, L. E., Fernandes, K, McEvoy, T. G., Butterwith, S. C., Gutierrez, C. G., Carolan, C., Broadbent, P. J., Robinson, J. J., Wilmut, I., and Sinclair, K. D. (2001). *Nature Genet.*, **27**, 153.
36. Brady, G., and Iscove, N. (1993). *Methods Enzymol.*, **225**, 611.
37. Moore, G. D., Ayabe, T., Kopf, G. S., and Schultz, R. M. (1996). *Mol. Reprod. Dev.*, **45**, 264.
38. Worrad, D. M., and Schultz, R. M. (1997). *Mol. Reprod. Dev.*, **46**, 268.
39. Oh, B., Hwang, S., McLaughlin, J., Solter, D., and Knowles, B. B. (2000). *Development*, **127**, 3795.
40. Salles, F. J., and Strickland, S. (1995). *PCR Methods Appl.*, **4**, 317.
41. Brevini-Gandolfi, T. A. L., Favetta, L. A., Mauri, L., Luciano, A. M., Cillo, F., and Gandolfi, F. (1999). *Mol. Reprod. Dev.*, **52**, 427.
42. Zimmerman, J. W., and Schultz, R. M. (1994). *Proc. Natl. Acad. Sci. USA*, **91**, 5456.
43. Zhang, J., and Byrne, C. D. (1999). *Biochem. J.*, **337**, 231.
44. Innis, M. A., and Gelfand, D. H. (1990). In: *PCR Protocols*, M. A. Innis, D. H. Gelfand, J. J. Sninsky, and T. J. White, eds., p. 3. Academic Press, New York.
45. Davies, T. C., Barr, K. J., Jones, H., Zhu, D., and Kidder, G. M. (1996). *Dev. Genetics*, **18**, 234.
46. Jones, D. H., Davies, T. C., and Kidder, G. M. (1997). *J. Cell Biol.*, **139**, 1545.

47. Birch, D. E. (1996). *Nature*, **381**, 445.
48. Borson, N. D., Strausbauch, M. A., Wettstein, P. J., Oda, R. P., Johnston, S. L., and Landers, J. P. (1998). *Biotechniques*, **25**, 130.
49. Moretti, T., Koons, B., and Budowle, B. (1998). *Biotechniques*, **25**, 716.
50. Williams, S. J., and Williams, P. M. (2001). *Methods Mol. Biol.*, **163**, 243.
51. Holland, P. M., Abramson, R. D., Watson, R., and Gelfand, D. H. (1991). *Proc. Natl. Acad. Sci. USA*, **88**, 7276.
52. Steuerwald, N., Cohen, J., Herrera, R. J., and Brenner, C. A. (1999). *Mol. Hum. Reprod.*, **5**, 1034.
53. Steuerwald, N., Cohen, J., Herrera, R. J., and Brenner, C. A. (2000). *Mol. Hum. Reprod.* **6**, 448.

15

D. R. NATALE
A. J. WATSON

Characterization of Novel Genes during Early Development by Application of Differential Display RT-PCR

1. USING GENOMICS AND PROTEOMICS IN UNDERSTANDING EARLY DEVELOPMENT

The introduction and development of genomics and proteomics during the past decade have significantly changed the way in which biomedical, agricultural, and biological research is conducted. Genomics and proteomics, which represent the identification and sequencing of genes and the systematic analysis and documentation of proteins in a biological sample, have been recently combined to give rise to functional genomics. Functional genomics encompasses assigning function to genes identified by genomics as well as investigating the organization and control of genetic pathways and the modification of their resulting proteins. This approach is breaking down the distinction between classic areas of study such as biochemistry, genetics, and physiology and bringing aspects of each field together to advance the understanding of biological systems. In this chapter we address the genomics aspect of functional genomics and its application to mammalian preimplantation-stage embryos by introducing and discussing methods designed to investigate and identify "new" genes involved in preimplantation development.

In the last several years, research on the molecular events that control early development has intensified. Specifically, there has been a concentrated effort directed at determining how transcript expression throughout preimplantation development supports the overall developmental program to the blastocyst stage (reviewed by Watson and Barcroft [1]). Characterization of these transcripts has largely relied on a gene-by-gene approach that arises from a specific result from which a hypothesis is generated based on individual genes that may be involved in driving the developmental event. An example is the study of blastocyst expansion and the related contribution of specific members of the Na^+/K^+-ATPase gene family in establishing an ionic gradient across trophectoderm cells to mediate accumulation of fluid supporting formation of the blastocoel cavity (reviewed by Watson et al. [2]).

The advent of highly efficient DNA sequencing methodology in the 1990s is one of the critical advances that made functional genomics a reality. This field has expanded in parallel with developments in the human genome sequencing project and other genome sequencing projects. As increasing amounts of sequence data are entered into databases around the world and techniques for large-scale analysis of gene expression are improved, it has become feasible to use these approaches to study gene expression in embryos (3–6). This has prompted a shift in the approach to experimental design from classic hypothesis-driven research and investigating genes on an individual basis to investigating questions in more general terms. By examining overall changes in gene expression surrounding a particular event or under specific treatments or conditions, inferences can be made about groups of genes required to be expressed in the experimental paradigm, the way in which different groups or families of genes may interact in the experimental paradigm, and previously uncharacterized genes playing a role in the experimental paradigm. Once this information has been generated, individual genes or gene families can then be examined and characterized more specifically in the context of the original experimental design. Investigators in many fields interested in gene expression underlying disease or specific cellular events are now able to study the expression of hundreds or thousands of genes in a single screen or assay, in contrast to investigating expression of individual genes. Using techniques such as DNA microarrays, differential display reverse transcription-polymerase chain reaction (DD-RT-PCR) and subtraction hybridization (SH), it is now possible to compare and contrast global patterns of gene expression between multiple tissues or samples. Although each of these techniques varies in the details in which they are applied and the type of information they provide, they are all similar in that they allow the direct comparison of patterns of gene expression between two or more representative pools of RNA and allow the identification of previously uncharacterized genes within the comparison groups. These types of approaches have eliminated the question of how to investigate gene expression and have instead largely replaced that question with another: how do we deal with the amount of information that is generated by these approaches? In response to this challenge, the field of bioinformatics has rapidly grown to handle the storage and analysis of information generated in gene and protein screens. Large databases of nucleotide and amino acid sequence information have been accumulated, and, in coordination with web-based software on these sites, unknown protein and nucleotide sequences can be screened for sequence similarity against previously identified/characterized sequences by the public (e.g., www.nbci.nlm.nih.gov/ and www.tigr.org).

In this chapter we focus specifically on the application of DD-RT-PCR to investigate global patterns of gene expression and identify novel gene products in preimplantation-stage mammalian embryos. This method is now widely applied to search for new genes and investigate gene expression in the preimplantation embryo (3, 5, 7–9). Its principal utility is that it can be applied to small starting mRNA samples, and it can also be applied to contrast mRNAs between more than two samples in a single experiment. For these reasons we believe that it offers an important approach for targeting functional genomics to the preimplantation embryo.

2. DIFFERENTIAL DISPLAY REVERSE TRANSCRIPTION-POLYMERASE CHAIN REACTION

DD-RT-PCR is a PCR-based technique that is an unbiased method to contrast mRNA pools from two or more samples (10) (see Figure 15.1 for overview). Introduced in 1992 (10), DD-RT-PCR uses short, arbitrary primers in the 5' position (10-mer) in coordination with one of four anchored oligo-dT primers in the 3' position. The basic principle is that by using a specialized or anchored oligo-dT to prime the reverse transcription reaction, the pool of messenger RNA can be subdivided into four representative parts. The same oligo-dT used in the RT reaction is then used in a PCR reaction in coordination with one of 26-5' primers. These 26 primers are short 10-mer oligos of specific

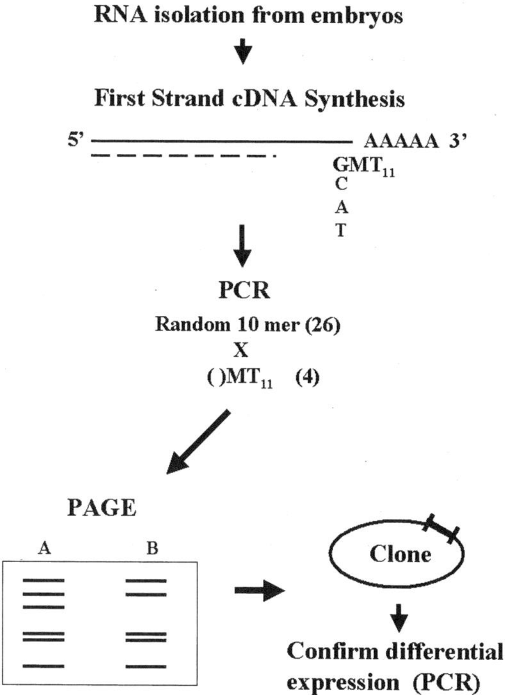

Figure 15.1. A schematic overview of the dd-RT-PCR approach as applied to preimplantation mammalian embryos. Briefly, total RNA is isolated from pools of embryos and undergoes first-strand cDNA synthesis using one of four anchored oligo-dTs. PCR is then carried out in the presence of a radiolabel using one of 26 arbitrary 10-mer 5' primers in coordination with the same oligo-dT used for first strand cDNA synthesis. Amplicons are visualized by standard PAGE followed by autoradiography. Differentially expressed cDNAs are then cloned, sequenced, and expression confirmed by PCR following the design of gene-specific primers against the differentially expressed amplicon.

sequence designed in various patterns of bases to theoretically prime all cDNA fragments present after RT arbitrarily with an anchored oligo-dT. By combining each 5' primer in PCR reactions with each anchored oligo-dT after RT with that same anchored oligo-dT, cDNA fragments should be generated that represent each transcript present in any RNA sample at least once. To visualize these products, the PCR reaction is carried out in the presence of a radiolabeled nucleotide, generally either ^{35}S or ^{32}P, and, after amplification, PCR products can be separated on a sequencing-grade polyacrylamide gel followed by drying of the gel and exposure to X-ray film. An anchored oligo-dT is a run of several T bases from the 5' to 3' direction that is anchored at the 3' end with either A, T, C, or G (random) at the second-to-last position and in the final 3' position, specifically by either A, T, C, or G, therefore giving the designations $T_{11}M$C, $T_{11}M$G, $T_{11}M$A and $T_{11}M$T, where M represents the random base.

After comparing banding patterns between samples and identifying differentially expressed cDNAs, these products can be easily isolated from the gels. After reamplification and cloning by standard T-A-cloning (T-Easy; Promega, Madison, WI), these products can then be sequenced and potentially identified by comparison of the nucleotide sequence to one of many available sequence databases. As a result, this method can lead to the identification of new gene products as well as the description of expression of previously characterized ones.

DD-RT-PCR has been used in a number of studies in our laboratory to assess patterns of gene expression in bovine embryos (5, 7). In one such experiment, we used this technique to investigate variance in gene expression after insemination of in vitro-matured bovine oocytes and during their progression through each cleavage division to the blastocyst stage (5). Briefly, pools of 50 in vitro-matured oocytes and embryos (at the 2–5-cell, 6–8-cell, 8–16-cell, morula, and blastocyst stages) were collected and analyzed by DD-RT-PCR using different primer sets. This approach allowed us to demonstrate that patterns of gene expression in mature oocytes are similar to patterns in early cleavage-stage embryos (2–5 cells). As the bovine embryo begins to undergo the major phase of genome activation at the 6–8-cell stage, the pattern dramatically shifted to establish a new pattern that was then conserved and further refined through the 8–16-cell, morula, and blastocyst stages. By examining gene expression in embryos in this fashion, we were able to gather data on how general patterns of gene expression varied with each cell division as well as to identify embryo stage-specific gene transcripts. To further enhance the study and to specifically examine gene transcripts actively transcribed from the embryonic genome, we also included treatment of the cultured embryos with a specific inhibitor of RNA polymerase II activity, α-amanitin, thereby preventing active production of mRNA species in these embryos.

Other experiments using DD-RT-PCR that have been carried out in our laboratory include the comparison of gene expression in bovine embryos cultured to the blastocyst stage in fully defined culture media and a reduced oxygen atmosphere versus embryos cultured in the presence of serum, oviductal co-culture, and 5% CO_2 in air atmosphere. We have also used this method to contrast gene expression in embryos produced by somatic cell cloning to embryos produced in vitro and in vivo as well as to the patterns of gene expression typical of the nuclei-donor cells used in the production of the somatic cell clones (7). As a final example, DD-RT-PCR has also been used to compare patterns of mRNA expressed in bovine oocytes collected from follicles of different sizes and stages (6).

A chief advantage of DD-RT-PCR is that it is possible to work with small amounts of starting material—for example, single preimplantation-stage embryos. In addition, using DD-RT-PCR, one can compare several samples at once in a single screen. In fact, the number of samples that can be compared in a single display is limited only by the researcher's ability to provide starting material, manage a large number of PCR reactions, and provide a gel apparatus large enough to display the resulting products.

3. APPLICATION OF DD-RT-PCR METHODS

3.1 Collection of Eggs and Embryos

We have investigated gene expression during preimplantation development for both murine and bovine species. Methods used to produce and collect preimplantation-stage embryos from either species are given in other chapters of this book. For bovine embryo production, cumulus–oocyte complexes (COCs) are harvested from slaughterhouse-collected ovaries, pooled, washed in oocyte collection medium, and placed in oocyte maturation medium for 22 h at 39°C under a 5% CO_2 in air atmosphere (11,12). Frozen bovine semen is thawed and prepared by a swim-up procedure (working concentration 1×10^6 sperm/ml) (13). Inseminated oocytes are cultured in 50 µl drops of modified synthetic oviduct fluid medium (SOFM) supplemented with 0.5 mM sodium citrate, 3 mg/ml PVA, and 1× nonessential and essential amino acids under a 5% CO_2, 7% O_2, and 88% N_2 atmosphere. Embryos are cultured for up to 8 days at 39°C.

Murine embryos are obtained from random-bred CF-1 or CD-1 females superovulated with pregnant mare serum gonadotrophin (PMSG) and human chorionic gonadotrophin (hCG) and mated with $CB6F_1/J$ males. COCs are collected from unmated female mice at approximately 13–15 h after administering hCG, whereas two-cell, four-cell, eight-cell, morula-, and blastocyst-stage embryos can be collected from mated females by standard oviduct and uterine flushing methods at approximately 48, 56, 72, 80, and

90 h after hCG, respectively. For culture of murine embryos, two-cell-stage zygotes are flushed from oviducts at 48 h after hCG administration and placed in 50-µl drops of KSOMaa medium (potassium simplex optimized medium plus amino acids [14]), covered by mineral oil, and cultured under a 5% CO_2, 7% O_2, and 88% N_2 atmosphere for up to 3 days at 37°C. These procedures produce the best culture environments for supporting bovine and murine preimplantation development in vitro.

3.2 RNA Isolation and Reverse Transcription

(For a detailed review of the methodologies currently applied for the isolation of total or Poly A+ mRNA from preimplantation staged embryos, refer to chapter 13.) We have used microisolation of RNA from pools of frozen, lysed embryos using messenger affinity paper (mAP), based on the method of Collins and Fleming (15) and described by De Sousa et al. (12). We used this method of RNA isolation in our experiments because it generated reproducible results in PCR amplification of specific gene products from small numbers of preimplantation-stage embryos. There are currently many RNA isolation methods used by different labs around the world, and any method that yields reproducible PCR results from small pools of embryos should be equally effective for RNA isolation in the context of DD-RT-PCR experiments.

Reverse transcription of total RNA from embryo samples employs an anchored oligodT$_{11}$MC primer (10). Reverse transcription reactions are conducted in a final volume of 20 µl consisting of 50 mM Tris HCl (pH 8.3), 75 mM KCl, 3 mM $MgCl_2$, 10 mM DTT, 750 µM dTNPs, and 300 U of Supercript RNase H (Gibco BRL, Gaithersburg, MD) for 90 min at 42°C. Reactions are terminated by heating for 5 min at 95°C and then placing on ice. The reverse-transcribed cDNA can be either used immediately for DD-RT-PCR analysis or stored at –20°C for weeks or even months. As a negative control for RNA isolation and reverse transcription, non-reversed transcribed samples must be included in the DD-RT-PCR experimental design.

3.3 Differential Display PCR

After reverse transcription of cDNA with the chosen anchored oligo-dT primer, the low-stringency PCR reaction is carried out. Although we have used [^{35}S]-ATP in our DD-RT-PCR reactions, alternative radioisotopes may be used to visualize PCR products. For example, ^{33}P-labeled nucleotides may be used with the benefit of enhanced detection of bands after autoradiography as well as a reduction in the potential for contamination of work space/equipment by aerosols that are associated with ^{35}S isotope. The DD-RT-PCR reaction mix set-up is detailed in Protocol 15.1.

3.4 Display of PCR Products by PAGE

Products of the DD-RT-PCR reaction are separated, according to size, by polyacrylamide gel electrophoresis on a standard sequencing-gel apparatus. Visualization of PCR products is then possible due to the incorporation of a radiolabel into the PCR amplicons by standard autoradiography/X-ray film methods (see Protocol 15.2).

3.5 cDNA Band Selection

As for any experimental, approach, careful consideration must be given to experimental design. It is important that non–reverse-transcribed controls are conducted along with reverse-transcribed embryo samples. In addition, we have adopted a conservative approach to designating individual bands as differentially expressed. All of our embryo samples are run simultaneously as triplicate PCR amplifications. Only those bands consistently represented in each of the triplicate lanes as positives are considered for further analysis. In addition to these criteria, when comparing cDNA banding patterns between different groups, we select only those bands that were clearly present in one group and absent in the comparison group. These steps are necessary to reduce the number of false positives that are selected for further analysis. The size of the original embryo pool used for RNA isolation affects these parameters in a dramatic way. It is possible to apply these methods to analyze mRNA pools in individual blastocysts, however, the use of single embryos from cleavage-stage embryos is likely not warranted due to the increased variability in banding pattern observed among triplicate reactions. It is possible to use DD-RT-PCR to examine changes in abundance of a message, but this method is not quantitative per se, and a more reliable quantitative assay must be used subsequent to the isolation of putative differentially expressed cDNA to confirm the differential gene expression (e.g., Northern blot, quantitative PCR).

3.6 Identification of Differentially Expressed Genes

One of the most compelling reasons for applying DD-RT-PCR methods to examine gene expression is the ability to easily identify the differentially expressed transcripts. Because the method is based on low-stringency PCR amplification, the technique results in the identification of both known transcripts and also of entirely novel gene transcripts. After selecting a differentially expressed band from the autoradiograph, the film is used as a template and positioned over the dried polyacrylamide gel and fastened down. The selected band is then cut out of the film along with the gel below. Using the following protocol, the band is eluted from the polyacrylamide, ethanol precipitated, and reamplified with the same primer set used in the DD-RT-PCR amplification series. The resulting PCR amplicon is then cloned using a commercially available TA-cloning or PCR-cloning vector/kit and can then be sequenced. The nucleotide sequence obtained from DD-RT-PCR products generally represents the 3' end of the message, as the technique relies on priming by a modified (anchored) oligo-dT. However, some of the products may not be generated as 5'–3' amplicons. Based on sequence similarity to sequences published in sequence databases, the differentially expressed transcript can then be putatively identified or deemed novel.

For excision and reamplification of differentially expressed cDNAs, see Protocol 15.3.

4. CONSIDERATIONS

In the application of DD-RT-PCR, there are several considerations at different steps of the protocol that should be taken into account to generate successful displays. First, one of the major limitations facing researchers attempting to characterize gene expression in early mammalian embryos is the limited amount of starting material. With only nanograms of RNA in a single embryo, many molecular techniques for the examination of gene expression are excluded, as they would require hundreds or thousands of embryos for a single experiment. This is both tedious and expensive. Differential display is not so dramatically affected by this problem, as PCR can be carried out successfully on single embryo equivalents. It is necessary, however, to consider the number of embryos from which RNA is harvested for use in the DD-RT-PCR reaction. To avoid minimizing sensitivity to low-abundance messages, we have found it useful to process embryos in groups of greater than 10. In fact, the more embryos that can be used for collection of RNA in a single pool and subsequent reverse transcription the better, as this pool can then be used for DD-RT-PCR with multiple primer sets. When applying DD-RT-PCR, it is important to consider the number of embryo equivalents of cDNA being used as a template in each reaction. Although it is important not to add too little cDNA, it is equally important to not add too much. We find it possible to achieve reproducible and consistent banding patterns using two embryo equivalents of cDNA in each PCR reaction.

This amount of cDNA works equally well for all stages of preimplantation development. It is possible to use less at some stages, for example, the blastocyst, but banding patterns may become weak and/or inconsistent. In terms of how much cDNA can be added, we find that the use of 5 or maximally 10 embryo equivalents per reaction is possible; however, more than this results in a dark smear on the gel.

Another important consideration is the selection of differentially expressed amplicons. A differentially expressed transcript may be identified and selected based on its intensity in one sample with respect to another, but it is important to consider that any given band on a differential display gel may actually represent different comigrating species of cDNA due to the nature of random priming by the short and arbitrary 5' primer. Some of these species will not be differentially expressed, therefore, upon excision/cloning and sequencing, it is possible that the amplicon cloned is one of those that is not differentially expressed. To reduce the likelihood of false positives in the selection of differentially expressed amplicons, a single-stranded confirmation polymorphism procedure can be applied to separate the different species of cDNA that may be present in what appears to be a single band in the differential display.

5. CONCLUSIONS

Elucidation of the genetic program controlling preimplantation development is an important endeavor. With the development of techniques that allow the investigation of gene expression on a large scale, this goal will likely be realized and specific mechanisms influencing gene expression eventually defined. While each technique has advantages and disadvantages, it is possible to successfully screen for new genes during preimplantation development using comparatively small embryo pools. As these techniques are refined and advanced, the only limitation may well be our imagination in establishing newer and more effective experimental paradigms for their application.

Protocol 15.1. Differential display PCR

1. When preparing samples for DD-RT-PCR, prepare a master mix including all reaction components except the cDNA to reduce pipetting error and minimize pipetting of [^{35}S]dATP. In addition, prepare mixture on ice adding Taq polymerase and [^{35}S]dATP last to minimize handling of and contamination with radioactivity. Set up DD-RT-PCR reaction mix (20 μl total minus volume of cDNA) as follows:

   2.0 μL 10× PCR buffer (Mg^{2+} free)
   2.0 μl 25 mM MgCl$_2$
   0.2 μl 200 μM dNTP
   (quantity depends on volume of cDNA) nanopure water
   2.0 μl 5' primer (5 μM stock)
   2.0 μl 3' primer (25 μM stock)
   0.5 μl Taq polymerase (5 U/μl)
   1.0 μl [^{35}S]-dATP (10 μCi/uL)

   Note: 5' primer is a random 10 mer; 3' primer is the same primer used in reverse transcription and preparation of cDNA.

2. Aliquot cDNA into 0.5-ml tubes in which DD-RT-PCR reactions will be carried out. The amount of cDNA required for each reaction depends on the embryo stage being used. DD-RT-PCR reactions were most reliable using the following minimum numbers of embryo equivalents (ee; where 1 ee = 1 μl of a 20-μl reverse transcription reaction in which RNA from 20 embryos/oocytes has been reverse transcribed) for each stage: oocyte to morula – 2ee, blastocyst – 1ee.

3. Add PCR mix to tubes containing aliquoted cDNA and overlay with light paraffin oil (1 drop/tube from a 200 μl pipettor) and cycle as follows: 40 cycles of 94°C for 30 s, 42°C for 60 s, 72°C for 30 s, followed by a hold at 72°C for 5 min and a 4°C dwell. *Note*: Annealing temperature is the same as annealing temperature used in reverse transcription reaction for preparation of cDNA being used.

4. Display DD-RT-PCR products by standard PAGE or store at –20°C for only a few days at most. Best results will be obtained if DD-RT-PCR products are displayed within the first half-life of the radioisotope used.

Protocol 15.2. Display of PCR products by PAGE

1. Prepare polyacrylamide (6%). As a guide, the sequencing apparatus used for differential display in our lab is 38 × 50 cm in size. Using spacers of 0.4 mm thickness, a gel volume of approximately 85 ml is required. For preparation of 100 ml of 6% polyacrylamide:

   42 g urea
   10 ml 10× TBE
   15 ml 40% acrylamide stock
   1.0 g Bio-Rad ion exchange beads
   ≤ 80 ml distilled water

   Cover and mix on stirrer under low heat to dissolve urea. When dissolved, top up to 100 ml with distilled water and filter through #1 Whatman paper (wet filter paper first with distilled water to reduce volume loss).

2. While urea is dissolving, clean sequencing plates carefully and thoroughly with a detergent such as Sparkleen and rinse. Dry using 95% EtOH and give a second rinse with 95% EtOH.

3. After drying, mix 200 μl of Kodak PhotoFlo in water (5 ml total) and apply half to gel surface of each plate. Spread using Kimwipes and wipe dry. Plates should be uniformly smooth after this treatment. This step is important because it influences the ease with which the gel can be removed from the gel plate after electrophoresis and before gel drying.

4. Clean spacers and comb and, when dry, put apparatus together.

5. To pour gel, add 100 μl 25% Ammoniun Persulfate (freshly prepared) and 100 μl TEMED to acrylamide and pour quickly. Gel will begin to polymerize within 10–15 min. Allow gel to polymerize 2 h. Gel may also be left to polymerize overnight. In this case, place wet paper towel over top of the wells, but not touching, and seal with plastic wrap to prevent well shrinkage.

6. Prepare 1× TBE running buffer and prewarm to 50°C before adding to apparatus. Add buffer to assembled apparatus, remove comb, flush wells using a syringe and needle (to remove urea and nonpolymerized acrylamide), and run at approximately 70 W for 2 h before loading (gel and plates should reach running temperature prior to loading, approximately 50–55°C).

7. For display, mix 5 μl of DD-RT-PCR products (from each sample; triplicate reactions are run for each experimental sample) with 2 μl of loading buffer. Flush wells a second time just before loading. Load DD-RT-PCR product carefully to prevent contamination of equipment with radioactivity and to prevent air bubbles forming in wells, as this may affect how distinct the bands appear after autoradiography.

8. Run gel at constant wattage to maintain a consistent gel temperature of 50–55°C. The wattage/temperature relationship will be determined by gel thickness, size, and so on, and simply requires trial and error. Start with a run time of 3 h and then judge from that the kind of product separation you desire. A three-hour run has allowed us to observe cDNA products from approximately 100–600 bp in size clearly.

9. After electrophoresis, drain buffer from gel apparatus and lay gel apparatus flat on the bench top. Gently separate the plate with the buffer chamber from the flat plate so that gel is left on the latter. Using a piece of filter paper just slightly larger than the gel, gently lower filter paper onto gel beginning at one end and following to the other. Make sure to avoid creases, folds, or bubbles. Once the filter paper is flat, the gel should stick to the paper and can now be removed from the flat plate in one swift motion, beginning from a corner. Lay paper flat on the bench, gel side up, and cover with plastic wrap. Place in gel dryer and dry under vacuum for 2 h.

10. When dry, transfer gel to cassette and put down film. Generally, if isotope is fresh, overnight exposure at room temperature is sufficient using Kodak BioMax film.

Protocol 15.3. Excision and reamplification of differentially expressed cDNAs

1. Align autoradiograph with the dried gel and cut out the bands of interest through the autoradiograph and the gel.
2. Place gel/paper into a 0.5-ml tube and rehydrate in 100 μl TE buffer (pH 8.0) for 10 min at room temperature.
3. Boil sample 15 min, spin 2 min in microfuge, and remove supernatant to new 0.5-ml tube.
4. Add 10 μl of 3 M Na$^+$-acetate, 5 μl glycogen (10 mg/ml), and 450 μl cold 100% EtOH and place at $-80°C$ for 30 min to precipitate the DNA.
5. Spin tube in a microfuge 10 min at 4°C to pellet DNA.
6. Remove supernatant and wash pellet with cold 85% ethanol.
7. Dissolve pellet in 10 μl of dH$_2$O and store at $-20°C$.

PCR master mix (35 μl)

4.0 μl 10× PCR buffer (Mg^{2+} free)
4.0 μl 25 mM MgCl$_2$
4.0 μl 200 μM dNTP mix
4.0 μl 5' DD-RT-PCR primer (5 μM)
4.0 μl 3' anchored primer (25 μM)
14.0 μl sterile H$_2$O
1.0 μl Taq (5 U/μl)

Aliqout 35 μl of master mix into 0.5-ml Eppendorf tubes containing 5 μl of eluted cDNA template giving 40 μl total volume. Cycle on thermocyler: 94°C for 30 s, 42°C for 30 s, 72°C for 30 s for 40 cycles; 72°C for 5 min, 4°C dwell.

References

1. Watson, A. J., and Barcroft, L. C. (2001). *Frontiers Biosci.*, **6**, D708.
2. Watson, A. J., Westhusin, M. E., De Sousa, P. A., Betts, D. H., and Barcroft, L. C. (1999). *Theriogenology*, **51**, 117.
3. Ma, J., Svoboda, P., Schultz, R. M., and Stein, P. (2001). *Biol. Reprod.*, **64**, 1713.
4. Robert, C., Gagné, D., Bousquet, D., Barnes, F. L., and Sirard, M. A. (2001). *Biol. Reprod.*, **64**, 1812.
5. Natale, D. R., Kidder, G. M., Westhusin, M. E., and Watson, A. J. (2000). *Mol. Reprod. Dev.*, **55**, 152.
6. Robert, C., Barnes, F. L., Hue, I., and Sirard, M. A. (2000). *Mol. Reprod. Dev.*, **57**, 176.
7. De Sousa, P. A., Winger, Q., Hill, J., Jones, K., Watson, A. J., and Westhusin, M. E. (1999). *Cloning*, **1**, 63.
8. Lee, K. F., Chow, J. F., Xu, J. S., Chan, S. T., Ip, S. M., and Yeung, W. S. (2001). *Biol. Reprod.*, **64**, 910.
9. Minami, N., Sasaki, K., Aizawa, A., Miyamoto, M., and Imai, H. (2001). *Biol. Reprod.*, **64**, 30.

10. Liang, P., and Pardee, A. B. (1992). *Science*, **257**, 967.
11. Betts, D. H., Barcroft, L. C., and Watson, A. J. (1998). *Dev. Biol.*, **197**, 77.
12. De Sousa, P. A., Westhusin, M. E., and Watson, A. J. (1998). *Mol. Reprod. Dev.*, **49**, 119.
13. Parrish, J. J., Susko-Parrish, J. L., Leibfired-Rutledge, M. L., Crister, E. S., Eyestone, W. H., and First, N. L. (1986). *Theriogenology*, **25**, 591.
14. Ho, Y., Wigglesworth, K., Eppig, J. E., and Schultz, R. M. (1995). *Mol. Reprod. Dev.*, **41**, 232.
15. Collins, J. E., and Fleming, T. (1995). *Trends Genet.*, **11**, 5.

16

K. E. LATHAM
G. NOURJANOVA
K. WIGGLESWORTH
J. J. EPPIG

Two-Dimensional Protein Gel Database Analysis of Embryos, Oocytes, and Oocyte-associated Granulosa Cells

1. ASSESSING CHANGES IN EMBRYONIC GENE EXPRESSION

A common objective of investigations of the mechanisms regulating early development is to characterize the effects of specific experimental or genetic manipulations on embryonic phenotype at the molecular level. In many circumstances, where molecular, cellular, or physiological mechanisms governing a process are well known, a marker gene approach can be adopted, in which the response of one or more specific genes to experimental manipulations is examined. In other circumstances, however, such marker gene approaches are of limited use, either because the nature of the biological response cannot be readily predicted or because the underlying biology of a given response is poorly understood, making the selection of the appropriate marker genes difficult.

In clinical medicine, the disease state needs to be characterized at a molecular level in order to understand the genetic basis of the disease and to design subsequently new therapies. In these cases, an alternative approach to characterizing the molecular responses of cells or embryos to manipulations is a global approach in which the effects on the expression of a large number of genes is examined, without any preconceptions or bias as to what specific genes should be examined. Two important advantages of using a global approach to assess changes in gene expression are, first, that unexpected relationships between diverse treatments and unexpected molecular responses can be seen and, second, that the large amount of gene expression data produced provides an unparalleled opportunity for quantitative and statistical analyses that are not possible with a simple marker gene approach. As a result, it becomes possible to measure the magnitude of change of overall gene expression patterns (i.e., one can quantify the degree of similarity between two cell types, something that is rarely achieved). It also becomes possible to observe incremental effects of treatment that may not be observed at the level of developmental potential or morphology and to identify by cluster analysis families of gene products that respond coordinately to experimental manipulations.

2. GLOBAL SCREENING METHODS

Two notable methodologies exist for global analysis of changes in gene expression. One exploits the recently developed methods for using cDNA microarrays, wherein hybridization of labeled cDNAs from two cell types to bound oligonucleotides or cDNAs can be performed, and the resulting pattern of hybridization can be interpreted to reveal genes expressed in both cell types or genes that are differentially expressed between the two cell types (1). These technologies provide substantial promise for more sensitive detection of differentially regulated genes and offer the important advantage that differentially expressed genes can be identified immediately. Their optimal application, however, requires the availability of arrays corresponding to the appropriate cell type being analyzed and the availability of significant amounts of mRNA from which to make the labeled cDNA probes. Either of these requirements can constitute a significant limitation in the applicability of the method.

Another approach to global analysis of gene expression, one which increasingly is being used in concert with cDNA microarrays (2), is the use of high-resolution, two-dimensional polyacrylamide protein gel electrophoresis (2D-PAGE) coupled to a system for computer-assisted gel image analysis. Molecular changes can at times be observed by 2D-PAGE where other methods of analyses fail, and 2D-PAGE can reveal significant effects of gene mutations on the expression of other gene products in the cell (3, 4). Thus, 2D-PAGE provides a powerful means of detecting gene products that undergo changes in expression.

3. HIGH-RESOLUTION 2D-PAGE

High-resolution 2D-PAGE can resolve up to 2000 individual polypeptides on a single gel, and as many as 1600 of these may be of suitable intensity and quality to permit accurate quantitation (5–8). Additional proteins can be detected and resolved using "zoom-in" gels (9, 10). For the preimplantation mouse embryo, the rates of synthesis of approximately 1200 polypeptides were followed from fertilization through the blastocyst stage (11, 12). High-quality gel images with a large number of well-resolved spots can be obtained with a small number of embryos (11) or granulosa cells (13), making this technique ideal for the study of preimplantation embryos, oocytes, or oocyte-associated granulosa cells. These gel images can be obtained for any number of stages or experimental treatments, entered into a cumulative computerized database, and the patterns of expression of proteins monitored through an almost limitless number of samples and experiments. This provides the ability to accumulate a large amount of information about the regulation of expression of the analyzed proteins as well as detailed knowledge of the cellular states of early embryos, oocytes, or oocyte-associated granulosa cells, such as cumulus cells.

Proteins detected within the 2D-PAGE gels can be identified through the inspection of size and isoelectric point in suitably calibrated gels and through such techniques as immunoprecipitation and immunodepletion (12), Western blotting (14), and sensitive methods for high-throughput mass spectrometry, coupled to expressed sequence tag and other genome sequence databases (2, 15–23). For studies of embryos and oocytes, the amount of material and labor required to identify systematically proteins within the 2D gel profile may limit the number of protein spots for which specific identities are established. It is possible, however, that as new and more sensitive methods are developed, the ability to obtain sequence information or spot identities for gels of early embryo proteins will be achieved. In addition, because comparisons of embryonic protein gel profiles to gel profiles for other cell types may reveal protein identities, as proteomic databases are developed for more cell types and species, it may become possible to assign protein identities through comparative image analysis.

4. 2D-PAGE AND DEVELOPMENT

Even in the absence of known identities for the majority of protein spots, however, the 2D-PAGE approach offers a powerful method for detecting proteins with developmentally regulated patterns of expression, detecting patterns of expression that respond to specific experimental manipulations, and for identifying groups of coordinately regulated proteins through the application of cluster analysis (8). This allows the investigator to identify novel arrays of molecular markers that can then be used in a manner reminiscent of the marker gene approach. Thus, even though the formal identities of the regulated proteins may not be known, their developmental properties nevertheless are clearly revealed by their expression patterns, making them suitable markers for experimental purposes. The ability to obtain quantitative data about the effects of a given treatment on the expression of a large number of regulated proteins provides more in-depth and informative results than can be achieved through the examination of one or few marker genes, and in fact can reveal how genetic polymorphisms in a single gene can affect the expression of multiple proteins. Because expression data are cumulative within the protein database, especially notable proteins will, over time, become evident, and proteins can be identified. In this way, the 2D-PAGE approach allows a global approach to evaluating biochemically cellular responses to specific manipulations, combined with the long-term pay-off of identifying specific gene products that exhibit biologically interesting patterns of expression, and can then be used for additional detailed studies.

5. APPLICATION OF 2D-PAGE TO STUDY EMBRYO PROTEIN PATTERNS

In general, 2D-PAGE has been applied to early mammalian embryos through the analysis of radiolabeled proteins after incubating embryos briefly with radiolabeled amino acids (11–13). Although hundreds or even thousands of embryos may be required to visualize cellular proteins by methods such as silver staining, only one or two dozen embryos may be required to obtain the necessary amount of radiolabeled protein to permit 1000 or more polypeptide spots to be visualized.

Embryos, oocytes, or oocyte-associated granulosa cells are cultured for a brief period (e.g., 1–3 h) in a high concentration of radiolabeled amino acid. The cells are then washed briefly in a protein-free medium or PBS and then lysed in buffer containing a reducing agent and a low concentration of SDS to denature proteins rapidly and dissociate protein complexes. The proteins are then resolved on isoelectric focusing (IEF) tube-gels and electrophoresed in a second dimension through a standard SDS PAGE gel (24). Each polypeptide migrates to a standard position within the second-dimension gel that is dictated by the combination of its isoelectric point and its molecular mass. This most common type of 2D-PAGE can be augmented through the use of nonequilibrium isoelectric focusing gels in the first dimension (14), which permit the visualization of more basic proteins. The proteins thus separated are detected either by fluorography using X-ray film followed by scanning densitometry or by using a phosphorimaging device (25).

Once images of the gels are acquired, a software program (5, 26) can be used to assemble the images into a protein database. A protein database consists of one or more gel images, matched to each other so that the expression of each polypeptide can be followed from gel to gel. Gels that have been matched to one another constitute a matchset (5). It is useful to organize matchsets to address specific experimental questions, rather than including a diverse array of gels within a given matchset. For example, matchsets can represent a series of samples collected at progressive times during development, a series of gels from samples in which the embryos or cells were treated in specific ways, or sets of gels in which proteins were isolated from specific subcellular compartments. Multiple experiments (i.e., matchsets) can then be linked to one another to form higher order matchsets and thus allow expression patterns to be followed between experiments as well as within experiments. This allows a wide variety of data to be collected for the analyzed proteins, and these data are then cumulative within the database. An artificial gel image can be created in which the positions of all spots detected in all gels of all experiments in the gel collection are marked. This standard or reference image then serves as a map for matching and tracking expression data for each polypeptide. The database can also include information about the individual experiments, samples, annotations of proteins including identities, isoelectric point, molecular mass, subcellular localization, and interesting attributes of protein expression. Recently, with the advent of proteomics, spot annotation provides a useful means to link 2D gels to expressed sequence tag and other genome sequence databases, thus allowing a comprehensive database to be developed.

6. TROUBLESHOOTING AND COMMON PROBLEMS

To apply 2D-PAGE to embryos, it is essential to obtain an adequate amount of incorporated radiolabel for analysis. In our earlier studies of mouse embryos, between 5×10^5 and 1×10^6 dpm was preferred per gel, with each embryo incorporating approximately 3×10^4 dpm. Although some gel images were obtained with less material, this required long exposure times using fluorography of 6 months or more. With the advent of phosphorimaging, we found that, although exposures can be obtained within a matter of days, suitable gels with a large number of detected spots require at least 2×10^6 dpm. Because samples of cells lysed in dSDS are lyophilized and resolubilized in an equal volume of sample buffer before IEF, the sample can not be concentrated, so it is important to lyse the embryos, oocytes, or cells in a volume that contains 2×10^6 dpm in 15 μl or less.

One must be careful not to exceed the protein-resolving capacity of the gel. For high-resolution analyses, thin analytical IEF gels are preferred, and these can only accept up to 10 μg of total protein (24). Because of the small size of oocytes and embryos, this is generally of no concern, but for cultured cells, efficient labeling is essential to avoid overloading the gel.

We have observed on occasion that, although a sufficient number of acid-precipitable disintegrations per minute is loaded to a gel, for some samples few or no spots may be detected. This most likely is the result of protein degradation or insufficient solubilization. It is essential to test each batch of dSDS and DNAse/RNAse solutions to ensure proper solubilization, nucleic acid digestion, and maintenance of protein integrity before using them for valuable experiments. For samples of embryos, culture conditions must support efficient uptake and incorporation of radiolabel. For cultured cells, it is essential that rapid and efficient solubilization and efficient DNase and RNase digestion be achieved; a noticeable decrease in sample viscosity should occur within a short time of DNase/RNase treatment on ice. Additional dSDS can be added if solubilization appears incomplete, and the DNase/RNase digestion can be prolonged if needed. It is also essential that no solutions containing urea be heated above 37°C.

Inconsistent or insufficient resolution of proteins on the IEF gel can occur. Substantial differences between gel runs in the resolution of proteins will prevent efficient matching of gel images. This can usually be overcome by changing ampholines or by using mixtures of ampholines. It is also important to apply consistent voltage and run times. Occasionally, IEF gels may break during removal from the IEF tubes. An experienced operator can often reassemble the IEF fragments when running the second dimension. However, a significant gap between the gel ends will prevent accurate matching of the gel images. IEF gels can also become folded when loading onto the second dimension. If this cannot be corrected, the gel will not be useful for analysis in a matchset.

7. CONCLUSION

As the ongoing genome initiatives progress, the field of proteomics should also continue to expand by establishing the identities of protein spots in a variety of cell and tissue types from a range of species, and the two areas of data collection should eventually come together so that detailed data for protein expression becomes linked to gene mapping and sequence data. Beyond the simple matching of expression patterns to gene sequence, however, it will be of great value to understand relationships between coordinately regulated genes and how such coordinately regulated genes work together to produce defined phenotypes in response to specific stimuli. Although DNA microarray technologies will reveal coordinate expression at the mRNA level, such approaches will not provide data at the level of protein expression or post-translational modifications, which are critical for determining the actual expression of the gene activity. Thus, global analyses undertaken using 2D-PAGE and databases as outlined in Protocols 16.1–16.4 housing the protein expression data should ultimately play an important role in the overall objective of assigning functions to newly identified genes.

Given the already well-documented power of combining 2D-PAGE with the use of other tools such as cDNA microarrays and genomic sequence databases, the application of 2D-PAGE analyses and database construction should not be viewed simply as a means to reach an immediate, short-term experimental objective, but rather as an investment toward the day when improved methods for protein spot identification will allow proteomic and genomic data to be incorporated into our specific embryo or oocyte databases and thus help us understand the mechanisms that underlie oogenesis and early embryogenesis. In the meantime, 2D-PAGE remains an valuable method of gaining insight into the molecular changes occurring within cells during normal development or in response to experimental manipulations and continues to offer the important advantage that it can be applied to very small amounts of material to which studies of oocyte, embryos, and granulosa cells are limited while yielding data for 1000 or more gene products.

Protocol 16.1. Sample labeling and lysis

1. Label preimplantation embryos, oocytes, or oocyte-associated granulosa cells in a suitable culture medium (e.g., CZB or Whitten's medium for embryos, or Whitten's embryo culture medium supplemented with 1 mg/ml polyvinyl pyrrolidone for developing oocytes or oocyte-associated cells) containing 1 mCi/ml of high specific activity, isotopically labeled amino acid.
2. Although mixtures of labeled amino acids can be used, we have generally used ^{35}S-methionine (~1000 Ci/mmol or greater) alone for labeling cells. This eliminates possible effects of variation in amino acid composition of labeled mixtures or effects related to differences in amino acid uptake.
3. After labeling, wash the cells once in PBS containing 0.4% PVP (molecular mass ~ 300,000).
4. Separate oocytes from cumulus cells by drawing the oocyte–cumulus cell complexes in and out of a micropipette with a diameter slightly smaller than that of the oocyte. The shearing action strips the cumulus cells off the oocytes.
5. Collect oocytes for analysis or discard. Pellet cumulus cells by centrifugation and resuspend in a small volume of lysis buffer. Remove cumulus cells from ovulated oocytes or embryos by treatment with hyaluronidase (50–100 U/ml, ~800 U/mg, Sigma, St. Louis, MO).
6. Collect oocytes and/or embryos in a minimal volume of PBS and transfer to a suitable volume (typically 20–40 µl) of dSDS lysis buffer preheated to 100°C in a boiling water bath, and then heated at 100°C for an additional 30 s.

7. Place the samples on ice for 1 min. Add one-tenth volume of DNase/RNase mixture and digest the sample for 1 min on ice. Freeze samples in liquid nitrogen and store at $-70°C$.

dSDS lysis buffer (100 ml)

0.3 g SDS
1.0 ml β-mercaptoethanol
0.44 g Tris-HCl
0.265 g Tris-base

Final composition: 0.3% SDS, 1% BME, 50 mM Tris, pH 8.0. Store at $-70°C$ in small aliquots.

DNase/RNase mixture (5 ml)

5.0 mg DNase I (Worthington Biochemical [Lakewood, NJ], code DPFF)
2.5 mg DNase A (Worthington, code RASE)
1.585 ml 1.5 M Tris-HCl
0.08 ml 1.5 M Tris-base
0.25 ml 1.0 M $MgCl_2$
2.96 ml Water

pH should be 7.0. Freeze in 50-µl aliquots at $-70°C$.

Protocol 16.2. 2D-PAGE

- Sample preparation and 2D-PAGE are performed exactly as described (24). In our studies, we have obtained high-quality gels from the gel laboratory at Cold Spring Harbor Laboratory (Cold Spring Harbor, NY) (11–13, 27).
- High-resolution gels can also be obtained by using a suitable commercially available apparatus that offers gel formats of approximately 20×20 cm. Other investigators have used larger gel formats successfully (28).
- "Zoom-in" gels, or the use of multiple, overlapping gels of more narrow pH range, can be used to increase the number of spots detected.
- The choice of ampholine and pH range to be incorporated in the IEF gel is important. A comparatively narrow range (e.g., pH 4.0–8.0) will produce enhanced resolution of spots within that range, but limits the number of spots that will be resolved as compared with a broader range (e.g., pH 3.5–10).
- A mixture of ampholines can be used to enhance resolution within a broad range. It is recommended that test gels be obtained for different combinations of ampholines and examined carefully to ascertain which combination and proportions of ampholines provide the most useful resolution for the specific cell type in question.
- Stringent quality control of electrophoresis reagents is necessary to ensure maximum reproducibility within gel runs, between gel runs, and over prolonged periods of time.
- Ideally, a set of gel reagents that produce optimum results is obtained in sufficient quantity to permit many gel runs. It is recommended that a standard gel be obtained from a readily available cell type and that whenever reagents are changed, a new sample of that cell type be electrophoresed and the new gels compared to the standard gel to ensure reproducibility of protein mobility.
- A list of commercial suppliers of gel reagents that have been successful for our gels is given below.

Sources of gel reagents

Reagent	Source
Acrylamide (30%), bisacrylamide (0.8%) (solution)	Genomic Systems (Ann Arbor, MI)
Ammonium persulfate	Genomic Systems
Temed	Genomic Systems
Tris-HCl	Genomic Systems
SDS	Genomic Systems
SDS/Tris/Glycine running buffer	Genomic Systems
Urea	Genomic Systems
Tris base	Sigma, St. Louis, MO
Igepal-630 (replaces NP-40)	Sigma
NaOH	Mallinkrodt (Hazelwood, MO)
Phosphoric acid	Millipore (Bedford, MA)
Acrylamide, bisacrylamide (powders)	Bio-Rad (Hercules, CA)
Dithiothreitol	Calbiochem (San Diego, CA)
Ampholine	Amersham Pharmacia Biotech (Piscataway, NJ)
Resolyte (ampholine), pH 3.5–10.0	British Drug House, Cater Chemicals (Bensenville, IL)
Ampholyte, pH 5–6	Crescent Chemical (Hauppauge, NY)

Protocol 16.3. Gel imaging

- Gel imaging is accomplished either by densitometric analysis of fluorographs or by phosphorimaging.
- Use of fluorography requires multiple exposures to be collected, due to the limited dynamic range of X-ray film. Shorter exposures thus permit quantitation of spots that are saturated on longer exposures. Exposure times for longest exposures can be 1 month or more.
- Phosphorimaging avoids the need for fluorographic enhancement of gels, offers a broad dynamic range, and can produce well-exposed images in as little as 3 days, with some increase in sensitivity with exposures of up to 7 days. Shorter phosphorimager exposures can also be collected if needed to quantify intense spots.

Protocol 16.4. Data analysis of 2D-PAGE images

The first step to analyzing 2D-PAGE images is to process the gel images for spot detection and quantitation. A merging step can be used to access quantitative data from multiple exposures to quantify intense spots as needed. Crop to select a common area of the gels for analysis. Gaussian modeling and other algorithms for subtraction of background (e.g., horizontal or vertical streaks, or "salt and pepper" background) can be used to create a synthetic image for each gel, to identify spot peaks, and to account for spot overlap in the quantification of individual spots (Figure P16.1). Depending on the parameters selected, some apparent spots visible to the eye may fall below the selected level of sensitivity. Other spots may be well separated from neighboring spots on some gels, but not others. Ideally, the analysis software permits manual editing to be used to resolve these deficiencies.

An additional part of spot quantitation that can be used if desired involves calibration of the gels to enable conversion of densitometric or phosphorimager units to more standard units, such as disintegrations per minute. Calibration strips, consisting of a series of gel segments containing known amounts and densities of ^{35}S, have been exposed alongside the gels for this purpose (24). Calibration chips of ^{14}C, for which the correspondence of imaging units for the chip to defined densities of ^{35}S has been empirically determined, can also be used (25). Alternatively, small paper disks containing small amounts of ^{35}S can be imaged by phosphorimaging with the gel and then quantified by

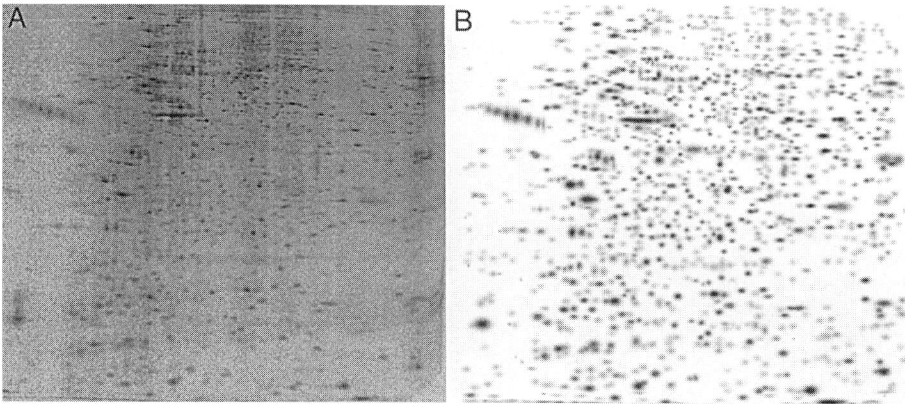

Figure P16.1. Representative two-dimensional PAGE images produced for analysis. (A) Raw scanned image produced by phosphorimaging of a gel of 16-day-old cumulus cells labeled in vitro. (B) Synthetic image of the same gel after background subtraction.

liquid scintillation counting at the end of the exposure. If the total number of acid-precipitable disintegrations per minute (i.e., incorporated radiolabel) loaded on each gel is known, then the disintegration per minute values for individual spots can be converted to units of parts per million to permit comparisons across a wide variety of gels with differing amounts of radiolabeled material applied (24).

The second step to analyzing 2D-PAGE data from multiple samples is to match the gel images to one another. A number of programs exist that permit this type of matching (see below). In general, the first step for comparing samples is to match proteins that are expressed in common in all gels in the matchset, which are known as "landmark proteins." Typically, several hundred of these landmarks will be entered. Matching algorithms then examine the spatial relationships between spots neighboring the landmarked spots, and this matching can be propagated for a certain distance with confidence. A larger number of landmarks improves the accuracy of automated matching. However, incorrectly landmarked spots will produce undesired effects, so only spots that are clearly the same on all gels should be selected as landmarks. The accuracy of the automated matching will depend on such features as the spacing of landmarks, degree of distortion in the gel, degree of resolution of individual spots, spot density in a given area, and whether specific spots are detectable in the selected standard gel. As a result, a number of spots typically remain unmatched or partially matched only between some gel images, and some spots may be matched incorrectly. Therefore, it is necessary to examine carefully all spot matches and correct or complete the matching as needed. In some cases, this can involve refinement of spot combining. It is important that consistency be applied throughout the gels to be analyzed during manual matching and editing of spots. It is also important to confine the analysis of the gels to a common area that encompasses a region in which spots are well resolved and detected on all of the gels being analyzed, while excluding regions, most notably gel margins, in which spots are less reproducibly resolved.

The selection of the appropriate gel image to serve as the standard or reference gel is important. This is the gel image from which a synthetic image is created; the positions of all spots in all gels are defined. Typically, for a single matchset experiment, the standard gel selected is often the one that appears to have the largest number of resolved spots. If an investigator plans to construct a database containing many linked matchsets, it is useful to select a cell type or condition to serve as a standard that is expected to resemble closely the gels that will be produced for each individual experiment or matchset. Because higher-order matchsets are created by matching standard gels from individual matchsets, including a specific cell type or condition as a constant in all experiments can be useful. This common cell type can serve as the standard gel for each individual matchset, and then matching these individual standard gels within the higher order matchset becomes easier, as all the gels are from a common cell type. Alternatively, a single, common gel image can be incorporated as a member into each matchset. For cases where very dissimilar cell types are to be compared, a gel receiving a mixture of lysates can be used to produce a standard gel to which either cell type can be matched efficiently.

After matching, spot intensity values can be normalized to correct for variations between gels. Although it is possible to normalize to the total number of acid-precipitable disintegrations per minute loaded to each gel, imprecision in measuring applied disintegrations per minute, variation in the amount of material entering each gel, and other factors may make this first level of normalization insufficient for detailed quantitative analysis. Two alternative approaches can be taken. One approach is to normalize to the total number of disintegrations per minute in all of the spots detected and analyzed. Another method is to identify a large set of constitutively expressed proteins and normalize to the total activity contained within that set of spots.

Options for quantitative analyses to identify spots with altered levels of expression among a set of gels includes pairwise comparisons between gel images. Such comparisons enable the investigator to create "spot sets" and then find overlap between sets. In this way, spots that change reproducibly by a specified amount between two or more conditions can be identified. A more detailed approach to analysis is to export the data for all fully matched spots into a spreadsheet program, such as Microsoft Excel, in which additional options for mathematical calculations exist for analysis, such as spot normalization (as described above), comparing mean spot values, creating spots sets, sorting of spots within spot sets by ratio of expression differences, and graphical representation of differences in expression for individual spots or for spot sets. By assigning each polypeptide a specific numerical identifier (standard spot number) from the standard gel, the developmentally regulated proteins identified through the more extensive spreadsheet analysis can be located in the standard gel map and annotated appropriately.

Another method of analysis that can be adopted is the use of cluster analysis to identify sets of coordinately regulated spots (8). Cluster analysis can be useful for discovering patterns of changes in expression that are not immediately apparent through simple pairwise gel comparisons (11, 12).

Once sets of regulated spots have been identified by one of the above methods, it is useful to reexamine the gel images to confirm that the spots thus identified are indeed differentially expressed. The spots are reexamined to ensure that they are well-resolved spots of high quality and density and that they are correctly matched. In the process, spots exhibiting valid differences can be entered into the database as spot sets, and that information can be accessed via spot annotation records.

One concern for accurate analysis of changes in protein synthesis detected by 2D-PAGE is to be certain that specific changes are reproducible. A small percentage of spots can exhibit significant changes in apparent intensity between gels of identical samples, or even between gels of the same sample (7). Therefore, it is essential that two or more gels representing two or more samples of a given cell type or condition be obtained. Although a greater number of gels is beneficial, the overall number of conditions to be tested, the cost and labor associated with obtaining the gels, and the eventual size of image matchsets to be processed must be taken into account. We have found that it is cumbersome to work with matchsets of more than 30 gel images. If necessary, however, this can be overcome by dividing the images into smaller matchsets that are then linked into higher order matchsets.

We have worked extensively with the PD-Quest program, initially supplied by Protein Database, Inc., and most recently by Bio-Rad. This program was formerly available on a Unix platform, but currently is supported on Macintosh or PC platforms only.

The program is useful for performing basic operations of spot detection and quantitation, calibration, matching, and editing. A large, high-resolution monitor (19" or larger) with a flat, low-glare screen is essential to permit the display and matching of large matchsets. Substantial disk storage space is required to store the gel images for a number of experiments. Each raw gel image may be approximately 10 MB in size, and several copies of each image may be made during the course of cropping, background subtraction, and spot detection, for a total of 40 MB or more for each gel image. Similarly, a generous allotment of RAM (512 MB or greater) is helpful to permit many gel images to be processed in batches for such operations as spot detection and for working with large numbers of gel images during matching. A high processor speed is highly advantageous to minimize the time required for matching calculations.

A suitable mechanism for image and matchset file backup and for long-term archiving is essential. It is recommended that a backup copy of matchset data be prepared weekly at minimum, preferably daily, and at least duplicate archive copies of completed matchsets maintained. A variety of phosphorimagers are available commercially. We prefer to have at least 100 μm resolution and the broadest linear range (e.g., four logarithms) available.

Acknowledgments   Work in the authors' labortories is supported in part by a grant from the NIH, NICHD (HD 21970).

References

 1. Iyer, V. R., Eisen, M. B., Ross, D. T., Schuler, G., Moore, T., Lee, J. C., Trent, J. M., Staudt, L. M., Hudson, J., Boguski, M. S., Lashkari, D., Shalon, D., Botstein, D., and Brown, P. O. (1999). *Science*, **283**, 83.
 2. Nelson, P. S., Han, D., Rochon, Y., Corthals, G. L., Lin, B., Monson, A., Nguyen, V., Franza, B. R., Plymate, S. R., Aebersold, R., and Hood, L. (2000). *Electrophoresis*, **21**, 1823.
 3. Klose, J. (1999). *Electrophoresis*, **20**, 643.
 4. O'Shaughnessy, R. F., Seery, J. P., Celis, J. E., Frischauf, A., and Wyatt, F. M. (2000). *FEBS Lett.*, **486**, 149.
 5. Garrels, J. I. (1989). *J. Biol. Chem.*, **264**, 5269.
 6. Garrels, J. I., and Franza, R. B. (1989). *J. Biol. Chem.*, **264**, 5283.
 7. Garrels, J. I., and Franza, R. B. (1989). *J. Biol. Chem.*, **264**, 5299.
 8. Garrels, J. I., Franza, R. B., Chang, C., and Latter, G. (1990). *Electrophoresis*, **11**, 1114.
 9. Gorg, A., Obermaier, C., Boguth, G., Harder, A., Scheibe, B., Wildgruber, R., and Weiss, W. (2000). *Electrophoresis*, **21**, 1037.
10. Hoving, S., Voshol, H., and Oostrum, J. (2000). *Electrophoresis*, **21**, 2617.
11. Latham, K. E., Garrels, J., Chang, C., and Solter, D. (1991). *Development*, **112**, 921.
12. Latham, K. E., Garrels, J., Chang, C., and Solter, D. (1992). *Appl. Theoret. Electrophoresis*, **2**, 163.
13. Latham, K. E., Bautista, F. D. M., Hirao, Y., O'Brien, M. J., and Eppig, J. J. (1999). *Biol. Reprod.*, **61**, 482.
14. Celis, J. E., and Gromov, P. (2000). *EXS*, **88**, 55.
15. Celis, J. E., Ostergaard, M., Jensen, N. A., Gromova, I., Rasmussen, H. H., and Gromov, P. (1998). *FEBS Lett.*, **430**, 64.
16. Chalmers, M. J., and Gaskell, S. J. (2000). *Curr. Opin. Biotechnol.*, **11**, 384.
17. Conrads, T. P., Anderson, G. A., Veenstra, T. D., Pasa-Tolic, L., and Smith, R. D. (2000). *Anal. Chem.*, **72**, 3349.
18. Gygi, S. P., and Aebersold, R. (2000). *Curr. Opin. Chem. Biol.*, **4**, 489.
19. Gygi, S. P., Rist, B., and Aebersold, R. (2000). *Curr. Opin. Biotechnol.*, **11**, 396.
20. Haynes, P. A., and Yates, J. R. (2000). *Yeast*, **17**, 81.
21. Keough, T., Lacey, M. P., Fieno, A. M., Grant, R. A., Sun, Y., Bauer, M. D., and Begley, K. B. (2000). *Electrophoresis*, **21**, 2252.
22. Yanagida, M., Miura, Y., Yagasaki, K., Taoka, M., Isobe, T., and Takahashi, N. (2000). *Electrophoresis*, **21**, 1890.

23. Yates, J. R. (2000). *Trends Genet.*, **16**, 5.

24. Garrels, J. I., (1983). *Meth. Enzymol.*, **100**, 411.

25. Patterson, S. D., and Latter, G. I. (1993). *Biotechniques*, **15**, 1076.

26. Kriegel, K., Seefeldt, I., Hoffmann, F., Schultz, C., Wenk, C., Regitz-Zagrosek, V., Oswald, H., and Fleck, E. (2000). *Electrophoresis*, **21**, 2637.

27. Latham, K.E., Beddington, R.S.P., Garrels, J.I., and Solter, D. (1993). *Mol. Reprod. Dev.*, **35**, 140.

28. Reik, W., Römer, I., Barton, S. C., Surani, M. A., Howlett, S. K., and Klose, J. (1993). *Development*, **119**, 933.

17

LISA C. BARCROFT
D. HOLSTEAD JONES-TAGGART
ANDREW J. WATSON

Antisense Disruption of Gene Expression in the Preimplantation Embryo

1. STRATEGIES FOR TARGETING GENE EXPRESSION

Development of techniques such as reverse transcription-polymerase chain reaction (RT-PCR), differential display RT-PCR, and, more recently, the use of gene-array chips has allowed a number of developmentally regulated gene products within the preimplantation embryo to be identified and characterized. Many of these gene products are involved in cell adhesion (1–10), inter- and intracellular signaling (11–19), cell cycle regulation (20–23), autocrine regulation (24–36), embryo metabolism (37–46), and ion transport (47–56). Despite extensive investigation of the timing of mRNA and protein expression of these gene products, there have been few studies examining the effect of disrupting the expression of a given gene. Most studies investigating gene function within somatic cells and preimplantation embryos have used one or two means of disrupting gene expression: antigene strategies or anti-mRNA strategies.

1.1 Antigene Strategies

A variety of strategies have been developed to interfere with expression at the level of the gene, including (1) synthetic oligodeoxynucleotides that bind to double-stranded (ds) DNA to generate triple-stranded DNA, thus preventing unwindase activity and/or transcription factor binding (57–61), (2) DNA decoy sequences meant to compete with the intact DNA template for transcription factor binding (62–66), (3) natural DNA binding molecules (e.g., polyamides and lexitropsins) to bind specific bases in the minor groove of dsDNA (67–69), and (4) physical disruption of the gene by gene targeting using homologous recombination to generate a gene null mutation in "knock-out" animals (70–72).

Homologous recombination is an effective means of ablating the expression of a gene of interest and has been used extensively to address the developmental implications of lost expression of a large number of genes. Generation and analysis of the phenotype of a null mutation has in some cases indicated that a given gene is not required for successful preimplantation development. In some cases, such as for gap junction proteins

(73) and β-catenin (74), functional redundancy between related proteins appears to compensate for lost function of a gene early in development. This approach has also resulted in the discovery of other genes having a key role in pre-/peri-implantation development (75–79). The generation of knock-out animals, however, is time consuming, expensive, and, as yet, not an efficient method for application to domestic animal models of early embryo development. Given these constraints, screening of a particular cell type or system for the importance of a given gene product is often more efficiently accomplished by using anti-mRNA strategies.

1.2 Anti-mRNA Strategies

Anti-mRNA strategies involve silencing a gene through destabilization of its mRNA template and include (1) the use of decoy oligonucleotides to provide an alternative binding site for protein-stabilizing elements, generally by interacting with a given mRNA (80, 81); (2) introduction of dsRNA molecules to produce gene-specific dsRNA (82–89); and (3) the use of antisense (reverse complement) nucleic acid sequences designed to duplex with mRNA and which can possess cleaving activity (90–94). The latter approach includes the use of synthetic antisense oligodeoxynucleotides, ribozymes, hammerhead ribozymes (95–99), hairpin ribozymes (100–103), and DNAzymes (93, 104–106). These methods are attractive because mRNA is easily accessible, and while inhibition is rarely complete, there is generally sufficient attenuation of gene expression to illuminate the gene product's role. Antisense methodologies using both antisense oligodeoxynucleotides (51, 107–125) and dsRNA (RNAi have been applied to investigate gene function during murine preimplantation development (85, 89). Antisense strategies also provide a valuable tool for determining interactions between two or more gene products, either in conjunction with established knock-out lines (108) or through simultaneous application of multiple antisense probes.

The purpose of this chapter is to review the potential application of antisense strategies to the analysis of gene function during preimplantation development in laboratory as well as in agricultural species. We discuss design and modifications of antisense oligonucleotides, methods for delivery of oligonucleotides within the preimplantation embryo, and experimental design and controls for successful antisense experiments.

2. ANTISENSE OLIGONUCLEOTIDE DESIGN AND SELECTION

Targeted downregulation of gene expression by using exogenous nucleic acids was first accomplished in cell-free systems in 1977 (126). Subsequent experiments demonstrated that in vitro viral replication could be inhibited by short antisense DNA sequences (127). These and other early experiments paved the way for substantial advances in the utilization of synthetic nucleotide sequences to modify gene expression in eukaryotes.

A key consideration for the application of antisense oligonucleotides is the design of antisense probes specific to the mRNA sequence of interest. In general, oligodeoxynucleotides have the potential for wider application than oligoribonucleotides because they are more stable (128). Although it is relatively easy to impart template specificity to nucleotide sequences, the efficacy with which a given sequence will decrease gene expression depends on mRNA secondary structures and the targeted region of the oligonucleotide within the mRNA sequence in relation to the functional domains of the protein. Oligonucleotide sequences of 15 nucleotides have been estimated to be sufficient to impart template specificity (129), although most published studies have used sequences ranging from 15 to 25 nucleotides. Strategies for selecting effective antisense oligonucleotides have been the subject of several review articles (130–133). In addition, an increasing number of companies offer custom antisense oligonucleotide design and synthesis. A quick search of the Internet web sites yields several options for antisense oligonucleotide design and synthesis. In addition to custom oligonucleotide design, some companies specializing in antisense technology now offer a catalogue of tested antisense oligonucleotides to selected gene targets. Before designing an oligonucleotide for experiments applied to preimplantation embryos, it is advantageous to search available commercial sources and published somatic cell studies for oligonucleotides directed against the target gene. Using previously tested oligonucleotides for embryo antisense experiments can greatly increase the efficiency of gene function characterization.

Another important consideration in oligonucleotide selection is the modifications of the nucleotides within the oligonucleotide sequences. Unmodified oligonucleotides are effective templates for RNase H-mediated RNA/DNA duplex cleavage; however, their susceptibility to endo- and exonuclease activity present in serum and within the cell (134, 135) limits their effectiveness. In an effort to increase stability, a number of modifications to the sugar backbone of nucleic acids have been made to impart greater nuclease resistance (137), although most 2'-sugar modifications prohibit RNase H activity. The most widely used modified oligonucleotides are the phosphorothioate oligodeoxynucleotides (sODNs); this modification not only imparts resistance to nuclease activity, but is also an effective substrate for RNase H-mediated cleavage of duplexed RNA/DNA (138–140). These oligos also have the advantage of downregulating gene expression using oligonucleotide concentrations often as low as 30 nM when used in combination with methods to enhance oligonucleotide uptake (141). Phosphorothioates are also reported to be less toxic than other modified oligonucleotides. A drawback of sODNs arises from their high affinity with various cellular proteins (142, 143) and inhibition of DNA polymerases and RNase H at high intracellular concentrations (130), resulting in nonspecific effects. More recently, morpholino oligonucleotides, which act via RNase H-independent mechanisms, have been generated and have the advantage of being relatively inexpensive and often more effective than standard phosphorothioate oligonucleotides (144–148).

3. OLIGODEOXYNUCLEOTIDE DELIVERY IN PREIMPLANTATION EMBRYOS

The majority of antisense experiments published for mouse embryos have generally not used a specific agent to mediate uptake of oligonucleotides (109–111, 113, 114, 116, 117, 120, 121), relying on cellular mechanisms leading to the uptake of charged nucleotides (149). The ability of murine preimplantation embryos to take up oligonucleotides from the culture medium was first demonstrated by Rappolee et al. (121). More recent experiments have used lysolecithin (51,108), Lipofectin (115), or Lipofectamine (112) to enhance uptake of oligonucleotides by the embryo.

The first report of lysolecithin (L-α-lysophosphatidylcholine) permeabilization in preimplantation embryos came from Dr. Armant's laboratory at Wayne State University (150). In this study morulae were treated with 0.05% lysolecithin for 2 min to mediate uptake of inositol triphosphate (IP_3) and α-amanitin. Measurements of Ca^{2+} release from intracellular stores after exposure to external IP_3 demonstrated that during incubation in the presence of lysolecithin and IP_3, Ca^{2+}-fluorescence was observed after 60 s and reached a maximum after 85 s. They were further able to demonstrate that lysolecithin-permeabilized morulae demonstrated increased permeability for up to 2 h after treatment. Lower concentrations of lysolecithin (0.001%) have been used over the same time interval in murine (51) and bovine (151) cleavage-stage embryos, as well as for longer time intervals (0.01% for 30 min [108]. In our experience, concentrations > 0.001–0.005% are generally toxic to murine four-cell embryos. The apparent difference in sensitivity of embryonic stages to a 2-min permeabilization with lysolecithin suggests either embryo stage-specific sensitivity to permeabilization or variability between different preparations of lysolecithin. These observations highlight the importance of determining the highest concentration of lysolecithin that can be tolerated by staged embryos for each vial of lysolecithin and ensuring that all subsequent lysolecithin preparations are made by diluting aliquots of a single starting stock solution in order to reduce variability. It is also critical to include a control group not treated with lysolecithin, as variation in embryo quality among different pools of embryos may reflect variations in their sensitivity to permeabilization.

A number of products have been designed to mediate transfection of somatic cell cultures with foreign nucleotide sequences. Again, it is important in preimplantation studies to determine the range of tolerance of the embryo. It is also useful to determine an optimal ratio between lipid and oligonucleotide for increased efficiency of transfection. MacPhee et al. (115) reported the effects of antisense ODNs directed against the $\beta1$ subunit of the Na/K-ATPase in murine embryos. Onset of cavitation was delayed and overall blastocyst diameter was decreased in embryos treated with 0.5 μM antisense ODNs for 5 h using 0.25 μg/μl Lipofectin (Invitrogen). Lipofectamine (Invitrogen, Burlington, ON) has also been used successfully to introduce antisense oligonucleotides into mouse preimplantation embryos (112). In this study, antisense ODNs directed against heat shock protein (hsp)70-1/hsp70-3 (2.5–10 μM ODN) were conjugated with a 1:100 dilution of Lipofectamine for 30 min followed by transfection of the embryos for 2 h. Many transfection agents come with a suggested protocol for transfection and generally require a preincubation period for complexing the product with the oligonucleotide before treatment of the cells.

4. ANTISENSE OLIGONUCLEOTIDE DISRUPTION OF GENE EXPRESSION IN MAMMALIAN PREIMPLANTATION EMBRYOS

Protocol 17.1 is a standard starting protocol for antisense oligonucleotide treatment of embryos using lysolecithin for permeabilization, which has been effective in our laboratories. This base protocol may have to be modified to accommodate a given oligonucleotide delivery strategy.

5. ASSESSMENT OF OLIGONUCLEOTIDE UPTAKE IN PREIMPLANTATION EMBRYOS

To determine whether a chosen antisense oligonucleotide gains access to the embryonic blastomeres, we suggest the use of a labeled oligonucleotide (same oligonucleotide type, generally a nonsense sequence) in conjunction with the oligonucleotide delivery strategy that will be used (Protocol 17.2). We have used a 21-nucleotide rhodamine-conjugated phosphorothioate ODN (R-ODN) in conjunction with several antisense oligonucleotide delivery methods in both murine and bovine cleavage-stage embryos to examine oligonucleotide uptake and nuclear localization. In both mouse and cow embryos, the following strategies have been examined for uptake of R-ODN: no agent; lysolecithin (L-α-lysophosphatidylcholine, from egg yolk; Sigma, St. Louis, MO); Lipofectin; Oligofectin G (Sequitur Inc., Natick, MA) and Lipofectamine in the presence of concentrations of R-ODN from 0 to 100 μM.

6. ANALYSIS OF mRNA DISRUPTION AFTER ANTISENSE TREATMENT

An important factor in confirming the antisense effect is the determination of disruption of mRNA expression for the targeted gene. Primer sets for analysis of the targeted gene should be designed to span the region homologous to the antisense ODN, such that semiquantitative analysis of intact mRNA can be made between treatments (51). In addition, it is important to consider the effect of ODN treatment on alternate isoforms of the targeted gene, and a housekeeping gene (e.g., actin or a metabolic marker) to examine nonspecific effects of the ODN within the embryo. To perform semiquantitative analysis of mRNA expression in mammalian preimplantation embryos, we carry out extraction of total RNA in the presence of an exogenous template (rabbit γ-globin; Invitrogen ([152, 153]) and reverse transcription using Oligo(dT)$_{18-22}$ and Superscript II reverse transcriptase (Invitrogen) using phenol:chloroform extraction of total RNA from small pools of embryos (36, 154). Wrenzycki et al. (Chapter 13), and De Sousa et al. (Chapter 14) outline these methods. Analysis of protein expression in both murine and bovine preimplantation embryos is performed on fixed, whole embryos using confocal laser scanning microscopy (3, 47, 49, 51) (Protocol 17.3).

7. SUMMARY AND IMPORTANT CONSIDERATIONS FOR ANTISENSE EXPERIMENTS IN PREIMPLANTATION EMBRYOS

Antisense strategies represent a powerful approach for investigating gene function during preimplantation development. However, using antisense molecules as effective agents against translation of specific RNAs thought to be important to preimplantation development requires considerable optimization. Consequently, each experiment should be designed with necessary controls in mind so that outcomes such as developmental delay, decreased target mRNA levels, and attenuation of the protein of interest can be interpreted as an antisense effect and not an effect of general oligonucleotide toxicity. The minimal experimental design should include a group treated only with the oligonucleotide delivery system in conjunction with a nontreated control and antisense and nonsense oligonucleotide treatment groups. All conditions, including permeabilization and oligonucleotide concentration must be determined for each targeting experiment. Generally, multiple oligonucleotides should be designed to each target mRNA and several delivery methods tried to ensure effective and specific downregulation. The nonsense oligonucleotide should be composed of the same nucleotide composition as the antisense oligonucleotide in a scrambled sequence to ensure effects do not stem from DNA toxicity or other nonspecific influences. It is also helpful to envision a possible phenotype stemming from the deletion of the target gene, as this will enable the detection of specific effects. If possible, a functional assay should be included so that the effects of the decrease in target protein on the embryo's physiological processes can be measured directly (e.g., dye coupling assays, measured uptake of ions/solutes, or an enzyme assay).

An important consideration for analyzing the effects of antisense oligonucleotides on embryos in culture is the issue of asynchronous development. In some cases, optimal antisense oligonucleotide effects are seen only when delivered at specific developmental stages, requiring the embryos be meticulously segregated before treatment. Oligonucleotide effects on morphological stages such as compaction and cavitation can be masked by asynchronous development, necessitating the use of appropriate controls as stated above. Oligonucleotides are certainly turned over with time, and thus if long-term experiments (e.g., >24 h) are conducted, it is necessary to examine oligonucleotide uptake and half-life to estimate oligonucleotide replenishment times to ensure continued inhibition of gene expression. This consideration can be particularly important in designing antisense experiments in preimplantation embryos from domestic animal models because they typically display extended developmental intervals compared to the murine embryo (in which the majority of antisense experiments have been conducted). Rarely, if ever, is antisense gene ablation complete. Therefore, these strategies promote downregulation of gene expression and not gene ablation. Estimates suggest that in some cases target genes can be downregulated by up to 80–90% of their control levels. It is likely that this is an extreme level of downregulation and that in most circumstances the downregulation is not as dramatic and may also be fairly short lived.

Reproducibility between experiments can also be a significant concern. We have observed variability in the reproducibility of results when using different synthesis batches of the same oligonucleotide sequences. The reason for this problem is unknown, but its existence certainly argues for great caution in interpreting results and also in obtaining large-scale synthesis of effective oligonucleotides that allow for a complete series of experiments to be conducted with the same oligonucleotide synthesis batch. However, reproducibility has been observed in the case of the glucose cotransporter studies (108, 109, 114). Experimental replicates are essential, and reproducibility must be established before valid conclusions regarding the effective targeting and phenotypic consequences from downregulating a specific gene transcript can be made.

Targeting specific developmentally regulated RNA transcripts in preimplantation embryos with antisense oligonucleotides is a powerful and readily available tool. Antisense technology now has the advantage of improved nucleotide stability as well as modifications for increased binding and functional characteristics. Balanced with these advantages are the challenges of oligonucleotide uptake, nonspecific effects and toxicity. Despite all these concerns, antisense strategies for targeting gene expression during preimplantation development are effective and have elucidated the function of many genes during this developmental interval. The recent application of dsRNA methods (85, 89) and the prospect of employing morpholino oligonucleotides to target gene expression in early embryos (156) ensures that this strategy for downregulating gene expression in early embryos will continue to become more robust and reliable well into the future. There is a great need to define simple, effective methods for targeting gene expression in preimplantation embryos, and this need will propel advances in this field. The aim of this chapter has been to provide the available information on antisense technology as it has been applied to studies in the mammalian embryogenesis in the hopes of encouraging further development in this field.

Protocol 17.1. Antisense oligonucleotide disruption of gene expression in mammalian preimplantation embryos

1. Prepare treatment media using a fresh dilution of lysolecithin stock solution in embryo culture medium and set up 20–50 µl drops for each treatment. Also prepare the embryo culture drops (10–20 µl +\- final concentration of ODN under light paraffin oil). Allow all the drops to equilibrate in the incubator that will be used for embryo culture. *Note*: Do not cover the lysolecithin drops with paraffin oil, as lipids will be absorbed by the paraffin oil; culture media should be serum-free. Also prepare and equilibrate 50–100 µl drops of culture media for each treatment to wash embryos in after transfection.

2. Collect embryos of the appropriate developmental stage and divide into pools for treatments (generally 20–30 embryos/treatment). In transferring embryos between drops, try to minimize the volume of medium transferred to less than 2 µl (embryos can be transferred to treatment/culture drops effectively using a 10 µl Eppendorf pipette set at the desired transfer volume).

3. Permeabilize the embryos with lysolecithin in the presence or absence of each concentration of ODN for 2 min.

4. Wash the permeabilized embryos in equilibrated culture media and transfer them (in < 2 µl) to the culture drop.

5. Culture embryos to assess antisense oligonucleotide effects on embryo development, protein, and mRNA expression under the desired culture atmosphere. For long-term cultures, refresh media with oligonucleotides every 24 h by replacing one-half of culture volume with an equal volume of media containing approximately 2× final oligonucleotide concentration.

Protocol 17.2. Assessment of oligonucleotide uptake in preimplantation embryos

1. Prepare for embryo permeabilization, oligonucleotide delivery, and embryo culture as described in Protocol 17.1.
2. Treat embryos in the presence or absence of tagged ODN at the desired concentration and length of time according to permeabilization strategy.
3. Wash oligonucleotide-treated embryos in fresh equilibrated culture media and transfer treatment groups to culture drops under oil for embryo culture.
4. Embryo samples for analysis can be collected at the end of oligonucleotide uptake interval and 24 h post-treatment to determine the cytoplasmic (0 h post-treatment) and nuclear (24 h post treatment) localization of the tagged ODN.
5. To assess oligonucleotide uptake, the zona pellucida (ZP) should be removed from the embryo by either acid tyrodes treatment or 1% Pronase treatment. Removal of the zona is essential because it is often a source of nonspecific binding of the fluorescent tag. Generally a few seconds in acid tyrodes solution (mice) or 1–2 min 1% Pronase (cow) is sufficient to remove the ZP.
6. Wash zona-free embryos 2× in fresh, equilibrated culture media to ensure zona pieces are washed away.
7. Transfer zona-free embryos to drops of 40% glycerol or FluoroGuard (antifade reagent; Bio-Rad, Mississauga, ON) on glass coverslips. Cover with glass coverslips and seal edges with nail polish or coverslip sealer.
8. Look for localization of oligonucleotide-induced fluorescence using either UV light or confocal laser scanning microscopy.

Protocol 17.3. Analysis of antisense effects on protein expression in whole preimplantation embryos

1. Collect embryos from treatment groups at 12 and 24 h post-treatment and/or treatment endpoint.
2. To fix the embryos for immunofluorescence, transfer the embryos from the treatment drops to 0.5-ml volumes of either 1× PBS or 1× PHEM buffer (2× PHEM: 60 mM PIPES, 25 mM HEPES, 10 mM EGTA, 1 mM MgCl$_2$·6H$_2$O, pH 6.9 [155]) in a multiwell tissue-culture plate to wash. Transfer the embryo pools to 0.5 ml of 1:1 buffer:methanol (MeOH) on ice and incubate for 2 min. Transfer the embryos from the 1:1 solution to a 1:2 solution of ice-cold buffer:MeOH for an additional 2 min before transferring to 100% ice-cold MeOH for 2 min. Remove the embryos from MeOH and transfer to antibody dilution/wash buffer (buffer + 0.005% Triton X-100 + 0.1 M lysine + 1% normal goat serum for approximately 5 min or until the embryos sink to the bottom of the well. Transfer the fixed embryos to 0.5 ml of fresh PBS/PHEM buffer in multiwell tissue-culture plates to wash. Embryos can be used immediately for immunofluorescence or stored for up to 2 weeks at 4°C. If embryos are to be stored at 4°C, they should be transferred into 0.5-ml volume of embryo storage buffer (PBS or 1× PHEM + 0.09% sodium azide) and the dish sealed with Parafilm to prevent buffer evaporation. Alternatively (depending on the antibody/antigen combination), fixation in 1–2% paraformaldehyde for 30 min to 1 h at room temperature may be preferred.
3. For whole-mount indirect immunofluorescence, remove the embryo pools from storage buffer and permeabilize/block in blocking buffer (buffer + 0.01% Triton X-100 + 0.1 M lysine + 1% normal goat serum; approximately 0.1–0.5 ml in well of glass-welled plate; Pyrex) for 30–45 min at room temperature. Wash the embryos once with fresh PBS/PHEM buffer. Meanwhile, prepare appropriate dilution(s) of antibody(s) in antibody dilution/wash buffer and place 30–100 µl of primary antibody in wells of glass plate for each treatment. Add blocked embryos to the appropriate wells and incubate in primary antibody for 3–4 h at room

temperature or overnight at 4°C in a humid box. In a multiwell tissue-culture dish, wash antibody-treated embryos 2 × 10 min in fresh antibody dilution/wash buffer (0.5 ml/well), followed by a final wash lasting at least 3 h. Dilute the fluorophore-conjugated secondary antibody in antibody dilution/wash buffer and add 30–100 µl to clean wells of glass plate and add washed embryos. Incubate embryos with secondary antibody for 1–2 h at room temperature in a light-proof humid box. In darkened room, wash embryos 3 × 10 min at room temperature and 1x overnight at 4°C in antibody dilution/wash buffer (multiwell culture plates covered with tin foil, 0.5 ml/well).

4. Prepare glass microscope slides by applying spots of nail polish at locations corresponding to the 4 corners of the coverslip size to be used to prevent crushing of embryos. Apply 15–20 µl of antifade reagent (e.g., FluoroGuard; BioRad) to glass slides and transfer embryos from final wash. Place coverslips over drops and seal edges with nail polish or coverslip sealer. Store slides in light-proof box at –20°C until able to examine protein localization patterns by confocal laser scanning microscopy.

References

1. Robson, P., Stein, P., Zhou, B., Schultz, R. M., and Baldwin, H. S. (2001). *Dev. Biol.*, **234**, 317.

2. Sheth, B., Fonraine, J., Ponza, E., McCallum, A., Page, A., Citi, S., Louvard, D., Zahraoui, A., and Fleming, T. P. (2000). *Mech. Dev.*, **97**, 93.

3. Barcroft, L. C., Hay-Schmidt, A., Gilfoyle, E., Caveney, A., Hyttell, P., Overstrom, E. W., and Watson, A. J. (1998). *J. Reprod. Fertil.*, **114**, 327.

4. Sheth, B., Fesenko, I., Collins, J. E., Moran, B., Wild, A. L., Anderson, J. M., and Fleming, T. P. (1997). *Development*, **124**, 2027.

5. Shehu, D., Mariscano, G., Flechon, J.-E., and Galli, C. (1996). *Zygote*, **4**, 109.

6. Collins, J. E., Lorimer, J. E., Garrod, D. R., Pidsley, S. C., Buxton, R. S., Fleming, T. P. (1995). *Development*, **121**, 743.

7. Fleming, T. P., and Hay, M. J. (1991). *Development*, **113**, 295.

8. Fleming, T. P., McConnell, J., Johnson, M. H., and Stevenson, B. R. (1989). *J. Cell. Biol.* **108**, 1407.

9. Damsky, C. H., Richa, J., Solter, D., Knudsen, K., and Buck, C. A. (1984). *Cell*, **34**, 455.

10. Pratt, H. P. M., Ziomek, C. A., Reeve, W. J. D., and Johnson, M. H. (1982). *J. Exp. Embryol. Morphol.*, **90**, 101.

11. Pauken, C. M., and Capco, D. G. (2000). *Dev. Biol.*, **223**, 411.

12. Domashenko, A. D., Latham, K. E., and Hatton, K. S. (1997). *Mol. Reprod. Dev.*, **47**, 57.

13. Davies, T. C., Barr, K. J., Jones, D. H., Zhu, D., and Kidder, G. M. (1997). *Dev. Genet.*, **18**, 234.

14. Wrenzycki, C., Herrmann, D., Carnwath, J. W., and Niemann, H. (1996). *J. Reprod. Fertil.*, **108**, 17.

15. Williams, C. J., Schultz, R. M., and Kopf, G. S. (1996). *Mol. Reprod. Dev.*, **44**, 315.

16. Worrad, D. M., Ram, P. T., and Schultz, R. M. (1994). *Development*, **120**, 2347.

17. De Sousa, P. A., Valdimarsson, G., Nicholson, B. J., and Kidder, G. M. (1993). *Development*, **117**, 1355.

18. Valdimarsson, G., De Sousa, P. A., and Kidder, G. M. (1993). *Mol. Reprod. Dev.*, **36**, 7.

19. Nishi, M., Kumar, N. M., and Gilula, N.B. (1991). *Dev. Biol.*, **146**, 117.

20. Kanatsu-Shinohara, M., Schultz, R. M., and Kopf, G. S. (2000). *Biol. Reprod.*, **63**, 1610.

21. Exley G. E., Tang, C., McElhinny, A. S., and Warner, C. M. (1999). *Biol. Reprod.*, **61**, 231.

22. Fleming, J. V., Fontanier, N., Harries, D. N., and Rees, W. D. (1997). *Mol. Reprod. Dev.*, **48**, 310.

23. Heikinheimo, O., Lanzendorf, S. E., Baka, S. G., and Gibbons, W. E. (1995). *Hum. Reprod.*, **10**, 699.

24. Ying, C., Hsu, W. L., Hong, W. F., Cheng, W. T., and Yang, Y. (2000). *Mol. Reprod. Dev.*, **55**, 83.

25. Raga, F., Casan, E. M., Kruessel, J., Wen, Y., Bonilla-Musoles, F., and Polan, M. L. (1999). *Endocrinology*, **140**, 3705.

26. Hiroi, H., Momoeda, M., Inoue, S., Tsuchiya, F., Matsumi, H., Tsutsumi, O., Muramatsu, M., and Taketani, Y. (1999). *Endocrinol. J.*, **46**, 153.

27. Stojanov, T., and O'Neill, C. (1999). *Biol. Reprod.*, **60**, 674.

28. Roelen, B. A., Goumans, M. J., van Rooijen, M. A., and Mummery, C. L. (1997). *Intl. J. Dev. Biol.*, **41**, 541.

29. Pantaleon, M., Whiteside, E. J., Harvey, M. B., Barnard, R. T., Waters, M. J., and Kaye, P. L. (1997). *Proc. Natl. Acad. Sci. USA*, **94**, 5125.

30. Winger, Q. A., de los Rios, P., Han, V. K., Armstrong, D. T., Hill, D. J., and Watson, A. J. (1997). *Biol. Reprod.*, **56**, 1451.

31. Sharkey, A. M., Dellow, K., Blayney, M., Macnamee, M., Charnock-Jones, S., and Smith, S. K. (1995). *Biol. Reprod.*, **53**, 974.

32. Rappolee, D. A., Basilico, C., Patel, Y., and Werb, Z. (1994). *Development*, **120**, 2259.

33. Palmieri, S. L., Payne, J., Stiles, C. D., Biggers, J. D., and Mercola, M. (1992). *Mech. Dev.*, **39**, 181.

34. Schultz, G. A., Hahnel, A., Arcellana-Panlilio, M., Wang, L., Goubau, S., Watson, A., and Harvey, M. (1993). *Mol. Reprod. Dev.*, **35**, 414.

35. Schultz, G. A., Hogan, A., Watson, A. J., Smith, R. M., and Heyner, S. (1992). *Reprod. Fertil. Dev.*, **4**, 361.

36. Watson, A. J., Hogan, A., Hahnel, A., Wiemer, K. E., and Schultz, G. A. (1992). *Mol. Reprod. Dev.*, **31**, 87.

37. Leppens-Luisier, G., Urner, F., and Sakkas, D. (2001). *Hum. Reprod.*, **16**, 1229.

38. Winger, Q. A., Hill, J. R., Watson, A. J., and Westhusin, M. E. (2000). *Mol. Reprod. Dev.*, **55**, 14.

39. el Mouatassim, S., Hazout, A., Bellec, V., and Menezo, Y. (1999). *Zygote*, **7**, 45.

40. Taylor, D. M., Ray, P. F., Ao, A., Winston, R. M., and Handyside, A. H. (1997). *Mol. Reprod. Dev.*, **48**, 442.

41. Lequarre, A. S., Grisart, B., Moreau, B., Schuurbiers, N., Massip, A., and Dessy, F. (1997). *Mol. Reprod. Dev.*, **48**, 216.

42. Houghton, F. D., Sheth, B., Moran, B., Leese, H. J., and Fleming, T. P. (1996). *Mol. Hum. Reprod.*, **2**, 793.

43. Morita, Y., Tsutsumi, O., Oka, Y., and Taketani, Y. (1994). *Biochem. Biophys. Res. Commun.*, **199**, 1525.

44. Aghayan, M., Rao, L. V., Smith, R. M., Jarett, L., Charron, M. J., Thorens, B., and Heyner, S. (1992). *Development*, **115**, 305.

45. Wiley, L. M., Wu, J. X., Harari, I., and Adamson, E. D. (1992). *Dev. Biol.*, **149**, 247.

46. Hogan, A., Heyner, S., Charron, M. J., Copeland, N. G., Gilbert, D. J., Jenkins, N. A., Thorens, B., and Schultz, G. A. (1991). *Development*, **113**, 363.

47. MacPhee, D. J., Jones, D. H., Barr, K. J., Betts, D. H., Watson, A. J., and Kidder, G. M. (2000). *Dev. Biol.*, **222**, 486.

48. Barr, K. J., Garrill, A., Jones, D. H., Orlowski, J., and Kidder, G. M. (1998). *Mol. Reprod. Dev.*, **50**, 146.

49. Betts, D. H., Barcroft, L. C., and Watson, A. J. (1998). *Dev. Biol.*, **197**, 77.

50. Betts, D. H., MacPhee, D. J., Kidder, G. M., and Watson, A. J. (1997). *Mol. Reprod. Dev.*, **46**, 114.

51. Jones, D. H., Davies, T. C., and Kidder, G.M. (1997). *J. Cell Biol.*, **139**, 1545.

52. Zhao, Y., Doroshenko, P. A., Alper, S. L., and Baltz, J. M. (1997). *Dev. Biol.*, **189**, 148.

53. Zhao, Y., Chauvet, P. J., Alper, S. L., and Baltz, J. M. (1995). *J. Biol. Chem.*, **270**, 24428.

54. Watson, A. J., Pape, C., Emanuel, J. R., Levenson, R., and Kidder, G. M. (1990). *Dev. Genet.*, **11**, 41.

55. Wiley, L. M., Lever, J. E., Pape, C., and Kidder, G. M. (1991). *Dev. Biol.*, **143**, 149.

56. Watson, A. J., and Kidder, G. M. (1988). *Dev. Biol.*, **126**, 80.

57. Faria, M., Wood, C. D., White, M. R., Helene, C., and Giovannangeli, C. (2001). *J. Mol. Biol.*, **306**, 15.

58. Praseuth, D., Guieysse, A. L., and Helene, C. (1999). *Biochim. Biophys. Acta*, **1489**, 181.

59. Vasquez, K. M., and Wilson, J. H. (1998). *Trends Biochem. Sci.*, **23**, 4.

60. Gunther, E. J., Havre, P. A., Gasparro, F. P., and Glazer, P. M. (1996). *Photochem. Photobiol.*, **63**, 207.

61. Maher, L. Jr. (1996). *Cancer Invest.*, **14**, 66.
62. Alper, O., Bergmann-Leitner, E. S., Abrams, S., and Cho-Chung, Y. S. (2001). *Mol. Cell. Biochem.*, **218**, 55.
63. Penolazzi, L., Lambertini, E., Aguiari, G., del Senno, L., and Piva, R. (2000). *Biochim. Biophys. Acta*, **1492**, 560.
64. Park, Y. G., Nesterova, M., Agrawal, S., and Cho-Chung, Y. S. (1999). *J. Biol. Chem.*, **274**, 1573.
65. Sharma, H. W., Perez, J. R., Higgins-Sochaski, K., Hsaio, R., and Narayanan, R. (1996). *Anticancer Res.*, **16**, 61.
66. Morishita, R., Gibbons, G. H., Horiuchi, M., Ellison, K. E., Nakama, M., Zhang, L., Kaneda, Y., Ogihara, T., and Dzau, V. J. (1995). *Proc. Natl. Acad. Sci. USA*, **92**, 5855.
67. Goodsell, D.S. (2000). *Stem Cells*, **18**, 148.
68. Kielkopf, C. L., Baird, E. E., Dervan, P. B., and Rees, D. C. (1998). *Nat. Struct. Biol.*, **5**, 104.
69. Walker, W. L., Kopka, M. L., and Goodsell, D. S. (1997). *Biopolymers*, **44**, 323.
70. Haber, J. E. (1999). *Trends Biochem. Sci.*, **24**, 271.
71. Marth, J. D. (1996). *J. Clin. Invest.*, **97**, 1999.
72. Bronson, S. K., and Smithies, O. (1994). *J. Biol. Chem.*, **269**, 27155.
73. De Sousa, P. A., Juneja, S. C., Caveney, S., Houghton, F. D., Davies, T. C., Reaume, A. G., Rossant, J., and Kidder, G. M. (1997). *J. Cell. Sci.*, **110**, 1751.
74. Haegel, H., Larue, L., Ohsugi, M., Fedorov, L., Herrnknecht, K., and Kemler, R. (1995). *Development*, **121**, 3529.
75. Voss, A. K., Thomas, T., Petrou, P., Anastassiadis, K., Scholer, H., and Gruss, P. (2000). *Development*, **127**, 13817.
76. Nichols, J., Zevnik, B., Anastassiadis, K., Niwa, H., Klewe-Nebenius, D., Chambers, I., Scholer, H., and Smith, A. (1998). *Cell*, **95**, 379.
77. Riethmacher, D., Brinkmann, V., and Birchmeier, C. (1995). *Proc. Natl. Acad. Sci. USA*, **92**, 855.
78. Larue, L., Ohsugi, M., Hirchenhain, J., and Kemler, R. (1994). *Proc. Natl. Acad. Sci. USA*, **91**, 8263.
79. Torres, M., Stoykova, A., Huber, O., Chowdhury, K., Bonaldo, P., Mansouri, A., Butz, S., Kemler, R., and Gruss, P. (1997). *Proc. Natl. Acad. Sci. USA*, **94**, 901.
80. Beelman, C. A., and Parker, R. (1995). *Cell*, **81**, 179.
81. Liebhaber, S. A. (1997). *Nucl. Acids Symp. Ser.*, **36**, 29.
82. Bass, B. L. (2000). *Cell*, **101**, 235.
83. Clemens, J. C., Worby, C. A., Simonson-Leff, N., Muda M., Maehama, T., Hemmings, B. A., and Dixon, J. E. (2000). *Proc. Natl. Acad. Sci. USA*, **97**, 6499.
84. Mette, M. F., Aufsatz W., van der Winden, J., Matzke, M. A., and Matzke, A. J. (2000). *EMBO J.*, **19**, 5194.
85. Svoboda, P., Stein, P., Hayashi, H., and Schultz, R. M. (2000). *Development*, **127**, 4147.
86. Zamore, P. D., Tuschi, T., Sharp, P. A., and Bartel, D. P. (2000). *Cell*, **101**, 25.
87. Sharp, P. A. (1999). *Genes Dev.*, **10**, 161.
88. Tuschl, T., Zamore, P. D., Lehmann, R., Bartel, D. P., and Sharp, P. A. (1999). *Genes Dev.*, **13**, 3191.
89. Wianny, F., and Zernicka-Goetz, M. (1999). *Nat. Cell Biol.*, **2**, 70.
90. Santiago, F. S., Lowe, H. C., Kavurma, M. M., Chesterman, C. N., Baker, A., Atkins, D. G., and Khachigian, L. M. (1999). *Nat. Med.*, **5**, 1264.
91. Joyce, G. F. (1998). *Proc. Natl. Acad. Sci. USA*, **95**, 5845.
92. Perman, R., and de Feyter, R. (1997). *Methods Mol. Biol.*, **74**, 393.
93. Santoro, S. W., and Joyce, G. F. (1998). *Biochemistry*, **37**, 13330.
94. Eckstein, F. (1996). *Biochem. Soc. Tans.*, **24**, 601.
95. Rossi, J. J. (1999). *Chem. Biol.*, **6**, R33.
96. Pieken, W. A., Olsen, D. B., Bensler, F., Aurup, H., and Eckstein, F. (1991). *Science*, **253**, 314.
97. Dahm, S. C., and Uhlenbeck, O. C. (1991). *Biochemistry*, **30**, 9464.
98. Vaish, N. K., Kore, A. R., and Eckstein, F. (1997). *Nucl. Acids Res.*, **226**, 5237.
99. Birikh, K. R., Heaton, P. A., and Eckstein, F. (1997). *Eur. J. Biochem.*, **245**, 1.
100. Hegg, L. A., and Fedor, M. J. (1995). *Biochemistry*, **34**, 15813.
101. Hampel, A. (1998). *Prog. Nucleic Acid Res. Mol. Biol.*, **58**, 1.

102. Earnshaw, D. J., and Gait, M. J. (1997). *Antisense Nucl. Acid Drug Dev.*, **7**, 403.

103. Perez-Ruiz, M., Barroso-DelJesus, A., and Berzal-Herranz, A. (1999). *J. Biol. Chem.*, **274**, 29376.

104. Wu, Y., Yu, L., McMahon, R., Rossi, J. J., Forman, S. J., and Snyder, D. S. (1999). *Hum. Gene Ther.*, **10**, 2847.

105. Zhang, X., Xu, Y., Ling, H., and Hattori, T. (1999). *FEBS Lett.*, **458**, 151.

106. Breaker, R. R., and Joyce, G. F. (1994). *Chem. Biol.*, **1**, 223.

107. Mohamed, O. A., Bustin, M., and Clarke, H. J. (2001). *Dev. Biol.*, **229**, 237.

108. Chi, M. M., Pingsterhaus, J., Carayannnopoulos, M., and Moley, K. H. (2000). *J. Biol. Chem.*, **275**, 40252.

109. Carayannopoulos, M. O., Chi, M. M., Cui, Y., Pingsterhaus, J. M., McKnight, R. A., Mueckler, M., Devaskar, S. U., and Moley, K. H. (2000). *Proc. Natl. Acad. Sci. USA*, **97**, 7313.

110. Jeong, S.-J., Kim, H.-S., Chang, K.-A., Geum, D.-H., Park, C. H., Seo, J.-H., Rah, J.-C., Lee, J. H., Choi, S. E., Lee, S. G., Kim, K., and Suh, Y.-H. (2000). *FASEB J.*, **14**, 2171.

111. Onohara, Y., Harada, T., Tanikawa, M., Iwabe, T., Yoshioka, H., Taniguchi, F., Mitsunari, M., Tsudo, T., and Terakawa, N. (1998). *J. Assist. Reprod. Genet.*, **15**, 395.

112. Dix, D. J., Garges, J. B., and Hong, R. L. (1998). *Mol. Reprod. Dev.*, **51**, 373.

113. Tsark, E. C., Adamsom, E. D., Withers, G. E., and Wiley, L. M. (1997). *Mol. Reprod. Dev.*, **47**, 271.

114. Pantaleon, M., Harvey, M. B., Pascoe, W. S., James, D. E., and Kaye, P. L. (1997). *Proc. Natl. Acad. Sci. USA*, **94**, 3795.

115. MacPhee, D. J., Barr, K. J., Watson, A. J., and Kidder, G. M. (1996). *Trophoblast Res.*, **11**, 87.

116. Blangy, A., Vidal, F., Cuzin, F., Yang, Y.-H., Boulukos, K., and Rassoulzadegan, M. (1995). *J. Cell Sci.*, **108**, 675.

117. Peters, J. M., Duncan, J. R., Wiley, L. M., Rucker, R. B., and Keen, C. L. (1995). *Reprod. Tox.*, **9**, 123.

118. Naz, R. K., Kumar, G., and Minhas, B. S. (1994). *J. Ass. Reprod. Genet.*, **11**, 208.

119. Brice, E. C., Wu, J.-X., Muraro, R., Adamson, E. D., and Wiley, L. M. (1993). *Dev. Genet.*, **14**, 174.

120. Xu, Y., Jin, P., and Warner, C. M. (1993). *Biol. Reprod.*, **48**, 1042.

121. Rappolee, D. A., Strum, K. S., Behrendsten, O., Schultz, G. A., Pedersen, R. A., and Werb, Z. (1992). *Genes Dev.*, **6**, 939.

122. Paria, B. C., Dey, S. K., and Andrews, G. K. (1992). *Proc. Natl. Acad. Sci. USA*, **89**, 10051.

123. Ao, A., Erickson, R. P., Bevilacqua, A., and Karolyi, J. (1991). *Antisense Res. Dev.*, **1**, 1.

124. Bevilacqua, A., Loch-Caruso, R., and Erickson, R. P. (1989). *Proc. Natl. Acad. Sci. USA* **86**, 5444.

125. Bevilacqua, A., Erickson, R. P., and Hieber, V. (1988). *Proc. Natl. Acad. Sci. USA*, **85**, 831.

126. Paterson, B. M., Roberts, B. E., and Sperling, R. (1977). *Proc. Natl. Acad. Sci. USA*, **74**, 4370.

127. Zamecnik, P. C., and Stephenson, M. L. (1978). *Proc. Natl. Acad. Sci. USA*, **75**, 280.

128. Wickstrom, E. (1986). *Biochem. Biophys. Meth.*, **13**, 97.

129. Marcus-Sekura, C. J. (1988). *Anal. Biochem.*, **172**, 289.

130. Lebedeva, I., and Stein, C. A. (2001). *Annu. Rev. Pharmacol. Toxicol.*, **41**, 403.

131. Smith, L., Andersen, K. B., Hovgaard, L., and Jaroszewski, J.W. (2000). *Eur. J. Pharm. Sci.*, **11**, 191.

132. Jen, K.-Y., and Gerwitz, A. M. (2000). *Stem Cells*, **18**, 307.

133. Galderisi, U., Cascino, A., and Giordano, A. (1999). *J. Cell Phsyiol.*, **181**, 251.

134. Leonetti, J. P., Rayner, B., Lemaitre, M., Gagnor, C., Milhand, P. G., Imbach, J. L., and Lebleu, B. (1989). *Gene*, **73**, 57.

135. Cazenave, C., Loreau, N., Toumle, J. J., and Helene, C. (1986). *Biochimie*, **68**, 1063.

136. Agrawal, S. (1999). *Biochim. Biophys. Acta*, **1489**, 53.

137. De Mesmaeker, A., Altmann, K. H., Waldner, A., and Wendeborn, S. (1995). *Curr. Opin. Struct. Biol.*, **5**, 343.

138. Crooke, S. T., Lemonidis, K. M., Neilson, L., Griffey, R., Lesnik, E. A., and Monia, B. P. (1995). *Biochem. J.*, **312**, 599.

139. Stein, C. A., Subsinghe, C., Shinozuka, K., and Cohen, J. S. (1988). *Nucl. Acid Res.*, **16**, 3209.

140. Walder, R. Y., and Walder, J. A. (1988). *Proc. Natl. Acad. Sci. USA*, **85**, 5011.
141. Cazenave, C., Stein, C. A., Loreau, N., Thuong, N. T., Neckers, L. M., Subainghe, C., Helene, C., Cohen, J. S., and Toulme, J. J. (1989). *Nucl. Acids Res.*, **17**, 4255.
142. Shoeman, R. L., Hartig, R., Huang, Y., Grub, S., and Traub, P. (1997). *Antisense Nucl. Acid Drug Dev.*, **7**, 291.
143. Guvakova, M. A., Yakubov, L. A., Vlodavsky, I., Tonkinson, J. L., and Stein, C. A. (1995). *J. Biol. Chem.*, **270**, 2620.
144. Heasman, J., Kofron, M., and Wylie, C. (2000). *Dev. Biol.*, **222**, 124.
145. Summerton, J. (1999). *Biochim. Biophys. Acta*, **1489**, 141.
146. Summerton, J., Stein, D., Huang, S. B., Matthews, P., Weller, D., and Partridge, M. (1997). *Antisense Nucl. Acid Drug Dev.*, **7**, 63.
147. Summerton, J., and Weller, D. (1997). *Antisense Nucl. Acid Drug Dev.*, **7**, 187.
148. Taylor, D. M., Ray, P. F., Ao, A., Winston, R. M., and Handyside, A. H. (1997). *Mol. Reprod. Dev.*, **48**, 442.
149. Stein, C. A. (1999). *Biochim. Biophys. Acta*, **1489**, 45.
150. Khidir, M. A., Stachecki J. J., Krawetz, S. A., and Armant D. R. (1995). *Exp. Cell Res.*, **219**, 619.
151. Barcroft, L. C., and Watson, A. J. (1998). *Biol. Reprod.*, **58** Suppl 1, 166.
152. De Sousa, P. A., Westhusin, M. E., and Watson, A. J. (1999). *Mol. Reprod. Dev.*, **49**, 119.
153. Watson, A. J., De Sousa, P. A., Caveney, A., Barcroft, L. C., Natale, D. R., Urquhart, J., and Westhusin, M. E. (1999). *Biol. Reprod.*, **62**, 355.
154. Offenburg, H., Barcroft, L. C., Caveney, A., Viuff, D., Thomsen, P. D., and Watson, A. J. (2000). *Mol. Reprod. Dev.*, **57**, 323.
155. Schliwa, M., and Van Blerkom, J. (1981). *J. Cell Biol.*, **90**, 222.
156. Siddal, L., Barcroft, L. C., and Watson, A. J. (2002). *Mol. Reprod. Dev.*, **63**, 413.

18

M. A. SIRARD
C. ROBERT

Gene Subtraction and Analysis

1. GENE SUBTRACTION TECHNIQUES

Research focused on investigating gene expression is required to better understand the complex mechanisms that control the formation of specialized tissue and organs during development. By increasing our current knowledge of the mechanisms controlling the expression of specific genes, novel approaches to the treatment of human diseases can be developed. One of the main benefits of studying gene expression in preimplantation embryos is that the precise nutritional and metabolic needs of the embryo will be defined, and this will lead to greater success in the production of normal offspring following embryo culture and embryo transfer. This outcome should also result in the generation of improved media better suited for human and animal embryo production in vitro.

Generally, the endpoint of gene expression is the production of functionally active proteins. The study of protein function is essential but is limited by the lack of effective antibodies and also by the requirement for amino acid sequence information, which limits investigation to known proteins. To circumvent these problems associated with protein analysis, the majority of screening methods have been applied at the mRNA transcript level. The rationale for these techniques is to screen the entire mRNA pool to isolate and characterize transcripts that display significant variation in the level of expression between specific cell types. This type of analysis results in the construction of an expression profile map that becomes associated with the specific tissue of interest and can be used to identify marker genes expressed during precise physiological events. The marker genes can also be used for diagnostic purposes. Therefore, targeting the characterization of variations in mRNA pools between specific cell types has become an efficient way to learn more about the cellular pathways that regulate physiological events in specific tissues or cell types.

Although this is not a strict principle, fluctuations in mRNA levels are often associated with corresponding variations in protein levels. The presence or absence of a specific mRNA is generally associated with the presence or absence of the protein in a given cell or tissue. Deducing the amino acid sequence from the nucleotide sequence of the candidate mRNA allows for further study at the protein level. In abundant tissues where

RNA availability is not a limiting factor, Northern blot analysis can be performed to quantitatively evaluate the mRNA level of candidate genes. However, when the tissue or cell sample is present in scarce quantities such as preimplantation embryos, methods that display a much greater sensitivity such as reverse transcription-polymerase chain reaction (RT-PCR) are required.

This area of research has been a very vigorous one in recent years, and several exciting new approaches have been developed to contrast gene expression in low abundance mRNA tissues. The differential display method and cDNA library analysis have been most commonly used to identify differentially expressed genes. Differential display analysis is a powerful technique using PCR amplification to compare side by side the banding patterns of cDNAs from several cell types. Since the typical mammalian cell contains about 10,000–30,000 mRNAs (1, 2), an exhaustive analysis of all the transcripts would require an extensive series of experiments that would require several sets of PCR amplifications.

Construction of cDNA libraries and large-scale sequencing on randomly picked clones is a powerful approach to discover genes (3). A functional gene expression map can be derived from this type of approach by measuring the relative abundance of each transcript by determining the number of times each cDNA is sequenced in the library. This provides valuable information regarding the relative abundance of each transcript in one specific tissue. Differentially expressed genes can be isolated by comparing expression maps derived from distinct cell types. A particular mRNA would be described as differentially expressed if its abundance was significantly greater in one condition versus another (4).

One of the first adaptations on this approach, the serial analysis of gene expression (SAGE) method was designed to overcome the time-consuming, large-scale sequencing of complete cDNAs. This method incorporates enzymatic digestions of the cDNA library and several ligation steps to produce concatenated small cDNA portions (expressed sequence tags, ESTs) cloned into plasmids (5). By sequencing the plasmids and compiling the frequency of each EST found, it is possible to quickly establish an expression map specific for each tissue. However, the ESTs are very small sequences, rending their identification difficult, which necessitates the use of complex techniques to lengthen the sequence. Overall, both approaches are very time consuming, expensive, and do not focus directly on the specific goal of isolating differentially expressed genes.

Subtracting two cDNA libraries is an efficient way to directly identify differentially expressed genes (6, 7). Standard procedures require up to 5–10 µg of poly A tail RNA (8). Preimplantation embryos on average contain a nanogram of total RNA, therefore these requirements would demand the extraction of RNA from tens of thousands of embryos. However, the development of PCR-based amplification has paved the way to applying these methods to samples as small as 10–100 ng of total RNA (9). The goal of the subtraction method is to eliminate all the transcripts expressed in both tissues, leaving only the specific ones behind. This subject is reviewed by Sagerström et al. (2).

The subtraction is performed by annealing the cDNA pools from the tissue of interest (often referred as the tester or tracer cDNAs) to an excess amount of cDNAs from the reference tissue to be subtracted (driver). The exceeding ratio of driver cDNAs will hybridize all the common transcripts, leaving the single-stranded cDNAs specific to the tissue of interest unhybridized in the sample. The double-stranded cDNAs are generally eliminated using hydroxyapatite or streptavidin-biotin interaction, thus leaving the single-stranded specific cDNAs to be cloned to build a subtracted cDNA library or to be individually used as probes in further screening procedures.

Many protocols are currently available to prepare cDNA or RNA for subtractive hybridization. Usui et al. (10) described a subtraction method based on directional tag PCR by cloning the tester and driver in opposite orientation in specific plasmids. Subtraction is conducted by hybridizing single-stranded antisense target RNA with a driver sense RNA. Hydroxyapatite eliminates the double-stranded RNA pool, leaving the single-stranded target ready for a PCR amplification step using the tag to prime the reaction.

To identify the differentially expressed genes, one can rapidly and efficiently apply the rapid subtraction hybridization (RaSH) method. This method involves the construction of digested cDNA libraries with specific linkers added to the driver cDNAs, which allow for PCR amplification. Different linkers are ligated to the tester and digested with specific restriction enzymes. The tester cDNA pool is annealed to the driver in a ratio of up to 1:30, respectively. At this point, only the cDNAs specific to the tester are able to reanneal with their complementary counterparts, producing tester-specific, double-stranded cDNA that can be cloned because they possess specific restriction sites on both sides, allowing for a cohesive ligation into the plasmid. Jiang et al. (11) used this technique to isolate differentially expressed genes in the growth arrest cells of human melanoma. A variation of the RaSH technique has been described (12). This method consists of eliminating the double-stranded templates by enzymatic restriction on an annealed, circular, single-stranded tester (phagemid) and linear, single-stranded driver derived from a reverse transcription reaction (first-strand cDNA synthesis). After the digestions, the remaining circular, single-stranded phagemids are specific to the tester. Transforming those into bacteria produces a subtracted cDNA library.

Suppression subtractive hybridization (SSH) is one of the recently developed techniques (13). It is completely PCR based and eliminates the step of single-stranded tester cDNA purification by streptavidin-biotin or hydroxyapatite. When the amount of starting material is limited, double-stranded cDNA of both tester and driver are first synthesized and amplified using Clontech's (Palo Alto, CA) SMART approach. For cDNA subtraction, the tester pool is divided in two fractions, and a different adaptor is ligated to each fractions. An excess of driver cDNA without linkers is denatured and hybridized with each tester cDNA pool (first hybridization). Both samples are mixed together with addition of more single-stranded driver (second hybridization). The resulting pool is a mixture of single stranded, double stranded with only one linker, double stranded like the original pools, and double stranded with both linkers corresponding to the tester specific fragments. Filling the ends of the linkers creates the templates to be amplified by PCR. The adaptors are created so that cDNAs possessing the same adaptor on both sides will form a hairpin preventing amplification. Only the ones possessing both linkers will be amplified exponentially. The resulting PCR product is enriched in tester specific cDNAs. The products are cloned and characterized to confirm their specificity by cDNA microarray.

In general, one of the major problems associated with specific cellular characterization is sample scarcity. Conventional methods for cDNA libraries are inappropriate because they require too large an amount of starting material (9). However, a restricted PCR amplification step before cDNA subtraction has eliminated most of the problems associated with tissue scarcity.

Many protocols are used for subtraction. Wan et al. (4) reported more than 300 examples of different subtraction techniques taken from different studies in various fields. Numerous procedural variations are reported in each publication, and it is up to the user to choose what will fit best his or her applications.

2. OVERALL METHOD COMPARISONS

The removal of double-stranded common cDNAs is one of the most common protocols (4). There are some advantages and disadvantages to either hydroxyapatite (HAP) or biotinylation and streptavidin methods.

Both biotinylation and HAP are effective methods that ensure the recovery of specific transcripts. HAP is very efficient but requires more refined conditions for its application (2). However, it is not commonly used in combination with cDNA driver. For successful subtraction, an excess of driver cDNA is needed, and the unannealed, single-stranded driver cDNAs will pass through the column and will contaminate the tester-specific, single-stranded cDNAs elution. To prevent this, the use of RNA driver is recommended as it is used in directional tag PCR subtraction (10). The elution can be treated by RNase to digest the contaminating driver.

Biotinylation is used for two purposes. When marking the driver molecules, it serves to remove completely the driver–driver and the driver–tester annealed molecules, leaving only the tester-specific cDNA. The choice of a DNA template is therefore appropriate for subtraction. DNA is not as sensitive to degradation as is RNA and is thus retrieved more easily. The biotin incorporation is conducted during the cDNA synthesis step or during the PCR reaction using biotinylated primers (14) or by RNA photobiotinylation. This last method is not recommended because the biotinylated nucleotides are rare (1/300 nucleotides) and insoluble in water. There are many ways to construct streptavidin columns to bind the biotinylated templates, and the method of choice depends on what is available in the laboratory.

As briefly described, it is today possible to combine PCR amplification of small samples of RNA for subtraction of cDNA libraries. By using the PCR technology, it is possible to start with only a few nanograms of total RNA and produce sufficient cDNA to subtract two specific cell populations (SMART, Clontech). These cDNA pools could be used as templates in standard subtraction procedures as described above.

The other techniques involve PCR amplification of the cDNAs used for the subtraction with a direct elimination of the annealed tester–driver and driver–driver cDNA populations. The plateau phase of the PCR amplification introduces a bias by skewing the relative abundance of the cDNAs. The Taq polymerases amplify preferentially smaller fragments, and some DNA secondary structures are more easily processed than others. Therefore, PCR amplification should only be used when RNA samples are limited or precious to allow for the application of more classical subtraction hybridization methods. RNA quality is essential for a good representation in the cDNA produced. There are two ways to control RNA quality. First, when a sufficient amount of RNA is available, it is important to analyze the total RNA on a denaturing agarose gel before the PCR amplification. Second, it is important to gauge the cDNA synthesis by amplifying and quantifying a housekeeping or known stable gene expressed in the sample.

RaSH is a simple technique that only requires a limited amount of starting RNA material. The PCR amplification is conducted for the synthesis of the cDNA before the subtraction. From the annealing step to the cloning of the subtracted library, the samples are not subject to any further PCR amplification. Because this method uses endonuclease digestions before cloning the subtracted cDNAs, the resulting library will be composed of variable length, incomplete cDNAs.

SSH employs a greater number of amplifying steps than the RaSH method, depending on the availability of the two samples to subtract. If a sufficient amount of poly A Tail RNA is available, a standard cDNA synthesis should be performed because it increases the probability of maintaining the relative representation of the cDNAs between each sample, which is important during the annealing step.

Using SSH, Robert et al. (15) produced subtracted libraries of oocytes at different stages of maturation. The total RNA content of oocytes in believed to be 2.4 ng (16, 17). It is well accepted that the ribosomal proportion of total RNA is about 95% of that found in somatic cells. However, because the oocyte stores RNA during its maturation, the proportion of ribosomal RNA is reduced to 23% (18), thus the mRNA content of the oocyte is estimated to be around 1.9 ng. Using only 50 oocytes (92 ng mRNA), Robert et al. (15) isolated specific cDNAs. Classic procedures would have required 5–10 µg of poly A RNA to obtain the same results. Other teams have also reported the use of similar technologies to amplify cDNAs from small tissue biopsy samples (19).

After the subtraction, specificity has to be confirmed because the resulting PCR product is only enriched in differentially expressed cDNA. To screen a large amount of candidates, the cDNAs are arrayed on nitrocellulose, nylon membrane, or glass and hybridized with probes made from the original tissues. In the case of low-expression genes, the use of subtracted probes may be useful because relative abundance is normalized during the suppressive PCR amplifications. Finally, Northern blot analysis or quantitative PCR standardized to the level of a stable, known housekeeping gene is needed to precisely measure the levels in differential expression between the samples.

3. METHODS FOR SSH USED ON OOCYTES AND PREIMPLANTATION EMBRYOS

3.1 Oocyte/Embryo Collection

During collection, oocytes are kept in follicular fluid for as long as possible to prevent spontaneous maturation and meiosis resumption. When all the oocytes are collected, they are washed in PBS buffer to remove the follicular fluid. This is done because the cumulus cells are stripped from the oocyte mechanically using a vortex, and the presence of follicular fluid causes the solution to foam. The denuded oocytes are recuperated and washed three to four times in PBS buffer to remove the cumulus cells in the solution. The oocytes are quickly transferred to a 0.5-ml Eppendorf tube. Most of the oocytes should drop to the bottom of the tube, and most of the remaining PBS buffer can be carefully removed. When small pools are used, it is important to confirm visually the number of oocytes in the tube because some oocytes may have been removed with the PBS. The preimplantation embryos are simply collected from the culture drops and transferred to a tube as described for the oocyte collection. It is not necessary to wash the embryos in PBS buffer, as the media will be removed from the tube. The tubes can be snap frozen in liquid nitrogen before storage at –80°C, but in our experience we did not have more RNA degradation when the liquid nitrogen dipping was skipped because only a few oocytes/embryos are remaining in the tubes, and they freeze instantly at –80°C. Although the stored maternal RNA should be fairly stable, it is important to process the cells as quickly as possible to reduce the effect of the stress associated with the oocyte/embryo manipulation.

3.2 RNA Extraction

In a previous study (15), total RNA was extracted using Trizol reagent (Gibco BRL, Burlington, ON). This is a fast and cost-effective method, but a fraction of the RNA is lost during the organic-phase collection. Furthermore, the resulting RNA is often contaminated with genomic DNA, necessitating DNase I treatment followed by a phenol/chloroform extraction or a column purification such as RNeasy (Qiagen, Mississauga, ON). These extended manipulations may result in the loss of RNA. Because the work with oocytes or preimplantation embryos is challenging due to tissue scarcity, it is important to be efficient during RNA extraction to prevent RNA loss. Other techniques of RNA extraction have recently been developed to circumvent the problem of RNA loss. We currently use the microextraction columns from Stratagene to extract the RNA and perform the DNase I treatment in the same column (see chapter 13). Generally, for the downstream reactions, it is better to use poly (A)$^+$ RNA when available. We do not extract the poly (A)$^+$ RNA from the cells because the oocytes/preimplantation embryos are rich in poly (A)$^+$ RNA, and we believe the RNA lost during poly (A)$^+$ RNA extraction would be more critical than using total RNA.

After the extraction, the RNA is precipitated using sodium acetate and isopropanol. Because the concentration of RNA is low, a coprecipitant must be added. Glycogen or linear acrylamide can successfully be used as coprecipitant, but the use of tRNA should be avoided because tRNA will interfere in the cDNA subtraction.

3.3 Reverse Transcription and Total Amplification

Even if an oligo-dT is used during first-strand synthesis to reverse transcribe specifically the RNA population with a poly A tail, sequences rich in poly A will allow the oligo-dT primer to bind, resulting in the reverse transcription of nonmessenger RNA. We use the superscript II enzyme (Gibco) to perform the reverse transcriptions because it is efficient and lacks the exonuclease activity of other moloney murine leukemia virus (MMLV) reverse transcriptases. Because the amount of starting material is usually insufficient to perform a standard subtraction when working with oocytes or preimplantation embryos, a total amplification of the RNA pool is needed. To do this, we use the SMART kit from Clontech, which is basically a regular reverse transcription reaction using two special oligonucleotides. The first one is an oligo-dT with a specific primer sequence in 5', whereas the second one is designed with the same specific primer sequence with the addition of a poly G tail. The first primer will bind to the poly A tail of the mRNA, while the second primer will bind to the poly C tail left by the reverse transcriptase leaving the RNA strand and thus creating another primed region for the reverse transcriptase to synthesize the second strand (Figure 18.1). The end product of this reverse transcription reaction is a double-stranded DNA pool bearing a known primer sequence at both ends. The primer sequence is then used to perform the PCR amplification of the entire DNA pool. To prevent over-cycling, which will introduce an important bias in the relative amount of each amplified transcript species, a fraction of the DNA pool is sacrificed to determine the optimal number of cycles. The amplification is performed on the sample for 12 cycles, and then, without stopping the reaction, an aliquot of the PCR product is taken out every 3 cycles until the amplification reach 24 cycles. The aliquots are run on an agarose gel, and the optimal number of cycles is determined visually according to the shape of the resulting DNA smear. For pools of 50 oocytes, we conduct on average 15 cycles.

3.4 SSH

For the SSH procedures (Figure 18.2), we follow the instructions as outlined in the user manual provided with the PCR Select Kit (Clontech). Briefly, the tester and driver DNA pools are digested with a specific restriction enzyme. The tester pool is divided in two fractions, and a different adaptor is ligated to each fraction. For the first hybridization, an excess amount of driver DNA is added separately to each tester-ligated fraction. The second and most important annealing step consists of mixing the two annealed fractions together and adding more denatured driver cDNAs. This increases the efficiency of the elimination of the common strands and allows the tester-specific cDNAs to anneal with their counterparts possessing the second linker sequence. Before the first subtracting PCR, the ends are filled with Taq polymerase, recreating the primer sites. An aliquot of this first PCR is diluted and used as template for the second round of subtraction by nested PCR. With these specific PCR amplifications, only the cDNAs bearing both linkers types that correspond to tester-specific cDNAs are amplified exponentially. The resulting PCR product is enriched in differentially expressed cDNAs of partial length.

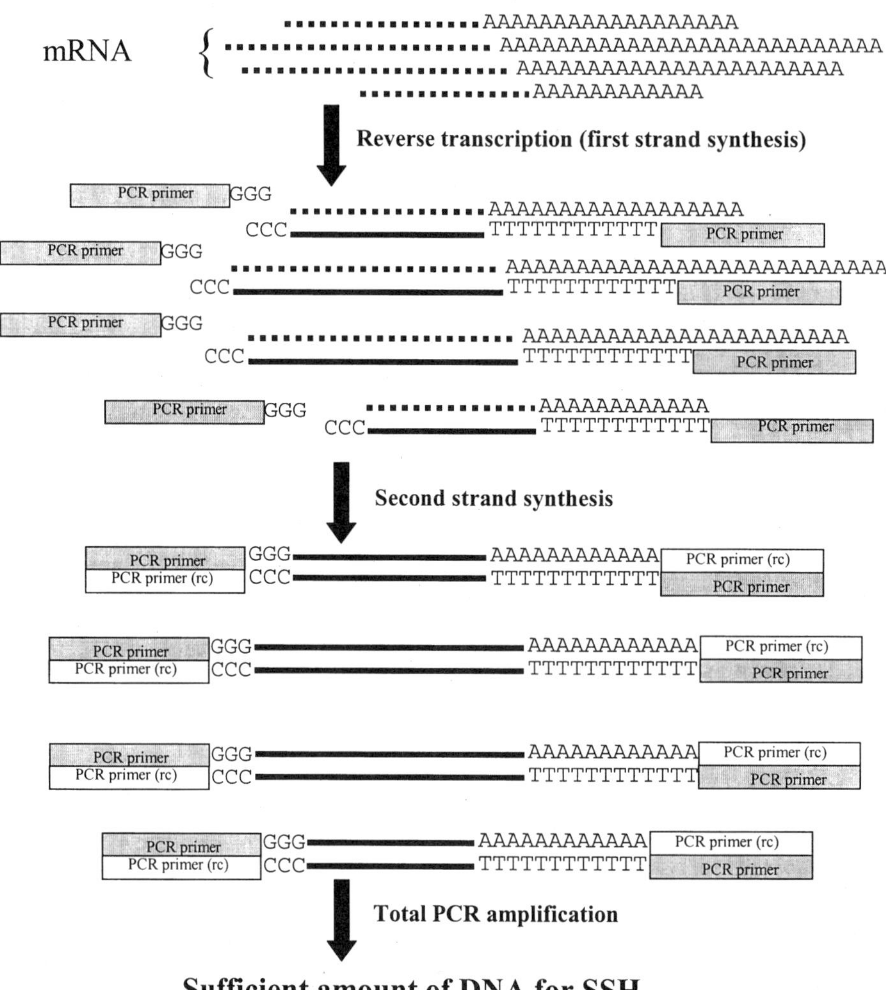

Figure 18.1. Reverse transcription and total cDNA amplification using the SMART approach from Clontech.

3.5 Cloning and Transformation

To clone the cDNAs we use the TA cloning kit from Invitrogen (Burlington, ON) because of its high efficiency with PCR products and because it offers the screening of the bacterial colonies with the lacZ system. After the PCR product is ligated into the plasmid, competent bacteria are transformed and plated onto agar plates containing ampicillin, X-Gal, and IPTG (isopropylthio-β-galactoside) using usual procedures. The white colonies are picked and grown individually overnight in 96-well plates.

3.6 Insert Amplification and Candidate Screening

The next day, 1 μl of each bacterial growth was subjected to PCR amplification using the nested PCR primers used in the second round of amplification during the SSH procedures. Each PCR product was run on an agarose gel to confirm the amplification and to evaluate the size of the insert. Some plasmids contain two inserts, and for an unknown reason some others do not show any inserts. To find the differentially expressed genes among all these candidates, the cDNAs are arrayed on a nylon membrane or a glass slide to perform high-throughput screening. Each candidate isolated by cDNA array is

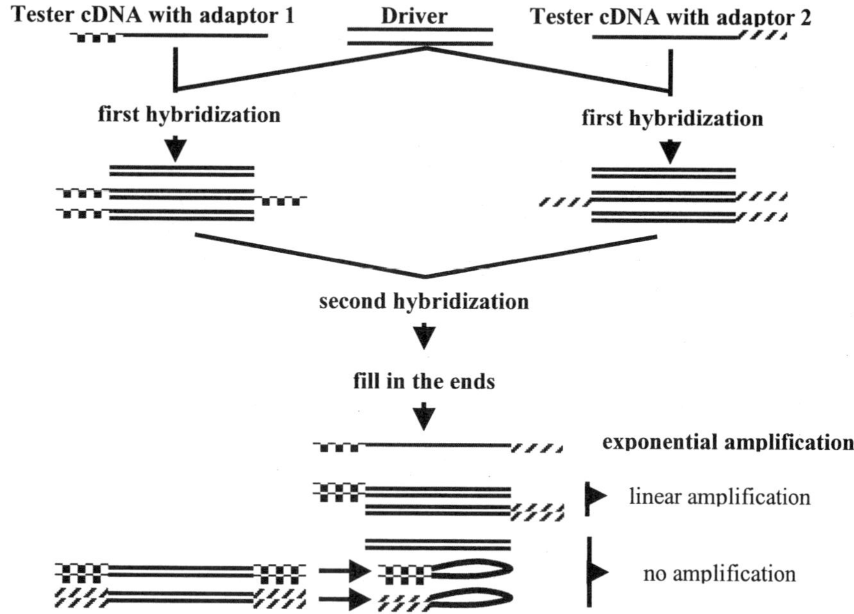

Modified from CLONTECH PCR-Select™ cDNA Subtraction Kit

Figure 18.2. Illustration of the PCR-based subtractive hybridization procedure modified from the Clontech kit.

then confirmed by quantitative RT-PCR because the techniques of cDNA amplification (SMART) combined with SSH procedures use several rounds of PCR that might skew the original relative abundance of some mRNA.

References

1. Davidson, E. H. (1976). *Gene Activity in Early Development.* 2nd edition. Academic Press, New York.
2. Sagerström, C. G., Sun, B. I., and Sive, H. L. (1997). *Annu. Rev. Biochem.,* **66**, 751.
3. Bonaldo, M., de, F., Lennon, G., and Soares, M. B. (1996). *Genome Res.,* **6**, 791.
4. Wan, J. S., Sharp, S. J., Poirier, G. M.-C., Wagaman, P. C., Chambers, J., Pyati, J., Hom, Y.-L., Galindo, J. E., Huvar, A., Peterson, P. A., Jackson, M. R., and Erlander, M. G. (1996). *Nature Biotechnol.,* **14**, 1685.
5. Velculescu, V. E., Zhang, L., Vogelstein, B., and Kinzler, K. W. (1995). *Science,* **270**, 484.
6. Martelange, V., De Smet, C., De Plaen, E., Lurquin, C., and Boon, T. (2000). *Cancer Res.,* **60**, 3848.
7. Hemberger, M., Himmelbauer, H., Ruschmann, J., Zeitz, C., and Fundele, R. (2000). *Dev. Biol.,* **222**, 158.
8. Sambrook, J., Fritsch, E. F., and Maniatis, T. (1989). *Molecular Cloning: A Laboratory Manual,* 2nd ed. Cold Spring Harbor Laboratory Press, Cold Spring Harbor, NY.
9. Lukyanov, K., Diatchenko, L., Chenchik, A., Nanisetti, A., Siebert, P., Usman, N., Matz, M., and Lukyanov, S. (1997). *Biochem. Biophys. Res. Commun.,* **230**, 285.
10. Usui, H., Falk, J. D., Dopazo, A., de Lecea, L., Erlander, M. G., and Sutcliffe, J. G. (1994). *J. Neurosci.,* **14**, 4915.
11. Jiang, H., Kang, D., Alexandre, D. and Fisher, P. B. (2000). *Proc. Natl. Acad. Sci. USA,* **97**, 12684.
12. Rivolta, M. N., and Wilcox, E. R. (1995). *Nucl. Acids Res.,* **23**, 2565.
13. Diatchenko, L., Lau, Y.-F.C., Campbell, A. P., Chenchik, A., Moqadam, F., Huang, B., Lukyanov, S., Lukyanov, K., Gurskaya, N., Sverdlov, E. D., and Siebert, P. D. (1996). *Proc. Natl. Acad. Sci. USA,* **93**, 6025.

14. Zhao, Z., Huang, X., Li, N., Zhu, X., Chen, S., and Cao, X. (1998). *J. Biotechnol.,* **73**, 35.
15. Robert, C., Gagné, D., and Sirard, M. A. (2000). *Mol. Reprod Dev.,* **57**, 167.
16. Bachvarova, R., De Leon, V., Johnson, A., Kaplan, G., and Paynton, B. V. (1985). *Dev. Biol.,* **108**, 325.
17. Bilodeau-Goeseels, S., and Schultz, G. A. (1997). *Biol. Reprod.,* **56**, 1323.
18. Piko, L., and Taylor, K. D. (1987). *Dev. Biol.,* **123**, 364.
19. Zanders, E. D., Goulden, M. G, Kennedy, T. C., and Kempsell, K. E. (2000). *J. Immunol. Methods,* **233**, 131.

19

DANIEL R. BRISON
ANTHONY D. METCALFE
DEBRA J. BLOOR
HELEN R. HUNTER
GERARD BRADY
SUSAN J. KIMBER

Analysis of Apoptosis in the Preimplantation Embryo

1. APOPTOSIS

1.1 Apoptosis in Development

A basic requirement in development is the proliferation of sufficient cells to form the structure of the organism. This is seen nowhere more clearly than during the earliest transition in mammalian development, from the single-celled zygote to the multicellular blastocyst. However, the generation and maintenance of a cell population is not just a function of cell division by mitosis and cytokinesis, but is counterbalanced by cell loss via apoptosis. Apoptosis is a physiological form of cell death and is characterized morphologically by nuclear condensation and blebbing, followed by shrinkage of the cell and formation of membrane-bound vesicles of degraded chromatin (1, 2). The cell does not release its soluble contents into intercellular spaces; rather, these apoptotic bodies are phagocytosed by macrophages or neighboring cells with phagocytic capacity. In contrast, necrotic cell death is accidental and occurs in response to trauma or injury such as osmotic shock. This results in loss of plasma membrane integrity, followed rapidly by cell and nuclear lysis and release of contents into intercellular spaces, frequently resulting in an inflammatory response.

Apoptosis has evolved to allow efficient and orderly disposal of cells without compromising the viability of the organism. It can be considered the primary driving force behind the establishment of cell number, against a background of cell division, and it is central to the regulation of animal size (3). In addition to regulating the number of cells, apoptosis also fulfills a crucial proofreading function in development by eliminating damaged or abnormal cells from an organism. It also eliminates cells that are normal but are no longer required in a particular place, such as interdigital webbing (4–6). Apoptosis usually occurs at specific times and places during development in response to a genetic program or specific signals—hence its original identification as "programmed" cell death (PCD) (4, 7, 8). Not all apoptosis is genetically programmed, and not all PCD occurs via apoptosis; however, the term apoptosis has now been generally adopted to refer to cell death with the features described above and is used throughout this chapter.

In mammalian development, apoptosis eliminates oocytes from the ovary (9) and defective sperm cells from the testis (10). In the postimplantation embryo, apoptosis performs a number of functions, including formation of cavities during development (11), as early as the proamniotic stage in the mouse (12). In the preimplantation embryo, loss of cells by fragmentation is common at the early cleavage stages. This process has been positively identified as apoptosis using a number of methods (see section 2, "Techniques for analyzing apoptosis") in mouse zygotes (13–15), human 2–8-cell embryos (16–20), and bovine 8–16-cell embryos (21). In mouse, human, bovine, and the amphibian *Xenopus* (22), the wave of apoptosis coincides with the earliest activation of the embryonic genome (EGA). This has led to suggestions that its function is to eliminate blastomeres or whole embryos that are not competent to undergo this developmental transition (13, 22, 23). The majority of embryos, however, develop normally without fragmentation and do not undergo apoptosis until they reach the blastocyst stage. This first major wave of apoptosis in development occurs in essentially all blastocysts, predominantly in the inner cell mass (ICM) (24, 25). Because this population of cells gives rise to the fetus and carries the germ line, the selective targeting of apoptosis to the ICM suggests strongly that it functions to protect the genomic integrity of the developing organism. Apoptosis could eliminate ICM cells that contain damage (23); induction of DNA strand breaks by X-irradiation induces apoptosis in the ICM within 2 h (26). Apoptosis may also eliminate cells that possess the "wrong" phenotype, since the incidence of apoptosis increases as the blastocyst expands, coincident with loss of ability of ICM cells to form trophectoderm (TE) (27). There is some evidence for this latter suggestion, as embryonal carcinoma cell lines that retain the potential to form TE are selectively killed in blastocoel fluid, possibly by the action of reactive oxygen species such as hydrogen peroxide (28, 29). Apoptosis may also regulate ICM cell number as this plateaus in the later blastocyst, although the rate of mitosis does not decrease (27, 30). However, apoptosis is not thought to play a role in size regulation of the mouse embryo after implantation (31), and neither is it responsible for formation of the blastocoel cavity. To date, apoptosis has been detected in blastocysts of a wide range of species including humans, rhesus monkeys, baboons, mice, rats, rabbits, pigs, and cows (21, 32–38).

1.2 Regulation of Apoptosis

Correct regulation of apoptosis is essential in development; failure to eliminate damaged cells can give rise to disease states such as cancer, whereas excess apoptosis can destroy viable cells and is implicated in neurodegenerative diseases (5). Regulation is especially critical in organisms such as the preimplantation embryo that contain few cells. Loss of ICM cells below a critical threshold can compromise fetal development (39, 40), and total blastocyst cell number is correlated with implantation potential (41, 42). Apoptosis can be regulated both positively and negatively, with pathways converging on the bcl2 family of intracellular regulators of apoptosis. The components of these pathways (known collectively as the "apoptosis machinery"; Figure 19.1) are expressed in almost all mammalian cells studied. Preimplantation mouse embryos express both pro- (bax, bad, bak, bid) and anti- (bcl2, bclxl, bclw) apoptotic members of the bcl2 family and several of the caspases (caspase 2, 3, 6, 7, 12, etc.) throughout development (13, 43; M. Kamjoo, D. R. Brison, and S. J. Kimber, unpublished). Human embryos express at least 10 members of the bcl2 family throughout preimplantation development (44–46). Thus, in mammalian cells the apoptosis machinery is present and ready to be triggered at any time. To ensure cell survival, apoptosis must be continually suppressed by extracellular survival signals. These can be derived from extracellular matrix via interactions with integrins, contact with other cells via the cadherin system, or soluble signals such as cytokines and peptide growth factors acting via their cognate receptors on the cell surface. In the absence of extracellular survival signals, apoptosis is the default cell fate (47).

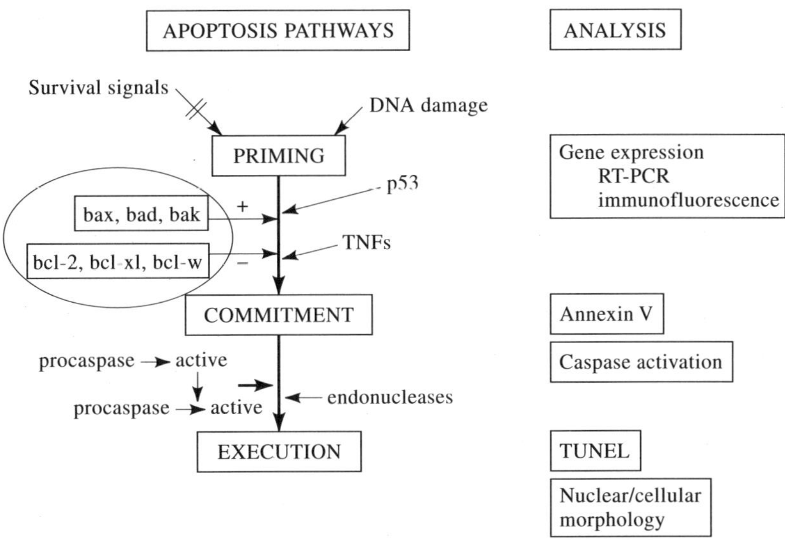

Figure 19.1. Model for regulation of apoptosis pathways and methods of analysis. A number of signals can prime a cell for apoptosis, including withdrawal of survival signals and detection of cellular DNA damage (e.g., by the tumor–suppressor gene p53). Priming leads to alterations in the expression and activity of the intracellular regulators of apoptosis, the bcl2 family of molecules, of which there are more than 20 mammalian family members (89). The balance and interaction between antiapoptotic members such as bcl2, bcl-xl, bcl-w (shown acting negatively on the apoptosis pathway) and proapoptotic members such as bax, bak, bad, bid, bcl-xs (shown acting positively) determines cell fate. This interaction occurs at the level of the mitochondrion (represented as an oval), with proapoptotic members such as bax and bid translocated from the cytosol to the mitochondrial outer membrane, possibly forming pores which allow release of cytochrome c into the cytosol (90, 91). Antiapoptotic members such as bcl-2 and bcl-xl effectively block this process. Regulation of these interactions can occur at the transcriptional and translational levels, with production of new protein altering the balance in favor of survival or death. Posttranslational modifications are also crucial. Survival receptor signaling via intracellular kinases such as protein kinase A causes the proapoptotic bad to be phosphorylated and sequestered away from the mitochondria, blocking the proapoptotic function of bax (92). p53 activation of apoptosis can act via the bcl2 family or directly via death factor signaling (93). After commitment to apoptosis, a cell is irreversibly destined to die, and the execution phase is carried out via the activation of a cascade of intracellular cysteine proteases known as caspases (94). Cytochrome c release causes the activation of the initiator caspase 9, together with an adaptor protein (Apaf-1) as part of a complex called the apoptosome. Caspase 9 then activates downstream caspases such as the effector caspase 3. Death factors such as tumor necrosis factor (TNF)α binding to its receptor (TNFR1) can bypass the bcl2 family by directly activating the initiator caspase 8, followed by caspase 3, etc. (95). A cascade of inactive procaspases are cleaved to yield active forms that kill the cell by cleaving cellular constituents, including specific known targets such as nuclear lamins, cytokeratins, and the DNA repair enzyme poly-ADP-ribose polymerase (PARP). This leads to breakup of nuclear structure and activation of endonucleases such as caspase-activated deoxyribonuclease (CAD), which cleaves DNA between nucleosomes, yielding fragments approximately 180 bp in length. The end result is the formation of membrane-bound apoptotic bodies, which are phagocytosed.

Exogenous survival signals can be blocked pharmacologically by incubation in staurosporine, a protein kinase inhibitor that blocks all signal transduction. Staurosporine induces apoptosis in all cells of the mouse blastocyst (48), but, interestingly, four-cell embryos are less sensitive. Cells can also be deprived of survival signals by culturing them singly in simple medium. Under these conditions, disaggregated and singly cultured four-cell mouse blastomeres are one of the few nucleated mammalian cells that do not undergo apoptosis (48). In contrast, mouse embryos grown singly in culture to blastocyst show three times the amount of apoptosis compared to blastocysts that developed in vivo, whereas embryos cultured in groups show intermediate levels (49). These data and other studies (50) suggest that survival factors are present in the maternal tract and also secreted by embryos in culture. One such survival factor is the peptide growth factor transforming growth factor α (TGFα). Embryos lacking the TGFα receptor, epidermal growth factor receptor erbB1, show a phenotype of failure of normal ICM development on at least one genetic background (51). TGFα added to embryo culture medium at 0.1 pM reduces apoptosis in whole mouse blastocysts and also in ICMs isolated and cultured singly. This rescue effect is specific because it is blocked by incubation with a control-blocking antibody to TGFα. Higher concentrations of TGFα increase apoptosis, presumably due to downregulation of erbB1 and loss of survival signaling (49). Loss of TGFα function also increases apoptosis. TGFα-null mutant mice are fertile and produce morphologically normal blastocysts, but with three times the rate of apoptosis seen in wild-type blastocysts (52). Insulinlike growth factor 1 (IGF1) is also a survival factor for rabbit (53), mouse (23), human (54), and cow (55) embryos. As with TGFα, high concentrations of IGF1 cause increased apoptosis in the blastocyst due to downregulation of the IGF1 receptor (IGF-1R) and loss of survival signaling (56). For a review of survival-factor regulation of apoptosis in the embryo, see Brison (23). Some cell types also express members of the endogenous inhibitor of apoptosis protein (IAP) family, which can block apoptosis (57). Maternal IAPs could be responsible for the survival of early mouse blastomeres, whereas failure in expression of these from the newly transcribed embryonic genome could explain the apoptosis seen in some early-cleavage embryos (13, 22, 23).

There is also evidence of positive regulation of apoptosis in mouse embryos. p53-null mutant mice develop normally but show an increased number of fetal abnormalities after paternal exposure to radiation (58). This confirms an important role for p53 in proofreading early mammalian development, probably by triggering apoptosis in response to DNA damage (59). This extends to the preimplantation period, as induction of DNA strand breaks by X-irradiation initiates apoptosis in the blastocyst within 2 h (26). Moreover, mouse embryos derived from DNA-damaged sperm arrest before the blastocyst stage with premature induction of apoptosis (60). Apoptosis is also positively regulated by death factors (see Figure 19.1), and in the mouse and rat embryo exposure to TNFα results in loss of cells preferentially from the ICM (61), decreasing embryo viability after implantation (40). High glucose concentrations associated with hyperglycemia in the diabetic state can also induce apoptosis in mouse and rat embryos (62, 63). This occurs via a bax-dependent pathway involving ceramide and caspase 1 (63), in response to downregulation of glucose transporters and decreased glucose transport (64). Glucose regulation of apoptosis has been recently reviewed (25, 65, 66).

1.3 In Vitro Development and Perturbation of Apoptosis

Apoptosis is seen during normal in vivo embryo development in a number of species and is not merely an artifact of in vitro culture (24). However, in vitro conditions do increase apoptosis (23). This can be a response to the absence of maternal or embryonic survival factors (49, 52), the presence or absence in culture medium of growth/survival factors including those in serum (23), and oxygen concentration (19). Culture medium composition can also influence the level of apoptosis (67, 68). IVF itself appears to cause dysregulation of apoptosis. Mouse IVF embryos show increased apoptosis at early cleavage stages, induced by a lack of autocrine growth factors (50), and in the blastocyst, with extensive cell death in the TE as well as the ICM, in mouse (69) and human (36) embryos. This gives rise to concerns about the appropriate level of apoptosis in human blastocysts replaced in clinical IVF programs (23). In particular, there is a danger that addition of growth factors to embryo culture medium could block apoptosis, which is required to eliminate damaged cells. In this context, it may be significant that there is interaction between IGF1 and p53-triggered apoptosis pathways (70).

2. TECHNIQUES FOR ANALYZING APOPTOSIS

A wide range of techniques are used in the analysis of apoptosis in cells. These have been described in detail elsewhere (71), and we only briefly review some here, concentrating on those that are especially applicable to preimplantation embryos (summarized in figure 19.1). These assays are categorized in terms of (1) expression of components of the apoptosis pathways, (2) activation of these components, and (3) endpoints or markers of apoptosis. The techniques can be further subdivided into biochemical assays that are carried out in solution and *in situ* assays that yield spatial information.

2.1 Expression of Apoptosis Pathway Genes

The role of apoptosis in development can be assessed by measuring the expression of genes involved in the regulation or execution of the pathways. This can be done at the level of mRNA or protein analysis.

2.1.1 mRNA Analysis

Expression of apoptosis genes can be carried out conveniently using RT-PCR, with the advantages that screening for a large number of genes is possible and transcriptional regulation of apoptosis can be studied. For species showing relatively homogeneous preimplantation development, such as the mouse, mRNA can be purified from groups of embryos using micro-scale methods (see chapters 13 and 14). For example, expression of the apoptosis survival factor genes TGFα and EGFR (erbB1) has been measured semiquantitatively in groups of embryos at all stages of preimplantation development (Figure 19.2) and in the ICM and TE (72). However, analysis of single embryos is often essential, especially for those species that develop heterogeneously in vitro such as the cow and human. Conventional RT-PCR protocols permit only a few genes to be analyzed in a single embryo, which is extremely limiting in the analysis of multifactorial pathways such as apoptosis. Thus, we have used a single-cell mRNA global amplification method (73, 74) which permits analysis of unlimited numbers of genes in a single embryo (see Figure 19.3 and Protocol 19.1 for PolyAPCR on single embryos). Using this technique, a large number of apoptosis genes have now been described in groups of mouse (13) and single human (46) embryos. Expression of the survival molecule bcl2 is low during mouse and human preimplantation development, suggesting that other family members may be more important (13, 43, 46, 75, 76). The prodeath molecule bax is strongly expressed and may be influenced by in vitro culture (46, 76). The relative expression of anti- and proapoptotic genes has also been measured in single embryos. Proapoptotic genes are associated with morphologically poor-quality embryos and antiapoptotic genes are associated with viable embryos (13, 45). Information on spatial expression of apoptosis-related mRNAs can also be obtained by disaggregating the blastocyst before RT-PCR (e.g., ICM versus TE [72]) or by using in situ hybridization (e.g., bcl2 mRNA in rat blastocysts [77]).

2.1.2 Protein Analysis

For most purposes, it is convenient to analyze the spatial distribution of genes in terms of their functional product, the protein, using immunocytochemistry. A number of apoptosis proteins have been studied in the embryo by immunofluorescence, using standard protocols for fixation and permeabilization (43, 44, 46). Precise conditions vary

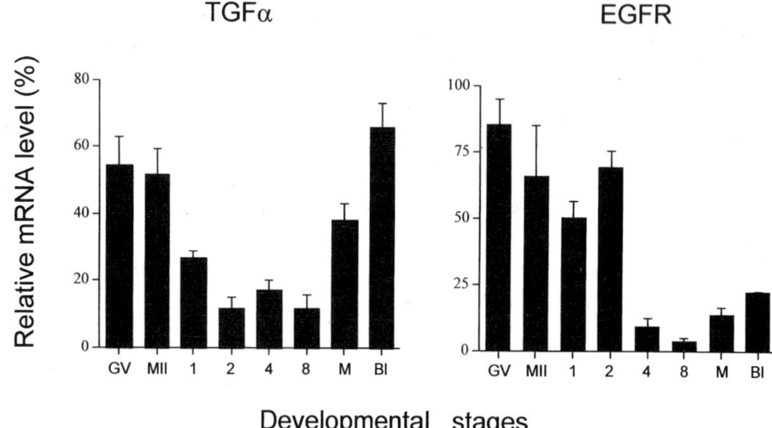

Figure 19.2. Developmental profile of relative levels of expression of mRNAs for TGFα and its cognate receptor, the epidermal growth factor receptor (EGFR; erbB1), in mouse germinal vesicle-intact (GV) and metaphase II (MII) oocytes, one-, two-, four-, and eight-cell stage embryos, morulae (M), and blastocysts (BI). mRNA was prepared from groups of 30 embryos at each stage, and levels were quantified against an exogenous β-globin standard (97). Both genes show a classic U-shaped profile of high expression in the oocyte and two-cell stages, low expression at the four-cell stage, and increased expression in the morula/blastocyst. This profile suggests that early transcripts are maternal in origin, whereas later transcripts arise from activation of the embryonic genome at the two-cell stage. The latter increase suggests a role for these genes in blastocyst formation. Data from D. R. Brison and R. M. Schultz (unpublished).

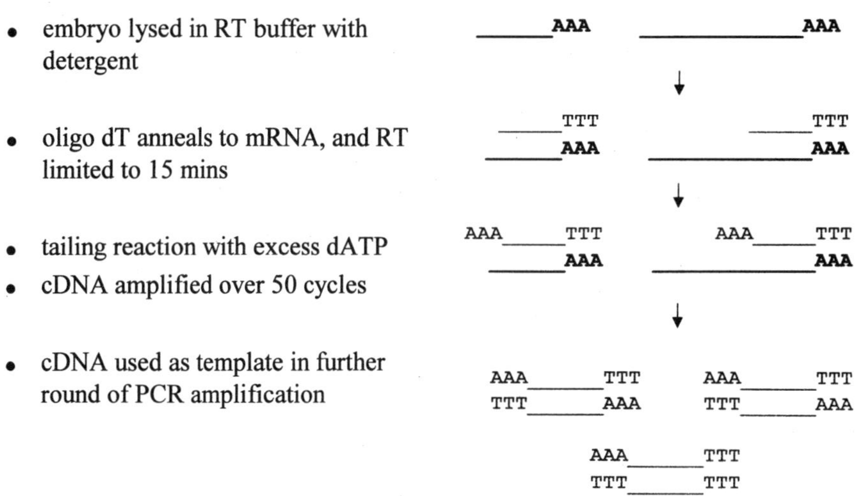

Figure 19.3. PolyAPCR on single embryos (adapted from Brady and Iscove [74]). This technique uses direct lysis without an mRNA purification step to increase the sensitivity of transcript detection. Reverse transcription is limited to 15 min, ensuring that first-strand cDNA synthesis is restricted to the most 3' 500–800 bp of the gene sequence. All mRNAs in the cell are therefore represented by amplicons of approximately the same size, ensuring that in subsequent PCR reactions there is no bias in favor of shorter mRNA transcripts. Similarly, rare mRNAs should be represented in proportion to their frequency. Thus the cDNA pools obtained reflect the mRNA populations in the intact embryo (although some discrepancies may still arise in determining relative frequencies of different mRNAs due to differences in GC content, as in any PCR-based technique). PolyAPCR has been used for qualitative analysis of mRNAs in single human embryos (76, 98) and quantitative analysis of small groups of mouse embryos (13, 99). This method has also been used to study gene regulation by polyadenylation of mRNAs during embryonic genome activation (100, reviewed by Brady [101]).

according to the antigen and antibody used, so it is inappropriate to give specific protocols here. However, fixation in paraformaldehyde and permeabilization with Triton X-100 according to the protocol given here for TUNEL (see Protocol 19.2 for blastocyst TUNEL) was found to be suitable for detection of TGFα in mouse blastocysts (72) and Bax, Bcl2, Bclx, Bclw, and activated caspase 3 in human embryos (46) (e.g., Bax; Figure 19.4). This information is essential in establishing a developmental role for particular proteins, especially in terms of subcellular localization. However, a major level of apoptosis regulation is post-translational (e.g., protein phosphorylation events). This requires an approach such as separation of proteins by gel electrophoresis followed by immunoblotting with, for example, antiphosphotyrosine antibodies.

2.2 Activation of Apoptosis Pathway Components

2.2.1 Caspase Activation

Caspase activation assays are relatively specific for apoptosis (excluding caspase-independent pathways) and make use of antibodies specific to the cleaved, active forms. Activation of the effector caspase 3 has been detected by immunofluorescence in mouse zygotes (15) and human embryos (A. D. Metcalfe, H. R. Hunter, S. J. Kimber, and D. R. Brison, unpublished data). This assay is highly specific to individual caspases but suffers from the disadvantage that activity is likely to be transient. Fluorogenic caspase substrates can also be used to assay caspase enzymatic activity. These work on the principle that a nonfluorescent substrate is cleaved by an active caspase to yield a fluorescent product. The substrate is added to live cells followed by fixation if desired and localization by fluorescence microscopy. This method is less specific for individual

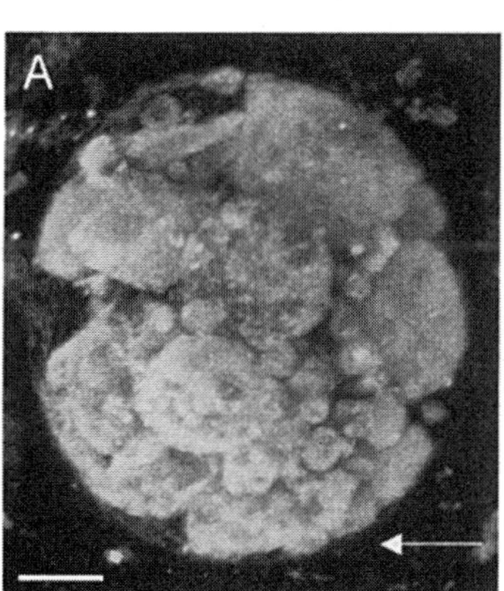
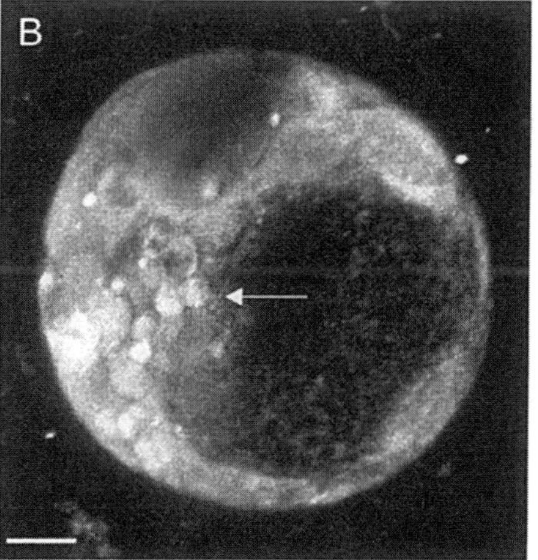

Figure 19.4. Confocal micrographs showing expression of Bax protein in a human fragmenting early cleavage embryo (A) and blastocyst (B) (23). Arrows indicate bright staining in cytoplasmic vesicles excluded from the embryo. In panel B, the ICM is located at the top left; the shadow is an artifact of embryo processing. Bars = 18 μm. Embryos were fixed in 1% paraformaldehyde for several days before processing, then permeabilized and washed according to the Protocol 19.2. The primary antibody used was a rabbit polyclonal antihuman Bax at a dilution of 1 in 10 in PBS/BSA (4 mg/ml), detected by a goat antirabbit FITC-conjugated secondary antibody diluted 1:50 in PBS/BSA. Negative controls omitting the primary antibody and with an irrelevant primary antibody showed no signal. Fluorescence detection was by confocal microscopy with a full Z series of optical sections collected for each embryo. Images can be visualized as full projections of all sections (A) or a selected region of the embryo, such as the ICM (B).

caspases but has been used to demonstrate general caspase activity in mouse oocytes (78) and embryos (43).

2.2.2 Annexin V labeling

One of the earliest events of apoptotic execution, the translocation of phosphatidyl serine to the outside of the plasma membrane, can be detected using binding of annexin V. Annexin V is conjugated to a fluorescent marker and localized by fluorescence microscopy. This technique has been used in oocytes (79) and in early-cleavage human embryos (18, 45). However, it must be applied to live, nonpermeabilized cells, making it unsuitable for studies of more complex structures such as the blastocyst. Annexin V also labels necrotic cells and must be used in conjunction with a method that discriminates between intact and lysed plasma membranes. The most commonly used method is exclusion of propidium iodide (PI), a DNA-intercalating dye, which does not cross intact cell membranes and will therefore label necrotic but not apoptotic cells.

2.3 Endpoints of Apoptosis

2.3.1 Endonuclease Degradation of DNA

The standard biochemical method of characterizing apoptosis in complex tissues is to isolate genomic DNA and subject it to electrophoresis in an agarose gel. Apoptotic DNA resolves into bands with a characteristic 180-bp internucleosomal repeat or ladder (80). This technique has been widely used, including in the study of apoptosis in granulosa cells during follicular development; however, it has so far not been used in preimplantation studies because it would require large numbers of embryos to generate sufficient DNA. A more useful technique that can be applied in situ, the TUNEL assay, is described below.

2.3.1 PARP Cleavage

Another classical marker of apoptosis, cleavage of the DNA repair enzyme PARP (poly (ADP-ribose) polymerase), can be measured biochemically by immunoblotting with an anti-PARP antibody. In general, biochemical assays of apoptosis such as DNA laddering and PARP cleavage are not useful for preimplantation embryo studies because they give no information on spatial distribution. However, antibodies have now become available that recognize cleaved, native PARP and may be useful for in situ immunocytochemical studies on embryos.

Other apoptosis endpoint assays include use of antibodies against cleaved cytokeratin 18 (81), single-stranded DNA (82), cytochrome c (15), and CAD (caspase-activated DNase) (83).

2.3.2 Cellular and Nuclear Morphology

The hallmarks of apoptosis remain the morphological endpoints of cellular and nuclear fragmentation (Figures 19.1, 19.4, 19.5, 19.6). Although a wide range of techniques are now available (see above), nuclear morphology in particular remains the most commonly used and reliable criterion of apoptosis. This has often led to controversy in defining apoptosis, not least during mammalian development (see Perez et al. [78] and van Blerkom and Davis [84] for interesting discussions of this in the oocyte). Cells with apoptotic morphology were first observed in preimplantation embryos by Wilson (32), and Wilson and Smith (33) and further studied by El-Shershaby and Hinchliffe (85) and Copp (86). These observations were made using light and electron microscopy on fixed, sectioned embryos. However, these approaches are difficult and highly labor intensive

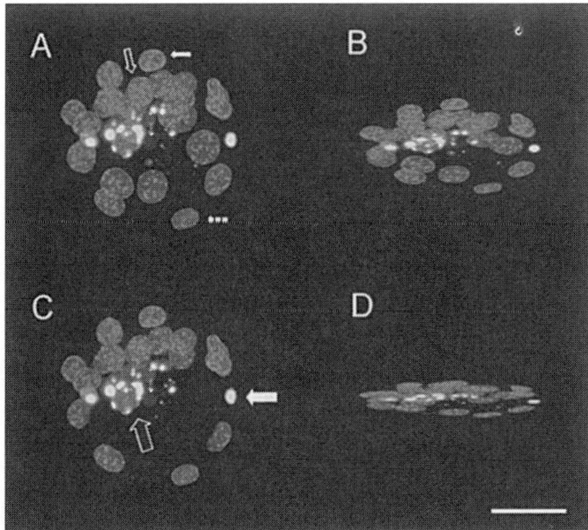
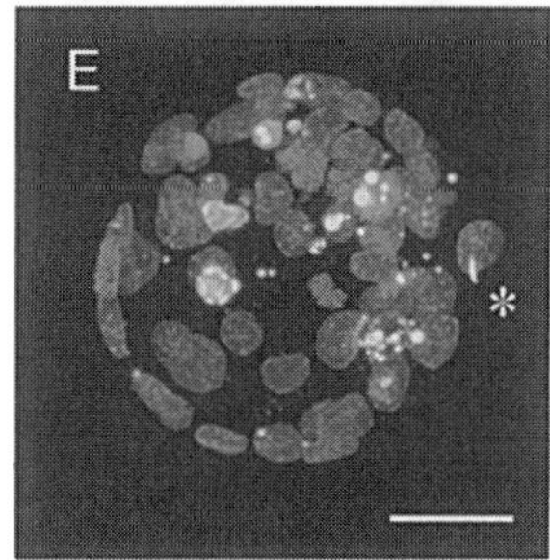

Figure 19.5. Apoptosis in whole mouse blastocysts. Fragmented apoptotic nuclei are labeled with TUNEL (yellow); healthy nuclei are labeled with propidium iodide (red). Computer-generated reconstructions of confocal Z series of the entire embryo show projections from (A) the front, (B) 60°, and (D) 90°, showing the position of apoptotic nuclei in the center of the embryo. (C) The middle sections through the ICM only. In A, the ICM is located at the top left of the blastocyst and ICM nuclei (small black arrow) can be distinguished from polar TE nuclei (small white arrow) and mural TE (small broken arrow). In C, TUNEL-positive nuclei show fragmented (large black arrow) and condensed (large white arrow) nuclear morphology. (E) A larger, expanded blastocyst with many TUNEL-positive, fragmented apoptotic nuclei in the ICM and loose in the blastocoel. The asterisk marks the position of a polar TE nucleus (red) beginning to hatch from the zona pellucida and a single sperm head (yellow). Bars = 40 μm. From Brison and Schultz (52). See color insert.

(85). Use of polynucleotide-specific fluorescent dyes combined with limited immuno-surgical lysis of the TE layer allowed TE to be distinguished from the ICM (30). Hoechst 33258 (bisbenzamide) was used to label the ICM and propidium iodide (PI) to label the TE only. Dead cells were identified by their fragmented nuclear morphology, represented as brightly staining vesicles of chromatin (27). This differential labeling method greatly simplified the quantitative study of cell proliferation, allocation, and death in the blastocyst (27, 30, 36, 49, 62, 87). However, differential labeling can be prone to artifacts because it relies on an intact permeability seal in the TE to exclude PI and conventionally also requires the labeled blastocyst to be disaggregated for cell counting, thus losing precious spatial information. It also requires the zona pellucida to be removed, increasing the risk of introducing artifacts by exposure of embryos to acid tyrodes solution.

2.3.3 Whole Blastocyst Analysis

The advent of confocal microscopy made it possible to overcome the problems discussed above by analyzing the spatial distribution of nuclei in whole, intact blastocysts (as described in the protocol 19.3 for whole mount confocal analysis). With this method, all nuclei in the blastocyst are labeled with a single fluorescent dye (e.g., PI, Hoechst 33258, DAPI). The embryo is scanned in whole mount, and a series of digital images (Z series) is captured of the whole embryo (52). Using computer software this is compiled as a stereo three-dimensional image, which can be analyzed from a number of different angles as a rotating movie, allowing full visualization of all nuclei in the blastocyst (52). Full reconstructions from different angles are shown in figure 19.5 (52). This analysis allows cells to be accurately counted and allocated to ICM, polar TE (pTE), and mural TE (mTE) on the basis of nuclear position, shape, and staining (see protocol 19.3).

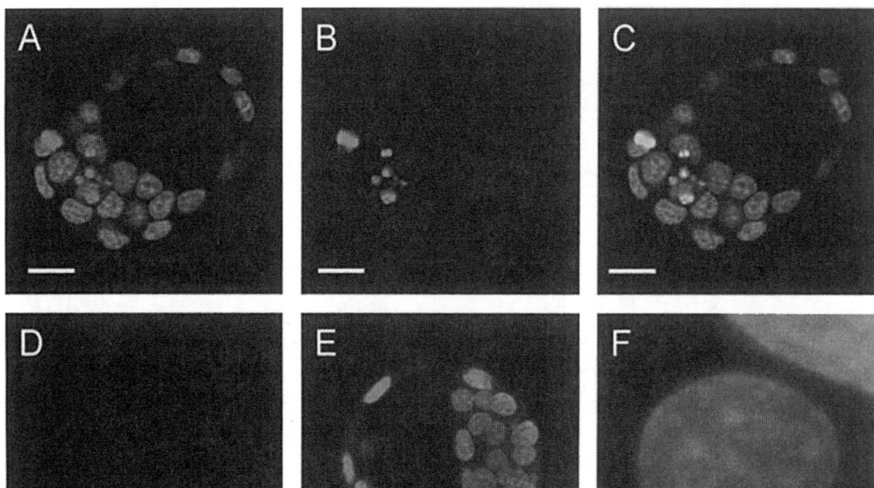

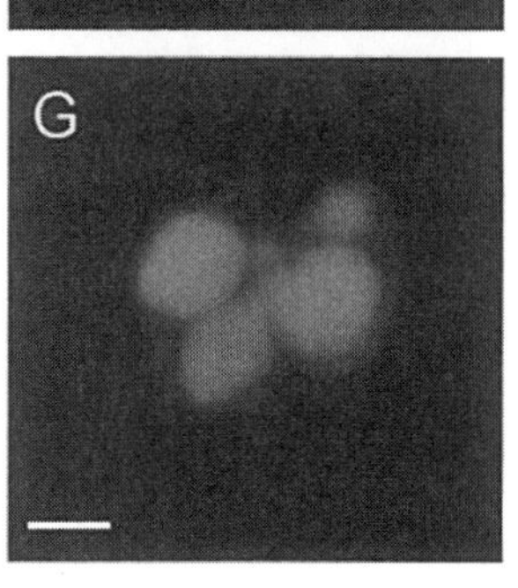

Figure 19.6. Detection of apoptotic nuclei in the blastocyst using TUNEL (see Protocol 19.2). The same blastocyst, labeled with (A) propidium iodide (PI; red) and (B) TUNEL (green), and (C) superimposed (yellow). Comparison between A and B shows that there is little bleedthrough of signal from the red to green channel. Controls are also performed using TUNEL in the absence of PI to demonstrate the converse (not shown). Negative controls lacking terminal transferase demonstrate an absence of nonspecific background labeling (D), and positive controls after DNase treatment show labeling of all nuclei in the blastocyst (E). (F, G) Morphology of healthy and fragmented apoptotic nuclei, respectively. (A–E) Bar = 20 μm (F, G) bars = 2 μm. From Brison and Schultz (49). See color insert.

A further advance was made with the use of TUNEL, which detects DNA strand breaks generated during apoptosis (88, 89). The TUNEL assay relies on the specificity of TdT (terminal deoxynucleotidyl transferase) for free 3'-OH ends of DNA generated by the action of endonucleases (e.g., CAD) activated during apoptosis. This technique is also known as in situ end labeling (ISEL) and is more specific to apoptosis than in situ nick translation (ISNT) assays, which use DNA polymerase. TdT catalyzes the addition of a polydeoxynucleotide to the 3'-OH end, and in the TUNEL assay this is generally dUTP, conjugated to a suitable marker. The marker can be biotin, in which case secondary detection is required to visualize the sites of dUTP addition. In the method described in Protocol 19.2, primary detection was adequately bright using dUTP directly conjugated to fluorescein. Although TUNEL preferentially detects apoptotic DNA, its use has been somewhat controversial because TdT can also label DNA breaks in late-stage necrotic cells and artifactual strand breaks induced by, for example, paraformaldehyde fixation and sectioning. This has made the use of TUNEL problematic in tissues that are complex or require extensive preparation. In a simple system with only a few cell types such as the preimplantation embryo, TUNEL is an extremely useful marker of apoptosis. However, even here it should be allied to a second marker, preferably nuclear morphology. TUNEL was first applied to fragmenting human embryos (17) and mouse embryos (48). We used TUNEL to provide further evidence of apoptosis in the blastocyst and to facilitate quantitation of the apoptotic index (expressed as TUNEL-positive cells per embryo cell number) using the confocal reconstruction method (21, 49, 52, 68).

2.4 Experimental Manipulation of Apoptosis

The levels of apoptosis can be manipulated in an in vitro model in a number of ways. Apoptosis can be most usefully blocked by incubation in caspase inhibitors, of which a number are now available (e.g., zVAD-fmk, DEVD-CHO). Incubation in the protein kinase inhibitor staurosporine will induce apoptosis in most mammalian cells including embryos (48) and can be used as a positive control. More physiological inducers of apoptosis are TNFα (62), high glucose (62, 63), and culture in the absence of survival signals, as single embryos (49, 52, 68) or blastomeres (48). Apoptosis can also be manipulated by culture media composition, (e.g., MF1 and C57Bl6 embryos both show increased apoptosis when cultured to blastocyst in M16 compared to KSOM [68]).

3. DATA ANALYSIS

3.1 TUNEL and Whole Blastocyst Confocal Analysis

The confocal reconstruction method allows blastocysts to be scored in each compartment (ICM, polar TE, and mural TE; see figure 19.5) for total cell (nuclear) number; number of mitotic figures; and number of TUNEL-positive fragmented nuclei, TUNEL-negative fragmented nuclei, and TUNEL-positive nonfragmented nuclei. TUNEL-positive and -negative fragmented nuclei are combined to give the number of apoptotic nuclei; we have rarely observed TUNEL-positive nonfragmented nuclei. Nuclear fragmentation and TUNEL labeling are coincident in the same nuclei with a correlation of approximately 0.9 (i.e., 90% agreement) in mouse and cow blastocysts (21, 49, 52, 68). One group has consistently shown that DNA fragmentation and nuclear fragmentation are rarely coincident in mouse blastocysts exposed to TNFα or high glucose and have suggested that the pathways leading to these two endpoints are regulated differently (62, 83). Nuclear condensation without fragmentation is occasionally seen and may also be considered an apoptotic morphology (figure 19.5A) (54, 83).

The number of nuclei undergoing apoptosis at any one time is most usefully expressed as a percentage of the total number of nuclei in each cell population, identified by PI labeling, for example. This gives an apoptotic index, similar to a mitotic index of cell proliferation. The apoptotic index is a useful parameter of the extent of cell death in an individual blastocyst and can form the basis of quantitative comparisons between populations of blastocysts or different experimental conditions. For example, embryos from TGFα homozygous null mutant (-/-) and wild-type (+/+) mice were cultured from the two-cell stage and fixed at the blastocyst stage (52). Data are presented in Table 19.1 and are discussed below.

Cell number and allocation to ICM versus TE did not vary between wild-type and homozygous null embryos (except for a slight difference in pTE cell numbers). Therefore, both the numbers of apoptotic nuclei and apoptotic indices could be compared between the two groups. These were significantly higher in homozygous null embryos in both mural and polar TE and, most strikingly, in the ICM, where nearly 16% of cells were undergoing apoptosis at any one time in the null embryos. There were also significantly more apoptotic cells loose in the blastocoel in null embryos. A number of caveats must, however, be borne

Table 19.1. Cell Numbers and Apoptotic Indices in TGFα +/+ and -/- Embryos In Vitro.*

		Whole Blastocyst	mTE	pTE	ICM	% ICM	Blastocoel
TGFα +/+	Cell number	61.1 ± 1.5	30.0 ±1.4	9.8[e] ± 0.5	21.5 ± 0.6	36.0	—
(*n* = 58)	No. apoptotic nuclei	2.7 ± 0.3[a]	0.4[c] ±0.1	0.4 ± 0.1	1.7[g] ± 0.2	—	0.2[i] ± 0.1
	Apoptotic index (%)	4.4 ± 0.5[a]	1.3[c] ±0.3	2.9[e] ± 0.7	8.9[g] ± 1.2	—	—
TGFα -/-	Cell number	61.5 ± 2.2	33.3 ±2.3	8.3[f] ± 0.5	19.7 ± 0.9	34.6	—
(*n* = 48)	No. apoptotic nuclei	5.9 ± 0.5[b]	1.3[d] ±0.2	0.9 ± 0.2	2.7[h] ± 0.3	—	1.0[j] ± 0.2
	Apoptotic index (%)	9.9 ± 0.1[b]	4.0[d] ±0.7	8.3[f] ± 1.5	15.8[h] ± 2.1	—	—

*Data are means ± SEM.
Different superscripts a–j in the same columns are significantly different between TGFα +/+ and -/- embryos, all at $\leq p < .01$. If the cell number differed between the two populations of embryos (e.g., pTE), then the apoptotic index but not the number of nuclei was compared. Data from Brison and Schultz (52).

in mind with these quantitative analyses. First, although expressing apoptosis as an index normalizes it to cell number, in order to make a valid comparison of apoptotic indices, the populations of blastocysts must have similar cell numbers (as in the TGFα +/+ and -/- example in table 19.1). A number of studies have shown that apoptosis varies with cell number, increasing as the blastocyst expands (27), then declining with increasing cell number before implantation (21, 27, 36). If populations of blastocysts are not matched for total cell number, a specific effect on apoptosis may be masked by an effect on the rate of development or cell proliferation. In cases where populations of blastocysts to be compared have different cell numbers, analysis of apoptotic indices must be stratified by cell number (68). Second, apoptosis endpoint assays represent only a snapshot in time. Thus, apoptotic indices in the blastocyst may appear to be low, but if phagocytosis and clearance of the corpse occurs very quickly, this could represent a high rate of ongoing cell death. The rate of clearance is not known for preimplantation embryos, and thus a major assumption when comparing apoptosis indices in different conditions is that the rate of clearance remains the same. In some blastocysts, large apoptotic nuclei can be observed floating in the blastocoel (e.g., Table 19.1 and Figure 19.5C; human blastocysts; H. R. Hunter, A. D. Metcalfe, S. J. Kimber, D. R. Brison, unpublished observations). These could be apoptotic early-cleavage blastomeres, which were not phagocytosed because they did not express appropriate cell-surface markers (17).

4. TROUBLESHOOTING AND COMMON PROBLEMS

4.1 PolyAPCR on Single Embryos

Common problems with polyAPCR include failure of lysis of the embryo and inconsistency in results between individual embryos. These problems can be minimized by taking the following steps:

- To avoid nucleic acid contamination, UV sterilize all PCR tubes (certified RNase/DNase-free) before beginning the procedure.
- Ensure that the embryo is transferred in a minimum of media (0.5 µl).
- Using a dissecting microscope after transfer, check that the embryo is in the tube undergoing lysis.
- The most common reason for failure to amplify has been traced to the batch of oligomer used. Once established, always compare new oligomers to proven existing oligomers.

- Purchase reasonable quantities of oligo dT24 primer so that the optimization step has to be done only at the outset.
- Once the buffers have been optimized, make large batches of the lysis and tailing buffers (minus enzymes and primer), aliquot, and store at $-80°C$.

On occasions, the RT negative control sometimes contains a DNA smear. This is believed to be due to concatamerization of primer amplicons accumulated during the amplification process or amplification of nucleic acids present in the Taq enzyme. On probing this apparent cDNA smear using Southern hybridization techniques or PCR, there should not be any detectable hybrids or PCR products. If there is a gene product or hybrid in the RT-negative sample, then the entire sample set created at the same time must be discarded.

4.2 TUNEL and Confocal Analysis

The use of TUNEL in embryos has proved consistently reliable in a number of laboratories. Most of the problems, namely poor permeabilization of the embryo and nonspecific labeling, will be familiar to those experienced with immunofluorescence techniques.

Positive and negative controls should be run with every assay (figure 19.6D, E) and, as with immunocytochemistry, are particularly important when labeling zona-intact embryos. Although the zona can easily be permeabilized using Triton to allow passage of reagents, its physical presence may create a microenvironment. This could potentially restrict access of reagents such as TdT to inner cells or result in persistence of unbound dUTP as background labeling. The positive control using DNase acts as a permeabilization control, while the negative control lacking TdT establishes that washing procedures are adequate.

Bleedthrough of signal between channels must be controlled for when establishing the assay. This is done by including embryos with PI only and TUNEL only and ensuring that a strong signal from each is seen only in the correct channel.

Occasionally embryos may be poorly fixed, with PI-labelled nuclei difficult to distinguish. This may require a fresh batch of paraformaldehyde.

With an optimized protocol, counterstained nuclei should be clearly delineated and easily distinguishable, and TUNEL-positive nuclei should be brightly labeled. One common problem concerns labeling of small fragments of DNA arising from fragmented nuclei. As with differential labeling of blastocysts (see above), small clusters of fragmented DNA in close proximity can be scored as a single apoptotic nucleus, while small, widely dispersed fragments should be ignored. As long as the analysis is performed blinded, this is unlikely to bias the results. In any case, with whole-mount analysis, these fragments tend to remain closely grouped rather than dispersed as when the embryo is spread. Alternatively, quantitative image analysis could be performed in an attempt to estimate the total signal arising from TUNEL. This, however, would negate some of the advantages of the in situ whole-mount analysis.

Polar bodies are often TUNEL positive, and this must be borne in mind when analyzing cleavage stage embryos because this is often the sole source of TUNEL signal.

5. CONCLUSION

There are a number of techniques available for the analysis of apoptosis in preimplantation embryos, but the assessment of nuclear morphology remains the most reliable single criterion. Future research will focus on the analysis of regulatory pathways and on the biological consequences of failure in this regulation. The preimplantation mouse embryo should prove to be an ideal model system for study of spatial and temporal aspects of apoptosis regulation during development.

Acknowledgments D.R.B. thanks Richard Schultz, in whose laboratory this work was started, and colleagues in the Department of Reproductive Medicine, St. Mary's Hospital. The work of the authors was supported by grants from the NIH (F32 HD 07957), the UK MRC, the Elstein fellowship, the NHS northwest region R&D fund, and the NHS R&D levy.

Protocol 19.1. PolyAPCR on single embryos

All lysis, tailing, and subsequent primary amplification are performed in the same tube (see Brady et al. [73, 74]).

Embryo lysis and cDNA amplification

1. Transfer a single embryo in a minimum volume of culture medium to 4.5 µl complete lysis buffer.
2. Add mineral oil overlay and heat to 65°C for 1 min before cooling to room temperature for 3 min. Add 25 µl of reverse transcriptase (Superscript RNAseH-, Life Technologies), incubate at 37°C for 15 min, heat inactivate at 65°C for 10 min, then cool on ice.
3. Add 1 volume of tailing buffer (includes TdT enzyme) and incubate at 37°C for 15 min, heat inactivate at 65°C for 10 min, then cool on ice.
4. Carry out PCR amplification of the resultant poly A tailed cDNA by adding 2 volumes of primary PCR reaction mix in the presence of a Not1dT$_{24}$ oligonucleotide primer (CATCTCGAGCGGCCGCTTTTTTTTTTTTTTTTTTTTTTTT).
5. Primary amplification: initial denaturation at 94°C for 2 min, 25 cycles of 1 min at 94°C, 2 min at 42°C, 6 min at 72°C, linked to a further 25 cycles of 1 min at 94°C, 1 min at 42°C, 2 min at 72°C. The optimal MgCl$_2$ concentration is determined empirically for every oligonucleotide primer preparation used.
6. Secondary amplification: 1 µl of primary PCR reaction is used as template in a 50-µl final reaction volume of secondary PCR mix; initial denaturation at 94°C for 2 min and 50 cycles of 30 s at 94°C, 30 s at 54°C, and 30 s at 72°C. Both primary and secondary amplifications should be subjected to electrophoresis and the resultant 500–800 bp cDNA smears visualized.
7. Controls: Negative controls should be used. Lyse embryos and subject them to the amplification protocol without reverse transcriptase to confirm the absence of contaminating genomic DNA. Negative controls without embryo material should also be processed in tandem with test samples to confirm absence of nucleic acid contamination. *Note.* Sperm and cumulus cells should be processed as controls in case of contamination.

Specific gene amplification protocol

For both test gene(s) used as internal standards and specific gene amplification, the following protocols are applied:

1. All gene-specific primers were designed to amplify target sequences within 3–500 bp of the polyadenylation tail using PRIMER version 0.5 (Whitehead Institute for Biomedical Research).
2. Use serial dilutions of secondary amplification products as templates in a standard PCR reaction to amplify β actin. For analysis of test genes, samples are then normalized relative to the expression of β actin. A general PCR profile would typically be an initial denaturation at 94°C for 1 min followed by 17 cycles (β actin) or up to 40 cycles (rarer test genes) of 30 s at 94°C, 30 s at the annealing temperature (established for each primer pair used), and 30 s at 72°C. *Note.* To assess overall global cDNA yields and the extent of sample to sample variation, it is useful to compare the level of a test gene to that of an abundant housekeeping gene. For the analysis of preimplantation embryos, we have typically used

β actin as the internal control. This methodology is sufficient for assessing relative levels of expression at different stages of development. If quantitative methodology is required, then gene-specific PCR can be used, keeping in mind that 1–50 ng of the globally amplified cDNA is necessary to produce a product.

3. After gene-specific PCR 10 µl of the amplification products are visualized after electrophoresis on a 2% agarose gel stained with ethidium bromide.

4. Partially sequence amplification products to verify identity using ABI Big dye technology.

Solutions for polyAPCR

Buffer	Components (optimal buffer concentrations determined empirically)
Lysis buffer	50 mM Tris.HCl, pH 8.3, 75 mM KCl, 3 mM MgCl$_2$, 20 mM DTT, 40 µg/ml BSA, 0.5% NP40, 9.7 pM dNTPs, 23 pM oligo dT$_{24}$, RNase inhibitor 0.8 U/µl (Roche Biochemicals). Equivalent to 1.04× First Strand Buffer (Life Technologies) + additional components
Tailing buffer	0.33 M potassium cacodylate, 6.7 mM CoCl$_2$, 0.67 mM DTT, 0.16 mM dATP, 0.45 µl rTdT (Life Technologies). Equivalent to 2.5× TdT Buffer (Life Technologies) + additional components
Primary PCR reaction mix	22 mM Tris.HCl, pH 8.3, 6.3 mM MgCl$_2$, 110 mM KCl, 2 mM dNTPs, 0.04 mg/ml BSA, 0.2% Triton X-100, 8.5 µM Not1dT$_{24}$ oligonucleotide primer, 0.16 U/µl Taq polymerase (Roche Biochemicals). Equivalent to 2.6× Taq Polymerase Buffer (Roche) + additional components
Secondary PCR reaction mix	10 mM Tris.HCl, pH 8.3, 1.5 mM MgCl$_2$, 50 mM KCl, 0.2 mM dNTPs, 2 µM Not1dT$_{24}$ oligonucleotide, 0.025 U/µl Taq polymerase. Equivalent to 1× Taq Polymerase Buffer (Roche) + additional components
Standard PCR reaction mix	10 mM Tris.HCl, pH 8.3, 1.5 mM MgCl$_2$, 50 mM KCl, 0.2 mM dNTPs, 1 µM forward primer, 1 µM reverse primer, 0.025 U/µl Taq polymerase. Equivalent to 1× Taq Polymerase Buffer (Roche) + additional components

Protocol 19.2. Blastocyst TUNEL

All steps of the procedure can be carried out in round-bottomed Nunc 96-well plates (Life Technologies), using one well for each group of embryos. This is particularly useful when simultaneously processing several experimental groups of embryos. Drawn Pasteur pipettes with a 45° bend are useful for handling embryos in these dishes. Propidium iodide is a DNA intercalating agent and a known mutagen, and the TUNEL (Tdt dUTP Nick End Labelling) buffer contains cacodylate, which is hazardous; appropriate precautions should be taken when using these reagents, including wearing gloves.

1. Remove embryos from culture drops and wash in PBS/PVP four times.

2. Fixation: incubate in 3.7% paraformaldehyde/PBS (freshly prepared) for 1–2 h at room temperature (RT) or overnight at 4°C. *Note*: DNA labeling is not particularly sensitive to fixation conditions. Paraformaldehyde can induce DNA strand breaks in some tissues, but in mouse embryos there was no difference in incidence of TUNEL labeling in embryos fixed at 1 h versus overnight (D. R. Brison and R. M. Schultz, unpublished data). However, this variable should still be controlled between experimental groups. It is not necessary to remove the zona for this protocol.

3. Wash in PBS/PVP two times.

4. Permeabilization: incubate in 0.5% Triton X-100/PBS for 1 h at RT. *Note*: Lower concentrations (0.1%) of Triton can be used for earlier embryos because there are fewer cell membranes to be permeabilized.

5. Wash in PBS/PVP two times.

6. Positive control: at this point, incubate a few embryos in DNase for 20 min at 37°C and wash (PBS/PVP 2×) before proceeding with TUNEL labeling as below (figure 19.6E, F). This acts as a control against inadequate permeabilization of the embryo. Great care must be taken not to contaminate experimental embryos or stock reagents with DNase.

7. Preincubate in 10–15 μl dUTP-FITC labeling mix 10 min at RT (optional). *Note.* The TUNEL buffer should be thawed on ice in the dark and refrozen.

8. TUNEL labeling: Incubate in 10–15 μl of TUNEL mix 1 h/ 37°C in the dark. *Note*: TdT is kept on ice, added in 1:9 ratio to TUNEL labeling mix.

9. Negative control: at this point, a few embryos are incubated in labeling mix minus TdT instead of TUNEL (figure 19.6D). This acts as a control against nonspecific background labeling. Then proceed with the rest of the protocol.

10. Wash with (0.5% Triton/PBS two times then with PBS/PVP once.

11. Nuclear counterstain (PI): wash in 1× RNase buffer. *Note*: The RNase treatment is necessary to remove cytoplasmic RNA which might be labelled by PI, obscuring nuclear labelling. The choice of nuclear counterstain is determined by the fluorescence filters or confocal lasers available. DAPI or Hoechst can also be used, with the advantage that they do not label cytoplasmic RNA and therefore RNase treatment is unnecessary.

12. Incubate in PI/RNase mix for 1 h at RT in the dark.

13. Wash with 0.5% Triton/PBS two times, then with PBS/PVP once.

14. Analysis for conventional fluorescence microscopy, embryos can be analyzed immediately in drops of medium in plastic dishes or on glass slides, either in whole mount or by disaggregating the embryo. For longer term storage and analysis by confocal microscopy, embryos should be mounted as described in the Protocol 19.3.

Protocol 19.3. Whole-mount confocal analysis

Fixed, permeabilized embryos can be mounted for long-term storage and analysis in whole mount by confocal microscopy.

1. Wash embryos in groups of up to 20–30 through a dilution series of 25%–50%–75%–100% Vectashield in PBS (10–15 min each step). *Note*: Vectashield is glycerol based and contains a fluorescence antifade agent. This nonaqueous environment minimizes free radical generation and therefore reduces the rate at which fluorescence fades. This allows the embryos to be stored for 2 weeks or more before analysis. However, blastocysts placed directly into Vectashield will collapse; gradual substitution of medium minimizes osmotic shock so that the blastocyst will reexpand. Before mounting, the blastocyst must be completely expanded with no folds in the TE for accurate confocal analysis of cell allocation. Great care must be taken with embryo handling because the embryos lose all refractivity in glycerol and become nearly invisible; in addition, they will float to the top of each dilution of Vectashield. When they have settled to the bottom of the drop, they can be transferred to the next dilution.

2. Mount in 20–40 μl Vectashield. At least two slides should be prepared for each experimental group.

3. Make vaseline posts at corners of coverslip; gently tap down so blastocysts are trapped but not collapsed. *Note*: It is important that the blastocyst is not distorted by too much pressure, but it must be trapped to minimize movement during confocal scanning.

4. Seal edges of coverslip with nail varnish.

5. Label slides and cover with tape to allow experimental groups to be analyzed blinded.

6. Store foil wrapped at 4°C. Slides should be read within 1 week but are readable for 3–4 weeks.

7. Confocal analysis: a complete Z series should be collected for each embryo, consisting of 20–60 optical sections at intervals of a maximum of 3 μm, to ensure that each nucleus in the blastocyst is sampled. A wide range of conventional confocal microscopes or multiphoton imagers can be used for this analysis. We have successfully used the Leica TCS 4D, MRC Bio-Rad 500, and a Bio-Rad 1024MP multiphoton imager with SpectraPhysics laser. Image processing can be carried out using Leica TCS software, Confocal Assistant, or similar.

8. The different compartments of the blastocyst (ICM, pTE, and mTE) can be distinguished on the basis of nuclear position, shape, and staining in blastocysts with intact morphology. ICM nuclei are positioned centrally and are large, rounded, and stain lightly with PI. TE nuclei are smaller, flattened, stain more intensely, and are arranged around the outside of the blastocyst in a single layer. pTE nuclei are defined as those in contact with an ICM nucleus; mTE nuclei are those positioned at least one cell away from the nearest ICM nucleus (see figure 19.5).

9. For each compartment, total nuclei, mitotic figures, and apoptotic nuclei can be counted. Apoptotic nuclei are identified on the basis of TUNEL and nuclear morphology; mitotic figures are a marker of cell proliferation.

10. After image capture, the slides can be carefully disassembled and the embryos can be individually disaggregated to perform manual nuclear counts by fluorescence microscopy.

Solutions for blastocyst TUNEL and confocal analysis

Solution	Components
TUNEL	In situ cell death detection kit (Roche Biochemicals)
PBS/PVP	PBS (pH 7.4) + 3 mg/ml PVP
DNase	2 μl RNase-free RQ1 DNase (Promega) in 40 μl PBS/PVP
RNase buffer	40 mM TRIS (pH 8.0), 10 mM NaCl, 6 mM MgCl$_2$
PI/RNase mix	1:1 RNase A (100 μg/ml) and 2× RNase buffer (above) + 1:20 of 1 mg/ml PI (equivalent to 0.05 mg/ml PI/50 μg/ml RNase A)
Vectashield	Vector Laboratories

References

1. Kerr, J. F., Wyllie, A. H., and Currie, A. R. (1972). *Br. J. Cancer*, **26**, 239.
2. Wyllie, A. H., Kerr, J. F., and Currie, A. R. (1980). *Int. Rev. Cytol.*, **68**, 251.
3. Raff, M. C. (1996). *Cell*, **86**, 173.
4. Sanders, E. J., and Wride, M. A. (1995). *Int. Rev. Cytol.*, **163**, 105.
5. Jacobson, M. D., Weil, M., and Raff, M. C. (1997). *Cell*, **88**, 347.
6. Meier, P., Finch, A., and Evan, G. (2000). *Nature*, **407**, 796.
7. Lockshin, R. A., and Williams, C. M. (1964). *J. Insect Physiol.*, **10**, 643.
8. Vaux, D. L., and Korsmeyer, S. J. (1999). *Cell*, **96**, 245.
9. Perez, G. I., Knudson, C. M., Leykin, L., Korsmeyer, S. J., and Tilly, J. L. (1997). *Nat. Med.*, **3**, 1228.
10. Woolveridge, I., and Morris, I. D. (1999). In: *Apoptosis in Toxicology*, R. Roberts, ed. Taylor Francis, London.
11. Metcalfe, A., and Streuli, C. (1997). *Bioessays*, **19**, 711.
12. Coucouvanis, E., and Martin, G. R. (1995). *Cell*, **83**, 279.
13. Jurisicova, A., Latham, K. E., Casper, R. F., and Varmuza, S. L. (1998). *Mol. Reprod. Dev.*, **51**, 243.
14. Liu, L., Trimarchi, J. R., and Keefe, D. L. (1999). *Biol. Reprod.*, **61**, 1162.
15. Liu, L., and Keefe, D. L. (2000). *Biol. Reprod.*, **62**, 1828.
16. Jurisicova, A., Varmuza, S., and Casper, R. F. (1995). *Hum. Reprod. Update*, **1**, 558.
17. Jurisicova, A., Varmuza, S., and Casper, R. F. (1996). *Mol. Hum. Reprod.*, **2**, 93.

18. Levy, R., Benchaib, M., Cordonier, H., Souchier, C., and Guerin, J. F. (1998). *Mol. Hum. Reprod.*, **4**, 775.
19. Yang, H. W., Hwang, K. J., Kwon, H. C., Kim, H. S., Choi, K. W., and Oh, K. S. (1998). *Hum. Reprod.*, **13**, 998.
20. Hardy, K. (1999). *Rev. Reprod.*, **4**, 125.
21. Byrne, A. T., Southgate, J., Brison, D. R., and Leese, H. J. (1999). *J. Reprod. Fertil.*, **117**, 97.
22. Stack, J. H., and Newport, J. W. (1997). *Development*, **124**, 3185.
23. Brison, D. R. (2000). *Hum. Fertil.*, **3**, 36.
24. Hardy, K. (1997). *Mol. Hum. Reprod.*, **3**, 919.
25. Pampfer, S., and Donnay, I. (1999). *Cell Death Differ.*, **6**, 533.
26. Kumazawa, T., Inouye, M., Hayasaka, I., Yamamura, H., and Murata, Y. (1998). *Teratology*, **57**, 146.
27. Handyside, A. H., and Hunter, S. (1986). *Roux's Arch. Dev. Biol.*, **195**, 519.
28. Pierce, G. B., Lewellyn, A. L., and Parchment, R. E. (1989). *Proc. Natl. Acad. Sci. USA*, **86**, 3654.
29. Parchment, R. E. (1993). *Int. J. Dev. Biol.*, **37**, 75.
30. Handyside, A. H., and Hunter, S. (1984). *J. Exp. Zool.*, **231**, 429.
31. Lewis, N. E., and Rossant, J. (1982). *J. Embryol. Exp. Morphol.*, **72**, 169.
32. Wilson, I. B. (1963). *J. Reprod. Fertil.*, **5**, 281.
33. Wilson, I. B., and Smith, M. S. (1970). In: *Ovo-implantation*, P. O. Hubinont, F. Leroy, C. Robyn, and P. Lelux, eds., p. 1. Karger, Basel.
34. Hurst, P. R., Jefferies, K., Eckstein, P., and Wheeler, A. G. (1978). *J. Anat.*, **126**, 209.
35. Papaioannou, V. E., and Ebert, K. M. (1988). *Development*, **102**, 793.
36. Hardy, K., Handyside, A. H., and Winston, R. M. (1989). *Development*, **107**, 597.
37. Enders, A. C., Lantz, K. C., and Schlafke, S. (1990). *Anat. Rec.*, **226**, 237.
38. Pampfer, S., De Hertogh, R., Vanderheyden, I., Michiels, B., and Vercheval, M. (1990). *Diabetes*, **39**, 471.
39. Tam, P. P. (1988). *Teratology*, **37**, 205.
40. Wuu, Y. D., Pampfer, S., Becquet, P., Vanderheyden, I., Lee, K. H., and De Hertogh, R. (1999). *Biol. Reprod.*, **60**, 479.
41. Lane, M., and Gardner, D. K. (1997). *J. Reprod. Fertil.*, **109**, 153.
42. Van Soom, A., Ysebaert, M. T., and de Kruif, A. (1997). *Mol. Reprod. Dev.*, **47**, 47.
43. Exley, G. E., Tang, C., McElhinny, A. S., and Warner, C. M. (1999). *Biol. Reprod.*, **61**, 231.
44. Antczak, M., and Van Blerkom, J. (1999). *Hum. Reprod.*, **14**, 429.
45. Liu, H. C., He, Z. Y., Mele, C. A., Veeck, L. L., Davis, O., and Rosenwaks, Z. (2000). *J. Assist. Reprod Genet.*, **17**, 521.
46. Metcalfe, A., Hunter, H., Bloor, D., Rutherford, A. J., Kimber, S. and Brison, D. R. (2000). *J. Reprod. Fertil. Suppl.*, **25**, 21.
47. Raff, M. C., Barres, B. A., Burne, J. F., Coles, H. S., Ishizaki, Y., and Jacobson, M. D. (1993). *Science*, **262**, 695.
48. Weil, M., Jacobson, M. D., Coles, H. S., Davies, T. J., Gardner, R. L., Raff, K. D., and Raff, M. C. (1996). *J. Cell. Biol.*, **133**, 1053.
49. Brison, D. R., and Schultz, R. M. (1997). *Biol. Reprod.*, **56**, 1088.
50. O'Neill, C. (1998). *Biol. Reprod.*, **58**, 1303.
51. Threadgill, D. W., Dlugosz, A. A., Hansen, L. A., Tennenbaum, T., Lichti, U., Yee, D., LaMantia, C., Mourton, T., Herrup, K., and Harris, R. C. (1995). *Science*, **269**, 230.
52. Brison, D. R., and Schultz, R. M. (1998). *Biol Reprod*, **59**, 136.
53. Herrler, A., Krusche, C. A., and Beier, H. M. (1998). *Biol. Reprod.*, **59**, 1302.
54. Spanos, S., Becker, D. L., Winston, R. M., and Hardy, K. (2000). *Biol. Reprod.*, **63**, 1413.
55. Byrne, A. T., Southgate, J., Brison, D. R., and Leese, H. J. (2002). *Mol. Reprod. Dev.*, **62**, 489.
56. Chi, M. M., Schlein, A. L., and Moley, K. H. (2000). *Endocrinology*, **141**, 4784.
57. Miller, L. K. (1999). *Trends Cell Biol.*, **9**, 323.
58. Armstrong, J. F., Kaufman, M. H., Harrison, D. J., and Clarke, A. R. (1995). *Curr. Biol.*, **5**, 931.
59. Heyer, B. S., MacAuley, A., Behrendtsen, O., and Werb, Z. (2000). *Genes Dev.*, **14**, 2072.
60. Haines, G., Brison, D. R., Hendry, J. H. D. P., and Morris, I. D. (1998). *Hum. Reprod.*, **13** (abstr. book 1), 210.

61. Pampfer, S., Vanderheyden, I., McCracken, J. E., Vesela, J., and De Hertogh, R. (1997). *Development*, **124**, 4827.
62. Pampfer, S., Vanderheyden, I., McCracken, J. E., Vesela, J., and De Hertogh, R. (1997). *Development*, **124**, 4827.
63. Moley, K. H., Chi, M. M., Knudson, C. M., Korsmeyer, S. J., and Mueckler, M. M. (1998). *Nat. Med.*, **4**, 1421.
64. Chi, M. M., Pingsterhaus, J., Carayannopoulos, M., and Moley, K. H. (2000). *J. Biol. Chem.*, **275**, 40252.
65. Moley, K. H. (1999). *Semin. Reprod. Endocrinol.*, **17**, 137.
66. Pampfer, S. (2000). *Placenta*, **21** (Suppl A), S3.
67. Devreker, F., and Hardy, K. (1997). *Biol. Reprod.*, **57**, 921.
68. Kamjoo, M., Brison, D. R., and Kimber, S. J. (2002). *Mol. Reprod. Dev.* **61**, 67.
69. Jurisicova, A., Rogers, I., Fasciani, A., Casper, R. F., and Varmuza, S. (1998). *Mol. Hum. Reprod.*, **4**, 139.
70. Butt, A. J., Firth, S. M., and Baxter, R. C. (1999). *Immunol. Cell Biol.*, **77**, 256.
71. McCarthy, N. J., and Evan, G. I. (1998). *Curr. Top. Dev. Biol.*, **36**, 259.
72. Brison, D. R., and Schultz, R. M. (1996). *Mol. Reprod. Dev.*, **44**, 171.
73. Brady, G., Barbara, M., and Iscove, N. N. (1990). *Meth. Mol. Cell. Biol.*, **2**, 17.
74. Brady, G., and Iscove, N. N. (1993). *Meth. Enzymol.*, **225**, 611.
75. Warner, C. M., Cao, W., Exley, G. E., McElhinny, A. S., Alikani, M., Cohen, J., Scott, R. T., and Brenner, C. A. (1998). *Hum. Reprod.*, **13** (Suppl 3), 178.
76. Metcalfe, A. D., Bloor, D. J., Lieberman, B. A., Kimber, S. J., and Brison, D. R. (2003). *Mol. Reprod. Dev.* **65**, 1.
77. Pampfer, S., Cordi, S., Vanderheyden, I., Van Der, S. P., Courtoy, P. J., Van Cauwenberge, A., Alexandre, H., Donnay, I., and De Hertogh, R. (2001). *Diabetes*, **50**, 143.
78. Perez, G. I., Tao, X. J., and Tilly, J. L. (1999). *Mol. Hum. Reprod*, **5**, 414.
79. Van Blerkom, J., and Davis, P. W. (1998). *Hum. Reprod.*, **13**, 1317.
80. Kaufmann, S. H., Mesner, P. W., Samejima, K., Tone, S., and Earnshaw, W. C. (2000). *Meth. Enzymol.*, **322**, 3.
81. Caulin, C., Salvesen, G. S., and Oshima, R. G. (1997). *J. Cell. Biol.*, **138**, 1379.
82. Kobayashi, S., Iwase, H., Kawarada, Y., Miura, N., Sugiyama, T., Iwata, H., Hara, Y., Omoto, Y., and Nakamura, T. (1998). *Breast Cancer*, **5**, 47.
83. Hinck, L., Van Der, S. P., Heusterpreute, M., Donnay, I., De Hertogh, R., and Pampfer, S. (2001). *Biol. Reprod.*, **64**, 555.
84. Van Blerkom, J., and Davis, P. W. (1998). *Hum. Reprod.*, **13**, 1317.
85. El-Shershaby, A. M., and Hinchliffe, J. R. (1974). *J. Embryol. Exp. Morphol.*, **31**, 643.
86. Copp, A. J. (1978). *J. Embryol. Exp. Morphol.*, **48**, 109.
87. Hardy, K., and Handyside, A. H. (1996). *Mol. Reprod. Dev.*, **43**, 313.
88. Gorczyca, W., Gong, J., and Darzynkiewicz, Z. (1993). *Cancer Res.*, **53**, 1945.
89. Gold, R., Schmied, M., Giegerich, G., Breitschopf, H., Hartung, H. P., Toyka, K. V., and Lassmann, H. (1994). *Lab. Invest.*, **71**, 219.
90. Ke, N., Godzik, A., and Reed, J. C. (2001). *J. Biol. Chem.* **276**, 12481.
91. Crompton, M. (2000). *Curr. Opin. Cell Biol*, **12**, 414.
92. Hengartner, M. O. (2000). *Nature*, **407**, 770.
93. Downward, J. (1999). *Nat. Cell Biol.*, **1**, E33.
94. Vousden, K. H. (2000). *Cell*, **103**, 691.
95. Earnshaw, W. C., Martins, L. M., and Kaufmann, S. H. (1999). *Annu. Rev. Biochem.*, **68**, 383.
96. Ashkenazi, A., and Dixit, V. M. (1999). *Curr. Opin. Cell Biol.*, **11**, 255.
97. Temeles, G. L., Ram, P. T., Rothstein, J. L., and Schultz, R. M. (1994). *Mol. Reprod. Dev.*, **37**, 121.
98. Adjaye, J., Daniels, R., Bolton, V., and Monk, M. (1997). *Genomics*, **46**, 337.
99. Rambhatla, L., Patel, B., Dhanasekaran, N., and Latham, K. E. (1995). *Mol. Reprod Dev.*, **41**, 314.
100. Wang, Q., and Latham, K. E. (2000). *Biol. Reprod.*, **62**, 969.
101. Brady, G. (2000). *Yeast*, **17**, 211.

20

D. P. BALUCH
C. M. PAUKEN
D. G. CAPCO

Cytoplasmic Signaling and Cell Cycle Control in the Mouse Egg and Embryo

1. ANALYSIS OF CYTOPLASMIC SIGNALING IN THE EARLY EMBRYO

Eggs have often been used for the analysis of cytoplasmic signal transduction events as well as the analysis of cell cycle events. This is because they have a distinct cell-cycle arrest point as they await fertilization by sperm. As a result of fertilization, there is a rapid transformation in the structure of the egg such that it becomes the zygote. These structural changes are often triggered by activation of cytoplasmic signaling pathways such as a calcium wave and/or propagation of a wave of kinase activity through the egg cytoplasm.

Fertilization can be induced in eggs relatively synchronously because all of the eggs are at the same, natural cell-cycle arrest point. However, at developmental stages after fertilization, the events become increasingly asynchronous as natural variations in cell-cycle progression accumulate within blastomeres of an embryo and between embryos. Preimplantation embryos can be selected as they cross from one major developmental state into another. For example, embryos that have just cleaved to form two-, four-, or eight-cell embryos can be grouped together for selective developmental study. However, such embryos do not have natural cell-cycle arrest points, and any manipulations that would halt the cell cycle progression at a particular stage would impede the developmental processes and damage the embryo.

Under these circumstances the best synchrony that can be obtained is through a combination of embryo selection as they pass from one cleavage state to another and through monitoring the time after cytokinesis to indicate when the embryos are at various stages of the cell cycle. At any of these points, the organization of DNA can be determined by staining with a membrane-permeable DNA dye, but the DNA staining procedure is best used immediately before the embryos are processed for experimental analysis because the dye could influence the progression of developmental events. Such a manipulation can be used if the researcher intends to collect embryos that have entered M phase (e.g., at the four-cell stage). The DNA dye would clearly indicate which embryos had chromosomes present and the investigator could hand select these embryos.

We have developed a number of techniques to analyze the action of different cytoplasmic signal transducers in the egg, zygote, and preimplantation embryo, and these procedures are presented in this chapter. A large number of cell cycle studies have been conducted on mammalian eggs arrested at meiotic metaphase II, in part because resumption of the cell cycle can be triggered synchronously by inducing a rise in intracellular free calcium.

Historically, many researchers investigating fertilization have focused on the well-known calcium signal that activates the egg, attempting to define how this signal is initiated by the penetrating sperm. A number of ideas have been put forward to explain how the calcium signal originates, such as GTP binding proteins (1), receptor–ligand interactions (2), and factors released by the penetrating sperm (3–5). Over time some of these hypotheses have been abandoned, but others have received increased interest. Although the calcium signal is the accepted initiator of egg activation events, the knowledge that calcium acts at the top of a hierarchy of signaling events does not explain the mechanistic basis of how the egg is converted into the zygote, nor does it define the subsequent signaling events that regulate the conversion of the egg to the zygote after sperm penetration.

Our own investigations have taken a quite different approach. We have sought to determine the mechanisms that lie downstream of the calcium signal and how are they regulated. We have successfully analyzed the regulation of several mechanisms that lie downstream of fertilization and the accompanying calcium signal and have provided evidence that these mechanisms are conserved among vertebrates that have the same cell-cycle arrest points (e.g., between the amphibian, *Xenopus*, and mammals). Our research on mammalian development has examined functions for protein kinase C (PKC) and calcium/calmodulin-dependent protein kinase II (CaM kinase II) that act downstream of the fertilization-induced calcium signal (6–9) and the functions of PKC during embryonic compaction (9), and has included a number of structural studies of mammalian fertilization and embryonic compaction (10–14). We have shown that these cytoplasmic signaling agents function within the three-dimensional space of the egg over time to remodel specific components within the egg that are essential for proper cell cycle resumption.

There is general agreement that the rise in intracellular free calcium inactivates MPF (maturation promoting factor) by targeting cyclin B1 for degradation through a ubiquitin-dependent pathway (5, 16, 17). PKC acts downstream of the calcium signal, and there have been several studies demonstrating this activation at a biochemical or immunocytochemical level in mouse and rat eggs (6, 18–21), although the specific isotypes of PKC that act appear to differ depending on the study and species. Once activated, PKC appears to have several roles. Some studies have shown that PKC is involved with the initial formation of the second polar body (22–24). Second polar body formation is a two-step process with initiation occurring first (i.e., initial out-pocketing of cytoplasm), followed by constriction of a contractile ring (23, 24). Gallicano et al. (6, 24) showed that PKC participates in the initiation of the second polar body, but that, in the absence of other signaling agents, the initiated polar body will be absorbed into the egg. Activation of PKC also causes cortical granule exocytosis in mammalian eggs (14, 21, 22, 25, 26); however, there is evidence suggesting that other calcium-dependent signaling pathways stimulate cortical granule exocytosis (27, 28) because PKC inhibitors cannot block the fertilization-induced exocytosis of cortical granules. Activation of PKC can drive the egg into an interphase state of the cell cycle as indicated by characteristics such as the formation of pronuclei, the Golgi apparatus, and the pattern of protein synthesis (18, 22, 23, 29, 30, 31), although this depends on the strain of mouse studied (32). Once activated, PKC subsequently is cleaved to form protein kinase M [PKM] and PKM can phosphorylate cytoskeletal elements in the interior of the activated egg (6).

Once the rise in intracellular free calcium occurs, another calcium-dependent pathway is activated. CaM KII is activated in mammalian eggs (7, 8, 32, 33), and its activation depends on the presence of both calmodulin and calcium (8). Although functions of CaM KII have been demonstrated in eggs from other classes of organisms, there are not many reports concerning CaM KII activation in mammalian eggs. One study with mammalian eggs used an inhibitor of calmodulin, which should inhibit CaM KII, and demonstrated delayed formation of the second polar body (34). This has been confirmed by two studies that used an inhibitor that acts on CaM KII (32, 35). CaM KII accumulates at the contractile ring of the forming second polar body (8), and at this location the kinase is in its active state (7). In addition, CaM KII action is essential in the transition from metaphase II to anaphase II (8, 32), and CaM KII also potentiates the action of MAP (mitogen activated protein) kinase (7).

2. GENERAL HANDLING AND MANIPULATION OF EGGS AND EMBRYOS

Mammalian eggs and preimplantation embryos are surrounded by an extracellular matrix, the zona pellucida. The mouse egg has a diameter of about 60 μm, and with the zona pellucida its diameter is about 80 μm. To handle such small specimens, we pull a Pasteur pipette to a fine tip. This is done by heating the narrow part of the Pasteur pipette over an alcohol lamp and pulling it to a much smaller bore at an angle of about at a 140° from the long axis of the pipette. Using a forceps, the tip of the pulled pipette is broken open in the area where it has a diameter of about 90–120 μm. The pipette is connected to flexible tubing that is connected to a tuberculin syringe. Suction or pressure is applied to the pipette by adjusting the position of the plunger within the syringe. Before transferring specimens, fluid is drawn into the pipette to wet the glass walls, which keeps the eggs or embryos from sticking to the walls. A small reservoir of media should remain in the tip to avoid exposing the samples to the fluid–air interface. Eggs or embryos should only be contained in the finely pulled portion of the pipette. Collecting fluid beyond the bend in the pipette risks losing many of the eggs or embryos. It is safer to transfer small lots of eggs or embryos rather than attempting transfer them all at once, especially if transferring more that 20 eggs or embryos. Eggs or embryos are cultured in 50–100 μl droplets under embryo-tested mineral oil in a tissue culture dish. Alternatively, they can be cultured in organ-culture plates. Because of the small volume contained in each of the droplets, the transfer pipette should be prefilled with media contained in the recipient droplet; this will reduce the occurrence of media dilution in the recipient dish when the eggs are placed there.

The next major consideration is media. The ingredients used to create the embryo handling media must be combined so that the ionic environment, osmolarity, and pH resemble the maternal environment from which they came. We primarily use KSOM (35–38), which is an optimized medium for the culture of the embryos in a CO_2/moisture incubator and to proceed beyond the two-cell block. We also use KSOM-H, which contains HEPES to buffer the pH of the medium for short-term handling outside of the incubator, including flushing of oviducts. The BSA in the KSOM-H is removed when calcium ionophore is applied in experiments that activate eggs. (The media types referred in this section and following sections are listed in the protocols of this chapter.)

3. ACTIVATION OF EGGS WITH CALCIUM IONOPHORE

The egg is naturally activated when it is penetrated by the sperm. When the sperm and egg membrane fuse, the mammalian egg undergoes a elevation of intracellular free calcium ($[Ca^{2+}]_i$) that approximates 1 μM, followed by a series of calcium oscillations of lower amplitude. Egg activation can be artificially induced by creating a rise in calcium within the egg. This rise in $[Ca^{2+}]_i$ activates the egg, and development proceeds normally for a period of time, albeit in the absence of a sperm nucleus. The reason that one might chose to activate eggs rather than to fertilize eggs is that egg activation occurs synchronously upon application of pharmacological agents that cause a rise in calcium, whereas fertilization, which relies on the speed with which the sperm penetrates the zona pellucida, is more asynchronous. Consequently, when one wants to investigate events related to cell-cycle resumption within the first 20 min of egg activation, and particularly in the first 5 or 10 min, the use of pharmacologic agents that cause a rise in calcium provide a synchronously developing group of eggs.

4. KINASE ASSAYS

Kinase activity within the cell serves to mediate rapid, short-term changes in cells. This is particularly true of cells that are in M phase of the cell cycle because rates of translation are reduced and the DNA is condensed in the form of chromosomes.

The action of kinases can be detected using biochemical assays where cells are lysed in the appropriate milieu and a known substrate is added to the lysate along with radiolabeled ATP. In such assays the amount of radiolabel can be measured after gel electrophoresis of the lysate. Such an assay does necessitate use of a specific substrate capable of distinguishing between different kinases. Because a large protein is likely to have many phosphorylation sites, a small peptide that has been designed to contain the consensus phosphorylation sites for the kinase of interest is ideal for such studies. There is an increasing number of these small peptide substrates that are commercially available. Another approach to aid in identifying the action of specific kinase would be to add a peptide or pharmacologic inhibitors that would block the action of kinases that are not of interest. Again, many such inhibitors are commercially available. However, the inhibitor approach could be difficult to interpret if the inhibitors have pleoitrophic effects. For this reason it is a good idea to conduct controls where both specific substrates as well as inhibitors are tested in lysate reaction mixtures. Moreover, two structurally different inhibitors to the kinase being investigated should also be used as separate controls. Both of these structurally different inhibitors should provide the same result.

Several companies offer assays kits for different kinases. A careful reading of the protocol supplied by the company will often indicate that the company has supplied components for an assay that activates all of a particular type of kinase in a cell (this is done by adding a excess of activators and cofactors to the kit's buffers). This is an acceptable approach if the goal is to determine if the kinase is present in the cell or if the goal is to compare the relative amounts of all of that kinase in different populations of cells. However, that approach will not inform the investigator about the change in activity at different points in the cell cycle. There is no reason to anticipate that when a kinase becomes active that all of that kinase in the cell becomes active. In fact, evidence indicates that only a small amount of the specific kinase becomes active (7, 41). If working with a kit purchased from a company, it is important to identify the function of all the components in the kit and omit the ones that activate all of the kinase so that the investigator measures only the proportion of kinase in the cell that was active at the time the lysate was prepared.

Recently it has become possible to purchase antibodies to a kinase when it is in its active state. To make such "anti-active antibodies," companies have taken advantage of the fact that many kinases become phosphorylated at a specific site when they are active, and they have thus designed the antibody to bind to that phosphorylation site on the kinase. Thus, either by Western analysis or by immunocytochemistry, an investigator can monitor the amount or spatial distribution (the latter by immunocytochemistry) of all of the kinase (i.e., the total kinase) with an antibody that binds a conserved location on all forms of the kinase. Then the investigator can use a different antibody that binds the phosphorylation site on the kinase and monitor the distribution of the active kinase (7, 41). Protocols 20.1–20.9 percent assays for the activity of the common isotypes of PKC and CaM KII. Protocol 20.10 describes media.

5. CONCLUSIONS

The procedures presented here provide approaches for investigating short-term changes in the cytoplasm of eggs and blastomeres that are mediated by cytoplasmic signaling events. These approaches are possible due to technological advances such as the development of peptide substrates, specific inhibitors, and sources of antibodies capable of recognizing signaling agents within the cell. Each of these procedures have been presented as a set of manipulations that can stand alone. This allows investigators to combine the protocols in the order appropriate to address a broad array of questions.

Protocol 20.1. Collection of mouse eggs and embryos

For experimental procedures we use the CD-1 strain of outbred mice. They are maintained on a 14-h light/10-h dark schedule.

1. Superovulate female mice between 6 and 8 weeks of age through the use of gonadotrophins. Superovulation allows the "scheduling" of ovulation so that one can estimate the time of ovulation and synchronize several females. In addition, through superovulation, a larger harvest of eggs or preimplantation embryos is usually obtained. Inject female mice intraperitoneally (i.p.) with 5 IU pregnant mare serum gonadotropin (PMSG), followed by an i.p. injection of 5 IU human chorionic gonadotropin (hCG) 46–48 h later. Deliver the hormones in 0.1 ml sterile sodium phosphate (0.1 M).
2. Remove unfertilized eggs within 13–15 h after hCG injection. As mouse eggs age beyond 16 h after hCG injection, they become more prone to spontaneous egg activation. Eggs arrested at meiotic metaphase II arrest are obtained from the oviduct.
3. To remove the eggs, euthanize the female by cervical dislocation, place the mouse on its back, and spray its ventral surface with 70% ethanol. Surgically open the abdominal cavity with one set of forceps and scissors by cutting through the skin and peritoneum at the pubic area, laterally toward the hip (on both sides) and then upward until the side of the rib cage is reached. This flap of skin and peritoneum is laid back over the thoracic cavity, and the intestinal viscera are pulled aside to reveal the uterine horns. Using another set of fine forceps and scissors, grasp the uterine horn near the oviduct and pull up and away from the body, exposing mesometrium tissue that surrounds the ovary and oviduct. Insert scissors between the uterine horn and fat to pull away the fat while moving the scissors toward the ovary. The procedure also gently separates the ovary from the oviduct. Cut the membrane that covers the oviduct and holds it in proximity to the ovary. Cut the uterine horn close to the opposite side of the oviduct. The oviduct is handled using this small piece of uterine horn. Place the oviduct in KSOM-H heated to 37°C until all oviducts are collected.
4. The meiotic metaphase II eggs are surrounded by cumulus cells to form the cumulus mass. This cumulus mass is visible in the ampullary region of the oviduct, when viewed with a dissecting microscope. Remove the cumulus mass by scoring the oviduct near the mass with a tuberculin syringe needle and allowing its extrusion with some gentle agitation of the oviduct. Using an unpulled Pasteur pipette, draw up the cumulus masses in a minimal amount of medium and place them in a dish containing 1.5 ml KSOM-H and 300 μg/ml hyaluronidase (Sigma Chemical Company, St. Louis, MO.; the stock solution is made by reconstituting

a vial with water to establish a 10 mg/ml solution and stored frozen) heated to 37°C. The cumulus masses can remain in the hyaluronidase for up to 5 min, but to decrease the chance of parthenogenetic activation it is best to monitor the cells as they come loose from the cumulus and transfer them immediately to the first of three washes in KSOM-H (37°C).

5. If embryos are needed for experiments, place the female mice with a male after injection with 5 IU hCG. Placement with the male can be delayed for 3–4 h to permit the female to reach a more receptive state. The following morning, check the female for a white mucous (vaginal) plug indicating that copulation has occurred. The time point of fertilization is assumed to be at the midpoint of the dark cycle and the morning in which the plug is detected is termed 0.5 days postcoitium (dpc).

6. Collect two-cell embryos on 1.5 dpc and four-cell embryos 8–12 h later. The eight-cell embryos are collected the morning of 2.5 dpc, and by evening a morula stage can be found.

7. Collect two-cell through eight-cell stage embryos by flushing the oviducts with 0.1 ml of KSOM. To do this a needle (26 or 30 G) is connected to a syringe filled with KSOM-H is placed in the infundibulum, a cufflike structure of the oviduct that before surgical manipulation is located next to the ovary. Flush the KSOM-H through the oviduct, expelling the embryos at the distal end of the oviduct.

8. For morula- or blastocyst-stage embryos, remove the uterine horns intact with the oviducts still attached. While holding the uterine horn over a dish of KSOM-H, insert a syringe needle (26 or 30 G) into the horn at the cervical end and inject the media, causing the horn to expand. The oviduct is then cut off, allowing the media to irrigate the horn flushing out the blastocysts.

Notes: The health of embryos can sometimes be detected by visual inspection. Embryos that appear grainy, pale, or ghostlike, have unevenly sized blastomeres, or have an inappropriate number of blastomeres for the stage expected should not be used. When working with unfertilized eggs, it is important that harvesting occurs before 16–18 h after hCG injection because of the increased risk of parthenogenetic activation (39–41).

Protocol 20.2. Activation of eggs with calcium ionophore

Some investigators use a brief treatment with ethanol to allow calcium to enter the egg from the surrounding medium. Presumably, ethanol destabilizes membrane structure, allowing calcium to leak in, but other components could also pass through the destabilized membrane. We use calcium ionophore A23187 in the free-acid form (Calbiochem. Inc., La Jolla, CA).

1. Dissolve the ionophore in ethanol as a 10 mM stock and immediately aliquot into microfuge tubes each containing 5 μl and store frozen until use. Titrate each new lot to determine the minimum time and concentration necessary to activate > 90% of the eggs.

2. Typically a 1 μM concentration of A23187, when exposed to eggs for 2 min, will activate > 90% of the eggs, but this should be tested because it can vary between supplier and lot. The treatment with calcium ionophore must be brief, otherwise the ionophore becomes toxic to the cells. Remove excess calcium ionophore by transferring the eggs through three washes of KSOM-H.

3. When conducting these manipulations to induce egg activation, the state of activation should be confirmed. Process a group of eggs and stain them with a DNA stain such as DAPI or Hoechst to confirm whether the chromosomes have been released from the arrest at meiotic metaphase II and are uniformly progressing into anaphase II and that the spindle has rotated. Alternatively, you can also assess whether the second polar body has been released, although sometimes this is more difficult to observe.

4. Allow a few eggs from the same group that has been processed to remain in KSOM-H. If, at the end of the experiment, these eggs still remain at meiotic metaphase II, then it is likely that the eggs that were used are healthy. In contrast, if the eggs in this control have spontaneously activated, the entire experiment should be discarded because this would indicate that the starting material was not in an optimal state.

Protocol 20.3. Mounting living or cytologically fixed eggs/embryos

To immobilize eggs surrounded by their zona pellucida, we use coverslips coated with polylysine. This treatment allows us to exchange the solution around groups of living or cytologically fixed eggs. We also use this treatment to immobilize eggs that are going to be viewed in a microscope so that they do not drift.

1. Clean slides and coverslips before use to remove excess debris that accumulates through storage or manufacturing. Sonicate coverslips three times for 5 min each in acetone and then sonicate again three times for 5 min each in deionizied water (by placing the coverslips in a glass beaker and placing the beaker in a bath sonicator). Allow the coverslips to air dry on filter paper within a laminar flow hood and place in a filter–paper–lined petri dish and cover. Wash slides by soaking in 75% ethanol overnight and dry with lint-free tissue and placed in a holding rack.
2. Make a 0.1% stock solution of polylysine (mw > 300,000; Sigma) in distilled water, aliquote into Nalgene plastic tubes, and stored at –20°C until use. For use the solution is thawed, briefly mixed, and 30 μl of 0.1% polylysine is thoroughly spread over the coverslip with the tip of a microliter pipette. Allow the coverslips to dry on a warmer set at 37°C. Coverslips left for a prolonged period should be covered. Once dried, and immediately before use, place a few drops of 1× intracellular buffer (ICB; without BSA or other sources of protein) over the polylysine surface.
3. Place eggs/embryos that have been prewashed in a protein/amino-acid–free medium (e.g., PBS or ICB) onto the coverslip. Within 1 min the egg/embryos will adhere to the slide/coverslip.
4. Treat specimens that are to be cytologically fixed through a two-step process: (a) pipette fixative onto the eggs/embryos; because of the higher density the fixative will displace the existing media and contact the specimens. (b) Remove excess media and add more fixative for a better-preserved specimen. It is best to remove as much excess media as possible so that the fixative is not diluted.
5. If the eggs/embryos are being processed for analysis by immunofluorescence mount the coverslip on a slide as follows. From a syringe filled with a wax/grease mixture (see Protocol 20.10 for composition and handling), place a small spot of the mixture in each of the four corners of the coverslip. Add a drop of ICB to the center of the coverslip over the eggs/embryos and gently place a clean slide over the coverslip. Gentle pressure flattens the wax/grease spots but still holds the coverslip apart from the slide so the eggs/embryos are not crushed. The space between the slide and coverslip should be free of air pockets and completely filled with ICB. Seal the coverslip to the slide with nail polish, which prevents movement and evaporation of the ICB.
6. In addition to cytological fixation, living eggs can also be processed through different solutions while immobilized on the polylysine-coated slides/cover slips. In this case, remove the medium by aspiration and add fresh medium by pipette. If the coverslip was lifted out of the solution, the shear force that developed as the specimen crossed the air–liquid interface would dislodge the specimen from the coverslip.

Notes: Polylysine will adsorb to glass, which reduces the concentration of free polylysine in the solution. To prevent this, store polylysine in nalgene tubes and handle with nonadsorptive pipette tips. Do not transfer polylysine with a Pasteur pipette.

Protocol 20.4. Preparation of the detergent-resistant cytoskeleton

The use of detergent extraction enables analysis of the cytoskeleton at a structural and biochemical level. Detergent extraction relies on the use of a nonionic detergent to remove the cell's membranes, permitting the release of detergent-soluble components from the cell. During biochemical analysis, the detergent-soluble fraction can be isolated from the detergent-resistant cytoskeleton and analyzed separately. Once treated with the detergent-extraction medium, the egg/embryo structural components can be viewed by various forms of microscopy or by gel electrophoresis. The following is a procedure for obtaining a detergent-resistant cytoskeleton and detergent-soluble fractions.

1. Wash eggs/embryos in PBS and transfer to the detergent-extraction media (ICB made 1% with Tween-20 and 200 μg/ml with the protease inhibitor 4-(2-aminoethyl) benzenesulfonil fluoride Sigma A8456 [AEBSF]). Allow the detergent extraction to proceed for 5 min at room temperature. (Prepare the AEBSF as a 20 mg/ml stock in water and store for up to 2 months frozen.)
2. If the detergent-resistant cytoskeleton is going to be processed solely for microscopy, wash the eggs through fresh extraction medium and then apply the cytologic fixative. If the eggs/embryos were transferred by pipette through these solutions, they can be adhered to a polylysine-coated coverslip at this time or they can continue to be processed by pipette through the various solutions (e.g., in solutions with antibodies) and mounted to the coverslip just before observation in the microscope (see Protocol 20.3 for specimen mounting).
3. If these fractions are going to be used for Western Blotting, pool fractions from approximately 200 eggs into detergent-soluble and detergent-resistant fractions. Chemiluminescence is needed to visualize a signal with this small number of eggs/embryos.
4. To collect and pool the detergent-soluble fraction, collect the extraction medium immediately after removal of the eggs/embryos and subject it to an organic precipitation by the addition of 4 volumes of ethanol (–20°C) and precipitate at least 24 h at –20°C. If replicate experiments need to be conducted to collect the 200 eggs, the soluble fractions can be pooled as precipitates, then transferred into a single microfuge tube, the ethanol removed, and the pellet solubilized in 1× sample buffer.
5. To collect and pool the detergent-resistant cytoskeletal fraction, transfer the eggs/embryos directly into a small volume of 1× sample buffer and vortex, which will solubilize the cytoskeletal components and the zona pellucida. As specimens are pooled into the sample buffer, if the volume of the sample buffer increases beyond the appropriate level, concentrate this fraction (even after solubilization in sample buffer) by ethanol precipitation described in step 4 and solubilize the precipitate in the appropriate volume of 1× sample buffer.

Notes: To obtain consistent results, apply Tween-20 well above its critical micelle concentration (we recommend use at 1%). Tween 20 is also light sensitive and should be stored in the dark. Tween-20 stock solutions are stored at –20°C, which prevents chemical degradation as well as bacterial growth in the stock solution. Studies in the somatic cell literature often use 1% Triton X-100 for detergent extraction. However, Triton X-100 has been found to destabilize the cytoskeleton of mouse eggs and embryos (11). Because of this we have replaced Triton X-100 with Tween-20.

Eggs/Embryos that are exposed to the detergent-extraction medium for prolonged periods of time will begin destabilization of the cytoskeleton as actin filaments and microtubules slowly begin to disassemble from their ends (in the absence of a pool of soluble monomer). Thus, specimens should either be cytologically fixed or solubilized in sample buffer soon after preparation at defined times for consistent results.

Detergent extraction works best on groups of cells that are only a few cell layers thick. In multiple cell layers the outer layers will be exposed to the nonionic detergent well in advance of the inner layers of cells, causing the extraction to be inhomogeneous.

The various solutions used during the detergent-extraction process have different densities and can cause the eggs/embryos to translocate in the dish. When placing the eggs/embryos into the detergent-extraction medium, it is best to place them on the bottom of the dish within the medium. Care should be taken that the eggs/embryos do not immediately float to the top because the air–medium interface may make them lyse. This problem can be circumvented by placing the eggs/embryos onto a polylysine-coated slide/coverslip before detergent extraction.

Protease inhibitors such as phenylmethyl sulfonyl fluoride (PMSF) or AEBSF block proteases by sulfonating serine residues at the active site of the proteins. These sulfonating groups can reduce or block enzyme function. If the purpose of the experiment is to perform an enzyme assay or assay a metabolic event, then PMSF or AEBSF should not be used. In this case we recommend the use of protease inhibitor cocktail that contains 1 μg/ml of each of the following protease inhibitors: pepstatin, aprotinin, chymostatin, leupetin, and trypsin-chymotrypsin inhibitor.

Protocol 20.5. Immunocytochemistry

Eggs/embryos can be processed for immunocytochemistry either after detergent extraction to assess components associated with the cytoskeleton or as intact specimens. If intact specimens are to be examined, the eggs/embryos must be permeabilized after cytologic fixation to permit the antibodies to enter the cells (step 2 below).

1. First fix eggs/embryos in 2% paraformaldehyde fixative (ICB made 2% with paraformaldehyde) for 30 min. In some cases the fixative is 2% paraformaldehyde and 0.1% glutaraldehyde; however, some antibodies will not bind if the glutaraldehyde is present in the fixative.
2. If the specimen is fixed intact (rather than detergent extracted), treat the sample with permeabilization medium (ICB made 2% with paraformaldehyde and 1% with Tween-20 in 10× ICB) for 30 min.
3. Place the eggs/embryos through 3 washes of ICB-BSA (ICB made 1% with BSA) at 15 min each.
4. Place the eggs/embryos in a primary antibody at the correct dilution (using ICB-BSA) within a covered, multiwell plate and place on a rocker at 4°C overnight.
5. The following morning rinse the eggs/embryos 4 times, 15 min each, in ICB-BSA, then place into a secondary fluorophore-conjugated antibody at the appropriate dilution (using ICB-BSA) and place on a rocker at room temperature for a minimum of 1 h (or at 4°C overnight).
6. Wash eggs/embryos 2 times, 15 min each, in ICB-BSA and then wash 1 time for 15 min in 1× ICB. The last ICB wash contains the DAPI (0.5 μg/ml) to enable visualization of the chromosomes.
7. Mount the eggs/embryos on polylysine-coated coverslips and seal as described in Protocol 20.3 and view with a scanning laser confocal microscope.

Note: We recommend that a control be used where the first antibody is omitted from the procedure. This should always be conducted to assure that the second antibody is not artifactually binding to the specimen.

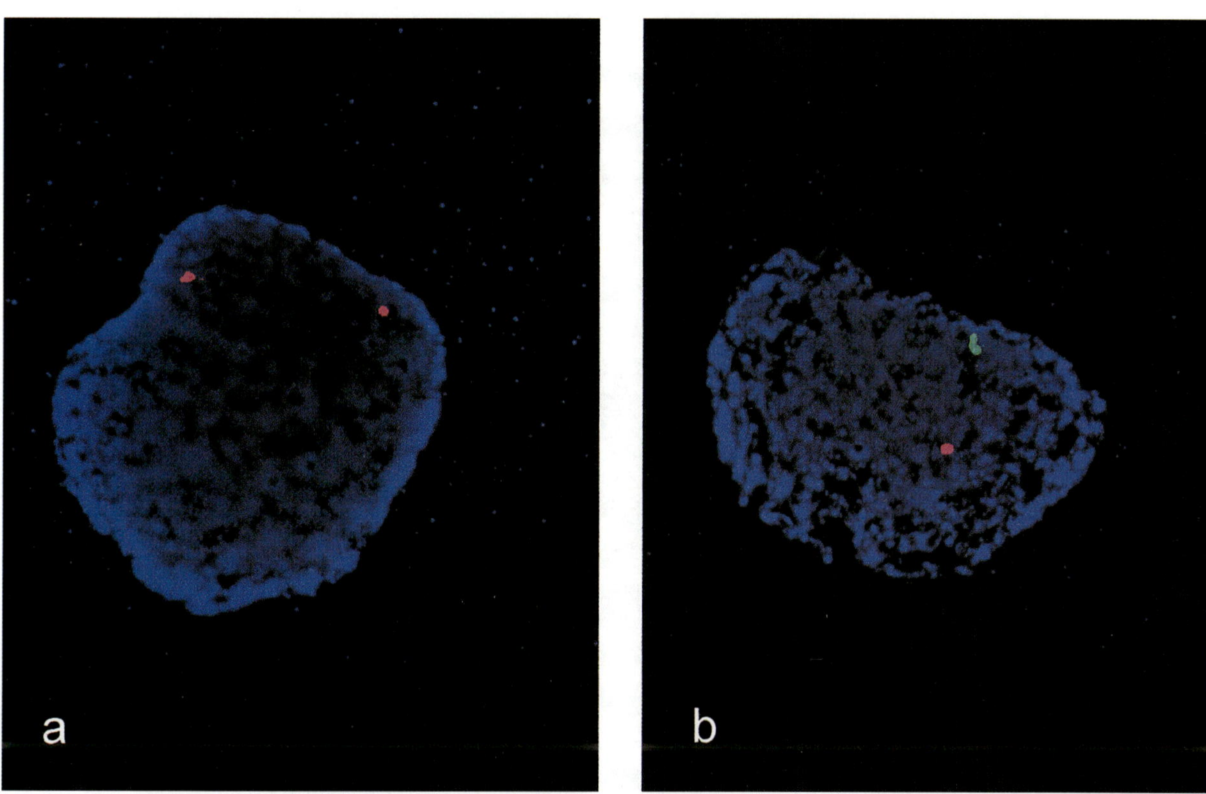

Figure 11.2. FISH image of blastomere nuclei counterstained with DAPI which shows the DNA in purple/blue. X chromosomes are labeled with red fluorochrome and Y chromosomes with green fluorochrome. (a) A nucleus with two red signals from a female embryo; (b) a nucleus with one red and one green signal from a male embryo.

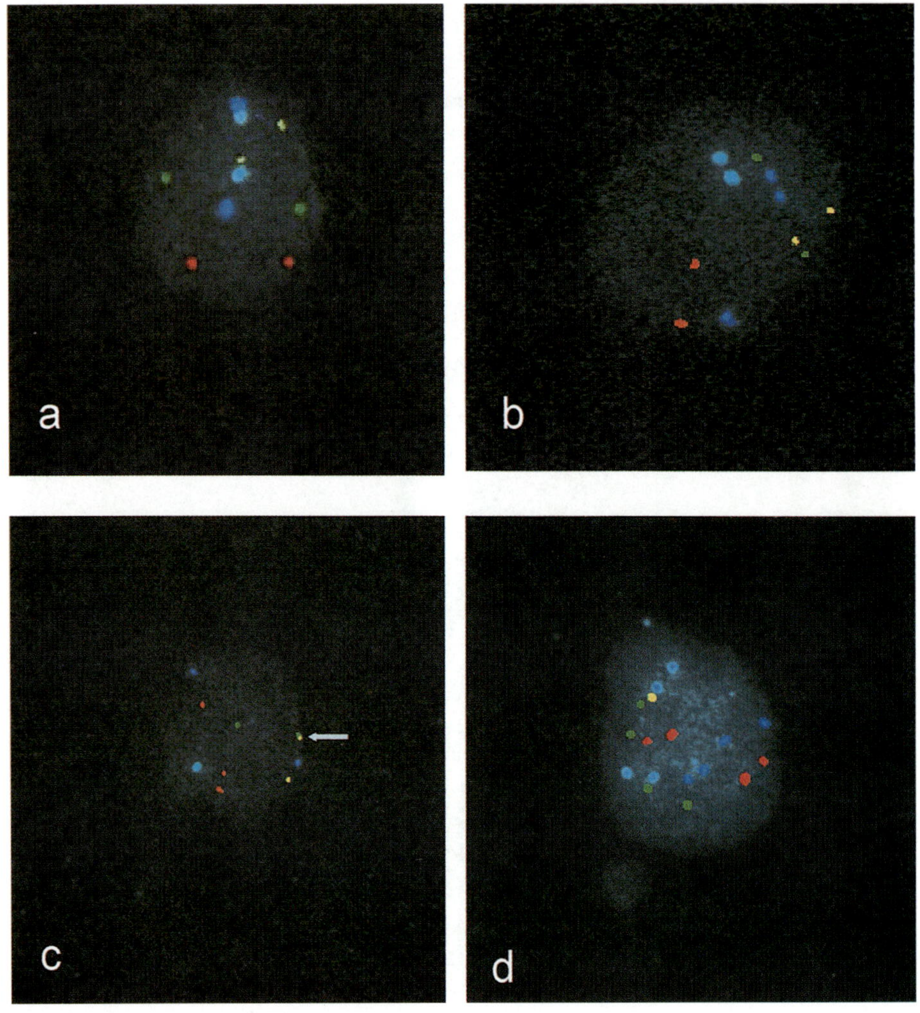

Figure 11.3. FISH images of blastomere nuclei counterstained in aqua. The probes used are for chromosomes 13 (red), 16 (aqua), 18 (blue), 21 (green) and 22 (yellow). (a) A normal nucleus with two signals of each color, (b) nucleus with an extra blue signal diagnosed as trisomy 18, (c) nucleus with an extra red signal diagnosed as trisomy 13. Note that one green and one yellow signal (arrow) are closely adjacent. (d) This nucleus approximates tetrasomy, as it has four signals of each color, with the exception of chromosome 22 (yellow), for which there is only one signal.

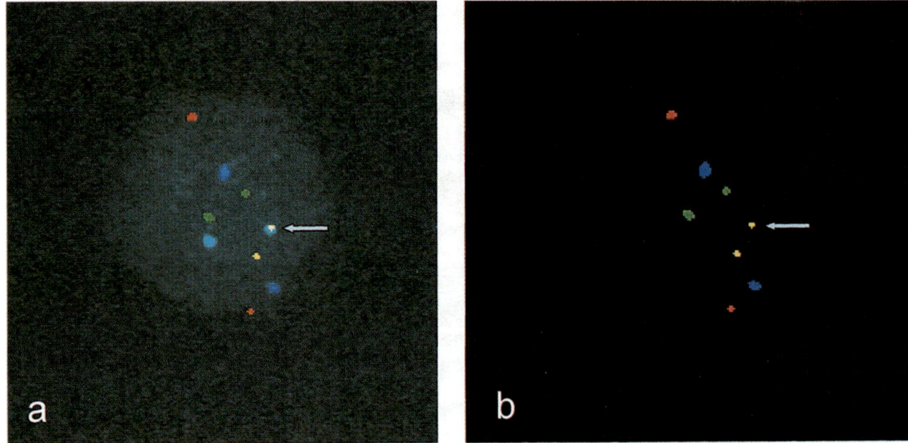

Figure 11.4. The advantages of using only monochromatically labeled probes and an imaging system. (a) This nucleus could easily be diagnosed as monosomy 22 with only one yellow signal. The arrow shows that one aqua signal may be overlying a yellow signal. (b) The same nucleus as in panel a, but the imaging system has been used to remove the aqua color plane, allowing unequivocal confirmation that a yellow signal is underneath the aqua.

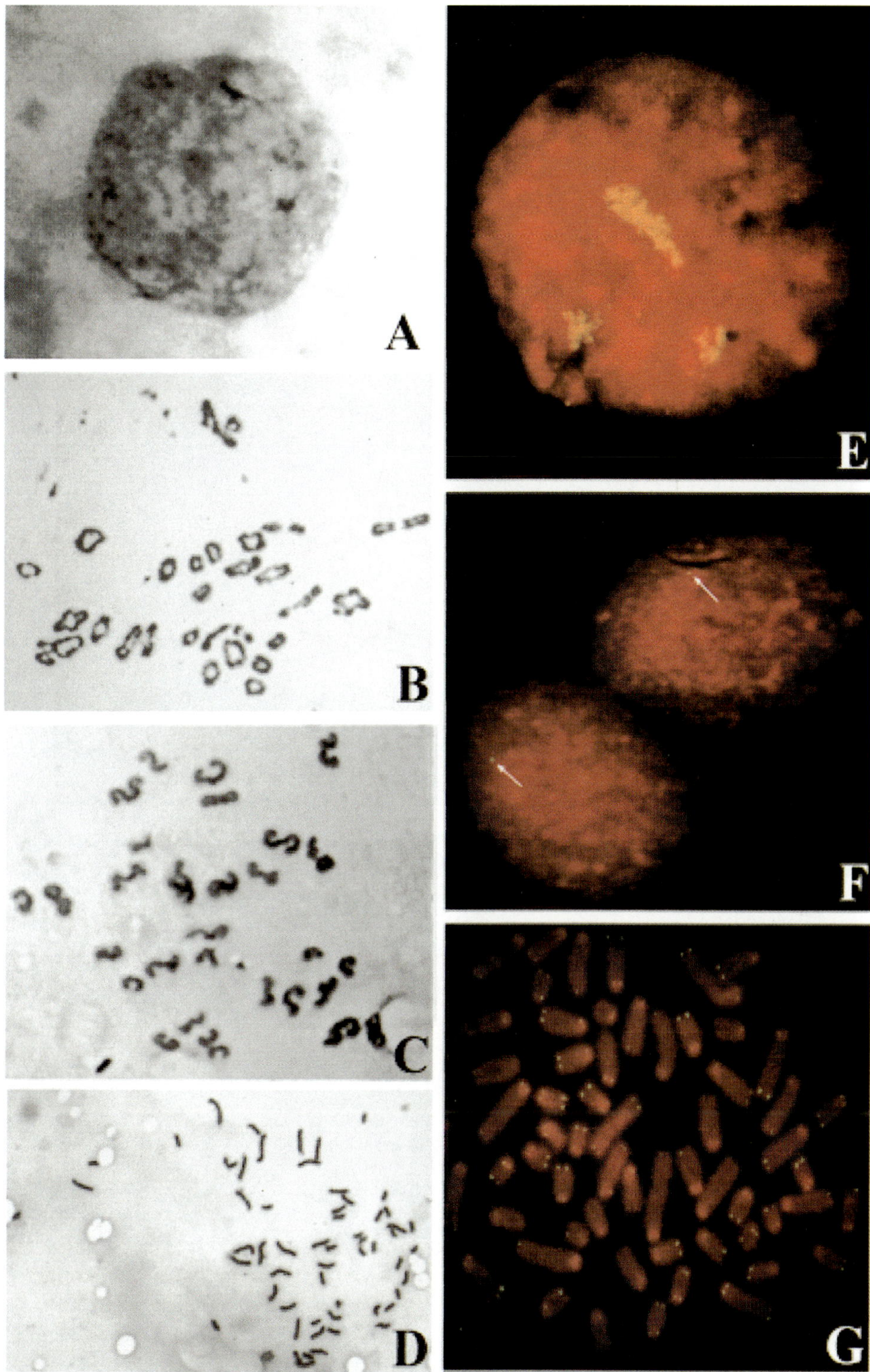

Figure 12.1. Methanol-acetic acid fixed preparations from bovine oocytes and two-cell and blasto-cyst stage embryos. Nuclei (interphase and metaphase) were stained with 4% Giemsa (A–D) or sub-jected to FISH using fluoresceinisothiocynate-labeled detection systems (green) and counterstained with propidium iodide (red). (A) Giemsa-stained germinal vesicle stage oocyte; (B) Giemsa-stained metaphase I; (C) Giemsa-stained metaphase II; (D) two Giemsa-stained metaphase spreads from a two-cell-stage embryo; (E) a blastomere nucleus after FISH using a repetitive DNA sequence specif-ic for chromosomes 14, 20, and 25; (F) two blastomere nuclei from a haploid parthogenetically acti-vated oocyte after FISH using a unique DNA sequence specific for chromosome 1 (signal is indicat-ed by the arrow); (G) a metaphase spread from a blastocyst after FISH using a PNA telomere repeat (TTAGGG) probe.

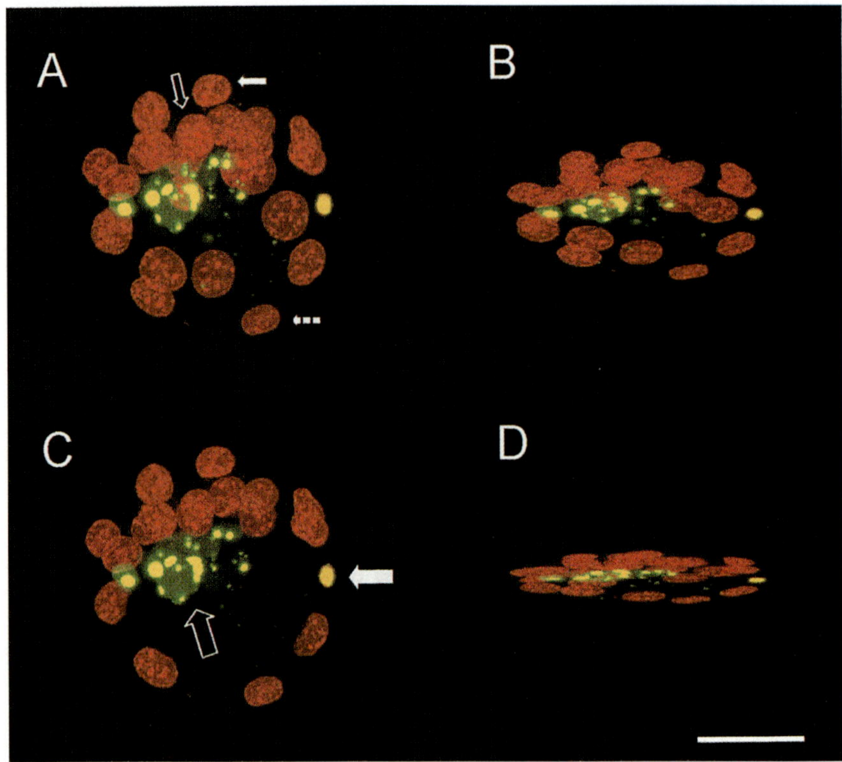
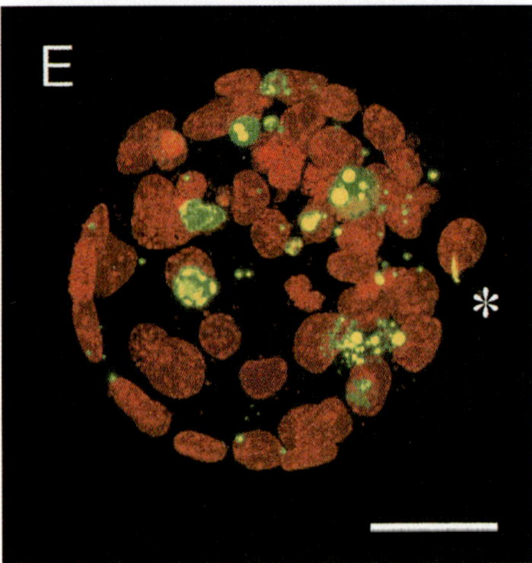

Figure 19.5. Apoptosis in whole mouse blastocysts. Fragmented apoptotic nuclei are labeled with TUNEL (yellow); healthy nuclei are labeled with propidium iodide (red). Computer-generated reconstructions of confocal Z series of the entire embryo show projections from (A) the front, (B) 60°, and (D) 90°, showing the position of apoptotic nuclei in the center of the embryo. (C) The middle sections through the ICM only. In A, the ICM is located at the top left of the blastocyst and ICM nuclei (small black arrow) can be distinguished from polar TE nuclei (small white arrow) and mural TE (small broken arrow). In C, TUNEL-positive nuclei show fragmented (large black arrow) and condensed (large white arrow) nuclear morphology. (E) A larger, expanded blastocyst with many TUNEL-positive, fragmented apoptotic nuclei in the ICM and loose in the blastocoel. The asterisk marks the position of a polar TE nucleus (red) beginning to hatch from the zona pellucida and a single sperm head (yellow). Bars = 40 µm. From Brison and Schultz (52).

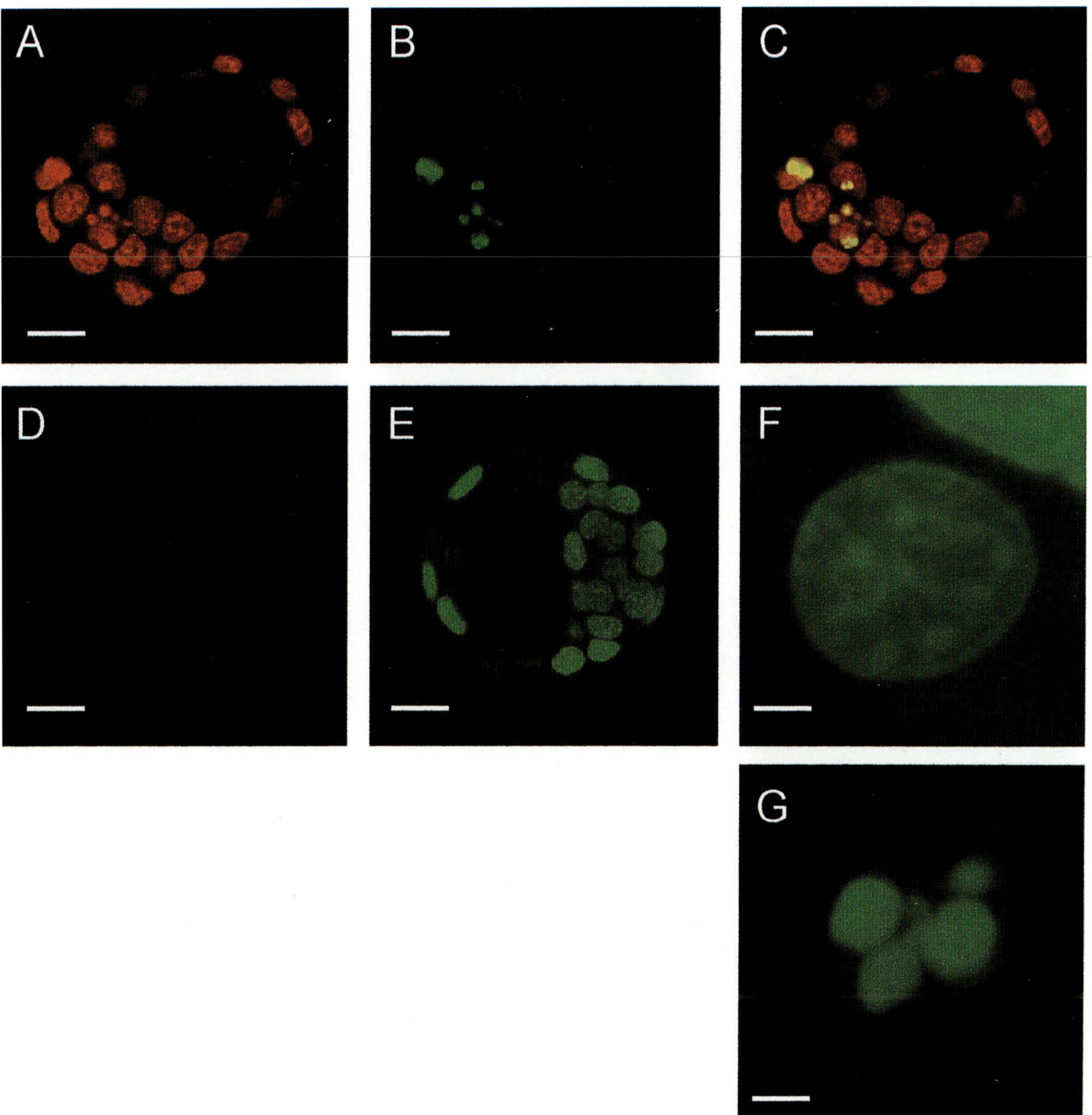

Figure 19.6. Detection of apoptotic nuclei in the blastocyst using TUNEL (see Protocol 19.2). The same blastocyst, labeled with (A) propidium iodide (PI; red) and (B) TUNEL (green), and (C) superimposed (yellow). Comparison between A and B shows that there is little bleedthrough of signal from the red to green channel. Controls are also performed using TUNEL in the absence of PI to demonstrate the converse (not shown). Negative controls lacking terminal transferase demonstrate an absence of nonspecific background labeling (D), and positive controls after DNase treatment show labeling of all nuclei in the blastocyst (E). (F, G) Morphology of healthy and fragmented apoptotic nuclei, respectively. (A–E) Bar = 20 μm (F, G) bars = 2 μm. From Brison and Schultz (49).

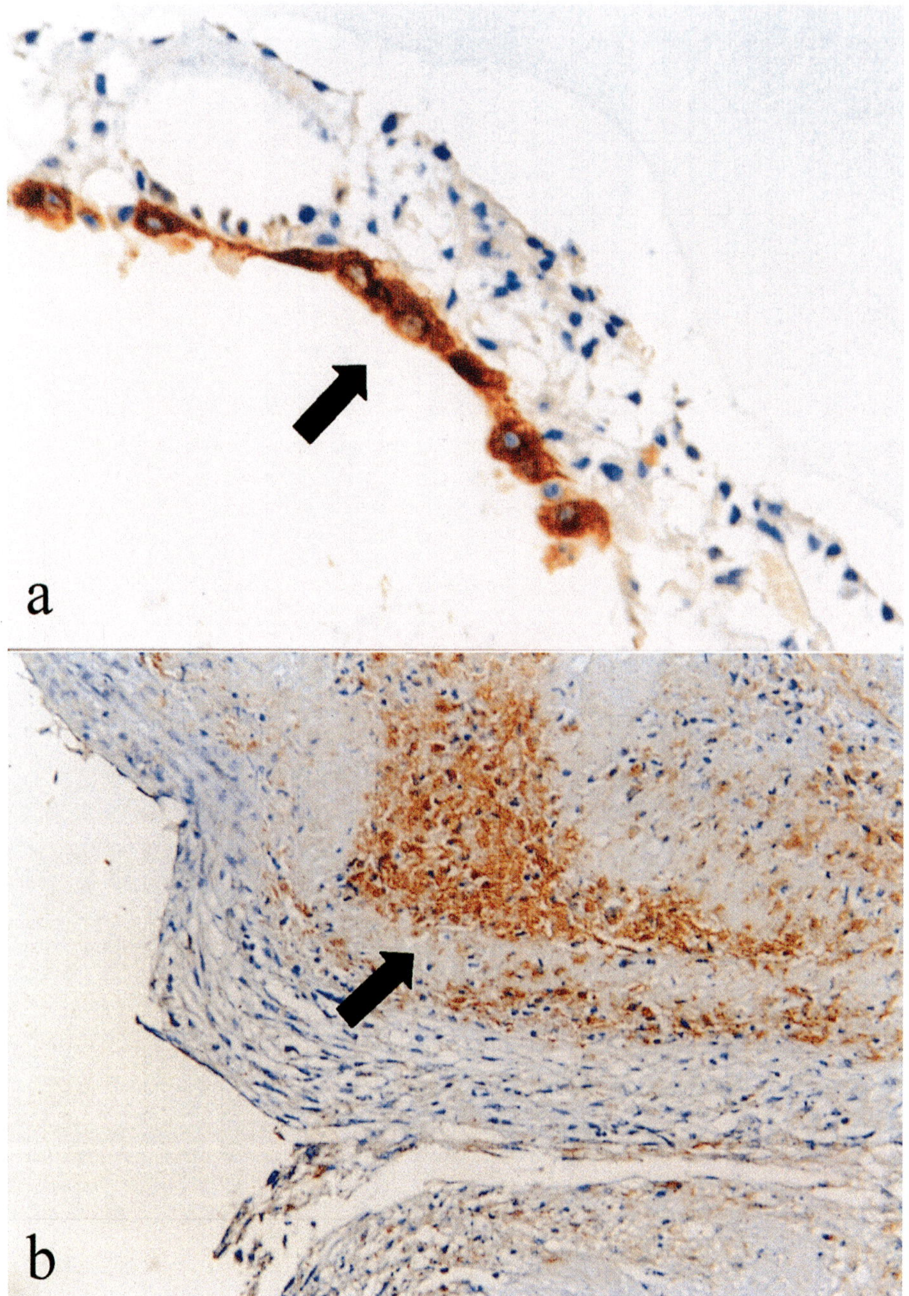

Figure 21.5. Light microscopic image of human embryonic stem cells after undirected differentiation in vitro. (a) Section of ES cells cultured 11 days without feeder layer, laveled with rabbit antihuman $\alpha-1-$fetoprotein (arrow), a marker for endoderm; original magnification $\times 400$. (b) Section of ES cells cultured 21 days without feeder layer, labeled with monoclonal mouse antihuman muscle actin (arrow) for indentification of an epitope present on muscle actin; original magnification $\times 200$.

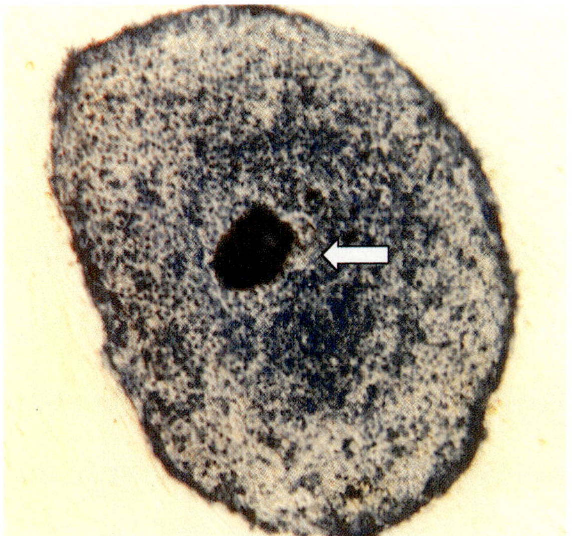

Figure 21.6. Human embryonic stem cell colony stained with alkaline phosphatase to show nondifferentiated areas (blue). An embryoid body is forming in the center of the colony (arrow). Original magnification × 100.

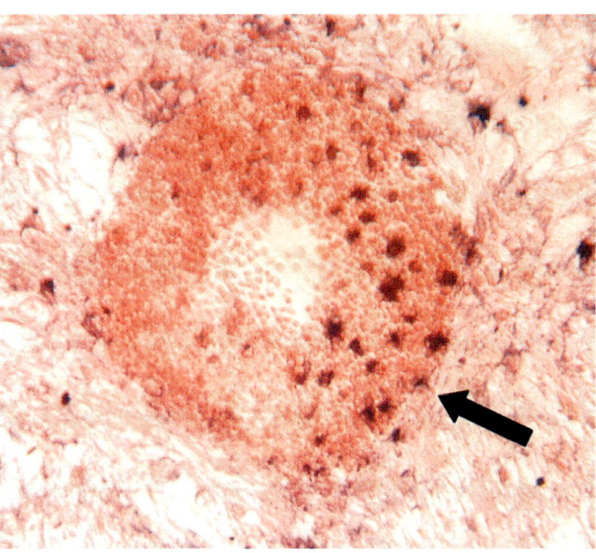

Figure P21.3. Human embryonic stem cell colony (arrow) stained positive for SSEA-4 showing that the cells are undifferentiated. Background cells (stained violet) are the primary mouse embryonic fibroblast. Original magnification × 100.

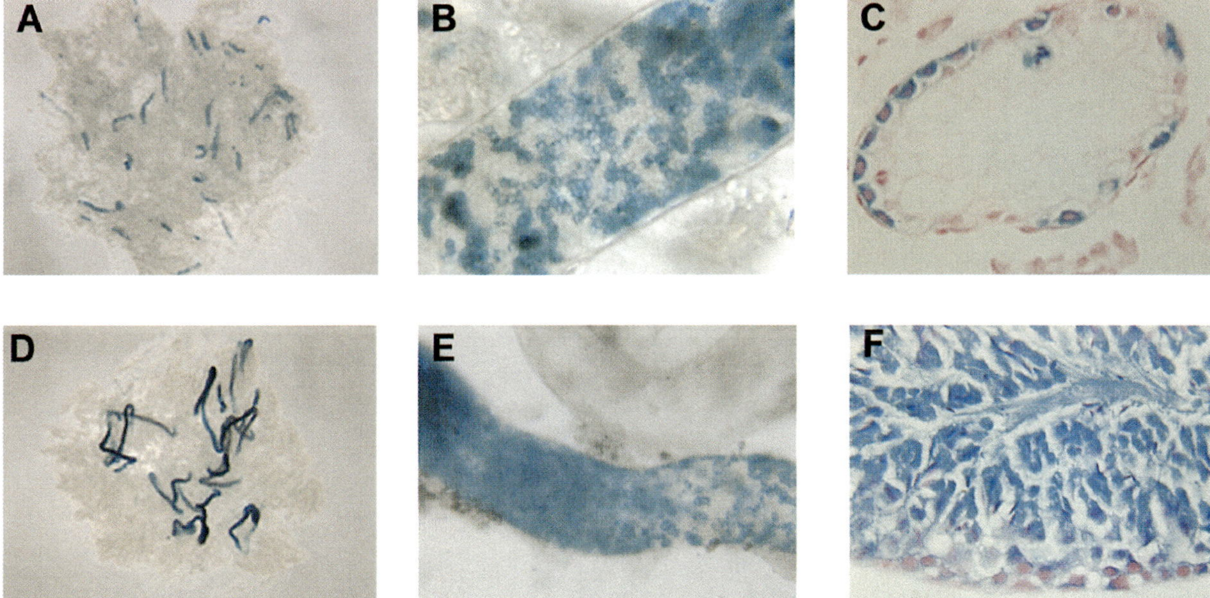

Figure 22.2. Donor germ cell colonization patterns 1 (A–C) and 2 (D–F) months after transplantation. Donor testis cells were obtained from ROSA26 mouse strain and transplanted into the testes of immunocompatible recipient mice. Recipient testes were stained blue with X-gal to visualize donor-derived spermatogenesis. (A) One month after transplantation, donor-derived spermatogenesis is observed as distinctive colonies. (B, C) The colonies at this time point are formed as a single cell layer, indicating that meiotic differentiation has not started. (D) Two months after transplantation, the colonies are well established and show a strong blue staining in the center, indicating development of spermatogenesis. (E) At the ends of colonies, donor cells show a similar colonization pattern observed at 1 month after transplantation, suggesting ongoing expansion of colonies. (F) A paraffin section showing that qualitatively complete spermatogenesis is reestablished at this time point (counterstained with nuclear fast red). Reproduced from Nagano et al. (18), with permission.

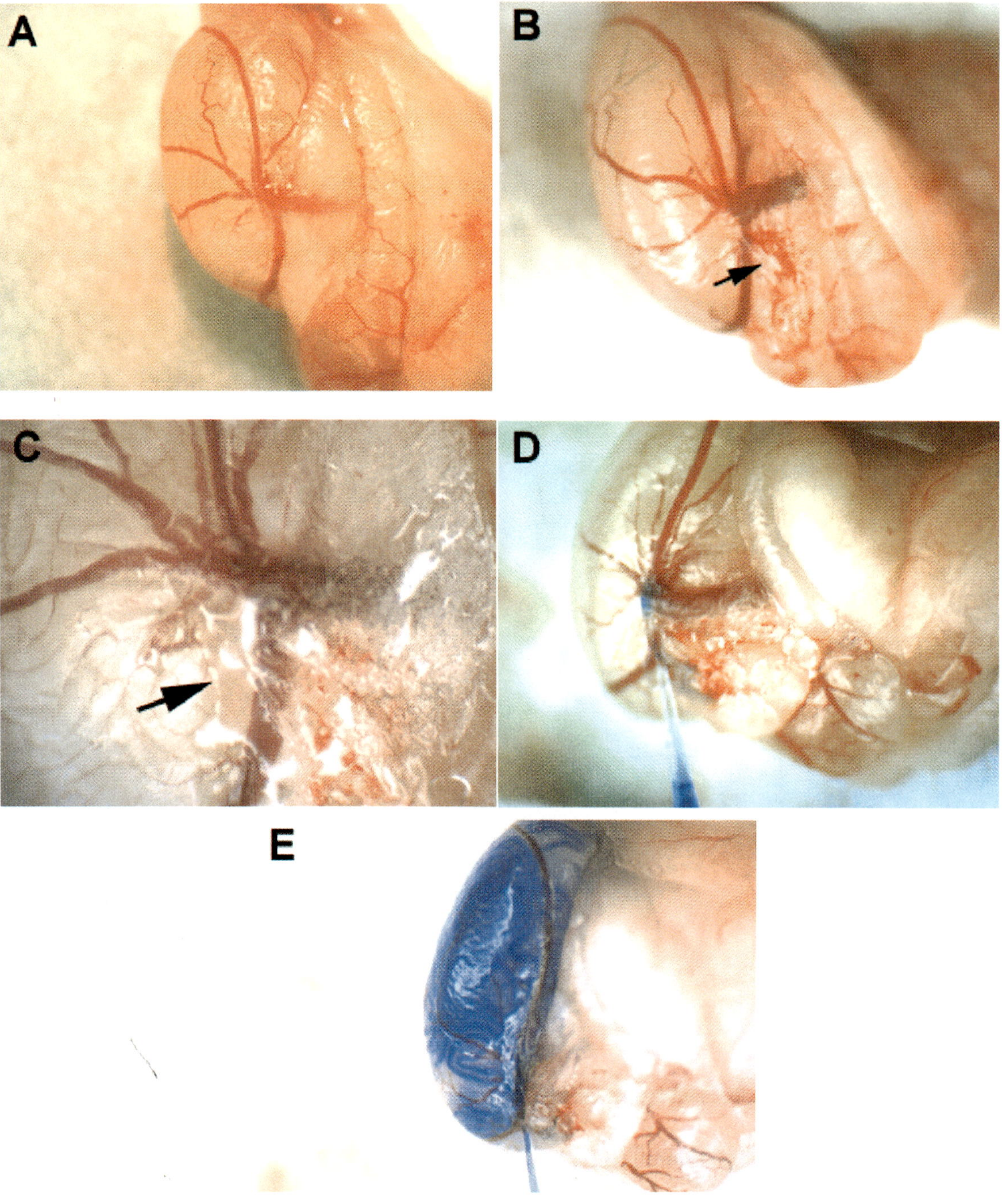

Figure 22.3. Spermatogonial transplantation procedure. Donor cell injection into a right testis. (A) A recipient testis is set in such a way that the cranial end of the testis faces the surgeon. The epididymis and fatty tissue are placed toward the left side of the recipient mouse. (B) A bundle of efferent ducts (arrow) can be found as a translucent line within opaque-colored fatty tissue between the epididymis and the testis. (C) The opening of the rete testis ("window") is found at the beginning of the efferent duct as an oval, shady hollow on the surface of the testis (arrow). (D) The injection pipette filled with donor cells (visualized with trypan blue) is inserted into the duct and reaches the rete testis, to which donor cell suspension is the first introduced. (E) Donor cells are further introduced into the seminiferous tubules through the rete testis.

Protocol 20.6. PKC assay

The activity of PKC can be assayed for at a biochemical level by providing both a substrate and radiolabeled ATP and measuring phosphorylation of this substrate.

1. Lyse 3 eggs in 2 μl of collection buffer (PBS containing 1 mg/ml PVA, 5 mM EDTA, 10 mM Na_3VO_4, and 10 mM NaFl) in a microfuge tube and flash-freeze in liquid N_2 to prepare a cell lysate.
2. Add the following reagents to individual microfuge tubes. The final concentrations in a volume of 10 μl are shown:

   - 100 mM KCl, 5 mM $MgCl_2$, 10 mM EGTA, 20 mM HEPES (pH 7.6) to adjust ionic conditions, pH, and chelate calcium
   - 54 mM glycerophosphate and 14.5 mM p-nitrophenylphosphate as phosphatase inhibitors from 100 mM and 50 mM stocks (kept frozen), respectively
   - 1 μg/ml of each of the following protease inhibitors: pepstatin, aprotinin, chymostatin, leupeptin, and trypsin-chymotrypsin inhibitor from 5 mg/ml stocks (kept frozen)
   - Kinase inhibitors: 2.4 μM PKI (protein kinase inhibitor KN-93), 10 μM ML-9, 75 μM genistein, to inhibit protein kinase A, myosin light-chain kinase, and tyrosine kinase.
   - 5 μCi [^{32}P]-ATP (s.a. 6000 Ci/mM, Redivue, Amersham, Arlington Heights, IL).
   - 5 μg MARCKS peptide (amino acids 151–175) as the peptide substrate
   - 1.5 μl of cell lysate which contains the presumptive PKC.

3. React the kinase assay for 30 min at 37°C and stop by adding an equal volume of 2× concentrated SDS sample buffer for electrophoresis. The MARCKS peptide can be viewed in 15% SDS-polyacrylamide gels after autoradiography. As controls, incubate an aliquot of the same cell lysate in a reaction mixture identical to that described for the experimental, except modify this reaction mixture to contain both of the following PKC inhibitors: 200 μM PKCγ (aa. 19–36) and 2 μM bisindolylmaleimide I. These two PKC inhibitors should suppress any phosphorylation due to PKC, consequently if any phosphorylation of MARCKS peptide is present in the control lane, it suggests that there is activity of a kinase other than PKC which can phosphorylate MARCKS peptide. PKCγ and BIM are two structurally unrelated PKC inhibitors that are highly specific for PKC. PKCγ is a peptide inhibitor that has the amino acids that constitute the pseudosubstrate domain of PKC, whereas BIM is a pharmacologic inhibitor that attaches to PKC at the ATP binding site on the kinase (42). When these PKC inhibitors are used, they should be preincubated with the lysate (or reaction mixture without substrate) for 15 min to permit binding of the inhibitor to PKC.

Notes: EGTA is present in the reaction mixture to chelate calcium in order to measure the endogenous levels of PKC activity. This is because once PKC is activated, it no longer requires calcium to remain active (43). Because homogenization of the cells is likely to release calcium (albeit in varying levels in different homogenates) from sequestered stores, a lysate made in the absence of a calcium chelator would be likely to artifactually activate more PKC than was active just before to homogenization. To measure the total amount of PKC, a specific amount of calcium in the form of calcium chloride can be added to the homogenate.

Protocol 20.7. CaM KII assay

1. For each reaction, wash 5 eggs/embryos briefly in PBS and add them in 2 ml PBS to the wall of a microfuge tube containing the kinase reaction mixture containing

the following components at their final concentrations after the addition of substrate and ATP (add the latter two components after making the lysate):

- 115 mM KCl, 5 mM 5,5'-diBrBAPTA (1,2,bis(-Aminophenoxy) ethane NNN'N' tetraacetic acid), 2.1 mM CaCl$_2$, 5.75 mM MgCl$_2 \cdot$ 6H$_2$O, 23 mM HEPES, pH 6.8. (This buffer adjusts the free-calcium level to 400 nM in the reactions with the calcium/diBrBAPTA buffer. The ionic conditions and pH are adjusted with the other components.)
- 240 mM β-glycerophosphate (Sigma), 120 mM *p*-nitrophenyl phosphate (Sigma) are present as phosphatase inhibitors
- 1 μg/ml of each of the protease inhibitors aprotonin, Bowman-Birk inhibitors (trypsin-chymotrypsin inhibitors), chymostatin, leupeptin, and pepstatin.
- 2.2 mM PKI (BIOMOL, Plymouth Meeting, PA), 75 mM genestein (BIOMOL), 10 mM ML-9 (BIOMOL), 200 mM PKCy are added as inhibitors of PKA, tryosine kinase, myosin light-chain kinase, and PKC, respectively

2. Immediately spin down samples and flash-freeze by immersion in liquid N$_2$. Keep frozen in liquid N$_2$ or store at –80°C until the kinase reaction is performed.
3. Thaw samples and add 10 mg Autocamtide-2 CaM KII peptide substrate (BIOMOL) and 0.25 mCi [^{32}P]-ATP (Amersham) to each tube.
4. Incubate complete reaction mixtures 30 min at 37°C in a heated microtube mixer (Vortemp 56, Labnet, Woodbridge, NJ). Add 10 mM KN-93, a CaM KII inhibitor, to otherwise identical reaction mixtures for controls at each time point. Pre-incubated control (KN-93) samples 15 min at room temperature with gentle agitation before adding of [^{32}P]-ATP and substrate.
5. Stop reactions by adding an equal volume of 2× tricene sample buffer (#161-0744, Bio-Rad, Hercules, CA) followed by mixing.
6. Run samples on precast Tris-tricine 16.5% polyacrylamide gels (#161-0922, Bio-Rad). Finally, fix the gels, dry and expose to a phosphor screen for 12 h. Exposed phosphor screens can be immediately scanned by a Molecular Dynamics Storm 840 Phosphorimager at 200 μm resolution. Bands corresponding to the molecular weight of Autocamtide-2 are then analyzed.

Protocol 20.8. Antiactive antibodies

Antibodies directed against the active forms of several different kinases now exist. Results from experiments with antiactive antibodies can be compared with results obtained using antibodies that can bind both active and inactive forms of the same kinase and can be viewed by Western analysis and immunocytochemistry.

When using immunocytochemistry, comparisons usually have to be performed in parallel batches of eggs/embryos because the active and inactive form of the antibodies have been made from mice, thus preventing double labeling in the same cell. For Western analysis we have found it useful to challenge with the antiactive antibody first, then strip the gel and challenge with the antibody that binds both forms of the kinase. For Western analysis we apply 75 eggs/embryos per lane, which can be monitored using chemiluminance detection system.

Our laboratory has developed a cytological treatment of eggs/embryos that essentially permeabilizes the structure yet keeps the basic cellular components intact. This allows various proteins, ions, inhibitors, and activators within the cell to be introduced (6). Below is a procedure using the permeabilized system to introduce antibodies for immunocytochemistry. *Note*: The permeabilized system is created to view an egg/embryo at a specific time in development. It is best to use various time points to get an accurate representation of cellular activity during development.

1. Permeabilize eggs/embryos by immersion in modified ICB (100 mM KCl, 5 mM MgCl$_2$, 5 mM BAPTA, 20 mM HEPES, pH 6.8) made 1% with Tween-20 and

1 µg/ml with each of the following protease inhibitors: aprotonin (Sigma), Bowman-Birk inhibitors (trypsin–chymotrypsin inhibitors, Sigma), chymostatin (Sigma), leupeptin (Sigma), and pepstatin (Sigma).

2. After permeabilization, briefly wash eggs/embryos in modified ICB without Tween-20 to remove the detergent. Place the eggs/embryos into droplets of medium containing the components to be flushed into the specimens. For example, in experiments in which the ratio of calmodulin to free calcium is 1:4, this medium contained 115 mM KCl, 5.75 mM $MgCl_2$, 10 mM diBrBAPTA (tetrasodium salt, Calbiochem, La Jolla, CA), 2.7 mM $CaCl_2$, 0.25 mM calmodulin (Calbiochem), 1 mM ATP, 1 mM phosphocreatine, 50 mg/ml creatine kinase, 1 mg/ml of the protease inhibitors listed above, and 23 mM HEPES, pH 6.8. We calculate the concentration of free calcium adjusted with a calcium/diBrBAPTA buffer using WinMaxChelator Ver. 1.0 (Chris Patton, Hopkins Marine Station, Pacific Grove, CA). Adjust the ratio of calmodulin to free calcium by varying the calmodulin concentration.

3. Apply kinase inhibitors, if used, by preincubation for 15 min in the modified ICB medium (lacking Tween-20) and also include them in the complete flushing mixtures during incubation.

4. Incubate permeabilized specimens for at least 30 min at 37°C, then cytologically fix with 2% paraformaldehyde in ICB, wash, and prepare for immunocytochemistry with antibodies or stained with Hoechst or DAPI to detect the configuration of the meiotic spindle or chromosomes, respectively.

Protocol 20.9. Electron microscopy

It is difficult to obtain high-contrast images of mammalian eggs and embryos through the conventional resin-embedded process for transmission electron microscopy of thin sections. We have found that the level of contrast and cellular detail can be enhanced by using one of the following procedures: (1) the addition of tannic acid (0.1%) to the glutaraldehyde fixation step to increase the electron density of the specimen or (2) *en bloc* staining with 2% aqueous uranyl acetate before embedding.

By adhering the egg/embryo sample to polylysine-coated slides/cover slips, the eggs/embryos can be processed through the necessary solutions for embedding. Through the dehydration steps, the solutions are transferred by rapidly pouring off most of the petri dish contents into a waste beaker and applying a new solution to the corner of the plate. Eggs/embryos are transferred off of the slide/coverslip by touching their zona pellucida with a blunt object to release them. They are then transferred to an embedding mold. For procedures that detail embedding in a removable medium and preparation of embedment-free sections, see Capco (44), and for those describing embedding in plastic resin, see Capco et al. (10).

Protocol 20.10. Media

KSOM and KSOM-H

	KSOM (g)[a]	KSOM-H (g)[a]	Concentration
NaCl	1.39	1.39	95 mM
HEPES	0	1.2 (0.02 M)	0.02 mM
$NaHCO_3$	0.525	0.084	25 mM/4.0mM
Na lactate	0.28 (214 µl)	0.28 (214 µl)	10 mM
Na pyruvate	0.0055	0.0055	0.2 mM
KCl	0.0465	0.0465	2.5 mM
$CaCl_2 \cdot 2H2O$	0.06275	0.06275	1.71 mM

KH_2PO_4	0.0119	0.0119	0.35 mM
$MgSO_4 \cdot 7H_2O$	0.0123	0.0123	0.20 mM
Glucose	0.009	0.009	0.20 mM
Penicillin	0.015	0.015	100 Ug/ml
Streptomysin	0.0125	0.0125	50 μg/ml
BSA	0.25	0.25	1 g/L
L-glutamine	0.0365	0.0365	1.0 mM
EDTA	0.0009	0.0009	0.01 mM
Phenol red	0.002	0.002	0.023 mM

From Lawitts and Biggers (35)

[a]Final volume 250 ml.

10× Intracellular buffer

37.3 g KCl (1000 mM)
5.1 g $MgCl_2$ (50 mM)
23.9 g HEPES (200 mM)

Add 125 ml of EGTA from a solution of 200 mM (pH 8.0) to make 50 mM. Bring to a final volume of 500 ml using distilled water (pH 6.8). Do not freeze this solution.

10× PBS

80 g NaCl
2 g KCl
11.5 g Na_2HPO_4
2 g KH_2PO_4

Bring to a final volume of 1 l using distilled water (pH 7.35).

10% Paraformaldehyde

Under a hood fill small Erlenmeyer with 17 ml distilled water and add stir bar. Warm water to 70°C and add 2.0 g paraformaldehyde and 1–2 drops of 10 N NaOH. Once dissolved, bring to a final volume of 20 ml. Filter and store at 4°C up to 6 months.

Paraformaldehyde fixative (5 ml final volume; 2% paraformaldehyde in ICB)

1.0 ml 10% paraformaldehyde
0.5 ml 10× ICB
3.5 ml distilled water

Permeabilizing medium for histochemistry (5 ml final volume; 2% paraformaldehyde, 1% Tween-20 in ICB)

1.0 ml 10% paraformaldehyde
250 μl 10% Tween-20
0.5 ml 10× ICB
3.25 ml distilled water

Detergent-extraction medium (5 ml final volume)

500 μl 10× ICB
500 μl 10% Tween-20
3.95 ml distilled water
50 μl AEBSF (light sensitive; add just before use)

ICB-BSA for antibody wash (50 ml final volume; 1% BSA in ICB)

0.45 g BSA
4.5 ml 10× ICB
45 ml distilled water

Store at 4°C up to 2 weeks.

Freezing medium for PKC assay (*10 ml final volume*)

1 ml 10× PBS
6 ml distilled water
0.01 g PVA (1mg/ml)
0.01861 g EDTA (5mM)
0.03804 g EGTA (10mM)
0.01839 g Na$_3$VO$_4$ (10mM)
0.00419 g NaF (10mM)

Bring to a final volume of 10 ml using deionized water.

Wax droplets

Wax droplets are made from a mixture of 1 part histological paraffin and 10 parts commercial petroleum jelly. The paraffin is melted in a histological oven until molten, at which point warmed petroleum jelly is slowly mixed in. While the mixture is still in a liquid state, draw up the wax mixture using a 10-ml syringe. When using the wax-filled syringes for mounting slides, flame the needle (18–20 G needles work best) to warm the wax, quickly wipe off the burned wax, then press out small droplets of wax onto the prepared coverslip. As the needle cools, it will become difficult to press the wax out. At this point flame the needle again; and repeat the above steps until all coverslips have been prepared.

References

1. Williams, C. J., Mehlmann, L. M., Jaffe, L. A., Kopf, G. S., and Schultz, R. M. (1998). *Dev. Biol.*, **198**, 116.
2. Cho, C., Bunch, D. O., Faure, J.-E., Goulding, E. H., Eddy, E. M., Primakoff, P., and Myles, D. G. (1998). *Science*, **281**, 1857.
3. Parrington, J., Swann, K., Shevchenko, V. I., Sesay, A. K., and Lai, F. A. (1996). *Nature*, **379**, 364.
4. Wu, H., He, C. L., and Fissore, R. A. (1998). *Mol. Reprod., Dev.* **49**, 37–47.
5. Yanagimachi, R. (1998). *Hum. Reprod.*, Suppl. 1, 87.
6. Gallicano, G. I., McGaughey, R. W., and Capco, D. G. (1995). *Dev. Biol.*, **167**, 482.
7. Hatch, K. R. and Capco, D. G. (2001). *Mol Reprod. Dev.*, **58**, 69.
8. Johnson, J., Bierle, B. M., Gallicano, G. I., and Capco, D. G. (1998). *Dev. Biol.*, **204**, 464.
9. Pauken, C. M., and Capco, D. G. (1999). *Mol. Reprod. Dev.*, **54**,135.
10. Capco, D. G., Gallicano, G. I., McGaughey, R. W., Downing, K. H., and Larabell, C. A. (1993). *Cell. Motil. Cytoskel.*, **24**, 85.
11. Gallicano, G. I., McGaughey, R. W., and Capco, D. G. (1991). *Cell. Motil.,* **18**, 143.
12. Gallicano, G. I., McGaughey, R. W., and Capco, D. G. (1992). *J. Exp. Zool.*, **263**, 194.
13. Gallicano, G. I., Larabell, C. A., McGaughey, R. W., and Capco, D. G. (1994). *Mech. Dev.*, **45**, 211.
14. Schwarz, S. M., Gallicano, G. I., McGaughey, R. W., and Capco, D. G. (1995). *Mech. Dev.*, **53**, 305.
15. Liu, L., and Yang, X. (1999). *Biol. Reprod.*, **61**, 1.
16. Lorca, T., Galas, S., Fesquet, D., Devault, A., Cavadore J.,-C. and Dorèe, M. (1991). *EMBO J.* **8**, 2087.
17. Moos, J., Xu, Z., Schultz, R. M., and Kopf, G. S. (1996). *Dev. Biol.*, **175**, 358.
18. Gallicano, G. I., Yousef, M. C., and Capco, D. G. (1997). Bioessays, **19**, 29.
19. Luria, A., Tennenbaum, T., Sun, Q. Y., Rubinstein, S., and Breitbart, H. (2000). *Biol. Reprod.*, **62**, 1564.
20. Raz, T., Eliyahu, E., Yesodi, V., and Shalgi, R. (1998). *FEBS Lett.*, **431**, 415.
21. Raz, T., Ben-Yosef, D., and Shalgi, R. (1998). *Biol. Reprod.*, **58**, 94.
22. Colonna, R., Tatone, C., Malgaroli, A., Eusebi, F., and Mangia, F. (1989). *Gam. Res.*, 24, 171.
23. Gallicano, G. I., Schwarz, S. M., McGaughey, R. W., and Capco, D. G. (1993). *Dev. Biol.*, **156**, 94.
24. Gallicano, G. I., McGaughey, R. W., and Capco, D. G. (1997a). *Mol. Reprod. Dev.*, **46**, 587.

25. Endo, Y., Schultz, R. M., and Kopf, G. S. (1987). *Dev. Biol.*, **119**, 199.
26. Sun, Q. Y, Wang, H.-H., Hosoe, M., Taniguchi, T., Chen, D.-Y., and Shioya, Y. (1997). *Dev. Grow. Diff.*, **39**, 523.
27. Ducibella, T., and LeFevre, L. (1997). *Mol. Repro. Dev.*, **46**, 216.
28. Raz, T., Skutelsky, E., Amihal, D., Hammel, I., and Shalgi, R. (1998c). *Mol. Reprod. Dev.*, **51**, 295.
29. Colonna, R., Tatone, C., Francione, A., Rosati, F., Callaini, G., Corda, D., and DiFrancesco, L. (1997). *Mol. Reprod. Dev.*, **48**, 292.
30. Endo,Y., Kopf, G. S., and Schultz, R. M. (1986). *J. Exp. Zool.*, **239**, 401.
31. Moses, R. M., and Kline, D. (1995). *Dev. Biol.*, **171**, 111.
32. Moore, G. D., Kopf, G. S., and Schultz, R. M. (1995). *Dev. Biol.*, **170**, 519.
33. Winston, N. J., and Maro, B. (1995). *Dev. Biol.*, **170**, 350.
34. Xu, Z., Abbott, A., Kopf, G. S., Schultz, R. M., and Ducibella, T. (1997). *Biol. Reprod.*, **57**, 743.
35. Lawitts, J. A., and Biggers, J. D. (1992). *Mol. Reprod. Dev.*, **31**, 189.
36. Erbach, G. T., Lawitts, J. A., Papaioannou, V. E., and Biggers, J. D. (1994). *Biol. Reprod.*, **50**, 1027.
37. Ho, Y., Wigglesworth, K., Eppig, J. J., and Schultz, R. M. (1995). *Mol. Reprod. Dev.*, **41**, 232.
38. Summers, M. C., McGinnis, L. K., Lawitts, J. A., Raffin, M., and Biggers, J. D. (2000). *Human Reprod.* **15**, 1971.
39. Moses, R. M., and Kline, D. (1995). *Dev. Biol.*, **171**, 111.
40. Moses, R. M., and Masui, Y. (1995). *Zygote,* **3**,1.
41. Shapiro, P. S., Vaisberg, E., Hunt, A. J., Tolwinski, N. S., Whalen, A. M., McIntosh, J. R., and Ahn, N. G. (1998). *J. Cell Biol.*, **142**, 1533.
42. Davis, P., Hill, C., Lawton, G., Nixon, J., Sedwick, A., Wadsworth, J., Westmacott, D., and Wilkinson, S. (1989). *FEBS Lett.* **259**, 61.
43. Ashendel, C. (1985). *Biochim. Biophys. Acta*, **822**, 219.
44. Capco, D. G. (1993). In *Polyethylene Glycol as an Embedment for Microscopy and Histochemistry,* K. Gao, ed., p. 97. CRC Press, London.

21

SUSAN E. LANZENDORF
CATHERINE A. BOYD
DIANE L. WRIGHT

Human Embryonic Stem Cells

1. ISOLATION AND PRODUCTION OF EMBRYONIC STEM CELLS

In 1981, Evans and Kaufman (1) described the technique for isolating embryonic stem (ES) cell lines from mouse blastocysts. In this procedure, the inner cell mass is used to give rise to an undifferentiated cell line capable of indefinite proliferation in vitro. ES cell lines are also pluripotent, meaning that under the appropriate conditions, they can undergo differentiation to give rise to any cell type in the body. This potential can be demonstrated in vivo by injection into mouse blastocysts to create chimeras or by grafting to an ectopic site, such as the testis, to form teratomas. In the absence of cytokines that prevent differentiation, ES cells form embryoid bodies in vitro, structures that contain the three primary germ layers.

ES cell lines have also been produced in other animal models, including the chicken (2), hamster (3), and pig (4). In primates, ES cell lines have been produced in the common marmoset (5), rhesus monkey (6), and human (7–9). The evolution of primate ES cell research raises hope that these cells may one day form the basis of revolutionary cell therapies for tissue degeneration in chronic disease (e.g., diabetes mellitus, Parkinson's disease) or after necrosis/apoptosis (e.g., ischemic heart disease, bone marrow irradiation).

The methods for producing ES cells involve development to the blastocyst stage followed by removal of the zona pellucida, isolation of the inner cell mass using immunosurgery, and maintenance of the cell line in vitro in an environment that prevents spontaneous differentiation. As in the mouse model, primate ES cell lines have been co-cultured with primary mouse embryonic fibroblasts (PMEFs) that have been treated to make them mitotically inactive. The PMEFs secrete leukemia inhibitory factor (LIF), a cytokine that has been shown to prevent differentiation of ES cell lines in the mouse, but the requirement needs further definition in other species (10).

To verify the pluripotential characteristics of an ES cell line requires rigorous testing of the phenotype. To ensure that the cell lines have not differentiated, samples must stain positive for alkaline phosphatase, and, in the primate, they must express the stage-specific embryonic antigen (SSEA)-4 (7–9). Investigators have also used other specific cell-surface markers such as SSEA-3, TRA-1-60, and TRA-1-81 to verify an undiffer-

313

entiated status for primate cell lines (5–7). In addition, as with other immortal cell lines, ES cell lines should express high levels of telomerase (7), the ribonucleoprotein involved in stabilizing telomeres and preventing senescence (11).

When removed from the presence of LIF, mouse ES cells begin the initial stages of differentiation to form embryoid bodies, which are cell aggregates containing the three germ layers, the mesoderm, endoderm, and ectoderm. There are reports (5, 6) that primate ES cells differ in their ability to form embryoid bodies during initial differentiation in vitro. In the nonhuman primates, embryoid body formation was not mentioned in the rhesus monkey (6), but it was reported in cell lines originating from the marmoset (5). Reubinoff et al. (8) reported that their human ES cell lines did not demonstrate embryoid body formation, but could, under extended culture, produce multicellular aggregates. Lanzendorf and co-workers (unpublished data), have also observed similar cellular aggregates and the possibility that these are the same as embryoid bodies should be investigated. These differences may be due to variations in culture techniques or reporting methods. In all primate species reported to date, the xenotransplantation of ES cells into immunodeficient SCID mice have produced teratomas histopathologically made up of tissues from all three germ layers, providing strong evidence of the pluripotentiality of undifferentiated cell lines (5–8).

In most ES cell models, the inner cell mass is isolated from the in vivo-fertilized embryo flushed from the uterine cavity. The exception to this is human ES cell lines, where the embryos are fertilized in vitro and inner cell mass isolation is performed on fresh embryos (7–9) or thawed embryos after cryostorage (7). It is unclear if cryopreservation of an embryo before use for generating an ES cell line decreases its potential, comparative studies of fresh and frozen embryos have not been performed. Most likely, cryopreservation decreases the numbers of viable embryos available for inner cell mass recovery due to cell death during freezing and thawing.

Currently, ES cells are propagated on fibroblast feeder-layers to prevent differentiation of the cell lines. The use of feeder-layers in the culture system poses many technical problems, including added quality control and quality assurance, difficulty assessing cell quantities and viability, and, because the feeder cells are from another species, risk of viral contamination across species.

PMEFs are used as the feeder layers because they synthesize and secrete LIF into the culture medium. PMEFs also provide an extracellular matrix that has been shown to associate with an immobilized form of LIF (12). However, studies have been performed that allow mouse ES cell lines to be established and propagated without a fibroblast feeder layer. Two early studies demonstrated the use of media previously conditioned with feeder layers to prevent mouse stem cell differentiation (13, 14). Another approach is to use an extracellular matrix produced by growing fibroblasts for a few days and then removing them with a lysis buffer (10). The ES cells are then grown on the extracellular matrix in the presence of soluble LIF.

Previous studies performed with primate ES cells demonstrate the need for some type of subcellular matrix to prevent differentiation as well as to optimize spontaneous differentiation. Reubinoff and co-workers (8) reported that the spontaneous differentiation of human ES cells could be accelerated by prolonging culture to up to 7 weeks without replacing the feeder layer. In both the human and nonhuman models, investigators report that removal of the feeder layer during differentiation will sometimes result in decreased survival and cell death (8, 12).

The techniques and protocols (Protocols 21.1–21.14) are those currently used for the initiation and propagation of human embryonic stem cell lines at our institution. We began using these methods in 1997 when little was known about human ES cell lines. Therefore, many of the protocols are modifications of those reported in the mouse model. Over the years and through trial and error, we have developed these protocols for our specific research needs. However, the field of ES cell research is growing rapidly, and new techniques and improved culture conditions are being reported everyday.

2. TECHNIQUES

2.1 In Vitro Fertilization and Embryo Culture

Previous reports of human ES cell line production have described the use of fresh embryos donated after in vitro fertilization (7, 8), embryos produced by donated gametes (9), or cryopreserved embryos donated by infertile couples (7). Human ES cell line production is unique from that of other species because the embryos are fertilized in vitro instead of flushed from the uterine cavity. Regardless of where the embryos originate, one could assume that embryos of higher quality will result in better or more ES cell lines, as in the case of pregnancy after embryo transfer. Therefore, optimization of in vitro conditions for development to the blastocyst stage should be performed as for clinical in vitro fertilization. If you are working in a laboratory that is not associated with a clinical program, you may want to evaluate the surrounding environment to ensure there are no contaminants that may adversely affect embryo quality (15, 16).

When donated by infertile couples, embryos will typically be cryopreserved, and the thaw procedure used will depend on how the embryos were frozen. Some clinical programs may also use specific culture requirements that will need to be followed when the embryos are thawed. If embryos are obtained from another facility, request copies of the thaw and culture procedures. It is our experience that the quality of the inner cell mass (ICM) will determine the initial quality of a cell line (Figure 21.1). Embryos are evaluated using the grading system presented in Table 21.1. With fresh embryos, we have found that blastocysts with compact, round ICMs (grades A and B) result in the highest numbers of human ES (hES) cell colonies after isolation. Also, when an ICM can be visualized before treatment, as with embryo grades 1 and 2, about 87% of the ICMs can be recovered after immunosurgery. When they cannot be visualized (grade 3), only about 10% of the embryos will contain a viable ICM. However, we have obtained a viable ICM from a grade-3 embryo, and after plating, the ICM did form a stem cell colony. Therefore, it may be worthwhile to perform immunosurgery on all blastocysts, even those without a visible ICM.

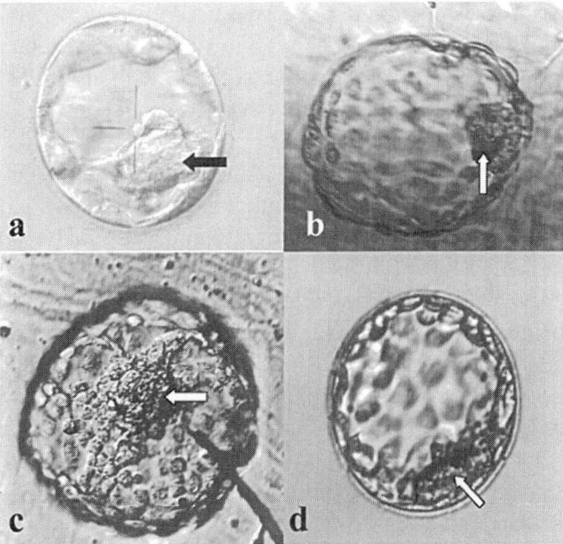

Figure 21.1. Grading system used to evaluate inner cell mass in expanded human blastocysts (see table 21.1). (a). Grade A ICM (arrow); (b). grade B ICM demonstrating small area of degeneration (arrow); (c). grade C ICM (arrow) photographed from above; (d). grade D ICM (arrow). All photographs: original magnification × 400.

Table 21.1. Grading System Used to Evaluate Human Blastocyst-Stage Embryos

Grade	Description
1	Expanded blastocyst, ICM visible, no areas of degeneration or bodies within blastocoel
2	Expanded blastocyst, ICM visible, areas of degeneration visible and/or bodies within blastocoel
3	Expanded blastocyst, ICM not visible, no or various amounts of degeneration or inclusions

Within the expanded blastocyst, the inner cell mass is graded as follows:

A	ICM is compact and clear, no dark cells; ICM is shaped like a ball
B	Same as A but with a small amount of dark cells
C	ICM is visible but is not compact; it is spread across the inside of the trophoblast; ICM appears flat, there are no dark cells
D	Same as C but with dark cells

2.2 Removal of the Zona Pellucida

The first step in isolating the ICM is to remove the zona pellucida from the blastocyst. This can be achieved by briefly exposing the embryo to acidified tyrodes medium or a Pronase solution or by allowing the embryo to complete hatching on its own. We have found that performing laser hatching on day 2 or 3 of culture facilitates hatching in vitro. We have not noted any difference in cell-line derivation between embryos treated with Pronase or acidified tyrodes medium or those that hatch unassisted. However, when possible, we prefer not to expose the embryos to any conditions that may affect their potential to develop into a stem cell line.

2.3 Isolation of the Inner Cell Mass

In mouse cells, the ICM can be isolated by allowing the trophoblast to attach and spread out on the bottom of the culture dish. The ICM forms on top of the trophoblast and can be pushed off with a pipette and used to generate a cell line (17). In human cells, it is often difficult to visualize the ICM within the attached and growing trophoblast and to recover it without also collecting trophoblast cells that may contaminate the cell line (Figure 21.2). For this reason, immunosurgery is used to isolate the ICM by destroying the cells of the trophoblast with an antibody (Figure 21.3). Using rabbit antimouse spleen serum, Solter and Knowles (18) successfully isolated the mouse ICM. Exposure of the blastocyst to the antiserum, followed by guinea pig complement, destroyed all of the tro-phoblastic cells but not the ICM. Because subsequent exposure of the isolated ICM to antiserum also resulted in its destruction, the authors suggested that the immunoglobulin molecules are too large to pass through the tight junctions between the trophoblastic cells (18). Human ES studies involving immunosurgery have used rabbit antiserum to BeWo cells (7, 9) and antihuman serum antibody (8).

In our laboratory, an antibody was prepared by intravenously injecting $1\text{--}2 \times 10^7$ BeWo cells, a human choriocarcinoma cell line (CCL-98; ATCC, Rockville, MD) into a New Zealand white rabbit in 0.5 ml PBS every 14 days for a total of three injections. With the exception of BeWo cell line expansion, all procedures were performed by Strategic BioSolutions (Newark, DE). A test bleed was performed 14 days after the last immunization, and the serum was evaluated for its ability to lyse BeWo cells in culture and the appropriate dilution for use determined. The animal was then euthanized and the serum collected and stored at $-80°C$.

2.4 Cell Line Expansion and Prevention of Differentiation

Once an hES cell line has been established, the next goal is to propagate the line to obtain sufficient numbers for experimentation and cryopreservation. To do this, one must pro-

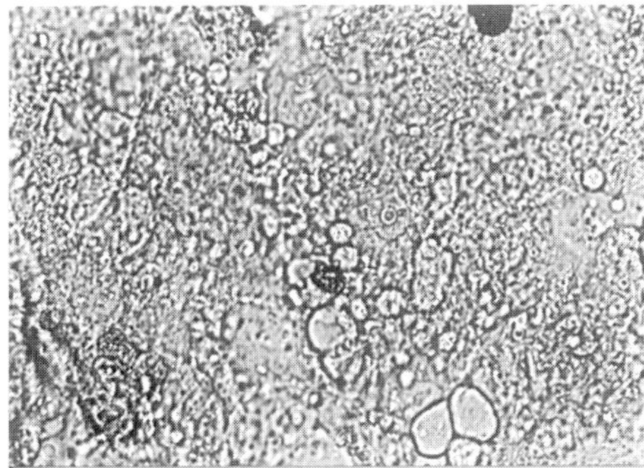

Figure 21.2. Photomicrograph of human ICM plated along with the trophoblast. The cells of the trophoblast have overgrown the colony, and it is no longer possible to distinguish the cells of the ICM. Original magnification × 400.

vide the cell lines with an environment that allows for cell growth but at the same time prevent the cells from differentiating. One method is to grow the cells on primary mouse embryonic fibroblasts (PMEFs) that have been mitotically inactivated using either γ-radiation or treatment with mitomycin-C (17). After inactivation, the PMEFs are seeded with the hES cells and should support their growth for about 7 days, at which time the hES cells should be replated to fresh PMEFs. It remains unclear if the cytokine LIF must be added to the medium when PMEFs are used as the substrate (19). The PMEFs synthesize and secrete LIF into the culture medium, and many researchers eliminate LIF from their media for ES culture (20, 21).

Another advantage the PMEFs provide is an extracellular matrix on which the hES cells can grow. However, the use of PMEFs to provide the matrix presents many problems, including the risk of cross-species viral contamination. Xu and co-workers (22) demonstrated that undifferentiated hES cells could be maintained on the artificial matrix Matrigel (Becton Dickinson, Bedford, MA) provided the culture medium was previously conditioned on PMEFs. Four human cell lines were evaluated, and the investigators reported that all lines could be maintained for a least 130 population doublings (22).

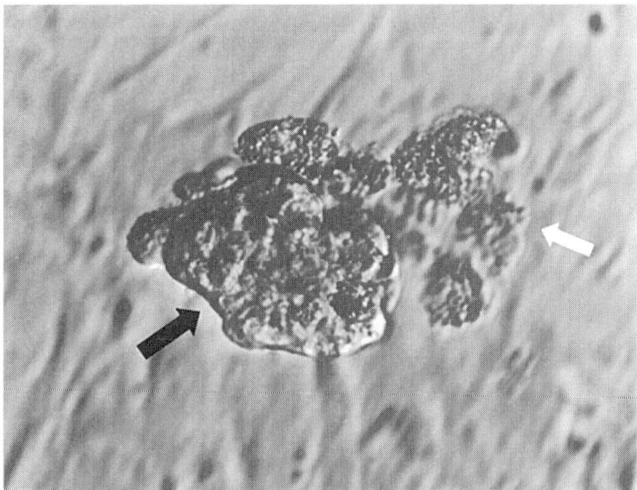

Figure 21.3. Isolated human ICM (black arrow) after immunosurgery. Degenerate cells of the trophoblast (white arrow) can be seen adhered to the ICM. Original magnification × 400.

The hES cell lines also maintained normal karyotypes, high telomerase activity, and expressed the appropriate cell-surface markers for undifferentiated human ES cells.

As in the mouse model (17), hES cell lines were initially cultured in DMEM (high glucose) supplemented with 20% FBS, 1 mM glutamine, 0.1 mM β-mercaptoethanol, and 1% nonessential amino acid stock. However, there has been a recent movement toward culture under more defined conditions and researchers now report the use of an optimized DMEM (KnockOut DMEM; Invitrogen, Carlsbad, CA) and a serum replacement (Knockout Serum Replacement, Invitrogen). The information provided by Invitrogen (www.invitrogen.com) states that the Knockout D-MEM is optimized for growth of undifferentiated ES cells and that the Knockout Serum Replacement is a serum-free formulation that has been evaluated for its ability to support the undifferentiated growth of ES cell colonies on PMEFs.

2.5 Assessment of Human ES Cell Lines

In the mouse model, a set of criteria has been developed to ensure that an ES cell line is truly an undifferentiated line of embryonic origin. When Thomson et al. (7) reported the production of the first human ES cell lines, this set of criteria was modified to fit the special needs of the human model (Table 21.2).

As in the other animal models, an hES cell line must be derived from a preimplantation embryo. There are other types of cell lines that demonstrate similar characteristics to ES cells lines, including embryonic germ (EG) cells, derived from the primordial germ cells of the fetus, and embryonic carcinoma (EC) cells, derived from testicular or ovarian teratocarcinomas. However, EG cell lines have a lower capacity for undifferentiated proliferation, whereas EC cell lines are more limited in what they can be directed to differentiate into and often contain subtle chromosomal abnormalities (23).

An hES cell line will display a characteristic morphology. The individual cells of a colony will have a high ratio of nucleus to cytoplasm, prominent nucleoli, and distinct cell borders. The colonies will have distinct boundaries around them, making them easy to differentiate from the surrounding PMEFs (Figure 21.4).

Embryonic stem cells of all species must be capable of prolonged, undifferentiated growth. For culture with PMEFs, hES cells will typically require replating to fresh PMEFs every 7–11 days to maintain the undifferentiated state. Overextending this timeframe on the same PMEFs will initiate spontaneous differentiation. Also, it is not uncommon for some colonies to display areas of differentiation, and, if this occurs, care should be taken to not include the differentiating cells when the colonies are replated.

Because ES cells are pluripotent and can theoretically develop into any cell type, an ES cell line should be capable of forming derivatives of all three embryonic germ layers—the endoderm, ectoderm, and mesoderm. Previous studies with hES cell lines demonstrated this ability by injecting the undifferentiated cells into severe combined

Table 21.2. Criteria for the Assessment of Human Embryonic Stem Cell Lines

1. Derived from a preimplantation stage embryo.
2. Morphology includes a high ratio of nucleus to cytoplasm, prominent nucleoli, and distinct cell borders.
3. Demonstrates numerous cell passages without undergoing differentiation.
4. Capable of forming derivatives of all three embryonic germ layers (endoderm, ectoderm and mesoderm).
5. Expresses high levels of telomerase activity.
6. Expresses cell-surface markers that characterize an undifferentiated line, such as alkaline phosphatase and SSEA-4.
7. Demonstrates normal karyotypes through frequent passages.
8. Capable of cryostorage.

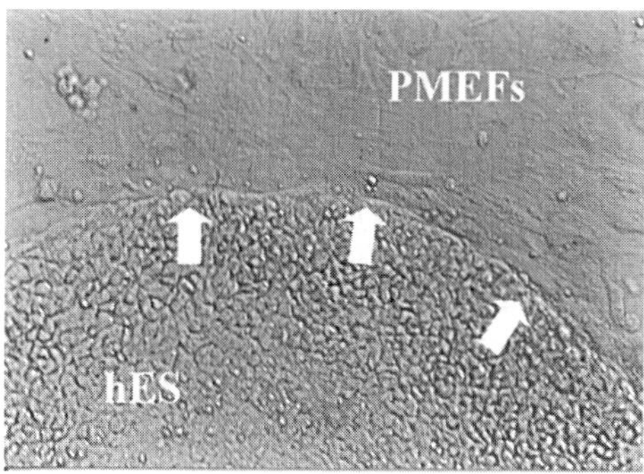

Figure 21.4. Human embryonic stem cell colony (hES) plated on primary mouse embryonic fibroblasts. Arrows point out the distinctive border between the hES cells and the feeder layer. Original magnification × 200.

immunodeficient mice to produce teratomas. Histological evaluation of these teratomas demonstrated the presence of many different cell types including gut epithelium (endoderm), muscle (mesoderm), and neural epithelium (ectoderm) (7, 8). Another method for demonstrating a cell line's potential for development in vivo is to allow the cells to begin the differentiation process in vitro and then evaluate the cells with markers for the germ layers. For example, polyclonal antibodies S-100 (DAKO Corporation, Carpintera, CA) will identify neural crest derivatives and ectoderm and α-1-fetoprotein (DAKO) identifies endoderm (Figure 21.5a), whereas the monoclonal antibody to muscle actin (DAKO) identifies mesoderm (Figure 21.5b).

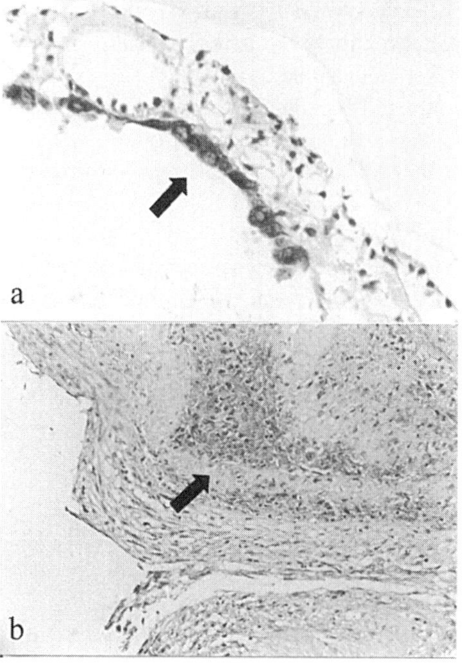

Figure 21.5. Light microscopic image of human embryonic stem cells after undirected differentiation in vitro. (a) Section of ES cells cultured 11 days without feeder layer, labeled with rabbit antihuman α-1-fetoprotein (arrow), a marker for endoderm; original magnification × 400. (b) Section of ES cells cultured 21 days without feeder layer, labeled with monoclonal mouse antihuman muscle actin (arrow) for identification of an epitope present on muscle actin; original magnification × 200. See color insert.

Another criterion for hES cell lines is their ability to express high levels of telomerase activity (7). Telomerase is a ribonucleoprotein that provides a mechanism for the stabilization of telomeres on the ends of chromosomes during cellular replication. DNA polymerase is unable to replicate these ends, and telomerase provides the template for the 6-bp repeating sequence. Without telomerase, a cell has a limited number of divisions possible because the telomeres shorten to a critical point at which the cell enters senescence. For this reason, high levels of telomerase activity have been found in cells types that continue to grow for many generations, such as embryonic tissue (24), hematopoietic cells (25), and tumor cells (26). We evaluate telomerase activity using the TRAPeze kit (Intergen, Inc., Purchase, NY), with the modifications described by Wright et al. (27).

In the mouse, another criterion includes the ability of the stem cells to contribute to normal development after injection into the blastocyst to form a chimera. The injected ES cells intermingle with the host ICM and may contribute to any tissue, including the germ cells (28). However, for ethical reasons, performing this same procedure with human ES cells and human blastocysts is not possible (7).

Until recently, another criterion unique to the mouse model was the ability to clone an ES cell line from a single cell. Amit and co-workers (29) reported the derivation of two clonally derived hES cell lines from an existing hES cell line that had already undergone continuous passage for 6 months. After 8 months of culture, the two cloned cell lines were also capable of meeting the criteria for hES cells described here.

Human ES cell lines should also express certain cell-surface markers to demonstrate that they are undifferentiated. The most common marker, alkaline phosphatase, is also the easiest assay to perform. When replating cell lines, one can periodically leave a few colonies in the plate and stain for alkaline phosphatase (Figure 21.6). It is also beneficial to perform this assay when first learning to work with hES cells to become acquainted with correct morphology. The assay kit is available from many different sources.

Undifferentiated hES cells should also stain positive for SSEA-4. To ensure that the assay is working properly, routinely evaluate mouse ES cells with SSEA-4, looking for a negative result. SSEA-1 can also be used, which should stain positive for mouse ES cells and negative for human ES cells. Other investigators working with hES cell lines also report the use of TRA-1-81 (7) and TRA-1–60 (7, 8).

It is also important that hES cell lines maintain a normal karyotype over repeated passages, particularly if they will be used for transplant therapies. Thomson and co-workers (7) reported normal karyotypes for all five of their established hES cell lines at passages 2–7. The methods for karyotyping and the number of cells examined in each cell line were not reported. Reubinoff et al. (8) reported the use of the standard G-banding procedure for evaluating their two hES cell lines, but the number of cells examined were

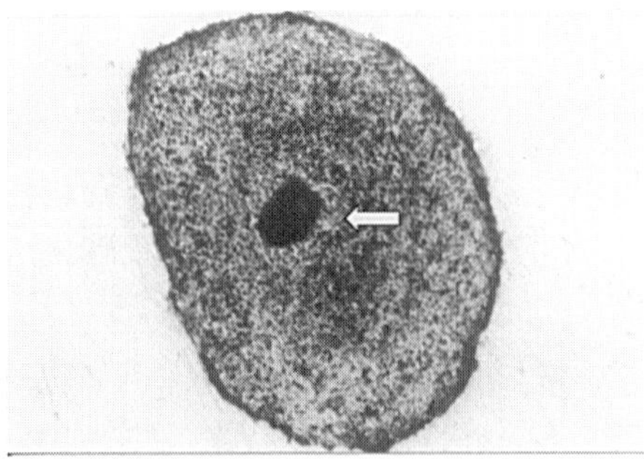

Figure 21.6. Human embryonic stem cell colony stained with alkaline phosphatase to show nondifferentiated areas (blue). An embryoid body is forming in the center of the colony (arrow). Original magnification × 100. See color insert.

not reported. When describing the clonal derivation of two hES cell lines, Amit and co-workers (29) reported the use of either the standard G-banding procedure or multicolor spectral karyotyping, as described by Schrock et al. (30). In that study, the evaluation of 20 chromosomal spreads per line examined appeared to be the standard.

An important criterion for a successful hES cell line is the ability to survive cryopreservation so that the line can be transported, saved for future use, and as insurance against problems in the laboratory. In the mouse model, ES cells survive well after cryopreservation (17). Unfortunately, human ES cells appear to be much more susceptible to cryodamage. A discussion of the methods used to date are found in the following section.

2.6 Cryostorage of Human ES Cell Lines

As each cell line is established, it should be cryopreserved at various passages. In the mouse, ES cells have been successfully cryopreserved with high rates of survival using different protocols, including the standard technique for tissue culture cells (31). This involves placing the cells in a 10% DMSO solution, followed by cooling between Styrofoam blocks for 24 h in a –80°C freezer before plunging and storage in liquid nitrogen (10, 17).

In the first reports of hES cell production, Thomson et al. (7) stated that three of the lines were successfully cryopreserved, but a method was not discussed. Reubinoff et al. (8) also reported successful recovery after cryopreservation, but, again, few details were provided. Unsatisfied with the low survival and high rates of differentiation after cryopreservation using the standard technique of tissue culture cells, Reubinoff et al. (32) evaluated vitrification, a method previously shown to be beneficial for freezing human blastocysts. The investigators found that vitrification did result in increased cell recovery and ability to form colonies after replating. In addition, although the vitrified cells did show evidence of cell death and differentiation, these deficiencies could be overcome by additional time in culture.

Our initial attempts at cryopreservation were performed using the standard tissue culture procedure of freezing cells using a solution of 90% FBS/10% DMSO in 1.8 ml cryotubes. The cells were added to 1.0 ml of this cryomedium, cooled for 24 h in a –80°C freezer between Styrofoam blocks, and then transferred to liquid nitrogen. Vials were thawed by placing them in a 37°C waterbath, followed by one wash in DMEM + 20% FBS. The pelleted cells were then added to PMEFs mitotically inactivated using γ-radiation 1 day earlier. Using this protocol, roughly 20% of the recovered colonies displayed an undifferentiated morphology over the next 3 days. The remaining colonies underwent differentiation and stained negative for alkaline phosphatase. We are currently investigating a method for cryopreservation that uses straws in a slow vapor-freeze protocol and results in about 50% of the colonies recovered in the undifferentiated state.

3. TROUBLESHOOTING

The level of quality assurance and control required for initiating and maintaining human ES cell lines is similar to that required by human in vitro fertilization. In both instances, the cells are very susceptible to toxins and inadequate culture conditions. In the future, hES cells may also be used therapeutically in human patients, so rigorous record keeping and the use of Good Laboratory Practices (GLPs) should be maintained at all times. For more information regarding the U.S. Environmental Protection Agencies policies for GLPs, visit the web site http://es.epa.gov/oeca/polguid/glp/glp.html. Outlined below are some of the common problems that our laboratory has experienced or problems that have been described to us by other investigators.

1. Colonies show signs of differentiation. The differentiation may be due to replating colonies that are too large. The hES cells do not respond well to being broken down into single suspensions, but at the same time, colonies that are too large tend to undergo spontaneous differentiation easily. Either mechanically break up colonies into smaller cell clumps or use dispase while drawing the cells in and out of a small-bore pipette to break up the colonies. Also, check media preparation log sheets to determine if a new lot of PMEFs, serum, or other media components could account for the spontaneous differentiation. We recommend that all new lots of serum or serum replacement be tested before using for routine culturing to demonstrate they support normal proliferation and do not initiate differentiation.

2. Cannot locate ICM after it has been added to the PMEFs. A viable ICM or early colony will increase in size daily. If you cannot immediately identify the colony, look for a group of cells that appear to be getting larger and measure it daily. It is not uncommon to lose a colony for a few days as it begins to take on the distinct morphology of a hES cell line.

3. Even though only one ICM was plated, there appear to be three or four colonies growing. Sometimes the PMEFs will form colonies that resemble hES cell colonies. The individual cells of these "false colonies" are rounder in appearance compared to hES cells, which tend to be more rectangular in shape. The false colonies will not undergo rapid growth because they are not capable of prolonged cell division. Before placing an ICM in a well, scan the well and mark on the lid any groups of cells that could be mistaken later for a HES cell colony.

4. CONCLUSIONS

The science of human ES cell production and its potential for treating a host of human diseases is rapidly evolving. Every day there are new reports on the directed differentiation of hES cells into precursor and/or adult cells that could one day be used for transplantation therapies. To date, most of the research involving human ES cells differentiation has been performed in vitro. Research has demonstrated the production of hematopoietic precursors cells (21), neural precursors (20, 33), and neurons (20) from human ES cells in culture. Zhang and co-workers (33) reported the transfer of neural precursor cells, derived from hES cells in vitro, into the neonatal mouse brain. The transplanted human cells were incorporated into different areas of the brains and further differentiated into neurons and astrocytes, further supporting the concept of using human ES cells as a source of transplantable cells for disease treatment.

Clearly, more research is needed to evaluate the use of human ES cells for disease treatments. In addition, studies are needed to evaluate the early stages of cell line production for optimization of the procedures and to further define the culture requirements. Other areas of concern involve the ethical issues surrounding the production and use of human ES cells and what governmental or agency oversight will be required as the field continues to grow.

Protocol 21.1. Isolation of primary mouse embryonic fibroblasts

This protocol is modified from the procedure of Abbondanzo et al. (10). It is reported that PMEFs can be made from any strain of mouse (10). We typically use outbred CD-1 females crossed with outbred CD-1 males.

1. Kill mice on day 13 of their pregnancy by cervical dislocation (day 1 = day copulation plug is observed). Prepare Falcon 150-mm petri dishes (Becton Dickinson) with 50 ml each of calcium, magnesium-free Dulbecco's phosphate-buffered serum (Ca^{2+},Mg^{2+}-free dPBS; Sigma, St. Louis, MO). Prepare three dishes for each pregnant mouse.

2. Remove the uterus and place it in a petri dish as described above. Dissect out the embryos and remove the yolk sac, amnion, and placenta (34). Move the embryos to a fresh dish to rinse. Transfer the embryos to the third fresh dish.

3. Using fine sterile forceps, remove the head and liver (dark red tissue in center of embryo).

4. Place 5–8 dissected embryos into the barrel of a 3- or 5-ml syringe and attach an 18-G needle (Becton Dickinson). Add 2 or 3 ml of Ca^{2+},Mg^{2+}-free dPBS and replace plunger.

5. While the tip of the needle is in a Falcon 50-ml conical centrifuge tube (Becton Dickinson), expel the contents of the syringe. Draw the embryos back into the syringe and repeat the expulsions 4–6 times until the tissue is broken up into single-cell suspension or small clumps of cells.

6. Add 25 ml of room temperature DMEM + 15% FBS (see below).

7. Split approximately 30 ml of PMEF cell suspension into three Falcon 175-cm^2 flasks (Becton Dickinson) containing 60 ml/flask of room-temperature DMEM + 15% FBS and place flasks in incubator (37°C, 5% CO_2). It should take 3–4 days for the flasks to become confluent, and a typical yield is $1–2 \times 10^7$ cells/embryo. Freeze the majority of PMEFs at this stage (see protocol 21.2. This is considered the first passage (P1).

8. To split the flask, rinse the flask with 10 ml Ca^{2+},Mg^{2+}-free dPBS. Add 3–5 ml of 0.25% trypsin-EDTA (Sigma) for a brief 15–30 s exposure and remove the trypsin. Replace with 8–10 ml of fresh trypsin-EDTA and place the flask in incubator for 3–5 min or until cells are rounding up and are dissociated from the flask surface. To inactivate the trypsin, add 15–20 ml of DMEM + 15% FBS, wash (10 min, 500 g at room temperature), and resuspend the cells and split three ways to new flasks for expansion.

Notes: PMEFs should be split by no more than a 1:3 ratio for expansion (10). PMEFs are only useful as a feeder layer through passages 5–6, afterward they cease proliferating. We have found that passages 2 and 3 provide the highest quality PMEFs for use with hES cells.

Preparation of DMEM-15% FBS

Media constituents

Component	Supplier	Catalogue No.
DMEM	Specialty Media (Lavalette, NJ)	SLM-120-B
FBS	Hyclone (Logan, UT)	SH300.70.02
Nonessential amino acids (100×)	Specialty Media	TMS-001-C
2β-Mercaptoethanol (100×)	Specialty Media	ES-007-E
Penicillin/Streptomycin (100×)	Specialty Media	TMS-AB2-C
L-Glutamine (100×) 200 mM	Specialty Media	TMS-002-C

Volumes

	Total volume		
	100 ml	200 ml	500 ml
DMEM	81 ml	162 ml	405 ml
FBS	15 ml	30 ml	75 ml
Nonessential amino acids	1.0 ml	2.0 ml	5 ml
Penicillin/Streptomycin	1.0 ml	2.0 ml	5 ml
2β-Mercaptoethanol	1.0 ml	2.0 ml	5 ml
L-Glutamine	1.0 ml	2.0 ml	5 ml

Mixing

1. Combine all components in tissue culture hood. Store flask of media at 4°C for 2 weeks.
2. Remove the amount needed daily under the tissue culture hood.
3. Equilibrate by gassing (5% CO_2/95% O_2 or clinical blood gas mixture) aliquoted flask: gas 1 min for 50-ml tissue culture flask or 2 min for 250-ml tissue culture flask. *Note*: Only fill equilibrating flask half full or less.
4. Warm to 37°C for a minimum of 1 h in incubator before use.

Protocol 21.2. Cryopreservation of early passage primary mouse embryonic fibroblasts

1. Prepare 30 ml of the freeze medium consisting of 10% DMSO (Sigma/90% FBS (Hyclone, Logan, UT).
2. Remove culture media from each flask and discard. To each flask add 2 ml of trypsin (Sigma) to rinse cells and remove any residual protein. Remove the trypsin and discard.
3. Add 8 ml of trypsin to flask and incubate for 3–5 min. Avoid handling the flask at this time. If the flask is disturbed, you will have clumps of cells instead of a single-cell suspension. Once cells are detached, add 2 ml of trypsin and vigorously pipette (using a 10-ml precooled pipette) the suspension in the flask to further break down the suspension into single cells. Add to the cell suspension 12 ml of the culture medium (DMEM + 15% FBS). The protein in the medium will inactivate the trypsin.
4. Place each cell suspension into a 50-ml centrifuge tube (Falcon) and spin the cells at 1500 g for 15 min.
5. Discard the supernatant.
6. To each 50-ml tube, add 5 ml of freeze media and vigorously pipette the cells up and down to break them up.
7. After each pellet is resuspended, combine them (15 ml), count the cells with a hemocytometer, and adjust the concentration with the freeze media to 3.0×10^6 cells/ml. Upon thaw, this should provide enough cells to prepare four 4-well plates. We recommend that several vials of 1.5×10^6 cells/ml be frozen in the event that you need to prepare two or six 4-well plates. *Note*: These numbers may differ in your lab, and the ratio of cells per well that you require may need to be determined.
8. Dispense 1 ml of cell suspension/freeze media into 1.8-ml cryovials (Corning, Corning, NY).
9. Label the vials with concentration, date, and cell type.
10. Using two Styrofoam 15-ml centrifuge racks, place the vials in the center of one rack and place the other on top. Tape shut and place in the –80°C freezer overnight.
11. The following day plunge vials into liquid N_2 and store until needed.

Protocol 21.3. γ-Radiation of primary mouse embryonic fibroblasts

1. Remove DMEM + 15% FBS from refrigerator and warm to 37°C.
2. Remove the appropriate number of gelatin plates (see below) and allow them to come to room temperature in the tissue-culture hood.
3. Thaw the required number of vials to make the number of plates needed (see Protocol 21.2). Vials are thawed by placing them in a 37°C water bath for 5 min or until all ice is melted.

4. Centrifuge thawed PMEFs in 5 ml of DMEM + 15% FBS (5 min, 500 g at room temperature). Use one thawed vial per 5 ml medium per centrifuge tube.
5. Decant supernatant and resuspend cell pellet in 7 ml DMEM + 15% FBS.
6. Pipette the 7 ml of cells into a sterile Falcon petri dish (100 × 15 mm; Becton Dickinson; do not use tissue culture-treated dishes). Place that dish in a larger dish for transport to the radiation facility and radiate the dish with 5 K rad. *Note*: Before using this procedure, we evaluated different cell concentration and radiation doses to determine the optimal conditions for mitotic inactivation. The conditions at other facilities may vary.
7. Remove the radiated cells from the dish, rinsing the bottom of the dish to remove settled cells. You should have a minimum of 6.4 ml at this point to fill the wells of two 4-well plates (Nunc, Naperville, IL).
8. Pipette 0.8 ml of cell suspension into each gelatin-coated well.
9. Incubate cells in 37°C, 5% CO_2 humidified incubator.
10. The PMEF plates are ready to accept hES cells after 24 h of culture. Before adding hES cells, the medium should be replaced with DMEM + 20% FBS (see below).

Gelatin coating of culture dishes

Use 0.1% gelatin in ultrapure water (Specialty Media, Lavallette, NJ). If stored in the refrigerator, allow the bottle of 0.1% gelatin to warm in a 37°C waterbath for 30 min before use. Once the bottle is opened, it should always be stored refrigerated. In a tissue culture hood, aliquot directly from the bottle enough 0.1% gelatin solution to cover the surface of the culture well. For a 4-well plate (16 cm^2), this is about 0.5 ml per well. Let sit on wells for a minimum of 20 min. Remove the excess gelatin solution from each well and seal each plate with parafilm. Return the plates to the original wrapping and seal with tape. Store at 4°C. The gelatin may be used for up to 2 months.

Preparation of DMEM- 20% FBS

Media constituents

Component	Supplier	Catalogue No.
DMEM	Specialty Media (Lavallette, NJ)	SLM-120-B
FBS	Hyclone (Logan, UT)	SH300.70.02
Nonessential amino acids (100×)	Specialty Media	TMS-001-C
2β-Mercaptoethanol (100×)	Specialty Media	ES-007-E
Penicillin/streptomycin (100×)	Specialty Media	TMS-AB2-C
LIF-human recombinant	Sigma, St. Louis, MO	L-5283
L-Glutamine (100×) 200 mM	Specialty Media	TMS-002–C
Fibroblast growth factor	Becton Dickinson, Franklin Lakes, NJ	40060

Volumes

	Total volume		
	50 ml	100 ml	200 ml
DME	38.0 ml	76.0 ml	152.0 ml
FBS	10 ml	20 ml	405 ml
Nonessential amino acids	500 µl	1.0 ml	2.0 ml
Penicillin/streptomycin	500 µl	1.0 ml	2.0 ml
2β-Mercaptoethanol	500 µl	1.0 ml	2.0 ml
LIF	50 µl	100 µl	200 µl
L-Glutamine	0.5 ml	1.0 ml	2.0 ml

Mixing

1. Combine all components in tissue culture hood. Store flask of media at 4°C for 2 weeks.
2. Remove amount needed daily under the tissue culture hood.
3. Equilibrate by gassing (5% CO_2/95% O_2 or clinical blood gas mixture) aliquoted flask: gas for 1 min for 50-ml tissue-culture flask or 2 min for 250-ml tissue culture flask. *Note:* Only fill equilibrating flask half full or less.
4. Warm to 37°C for a minimum of 1 h in incubator before use.
5. Just before use, add 8 μl FGF per 10 ml of culture medium (see below).

Preparation of fibroblast growth factor (FGF)

1. Use 10 mg/vial FGF = (Becton Dickinson)
2. Reconstitute vial with 2 ml of calcium-, magnesium-free Dulbecco's phosphate-buffered serum (Sigma, St. Louis, MO).
3. Prepare 20 μl aliquots and store at –80°C until ready to use.
4. Add FGF to the culture media just before use.

Note: We are currently adding 8 μl/10 ml of culture medium for a working concentration of 4 ng/ml.

Protocol 21.4. Mitomycin-C treatment of primary mouse embryonic fibroblasts

This protocol is modified from the protocol by Abbondanzo et al. (10).

1. Thaw centrifuge tubes of mitomycin-C dilution (see below) in a 37°C water bath.
2. Apply 0.8 ml of mitomycin-C solution to each well of feeder cells in 4-well plates that were prepared one day earlier. *Note:* Use the Protocol 21.3 irradiation, but do not perform the γ-radiation step. Also, decrease the number of cells plated so that they will reach confluency over the next 24 h.
3. Return plates to the incubator and culture for 3 h.
4. Remove the mitomycin-C solution and discard within specimen container in bio-hazard trash.
5. Rinse each well twice with DMEM + 20% FBS (see Protocol 21.3).
6. Seed plates with hES cells.

Preparation of Mitomycin-C

1. Use 2-mg vials of mitomycin-C (Sigma *toxic, light sensitive*)
2. Use respirator mask during initial dilution. Conduct dilution in a chemical hood on top of blue pad in case of a spill. Wear two pair of gloves and lab coat.
3. Inject 5 ml sterile calcium-, magnesium-free Dulbecco's phosphate-buffered serum into the vial. Pull the 5-ml solution back into the syringe and recap with one hand. Remove outer gloves.
4. In a tissue culture hood, add the 5-ml solution into 195 ml DME + 5% FBS filter sterilized 190 ml DMEM + 10 ml FBS; (wrap 250-ml tissue culture flask with foil to minimize light exposure). Mix.
5. In a tissue culture hood, aliquot 10-ml portions into 14-ml centrifuge tubes, being careful to avoid significant light exposure of the mitomycin solution.
6. Store at –20°C.

Protocol 21.5. Removal of the zona pellucida using acidified tyrode's solution

1. Transfer the acidified tyrode's solution (Irvine Scientific, Santa Ana, CA) to a culture dish at room temperature.
2. Using a pipette that is slightly larger than the embryo, transfer the blastocyst to the acidified tyrode's solution. Using a dissecting microscope, visualize the em-

bryo while gently pipetting it in and out. Care should be taken not to let the embryo settle to the bottom of the dish, as exposed areas of the embryo will stick.

3. Remove the blastocyst from the acidified tyrode's solution as soon as the zona pellucida begins to thin and lose its shape. Finish removing the zona by pipetting it in and out while in a dish of culture medium.

4. Wash the embryo three times in fresh culture medium and continue culture until needed for ICM isolation. If the blastocoel collapses during zona removal, it is best to culture the embryo until it has reexpanded before performing ICM isolation.

Protocol 21.6. Removal of the zona pellucida using Pronase

1. Prepare the Pronase solution as detailed below.

2. Using a pipette slightly larger than the embryo, transfer the blastocyst to the Pronase solution. Using a dissecting microscope, visualize the embryo while it is being treated. The Pronase solution takes between 3 and 5 min to thin the zona pellucida sufficiently.

3. Remove the blastocyst from the Pronase solution as soon as the zona pellucida begins to thin and lose its shape. Finish removing the zona by pipetting it in and out while in a dish of culture medium.

4. Wash the embryo three times in fresh culture medium and continue culture until needed for ICM isolation. If the blastocoel collapses during zona removal, it is best to culture the embryo until it has reexpanded before performing ICM isolation.

Preparation of Pronase solution for zona pellucida removal

Use 100 mg /vial Pronase (Sigma, 4.4 IU/mg). Weight out 2.2-mg aliquots of Pronase and store in snap-top tubes at –20°C. When needed, add 1.0 ml of embryo culture medium to one aliquot and invert to mix. This produces a solution of 10 IU/ml. Use immediately.

Protocol 21.7. Isolation of the inner cell mass

1. Prepare a 4-well plate as follows:

   - Well 1—antibody to the trophoblast. *Note*: You will need to determine the appropriate dilution for your antibody.
   - Well 2—standard guinea pig complement (Cedarlane, Westburg, NY), we use a dilution of 1:10.
   - Wells 3 and 4—0.6 ml DMEM-20% FBS.

   Cover all four wells with equilibrated mineral oil (Squibb, Princeton, NJ).

2. Pipette zona-free blastocyst into well 1 containing the antibody and incubate at 37°C, 5% CO_2 for 15–30 min (time will vary depending on the type of antibody you are using).

3. Rinse the embryo in well 3 for about 1 min.

4. Transfer the embryo to well 2 containing the complement. Incubate for about 5–10 min at 37°C, 5% CO_2. Evaluate the embryo while it is in the complement. The cells of the trophoblast should appear dark and degenerate. At this time, begin pipetting the embryo through pipets of decreasing diameters. As you do so, the degenerate cells of the trophoblast should fall away. It is important to get all of the trophoblast cells off of the ICM so that they do not interfere with attachment to the substrate. An ICM is about 50 μm in diameter, so a pipette just slightly larger will facilitate the removal of trophoblast cells during the final stages of isolation.

5. As you are removing the trophoblast cells, transfer the ICM into well 4 to complete the process. Once all of the trophoblast cells have been removed, transfer

the ICM to the substrate (i.e., PMEFs, Matrigel, etc.) on which you want to initiate the colony. Using a computer scale (we use the ZILOS computer scale from Hamilton Thorne Biosciences, Beverly, MA) or ocular micrometer, measure the ICM to aid in its identification later (Figure 21.3).

6. Periodically evaluate the well for attachment of the ICM and mark its location with a pen on the plate lid. This will be helpful later when trying to identify the ICM/colony within the first few days of culture.

Protocol 21.8. Initiation of human ES cell colonies

1. After plating an ICM, a healthy colony will begin to increase in size over the next 7–9 days and retain a characteristic morphology (Figure P21.1). It is our experience that many colonies will initially grow upward for the first 3–5 days. The colony will then flatten out and form a more typical cell colony.
2. Feed cultures every 48 h with DMEM-20% FBS (Protocol 21.3) or every 24 h with conditioned medium (see Protocol 21.11).
3. During the first week of growth, check colonies for signs of differentiation. This will sometimes occur to a portion of the colony, and, if caught quickly, the area of differentiation may be removed with a 30.5-G needle (Becton Dickinson).
4. Allow the colony to remain on the PMEFs for 7–9 days before replating to fresh PMEFs prepared 24 h earlier.
5. Remove the colony mechanically with a 30.5-G needle, using the needle to gently separate the colony from the PMEFs. Transfer the colony to a sterile watch glass, and, using two needles, break apart the colonies into clumps containing about 50 cells. Transfer these cell clumps to the fresh PMEFs.

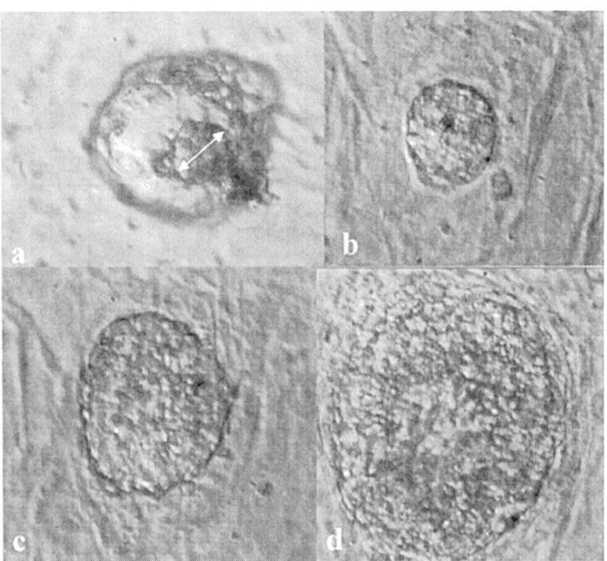

Figure P21.1. Photomicrographs of stages of human ES cell colony production from the same embryo. (a) Human blastocyst on day 6 of culture (day 0 = day of insemination). ICM measures 52 μm in diameter (arrow). This embryo (grade 2; ICM grade B) collapsed in acid tyrodes and was cultured an additional 3 h for reexpansion. (b) Human ES colony from same embryo on day 3 of culture (day 1 = day ICM plated on primary mouse embryonic fibroblasts). Colony measures 74 μm in diameter and is growing up to form a columnar-shaped structure. (c) Day 6 of culture; colony measures 129 μm in diameter. (d) Day 8 of culture; colony has flattened and is 260 μm in diameter. The colony was removed and replated onto new primary mouse embryonic fibroblasts on this day.

6. Alternatively, add dispase (Becton Dickinson) to the well and, using a finely drawn pipette, remove the colony. The dispase works quickly, so be ready to remove the colony as soon as you see the cells separating from the PMEFs. Transfer the colony to the next well and rinse. Continue through the other 3 wells of the plate, rinsing the dispase from the colony. Transfer the colony to a sterile watch glass, and, using two needles, break apart the colonies into clumps containing about 50 cells. Transfer these cell clumps to the fresh PMEFs.

Protocol 21.9. Passage of human embryonic stem cell lines onto fresh primary mouse embryonic fibroblasts

1. Human ES cells grown on PMEFS should be replated onto fresh PMEFs every 7–10 days to prevent spontaneous differentiation. The hES cell colonies can be moved mechanically or the colonies separated from the PMEFs using dispase (8) or collagenase (7,20,22).
2. For mechanical replating, the colonies are teased off with a 30.5-G needle and broken up into smaller cells clumps by

   - using two needles to break the cells apart,
   - drawing the colonies in and out of a 30.5-G needle or finely drawn pipette, or
   - exposing the cells to trypsin or collagenase while pipetting them through a finely drawn pipette. Be sure to rinse the cells free of trypsin or collagenase before replating.

3. Alternatively, add dispase (Becton Dickinson) to each well, and, as the colonies separate from the PMEFs, pull them off the wells with a drawn pipette. You may have to tease them off the bottom a little. Add the colonies to 3.0 ml of DMEM-20% FBS (Protocol 21.3) and centrifuge the cells and remove the supernatant. Resuspend in 1.0 ml of DMEM-20% and, if needed, break the colonies into smaller cell clumps by using two needles to break the cells apart or drawing the colonies in and out a 30.5-G needle or finely drawn pipette.
4. Add the cell suspension to wells of PMEFs prepared the day before.

Protocol 21.10. Preparation and use of MatriGel basement membrane matrix plates

1. Perform all work in a laminar flow hood.
2. The day before coating, thaw MatriGel basement membrane matrix (Becton Dickinson) overnight on ice at 4°C and precool a variety of sterile pipettes at –20°C for use in aliquotting. MatriGel matrix will gel rapidly at 22°–35°C, so all items must be kept as cold as possible throughout the aliquotting procedure.
3. On the day of plate preparation, prepare a bucket of ice to hold 50-ml centrifuge tubes (Becton Dickenson) and pipettes. Use precooled pipettes to mix the MatriGel matrix.
4. In a cooled tube, dilute MatriGel to the desired concentration with a serum-free DMEM (use medium preparation from Protocol 21.3, omitting the FBS). We currently use 5% MatriGel coating on our plates.
5. Using a precooled 10-ml pipette, enough MatriGel is added to each well of a 4-well plate to coat the bottom.
6. Incubate plates at room temperature for 1 h in the laminar flow hood.
7. Aspirate off the unbound MatriGel. Gently rinse the wells with serum-free DMEM. Plates are now ready for use.
8. Pack unused plates in groups of four, wrap in parafilm, and place back in their original wrapper, taped and dated. They may be stored at 4°C for up to 2 months.
9. Seed plates with hES cells in the same manner as when PMEFs are used. As reported by other investigators (22), cells are fed every day with conditioned me-

dium (see Protocol 21.11). When initiating colonies on MatriGel the cells of the ICM will often disperse to form a flat colony much more quickly than on PMEF (Figure P21.2). Use the same procedure described previously for replating of colonies (see Protocol 21.8.).

Protocol 21.11. Preparation of conditioned medium

1. Conditioned medium may be used for culturing on MatriGel or PMEFs. We use conditioned medium fresh each day, however, Xu et al. (22) report that it can be frozen and stored at –20°C for 1 month. Before using, it is important to add the fibroblast growth factor fresh each day. Depending on your needs, the conditioned media can be produced without serum or with a serum substitute and with or without LIF.

2. The conditioned media is produced by seeding a 75-cm^2 flask (Falcon) with 3×10^6 PMEFs (passages 3 or 4) in 20 ml of DMEM-15% FBS (Protocol 21.1). All work is done under a laminar flow hood. Twenty-four hours later, draw off all of the medium and replace with 20 ml of DMEM- 20% FBS (Protocol 21.3). After an additional 24-h culture, the flask is ready to provide the conditioned medium.

3. To use, draw off the 20 ml of conditioned medium and replace it with 20 ml of fresh DMEM-20% FBS (for use the next day). Filter the conditioned medium through a 0.2-μm acrodisc syringe filter (Pall Co., Ann Arbor, MI) that has been flushed with about 5 ml of DMEM-20% FBS. This sterile filtered media is then used to feed the ES cells. When using fibroblast growth factor in the culture medium, be sure to add it to the conditioned medium at this time. We use 4 ng/ml of DMEM- 20% FBS (Protocol 21.3).

4. The next flask of PMEFs must be started 2 days before needing the conditioned media to allow for the change over from 15% serum to 20% serum + LIF. The flask is then used for an additional 7 days to provide conditioned medium.

Protocol 21.12. Alkaline phosphatase staining of human embryonic stem cells

We use the Vector Blue Staining Kit from Vector Laboratory (Burlingame, CA). Store kit at 4°C. The kit contains 3 dropper bottles and 1 empty bottle. You will not need the empty bottle. Each bottle has a number.

1. For staining, prepare the working solution just before use. To prepare working solution:

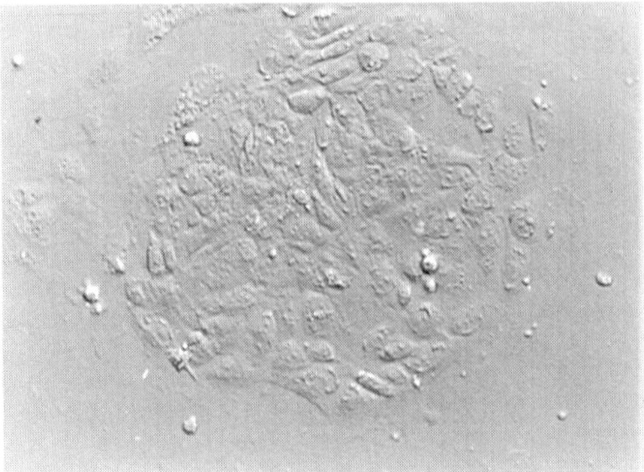

Figure P21.2. Human inner cell mass plated on MatriGel on day 4 of culture. Original magnification × 200.

- Add 2 drops of reagent 1 to 5ml of 100 mM Tris-HCL, pH 8.2 buffer (see below) Vortex.
- Add 2 drops of reagent 2. Vortex.
- Add 2 drops of reagent 3. Vortex.

2. Rinse the wells to be stained with the 100 mM Tris-HCl buffer.
3. Add 1.0 ml of the working stain solution to each well. Do not heat the working solution.
4. Incubate the dishes in a drawer, out of the light, for a minimum of 20–30 min. Increasing the incubation time in the working solution up to 4 h may increase staining sensitivity 2–4 times. However, with extended incubation, background staining may increase, and the stain may begin to precipitate over time.
5. Positive cells will show an intense blue staining (Figure 21.6). Always stain a well without the hES cells to serve as a negative control.

Preparation of 100 mM Tris-HCl buffer for alkaline phosphatase assay

Use Trizma HCl (Sigma), and 10 M NaOH (Sigma).

1. Add 1.576 g of Trizma HCl to 100 ml Ultrapure water.
2. Adjust pH to 8.2 (original pH is about 4) using 10 M NaOH (about 15 drops).
3. Filter with disposable filter unit.
4. Store at room temperature in sterile bottle.

Protocol 21.13. Stage specific embryonic antigen -4 assay

1. Fix cells in 4% paraformaldehyde fixative (Electron Microscopy Sciences, Ft. Washington, PA.).
2. Wash wells in PBS for 5 minutes.
3. Incubate for 20 min with normal goat serum consisting of 150 ml goat serum (Vector Laboratories) diluted in 10 ml PBS.
4. Remove serum from the wells.
5. Incubate for 1 h with the SSEA-4 primary antibody (Developmental Studies Hybridoma Banks, Iowa City, IA, cat. No. MC-813-70) diluted 1:2 in PBS.
6. Wash cells in PBS for 5 min.
7. Incubate cells for 30 min with biotinylated secondary antibody anti-mouse IgG (Vector Labatories) diluted 1: 200 in PBS.
8. Wash cells in PBS for 5 min.
9. Incubate cells for 30 min with VECTASTAIN ABC Reagent (Vector Laboratories).
10. Wash cells in PBS for 5 min.
11. Incubate cells in TMB (3, 3', 5, 5'-tetva methylbenzicline) substrate solution (Vector Laboratories, TMB Substrate Kit) until the desired stain intensity develops.
12. Wash cells for 2–3 minutes in PBS and then rinse briefly in water.
13. Counterstain the cells with Nuclear Fast Red (Vector Laboratories) for 1–2 min.
14. Wash cells in water.
15. Evaluate cells using light microscopy (Figure P21.3).

Protocol 21.14. Cryopreservation of human embryonic stem cells in straws

1. The cryopreservation medium consists of 10% DMSO and 90% DMEM-20% FBS (Protocol 21.3). Prepare the cryopreservation medium fresh before each freezing procedure.
2. Perform cryopreservation in liquid nitrogen vapors using a Taylor-Warton Freezing Tray that fits in the opening of the storage container. You will need to predetermine the levels on the tray which correspond to –2°C, -40°C, and –90°C.

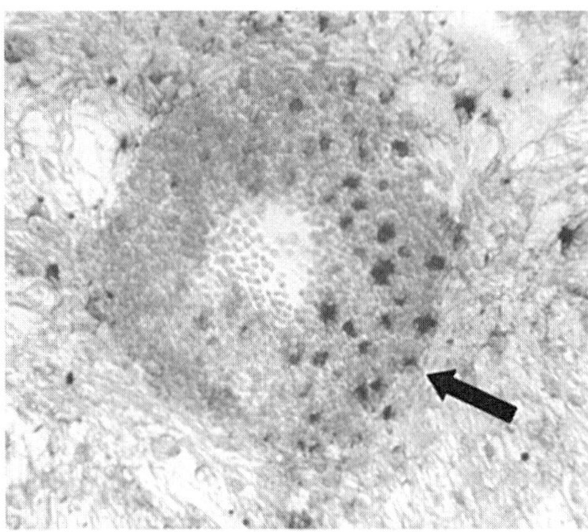

Figure P21.3. Human embryonic stem cell colony (arrow) stained positive for SSEA-4 showing that the cells are undifferentiated. Background cells (stained violet) are the primary mouse
embryonic fibroblasts. Original magnification × 100. See color insert.

3. Remove colonies from wells as described in Protocol 21.9. Do not break the
   colonies down into smaller cell clumps. Using a drawn pipette, transfer the colonies to a dish containing the cryopreservation medium. Try to keep the colonies
   close together in the dish.
4. Attach a 1-ml syringe to the end of a 0.25-cc sterile straw (TS Scientific, Perkasie,
   PA). Draw the cryomedium and ES colonies into the straw. Do not allow the
   medium to reach the cotton plug. Leave about 0.5 inch of air at the end of the
   straw, and plug straw with Critoseal.
5. Place the straw in the freezing tray and place the freezing tray in the refrigerator
   (4°C) for 1 h.
6. Transfer the straw to the freezing tray that has been preequilibrated in the storage tank at –2°C. Let the straw remain at this position for 25 min.
7. Turning the crank on the freezing tray, lower the straw to the position that corresponds with –40°C and let the straw remain at this position for 10 min. Do not
   remove the freezing tray from the mouth of the storage tank at any time.
8. Lower the freezing tray to the position that corresponds with –90°C and allow
   the straw to remain at this position for 10 min.
9. Quickly remove the tray from the tank and drop the samples into a Dewar flask
   containing liquid nitrogen. Place the straw in a precooled cane and store in liquid nitrogen.
10. Thaw straws by placing them at room temperature until all ice disappears. Empty
    the straw into a dish containing DMEM-20% FBS (Protocol 21.3). With a drawn
    pipette, transfer the colonies to fresh medium and break the cell colonies up into
    smaller clumps by using two needles to break the cells apart or drawing the
    colonies in and out a 30.5-G needle or finely drawn pipette.
11. Add the cell suspension to wells of PMEFs prepared the day before.

References

1. Evans, M. J., and Kaufman, M. H. (1981). *Nature*, **292**, 154.
2. Pain, B., Clark, M. E., Shen, M., Nakazawa, H., Sakuri, M., Samarut, J., and Etches, R. J.
   (1996). *Development*, **122**, 2339.
3. Doetschman, T., Williams, P., and Maeda, N. (1988). *Dev. Biol.*, **127**, 224.
4. Wheeler, M. B. (1994). *Reprod. Fertil. Dev.*, **6**, 563.
5. Thomson, J. A., Kalishman, J., Golos, T. G., Durning, M., Harris, C. P., and Hearn, J. P.
   (1996). *Biol. Reprod.*, **55**, 254.

6. Thomson, J. A., Kalishman, J., Golos, T. G., Durning, M., Harris, C. P., Becker, R. A., and Hearn, J. (1995). *Proc. Natl. Acad. Sci. USA*, **92**, 7844.

7. Thomson, J. A., Itskovitz-Eldor, J., Shapiro, S. S., Waknitz, M. A., Swiergiel, J. J., Marshall, V. S., and Jones, J. M. (1998). *Science*, **282**, 1145.

8. Reubinoff, B. E., Pera, M. F., Fong , C. Y., Trounson, A., and Bongso, A. (2000). *Nature Biotech.*, **18**, 399.

9. Lanzendorf, S. E., Boyd, C. A., Wright, D. L., Muasher, S., Oehninger, S., and Hodgen, G. D. (2001). *Fertil. Steril.*, **76**, 132.

10. Abbondanzo, S. J., Gadi, I., and Stewart, S. L. (1993). *Methods Enzymol.*, **225**, 803.

11. Morin, G. B. (1989). *Cell*, **59**, 521.

12. Rathgen, P. D., Toth, S., Willis, A., Heath, J. K., and Smith, A. G. (1990). *Cell*, **62**, 1105.

13. Smith, A. G., Heath, J. K., Donaldson, D. D., Wong, G. G., Moreau, J., Stahl, M., and Rogers, D. (1988). *Nature*, **336**, 688.

14. Williams, R. L., Hilton, D. J., Pease, S., Wilson, T. A., Steward, C. L., Gearing, D. P., Wagner, E. F., Metcalf. D., Nicola, N. A., and Gough, N. M. (1988). *Nature*, **336**, 684.

15. Cohen, J., Gilligan, A., Esposito, W., Schimmel, T., and Dale, B. (1997). *Hum. Reprod.*, **12**, 1742.

16. Hall, J., Gilligan, A., Schimmel, T., Cecchi, M., and Cohen, J. (1998). *Hum. Reprod.*, **13 Suppl 4**, 146.

17. Robertson, E. J. (1987). Embryo-derived stem cell lines. In: *Teratocarcinomas and Embryonic Stem Cells: A Practical Approach*, E. J. Roberson, ed., p. 71. IRL Press, Oxford.

18. Solter, D., and Knowles, B. B. (1975). *Proc. Nat. Acad. Sci. USA*, **72**, 5099–5.

19. Eiges, R., Schuldiner, M., Drukker, M., Yanuka, O., Itskovitz-Eldor, J., and Benvenisty, N. (2001). *Curr. Biol.*, **11**, 514.

20. Carpenter, M. K., Inokuma, M. S., Denham, J., Mujtaba, T., Chiu, C., and Rao, M. S. (2001). *Exp. Neurol.*, **172**, 383.

21. Kaufman, D. S., Hanson, E. T., Lewis, R. L., Auerbach, R., and Thomson, J. A. (2001). *Proc. Natl. Acad. Sci. USA*, **98**, 10716.

22. Xu, C., Inokuma, M. S., Denham, J., Golds, K., Kundu, P., Gold, J. D., and Carpenter, M. K. (2001). *Nature Biotech.*, **19**, 971.

23. Thomson, J. A., and Odorico, J. S. (2000). *TIB Tech.* **18**, 53.

24. Wright, W. E., Piatyszek, M. A., Rainey, W. E., Byrd, W., and Shay, J. W. (1996). *Dev. Genet.*, **18**, 173.

25. Ogashi, M., Takashima, A., and Taylor, R. S. (1997). *J. Immunol.*, **158**, 622.

26. Kim, N. W., Piatyszek, M. A., Prowse, K. R., Harley, C.B., West, M. D., Ho, P. L. C., Coviello, G. M., Wright, W. E., Weinrich, S. L., and Shay, J. W. (1994). *Science*, **266**, 2011.

27. Wright, D. L., Jones, E. L., Mayer, J. F., Oehninger, S., Gibbons, W. E., and Lanzendorf, S. E. (2001). *Mol. Hum. Reprod.,* **7**, 947.

28. Jones, J. M., and Thomson, J. A. (2000). *Sem. Reprod. Med.*, **18**, 219.

29. Amit, M., Carpenter, M. K., Inokuma, M. S., Chiu, C., Harris, C. P., Waknitz, M. A., Itskovitz-Eldor, J., and Thomson, J. A. (2000). *Dev. Biol.*, **227**, 271.

30. Schrock, E., du Manoir, S., Veldman, T., Schoell, B., Wienberg, J., Ferguson-Smith, M. A., Ning, Y., Ledbetter, D. H., Bar-Am, I., Soenksen, D., Garini, Y., and Ried, T. (1996). *Science*, **273**, 494.

31. Freshney, R. I., ed. (2000). *Culture of Animal Cells: A Manual of Basic Technique,*. 4th ed., Wiley-Liss., New York.

32. Reubinoff, B. E., Pera, M. F., Vajta, G., and Trounson, A. O. (2001). *Hum. Reprod.*, **16**, 2187.

33. Zhang, S. C., Wernig, M., Duncan, I. D., Brustle, B., and Thomson, J. A. (2001). *Nature Biotech.*, **19**, 1129.

34. Mudgett, J. S., and Livelli, T. J. (1995). In: *Methods in Molecular Biology, vol 48; Animal Cell Electroporation and Electrofusion Protocols*. J.A. Nickoloff, ed., Humana Press, Totowa, NJ.

22

MAKOTO C. NAGANO

Spermatogonial Transplantation

1. SPERMATOGENESIS

Spermatogenesis is a highly active and complex process that continuously produces male gametes, the spermatozoa. A technique has been developed in which germ cells derived from the testis of a fertile male are transferred to the testis of an infertile male and donor-derived spermatogenesis is reconstituted in recipient testes (1, 2). This technique, called "spermatogonial transplantation," provides unique opportunities not only to understand the process of spermatogenesis but also to manipulate male germ cells for medical and agricultural applications. In this chapter I describe the general concept and applications of the transplantation technique and standard procedures using the mouse as a model species. An excellent technical review has been published that complements this chapter (3).

A high rate of spermatozoa production is one of the remarkable characteristics of spermatogenesis. For example, around 10^7 spermatozoa/g of testicular tissue/day are produced in the rat, and daily production of spermatozoa in the human and boar is approximately 10^8 and 10^9, respectively (4–6). Furthermore, this high output process continues from puberty throughout the life of an individual male. This contrasts markedly with female gametogenesis, which produces only a limited number of eggs in mammals until the ovarian reserve is exhausted. Spermatogenesis is also a highly organized process, which can be divided into three phases (7). Each of the three phases is represented by a unique cell type: diploid spermatogonia in the first phase, meiotic spermatocytes in the second phase, and haploid germ cells (spermatids and spermatozoa) in the third phase. In the first phase, the proliferation phase, spermatogonia replicate and undergo a series of differentiation on the basement membrane at the base of the seminiferous tubules. These spermatogonia amplify their population to supply germ cells for the terminal differentiation stages that follow. In the second phase, the meiotic phase, spermatogonia move from the basement membrane toward the lumen of the seminiferous tubules, differentiating into spermatocytes that undergo meiosis to produce spermatids. It is during this second phase that genetic recombination takes place. The third phase is called spermiogenesis, in which spermatids proceed through a complex morphological differentiation to form spermatozoa. The morphological architecture of spermatogenesis follows these three germ cell development processes. While spermatogonia form a single cell layer at the base of the tubules, more advanced germ cells are formed in multiple layers toward the lumen, and terminally differentiated spermatozoa reside at the top of the layer, facing to the luminal space. Diploid spermatogonia are also structurally separated from meiotic and postmeiotic germ cells by junctions between somatic Sertoli cells, called the Sertoli cell barrier; spermatogonia reside in the basal compartment of the seminiferous epithelium, whereas terminally differentiating cells are found in the adluminal compartment. Although the Sertoli cell barrier permits the transport of a variety of ions and other small molecules, it prevents larger molecules, such as protein, from penetrating the seminiferous epithelium into the lumen (8). Consequently, germ cells in the adluminal compartment depend significantly on Sertoli cells for support in nutrition, growth factors, and waste disposal (8). Therefore, appropriate communication between male germ cells and Sertoli cells is critical for the maintenance and regulation of spermatogenesis.

The long-term and vigorous activity of spermatogenesis is founded by the male germline stem cells (GSCs), also called spermatogonial stem cells. The male GSCs are believed to be only a small fraction of spermatogonia and estimated to be approximately 0.02% of total testis cells in the mouse (9). These stem cells self-renew and produce progenitor spermatogonia that are committed to differentiation. This dual biological function is the characteristic that defines stem cells in general. Compared to the stem cells of somatic self-renewing organs, such as those in the intestine and bone marrow, male GSCs have unique properties. Although somatic stem cells are vital for the survival of an individual, GSCs are dispensable, and their absence does not significantly affect the normal life of an individual. However, GSCs are essential for survival of species and for diversifying the gene pool in a given species through meiotic recombination. Another distinct characteristic is that the GSCs in postnatal mammals exist only in males because female germ cells enter meiosis and lose their self-renewing capacity before birth. Therefore, male GSCs are the only cell population in mammals that self-renews throughout postnatal life and transmits the appropriate genetic information to the next generation. In addition, male GSCs are also characteristic in that the differentiation cascade in spermatogenesis is unidirectional toward production of spermatozoa, and male GSCs generate only a single type of committed progenitor cells in this cascade. In contrast, differentiation originating from somatic stem cells is often multidirectional, and the somatic stem cells produce various types of committed progenitors, as seen in hematopoiesis (10).

Because a stem cell is defined by its dual function regardless of its tissue of origin, a biological assay system that detects stem cell activity is a prerequisite for the study of stem cells. This is certainly the case for the male GSCs, as these cells cannot be identified to date using morphological, genetic, or biochemical markers. In 1994, the Brinster's group at the University of Pennsylvania developed the spermatogonial transplantation technique using the mouse as a model species (1, 2). In this technique, donor testis cells derived from a fertile male are transplanted into the seminiferous tubules of an infertile male. The stem cells injected into the lumen of the tubules reach the basement membrane and initiate proliferation and differentiation to reestablish donor-derived spermatogenesis in recipient testes. Recipients produce donor-derived spermatozoa and, following mating with females, offspring carrying the donor haplotype (2, 11) (Figure 22.1). Since such long-term and complete spermatogenesis can originate only from GSCs, the transplantation technique is the unequivocal method for detection of male GSCs. Even if nonstem germ cells were transplanted together with stem cells, these nonstem cells would not survive or establish continuous spermatogenesis, but rather would be exhausted during the unidirectional differentiation of spermatogenesis. Therefore, only definitive stem cells can reconstitute donor-derived spermatogenesis after transplantation. In turn, the technique can also be used to analyze the effects of the testicular microenvironment

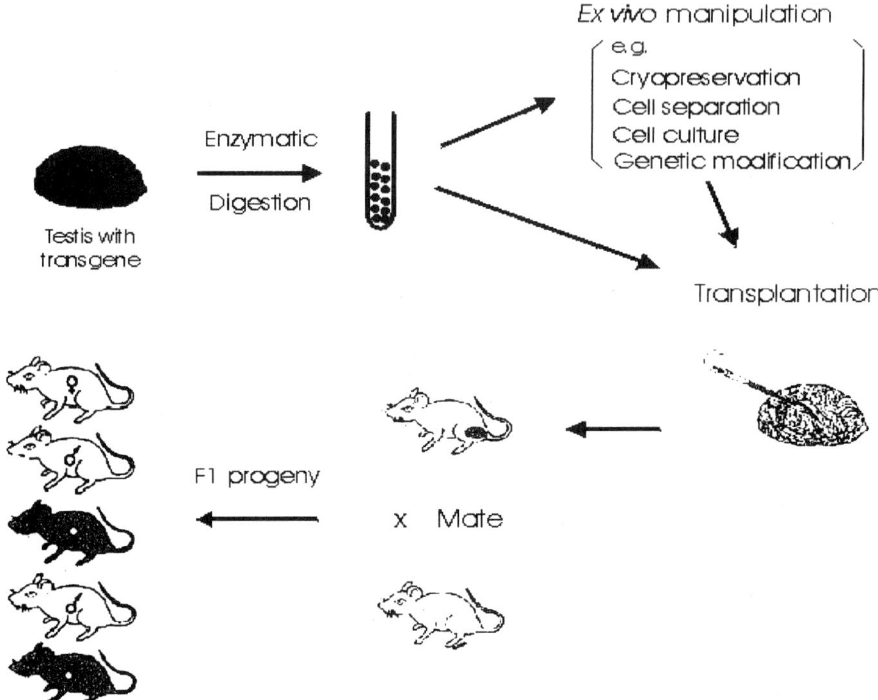

Figure 22.1. Schematic representation of the spermatogonial transplantation procedure. In this scheme, donor cells are isolated by enzymatic digestion from the testis of a transgenic mouse strain that expresses *lacZ* in germ cells. Before transplantation, these cells can be subjected to ex vivo manipulation, such as cryopreservation, cell separation, in vitro culture, or genetic manipulation. Donor cells are injected into the seminiferous tubules of infertile mouse testes, and stem cells derived from donor testes colonize recipient testes. Donor-derived spermatogenesis can be detected in recipient testes by blue staining. Offspring carrying the donor haplotype can be produced after mating with females. Modified from Nagano and Brinster (11), with permission.

on germ cell development. Furthermore, this technique has the potential to be a key method in restoring male fertility and genetically modifying offspring through the male germ line.

2. DESCRIPTION OF TECHNIQUE

2.1 Principles

Here I focus on the spermatogonial transplantation technique using the mouse as a model species, for which the technique has been best developed. Applications of the transplantation technique in other species are under investigation (12). The transplantation procedure is composed of three steps: preparation of recipients, donor cell preparation and transplantation, and analysis of recipient testes (Figure 22.1).

First, infertile recipient mice are prepared before transplantation. As the seminiferous tubules in normal testes are filled with multiple layers of germ cells, it is ideal to eliminate the endogenous germ cells to provide access of donor stem cells to their designated niche on the basement membrane and to reduce competition with endogenous spermatogenesis. Infertility in recipients can be chemically induced using an alkylating agent, busulfan (3, 13). It should be noted, however, that a degree of recovery of endogenous spermatogenesis can be seen after busulfan treatment. Alternatively, genetically infertile mouse strains can be used as recipients (1, 14). It is also ideal to use recipient mice with a genetic background that is immunologically compatible with that of the donor mice. Although the testis is known to be an immunoprivileged organ (15), and complete donor-derived spermatogenesis can be established after transplantation into the testis of partially compatible recipients (16), the effects of histocompatibility between donor and recipient have not been rigorously analyzed. In this regard, immunodeficient mice, such as SCID or nude mice, could be universal recipients for transplantation.

Second, a single cell suspension of donor cells is prepared from the testes of fertile mice using the conventional two-step enzymatic digestion method: digestion of the testis using collagenase followed by trypsin (3, 17). The donor cells are then transferred into a microinjection pipette inserted into the bundle of efferent ducts, which run into the rete testis. Finally, donor cells are introduced into the recipient seminiferous tubules through the rete testis. Therefore, in the spermatogonial transplantation procedure, donor germ cells travel in the opposite direction to the normal route that spermatozoa take through the male reproductive tract. Before transplantation, donor cells can be subjected to ex vivo manipulation, such as cryopreservation, cell separation, tissue culture, and genetic modification.

The third step of the transplantation technique, analysis of recipient testes, is significantly affected by the choice of the donor strain. For example, if donor cells are derived from a transgenic strain that expresses a marker gene in germ cells, such as bacterial *lacZ*, donor-derived spermatogenesis can readily be visualized as a distinct "colony" by staining recipient testis for the marker gene activity (Figure 22.2) (18). If visualization of donor cells cannot be achieved, the origin of established spermatogenesis in re-

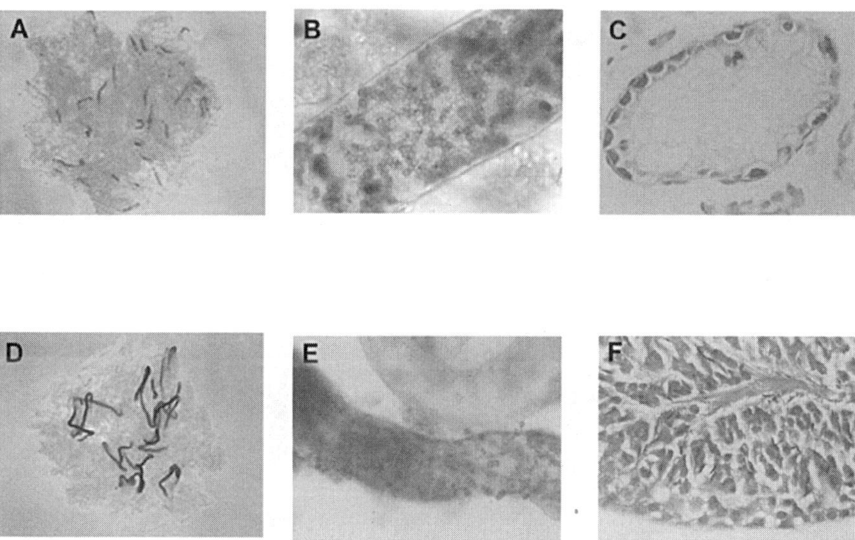

Figure 22.2. Donor germ cell colonization patterns 1 (A–C) and 2 (D–F) months after transplantation. Donor testis cells were obtained from ROSA26 mouse strain and transplanted into the testes of immunocompatible recipient mice. Recipient testes were stained blue with X-gal to visualize donor-derived spermatogenesis. (A) One month after transplantation, donor-derived spermatogenesis is observed as distinctive colonies. (B, C) The colonies at this time point are formed as a single cell layer, indicating that meiotic differentiation has not started. (D) Two months after transplantation, the colonies are well established and show a strong blue staining in the center, indicating development of spermatogenesis. (E) At the ends of colonies, donor cells show a similar colonization pattern observed at 1 month after transplantation, suggesting ongoing expansion of colonies. (F) A paraffin section showing that qualitatively complete spermatogenesis is reestablished at this time point (counterstained with nuclear fast red). Reproduced from Nagano et al. (18), with permission. See color insert.

cipient testes needs to be determined using histological sections and genetic markers of donor cells (19). In such a case, the use of genetically infertile recipients would ensure the origin of spermatogenesis because endogenous spermatogenesis cannot occur in such mutant recipients (1, 14). Donor-derived spermatogenesis can be analyzed in detail using histological sections. Recipient animals can be maintained and mated with females to obtain offspring carrying donor haplotype.

2.2 Applications

Since its development, the spermatogonial transplantation technique has been applied to investigate the biology of GSCs and spermatogenesis. The technique is amenable both for qualitative and quantitative analyses of the stem cells. When donor cells are obtained from transgenic mice that express β-galactosidase in germ cells, donor-derived spermatogenesis can be visualized as a morphologically distinct colony, and the number and size of colonies can be measured (figure 22.2) (18). Because one colony is estimated to originate from one stem cell (20), and long-term spermatogenesis can be established only from GSCs, the number of colonies reflects the number of stem cells.

Basic characterization of post-transplantation development of donor-derived spermatogenesis has been described in the mouse (18, 21). After transplantation, the GSCs reach the basement membrane and apparently initiate proliferation during the first week. Daughter spermatogonia produced by GSCs proliferate laterally on the basement membrane during the first month after transplantation and produce a single cell layer that will serve as a mother population for regeneration of spermatogenesis (Figure 22.2). Meiotic differentiation can be observed beginning approximately 1 month after transplantation and spermatozoa can first be found around 2 months after transplantation. A recent study using the transplantation technique showed an age-related difference in biological properties of GSCs and recipient testis microenvironment (22). The number of GSCs in immature mouse testes is significantly smaller than that in adult testes. However, environment of immature recipient testes allows more efficient colonization of donor stem cells than that of adult recipient testes.

Because the number of GSCs is minute, stem cell enrichment is often required in biological studies of GSCs, and spermatogonial transplantation is an indispensable technique to evaluate various enrichment methods. Using the transplantation technique, two enrichment procedures have been developed for male GSCs in mice. First, GSCs can be significantly enriched by experimental cryptorchidism (23). Cryptorchidism, produced by suturing the testis to the abdominal wall, exposes the testis to high body core temperature that eliminates differentiated germ cells. Consequently, only somatic Sertoli cells and undifferentiated spermatogonia remain in the tubules, resulting in a 20- to 25-fold enrichment for stem cells (23). Second, molecules of the integrin family of extracellular matrix receptors can be used as markers to enrich male GSCs. Immunological cell separation using antibodies against α6- and β1-integrins leads to approximately eight- and four-fold enrichment for stem cells, respectively (24). Combing the experimental cryptorchidism and immunological cell separation, the GSCs can be enriched 150-fold or more (25). Thus, the transplantation technique is essential to develop efficient stem cell enrichment procedures and further to determine GSC identification markers.

The transplantation technique (Figure 22.3) has been applied to dissect the mechanism of spermatogenesis. After transplantation of rat germ cells into testes of immunodeficient nude mice, spermatogenesis derived from rat stem cells in mouse testes shows the temporal pattern specific to rat spermatogenesis, indicating that the species-specific cycle of spermatogenesis is predominantly controlled by germ cells (26). This study was based on the fact that interspecies germ cell transplantation is feasible (27). It has been shown that rat and hamster spermatogenesis can be completed in the testes of nude mice (27, 28), while GSCs of all other species thus far studied survive more than

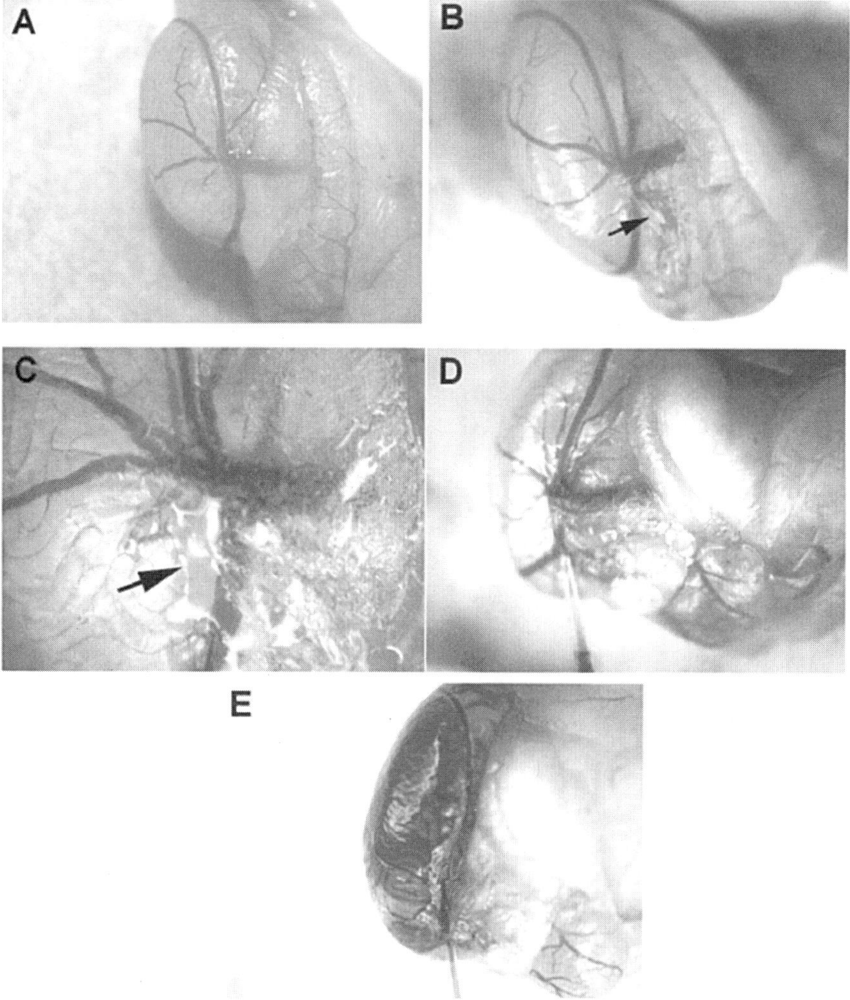

Figure 22.3. Spermatogonial transplantation procedure. Donor cell injection into a right testis. (A) A recipient testis is set in such a way that the cranial end of the testis faces the surgeon. The epididymis and fatty tissue are placed toward the left side of the recipient mouse. (B) A bundle of efferent ducts (arrow) can be found as a translucent line within opaque-colored fatty tissue between the epididymis and testis. (C) The opening of the rete testis ("window") is found at the beginning of the efferent duct as an oval, shady hollow on the surface of the testis (arrow). (D) The injection pipette filled with donor cells (visualized with trypan blue) is inserted into the duct and reaches the rete testis, to which donor cell suspension is first introduced. (E) Donor cells are further introduced into the seminiferous tubules through the rete testis. See color insert.

6 months in mouse testes. Spermatogonia derived from the rabbit, pig, and baboon are able to proliferate in mouse testes (28–30), whereas those from the dog, bull, and horse do not (28, 29). No meiotic differentiation occurs in mouse testes after xenogeneic transplantation of germ cells derived from all the species described above, except those from rats and hamsters. These results indicate that the mechanism for stem cell survival in the testicular microenvironment has been well conserved during evolution, but that for further germ cell development has markedly diverged.

Using the transplantation technique, germ cell–somatic cell communication has been examined. A number of spontaneous and induced mutant mouse strains are currently available that exhibit defective spermatogenesis. One of these mutant mouse strains, the juvenile spermatogonial depletion (*jsd*) mouse, shows a single wave of spermatogenesis after birth but spermatogenesis ceases thereafter, resulting in male infertility in the adult (19, 31). When germ cells derived from wild-type mice are transplanted into the testis of *jsd* mice, complete spermatogenesis is regenerated. In contrast, *jsd* germ cells transplanted into wild-type mouse testes fail to establish donor-derived spermatogenesis (19). The study thus indicates that the mutant phenotype is not caused by defects in somatic supporting cells, but by defects in germ cells themselves, which may have an intrinsic deficiency in spermatogenesis and/or be incapable of responding to extrinsic signals to undergo spermatogenesis. Similarly, transplantation experiments have demonstrated that estrogen receptor-α and testosterone receptor are not required in the germ cells to complete spermatogenesis (32, 33).

Spermatogonial transplantation also has the potential to be a new tool in male fertility preservation and transgenesis. As the survival rate of cancer patients increases, sterilizing cancer therapies have become a major concern for the quality of life of these patients. Male germ cells replicate at a rapid rate and are thus highly sensitive to cancer therapy, which causes more infertility in men than in women (34, 35). This is a particularly serious problem for childhood/adolescent cancer patients because cryopreservation of sperm is not an option for young patients. Thus, a future potential application of transplantation includes restoring male fertility following autologous transplantation of GSCs collected before sterilizing treatments. It is encouraging in this respect that male GSCs can readily be cryopreserved for a long time while retaining their full biological activity to regenerate complete spermatogenesis upon transplantation (36). In addition, a study has demonstrated that male GSCs that remain in the testis of genetically infertile mice are able to reestablish spermatogenesis after transplantation into a normal somatic environment, suggesting the application of the transplantation technique to male infertility treatment (14). For agricultural technologies, cryopreservation and transplantation of male GSCs would enable virtually indefinite preservation of important animal traits, while retaining possibilities of genetic diversification through recombination.

Finally, spermatogonial transplantation provides an opportunity to modify the genome of offspring through the male germ line by animal transgenesis and, in the future, to cure genetic infertility of men and to develop germ-line gene therapy techniques. In mice, male GSCs can be cultured in vitro without loss of their biological activities (37). When GSCs are cultured and exposed to the retrovirus that contains a marker gene, the viral gene is integrated in the stem cell genome, and spermatogenesis with expression of viral marker gene can be established in recipient testes after transplantation (38). These recipient mice can further produce transgenic offspring (16). Transgenic techniques developed to date (pronuclear injection of DNA and embryonic stem cell technologies) are based on female-derived cells as a target (egg and embryo) and require manipulation of each individual offspring. Since spermatogenesis has a life-long, high output activity, transgenesis through the male germ line could be a more efficient method of producing a large number of transgenic offspring. In addition, because the retrovirus requires host cell proliferation for its normal life cycle, the results described above demonstrate that the male GSCs can replicate in vitro. If the proliferation activity of GSCs can be maintained in vitro, it will be possible to amplify the GSC, precisely introduce a gene at a designated locus in its genome, and select appropriately modified stem cells. Therefore, the spermatogonial transplantation technique has the potential for applications to manipulate male GSC genome in a site-specific manner.

3. PROTOCOLS

Protocols described here (Protocols 22.1–22.5) use a transgenic mouse strain that expresses the *lacZ* marker gene in germ cells as the donor and chemically sterilized immunocompatible mice as recipients. For transgenic donor mice, Gtrosa26 (ROSA26 hereafter) mice (18, 39), which express *lacZ* in virtually all cell types including germ cells, and Syx4 mice (1,2), which express *lacZ* from the spermatid stage, are commercially available (The Jackson Laboratory, Bar Harbor, ME). Solutions and media required for these procedures are listed in Protocol 22.1.

4. EXAMPLES OF DATA ANALYSIS

4.1 Quantification of Male Germ Line Stem Cells and Measurement of Donor-derived Colonies

A typical staining pattern after transplantation of testis cells derived from ROSA26 mice is illustrated in Figure 22.2. Although still mostly monolayer, donor spermatogenesis 1 month after transplantation can be recognized as distinctive colonies (Figures 22.2A–C). As the number of colonies per testis does not change from 1 to 4 months after transplantation (Figure 22.4A) (18), these monolayer colonies represent definitive colonies of donor-derived spermatogenesis that are still in process of reestablishment. Well-established colonies can be observed at 2 months after transplantation using ROSA26 donor mice (Figure 22.2D, E). In Figure 22.2D, there are 13 colonies observed after injection of a donor cell suspension at a final concentration of 100×10^6 cells/ml. As approximately 10 µl of donor cell suspension can be injected into one recipient testis, it is estimated that 10^6 cells were injected into this recipient testis (100×10^6 cells/ml $\times$ 10 µl/1000 µl). Therefore, 13 colonies are estimated to be obtained from 10^6 donor cells. Since 1 colony is derived from 1 stem cell (20), the staining result shown in Figure 22.2D indicates that 13 stem cells colonized after transplantation of 10^6 donor cells per testis.

In addition, the colonies shown in Figure 22.2 are located apart from other colonies, and each single colony is clearly identified. Therefore, measurement of the length of a colony is feasible using an eyepiece micrometer. Figure 22.4B shows the results of the length measurement of 1021 colonies observed from 1 to 4 months after transplantation. The results indicate the linear growth in the average size of donor-derived colonies from 1 to 4 months after transplantation, even though a large variation is also apparent.

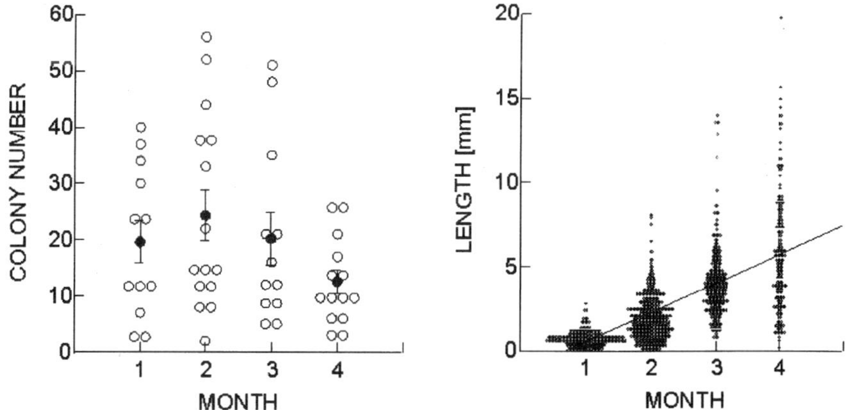

Figure 22.4. Kinetics of donor cell colonization after transplantation. Donor cells were obtained from the testes of ROSA26 mice. (A) The number of colonies per testis does not significantly change over the 4 months after transplantation, indicating that the colony number can be determined 1 month after transplantation when ROSA26 mice are used as donors. (B) Donor-derived colonies linearly increase in size during the 4 months after transplantation. The length of colonies can be measured under a stereomicroscope using an eyepiece micrometer.

4.2 Estimation of Donor Stem Cell Maintenance in Culture

In vitro culture of stem cells is important to characterize their biological properties and to manipulate male GSCs. To this end, it is necessary to compare different culture conditions that would affect *in vitro* GSC maintenance and behavior. Transplantation of donor cells cultured in vitro allows such an evaluation. Table 22.1 shows an example of the evaluation procedure of stem cell culture conditions using the transplantation technique (38).

In a series of experiments shown in Table 22.1 (culture condition A), a total of 8.2×10^6 donor testis cells were placed in culture. After a 7-day culture, all cultured cells were recovered by trypsin digestion. The cells were resuspended in 186 µl of germ cell culture medium and injected into recipient testes. As approximately 10 µl of donor cell solution can be injected into one recipient testis, it is estimated that an average of 0.44 $\times 10^6$ cells originally placed in culture were injected per recipient testis ($8.2 \times 10^6 \times 10/186$). A total of 62 recipient testes were injected, which resulted in a total of 1068 colonies established. Therefore, 39.1 colonies were obtained after transplantation of 10^6 cells originally placed in culture ($1068/62/0.44 \times 10^6$). In the same manner, after donor cell culture under the condition B, 2.5 colonies were obtained as a consequence of transplantation of 10^6 cells originally placed in culture ($27/14/0.78 \times 10^6$). Thus, these results indicate that 15.6-fold ($39.1/2.5$) more stem cells remained under condition A than under condition B.

It should be noted that the denominator used to calculate the relative colonization efficiency was the number of donor cells originally placed in culture and not the number of cells recovered after cell culture. The testis cell culture described above includes a number of unknown factors. The stem cells may proliferate or be lost by cell death and/or differentiation to committed progenitors in culture. The donor cells transplanted contain not only germ cells but also somatic testis cells, which may well affect GSC maintenance in culture. However, it is currently impossible to directly evaluate the stem cell kinetics and behavior during the culture period because GSCs cannot be identified in vitro. Therefore, the number of donor cells originally placed in culture must be used as a denominator to involve all these unknown events taking place in culture. In other words, the use of this denominator leaves all unknowns in culture as a blackbox and simply provides the end-point outcome; thus it is the most appropriate method to normalize the results and compare different culture conditions.

5. COMMENTS AND TROUBLESHOOTING

In this chapter, a manual transplantation procedure was described. Manual transplantation does not require a sophisticated microinjection apparatus and allows easier guidance of the injection pipette than does mechanical guidance of the pipette. However,

Table 22.1. Evaluation of In Vitro Maintenance of Male Germ Line Stem Cells

Culture Conditions	No. of Testis Cells Cultured ($\times 10^6$)	Vol. of Donor Cell Suspension (µl)	No. of Cells Injected per Testis ($\times 10^6$)	No. of Colonies	Relative Colonization Efficiency
A	8.2	186.0	0.44	1068	39.1
B	7.3	93.5	0.78	27	2.5

Adapted from Nagano et al. (38) with permission. A total of 62 (A) and 14 (B) recipient testes were analyzed. See text for details.

major drawbacks of the manual procedure are difficulty in the regulation of pressure to inject the donor cell suspension into recipient testes and limitations in manual handling of the injection pipette. These issues rarely become problems when recipients are adults. However, when immature mice are used as recipients, a procedure using a mechanical microinjection instrument and automatic microinjector may be superior to the manual procedure.

Busulfan treatment of recipient mice kills most endogenous GSCs. However, a small number of endogenous GSCs remain and regenerate recipient-derived spermatogenesis. Thus, if visualization of donor-derived spermatogenesis is not feasible, genetic determination of the origin of reestablished spermatogenesis is necessary (19).

A technical problem during the transplantation procedure may occur when cells aggregate at the tip of the injection pipette and block the donor cell flow. Cell aggregates may remain even after filtration or may form while the cells are placed on ice during injection. Careful pipetting of the cell suspension before injection prevents this problem. However, if the blockade of the pipette still occurs, pull the pipette out of the ducts, break the tip of the pipette with the forceps at the location of the cell aggregate, and reinsert the pipette at the initial insertion point. Multiple insertion is not difficult because traces of trypan blue that remain in the ducts allow easier identification of the initial location and route of the previous insertion.

During the injection procedure, care should be taken not to insert the injection pipette too deeply into the rete testis. Once the rete testis is punctured by the pipette, donor cells will leak out to the interstitial tissue and will not flow into the seminiferous tubules: donor cell leakage can readily be recognized because the area indicated by trypan blue does not show a clear lining of tubules. Donor cell leakage into the interstitial tissue may also occur due to partial rupture of the rete testis or the tubules. However, problems specifically caused by leakage have not been recognized. Recipient testes occasionally become fibrous or even atrophic. The cause of these problems has not been clearly identified, but they seem unlikely to be associated with donor cell leakage. Similarly, no adverse effects have been observed due to accidental injection of air into seminiferous tubules. At present, it is assumed that fibrous or atrophic testes result from interference with blood supply to recipient testes.

After *lacZ* staining, elongated spermatids and spermatozoa are recognizable on histological sections. However, more detailed identification of each germ cell type by nuclear morphology is difficult due to the blue precipitate formed after *lacZ* staining.

6. CONCLUSIONS

The spermatogonial transplantation technique is currently the only biological assay system that unequivocally detects male GSCs. The technique is amenable not only for qualitative but also for quantitative analyses of GSCs and spermatogenesis. Since its development, the transplantation technique has facilitated investigations into the characterization of GSCs and a better understanding of the process of spermatogenesis. Furthermore, ex vivo manipulation of the GSC genome during the transplantation procedure has provided a new approach to producing transgenic animals.

However, similar to research on all types of stem cells, the major drawback of the biological assay of GSCs, spermatogonial transplantation, is that GSCs are detected only retrospectively. Because we have to rely on this retrospective assay system, the study of male GSCs and spermatogenesis is time consuming and limited in its extent. To solve these problems, GSC-specific markers need to be identified that are tightly linked to the dual biological functions of stem cells. Stem cell markers will allow a prospective assay of GSCs and further enable us to isolate GSCs. Efficient GSC manipulation will be achieved with the establishment of in vitro culture systems in which

GSCs expand their population. Efforts to determine stem cell markers and stem cell expansion culture conditions will complement each other and contribute to a better understanding of male germ cell biology and manipulation of male germ cells.

Extension of the transplantation technique, including cryopreservation, purification of GSCs, and manipulation of the male germ line, should be pursued in other animal species, in particular agricultural animals. The testis is a powerful bioreactor that produces numerous gametes over the lifetime of a male. Manipulation of the male germ line of agricultural species is certainly a valuable approach to efficient production of transgenic animals for agricultural and medical applications.

It is also necessary to improve the efficiency of GSC colonization and regeneration of spermatogenesis upon spermatogonial transplantation into recipient testes. It has been reported that administration of gonadotropin-releasing hormone (GnRH) agonist to recipient mice can improve the regeneration of spermatogenesis after transplantation (41, 42). Such an approach to improvement of the transplantation technique will facilitate development of practical and clinical applications of this technique.

As described earlier, xenogeneic transplantation is feasible but has had limited success. However, xenogeneic germ cell transplantation into testes of experimental animals is expected to facilitate studies of male germ cells in various species, particularly in humans. Human hematopoietic stem cells (HSCs) can be transplanted into immunodeficient mice and regenerate human hematopoiesis in the mouse (43, 44). Supported by xenogeneic transplantation systems, biological studies and gene therapy technologies for human HSCs have greatly advanced (10). For example, the stem cell markers for human HSCs are different from those for mouse HSCs (10). Therefore, to understand the biology of human male germ cells and GSCs as well as to extend the transplantation technique to clinical situations, the possibility of regeneration of human spermatogenesis in the testes of mice or other appropriate animal species needs to be further pursued. It should also be noted that xenogeneic transplantation will reduce ethical concerns associated with human germ cell manipulation and experimentation.

Finally, together with development of various assisted reproductive technologies, the spermatogonial transplantation technique will pave new avenues toward male fertility preservation and genetic modification through the male germ line in the future.

Acknowledgments   I am deeply indebted to Dr. Ralph Brinster for introducing me to the exciting research field of male GSC biology and for providing me with the opportunity to learn his innovative technique of spermatogonial transplantation. I am grateful to Drs. Hugh Clarke, Takehiko Ogawa, Teruko Taketo, and Jacquetta Trasler for critical reading of the manuscript and for helpful comments.

Protocol 22.1. Standard operating procedure

Preparation of recipient mice

1. Use recipient mice at 4 weeks of age or older and immunocompatible to the donor strain.
2. Prepare busulfan solution first in DMSO and then add an equal volume of water to make a desired final concentration.
3. Weigh each mouse and intraperitoneally inject the well-mixed busulfan solution according to the body weight. See (18) for dose regimens for specific mouse strains.
4. Maintain the busulfan-treated recipient mice for at least 4 weeks before transplantation to destroy endogenous spermatogenesis.

Donor cell preparation

1. Remove testes and peel off the tunica albuginea. Mechanically loosen the seminiferous tubules.
2. Incubate testes in ~10 volumes of collagenase type IV solution (1 mg/ml in Hank's balanced salt solution; HBSS) at 32°–37°C for 10–15 min with occasional agitation to release interstitial cells from the seminiferous tubules.
3. Sediment the tubules by a unit gravity and remove supernatant.
4. Add HBSS and wash off the interstitial cells by removing supernatant. Repeat once.
5. Resuspend the tubules in 0.25% trypsin/1 mM EDTA and incubate 10 min.
6. Confirm the complete digestion of tubules and add 10% volume of FBS to stop digestion. Cell aggregations can be further digested by addition of DNase I at 1 mg/ml.
7. Filter the donor cells using nylon mesh to remove remaining debris.
8. Centrifuge the cells at 500 g and resuspend at an appropriate cell concentration with the germ cell culture medium. Addition of trypan blue enables visualization of donor cell introduction into recipient seminiferous tubules.

Injection procedure

1. Load the cells into a silicone tubing that is connected to a 1-ml syringe.
2. Attach a siliconized glass microinjection pipette to the tubing and transfer donor cells with pressure applied to the plunger of the syringe.
3. Make a small incision on the midline of the recipient's abdomen. Take out a testis, and place the epididymis and the fatty tissue in such a way that the opening of the rete testis can be observed.
4. While pulling the epididymis and fatty tissue, find the efferent ducts.
5. While holding and gently pulling the efferent ducts, remove fatty tissues from the ducts.
6. While pulling the ducts, insert the injection pipette into the ducts at around one-quarter of the distance between rete testis and epididymis.
7. Introduce the pipette following the duct until the tip of the pipette reaches the rete testis.
8. Gently apply the pressure to the plunger of the syringe, while holding the pipette. The donor cell suspension first goes into the rete testis and then into seminiferous tubules.
9. Return the testis to the abdomen and repeat the same procedure for the other testis.
10. Suture the incision and maintain the recipient mice until analysis.

Solution/buffer/medium

Collagenase IV (Sigma, St. Louis, MO): final concentration 1 mg/ml in HBSS. Prepare fresh.
DNase I (Sigma): final concentration 1 mg/ml in HBSS. Prepare fresh.
Germ cell culture medium—DMEM (high glucose, Gibco BRL, Gaithersburg, MD) containing
   10% FBS
   6 mM glutamine
   30 µg/ml penicillin
   50 µg/ml streptomycin
lacZ rinse buffer
   200 mM phosphate buffer (pH 7.3)

2 mM $MgCl_2$
0.01% sodium deoxycholate
0.02% Nonidet P-40 (NP40)
X-gal (5-bromo-4-chloro-3-indolyl-β-D-galactopyranoside): 25 mg/ml stock solution
  in DMSO
lacZ staining buffer. Prepare fresh. In the *lacZ* rinse buffer
  5 mM potassium ferricyanide
  5 mM potassium ferrocyanide
  X-gal solution at a final concentration of 1 mg/ml

Protocol 22.2 Preparation of recipient mice

1. Use mice immunocompatible with a donor strain at 4 weeks of age or older as recipients.
2. Determine empirically an appropriate dose of busulfan for a recipient strain of choice. Previous studies show that 44 mg/kg body weight is optimal for C57BL/ 6 (B6) × SJL F_1 hybrids and Ncr nu/nu mice, and 50 mg/kg body weight is optimal for B6 × 129 F_1 hybrids (3, 18, 28). Histological sections can be made to evaluate the level of germ cell elimination after busulfan injection.
3. Weigh and dissolve busulfan (Sigma, St. Louis, MO) in DMSO. The busulfan solution can be prepared at 4.0 mg/ml (final, after adding water; see step 4) and an appropriate volume of the solution injected according to the body weight of an individual mouse to provide the dose of busulfan predetermined in step 2. *Note*: Busulfan is a carcinogen and should be handled accordingly.
4. Add an equal volume of sterile distilled water immediately before injection. The water solubility of busulfan is low; thus, after adding water, busulfan may precipitate if the injection procedure is lengthy. To prevent precipitation, the busulfan solution can be maintained at 35°–40°C (3). Alternatively, multiple vials of a small amount of busulfan solution dissolved in DMSO can be prepared and water added to each vial immediately before each set of injections.
5. Mix the busulfan solution well and intraperitoneally inject the appropriate volume.
6. Maintain recipient males for at least 4 weeks before use. During this period, most endogenous germ cells disappear, and Sertoli cells and a small number of spermatogonia are found in the seminiferous epithelium.

Protocol 22.3 Donor cell preparation

Donor cells are prepared using a two-step enzymatic digestion (3, 17).

1. Sacrifice donor mice and remove the testis. Remove the tunica albuginea using fine forceps. The testis can be mechanically loosened to allow efficient enzymatic digestion.
2. Prepare 1 mg/ml collagenase type IV (Sigma) in HBSS without calcium and magnesium in a conical tube. Incubate testes in 10 volumes of collagenase solution. For immature testis, addition of 1 mg/ml collagenase type I (Sigma) allows more efficient digestion.
3. Incubate at 32°–37°C with occasional agitation for 10–15 min or until the tubules are dissociated.
4. Let the tube stand to allow the tubules to sediment.
5. Remove the supernatant to eliminate dissociated interstitial cells. Add HBSS without collagenase and let the tubules sediment.
6. Repeat the washing step (step 5).
7. If removal of the peritubular myoid cells is desired, treat the tubules with 1 mg/ ml hyaluronidase dissolved in HBSS in the same manner as in steps 3–6 (40).

8. Incubate the tubules in HBSS containing 0.25% trypsin and 1 mM EDTA at 32°–37°C for 5–10 min. Occasional pipetting allows efficient digestion.

9. Add HBSS containing DNase I to a final concentration of 1 mg/ml. Pipette the cell suspension to make a single cell suspension and to completely disrupt cell aggregates that may develop during the preceding digestion processes. Verify the completion of enzymatic digestion under a microscope.

10. Add 10% volume of FBS to stop trypsin activity.

11. Filter the cell suspension through a nylon mesh to remove any large pieces of undigested materials, which may plug a pipette during transplantation. A 60- to 70-μm pore size nylon mesh is appropriate, but a different pore size may be preferred depending on the type of application. For example, donor cell preparation for fluorescence-activated cell sorting requires a finer filter (25).

12. Wash the nylon mesh with HBSS.

13. Centrifuge the filtered cell suspension at 500 g for 5 min at 16°C. Remove the supernatant.

14. Resuspend the cell pellet in germ cell culture medium. Count cells. Cell viability after digestion of freshly dissected testes is usually ≥ 95%. These cells can be cryopreserved using a regular cell-freezing medium required for somatic cell cryopreservation (36).

15. Adjust the cell concentration as desired. When intact adult testes of B6/129 ROSA26 transgenic mice are used as the source of donor cells, transplantation of 100×10^6 cells/ml cell suspension results in approximately 20 colonies/testis (18). Trypan blue solution (0.4%) can be added at approximately 10% of the final volume to visualize donor cell suspension flowing into the tubules during the injection procedure.

16. Place the cell suspension on ice before and during injection.

17. Donor cells prepared in this manner can be used for ex vivo manipulation of germ cells, such as cell separation and cell culture. If donor cells are cultured in a monolayer, the cells can be digested with 0.25% trypsin–1 mM EDTA solution and prepared for transplantation as described in steps 8–16.

Protocol 22.4 Injection procedure

Three approaches have been reported for the donor cell injection procedure (3). The difference in these approaches is where the injection pipette is inserted: (1) directly into the seminiferous tubules, (2) into the efferent duct or the bundle of the ducts, or (3) into the rete testis. Although all three methods are equally effective in spermatogonial transplantation, the most versatile approach is the second, injection through the efferent duct or the bundle of the ducts, which is described below. The other two approaches have been described in Ogawa et al. (3).

The injection procedure can be performed either using a microinjection apparatus (1–3) or manually. The setting and injection procedure with a microinjection apparatus have been described previously (1–3), so only a manual injection procedure is described here.

Injection pipette

Prepare the injection pipette using a 3-inch length of borosilicate glass with internal diameter of 0.75 mm and external diameter of 1 mm. The glass tubing, which is siliconized to minimize cell loss, is drawn on a pipette puller. Break the tip of the pipette with a pair of watchmaker's #5 forceps to adjust the size of the tip to that of the bundle of efferent duct. Connect the other end of the pipette to silicone tubing (0.03 inches internal diameter × 0.065 inches outer diameter), which is attached to an edge-polished

20-G needle connected to a 1-ml syringe. Before connecting to the needle, the donor cell suspension is loaded into the silicone tubing by negative pressure applied through the plunger of the syringe, then transferred into the pipette.

Efferent duct injection

The efferent ducts run between the head of the epididymis and the rete testis, from which all seminiferous tubules originate. Insertion of the injection pipette into the efferent ducts allows the transfer of donor cells through the rete testis into virtually all the seminiferous tubules.

1. Anesthetize a recipient mouse and make a small incision (~ 3 mm) on the midline on the abdomen. Place the mouse under a dissection microscope set at ~12× magnification.

2. Set a testis on a sheet covering the incision in such a way that the cranial end of the testis faces the surgeon (Figure 22.3A). Place the epididymis and the fatty tissue toward the opposite direction to the side of the testis; when the right testis is injected, place the epididymis toward the left side of the recipient mouse (Figure 22.3A).

3. Identify the efferent ducts and clear them from the fatty tissue under a dissection microscope. In mice, multiple efferent ducts are bundled together into one that runs between the rete testis and the head of the epididymis. The bundle of the ducts, which is encapsulated with a thin membrane and surrounded by the fatty tissue, can be found cranial to the vascular pedicle as a transparent line in contrast to opaque fatty tissue (Figure 22.3B). Following the ducts toward the testis, the opening of the rete testis ("window") can be found as a shady spot with an oval shape on the surface of the testis (Figure 22.3C). In rats, multiple windows can be found according to the number of ducts. While gently pulling the ducts with the forceps, clear excessive fatty tissue away from the ducts to expose the bundle near the testis at a length one-fourth to one-third the distance to the epididymis. The dissection should be done carefully in the area close to the rete testis so that the ducts are aligned in one line toward the rete testis. However, complete isolation of the ducts from the fatty tissue is not necessary if the fatty tissue is cleared enough to expose the ducts and to enable alignment of the window, ducts, and needle.

4. While gently pulling the ducts with the forceps, insert the tip of the injection pipette into the area of the ducts bundle cleared in step 3. Follow the duct until the tip of the pipette reaches the rete testis. As the donor cell suspension includes trypan blue, the location of the pipette is relatively easily identified. When the pipette reaches the rete testis, release from the friction can be felt or the seminiferous tubule fluid can be observed to flow into the injection pipette from the testis. The edge of the opening of the rete testis may also be felt when the tip of the pipette is laterally moved.

5. While holding the injection pipette with one hand, apply pressure by gently pushing the plunger of the syringe with the other hand. The cell suspension, visualized by trypan blue, first accumulates in the rete testis, then runs into the tubules (Figures 22.3D, E). Inject 5–10 μl of cell suspension into one recipient testis that was pretreated with busulfan.

6. Pull out the injection pipette and observe the surface of the testis visualized by trypan blue included in the donor cell suspension. Although donor cells can be injected until the cell suspension fully covers the whole surface of a recipient testis, it is optimal to inject at a level where 70–85% of the testis surface is covered with the dye. Care should be taken not to inject so much of the donor cell suspension that blood vessels become very thin or disappear from the sur-

face of the testis. If the recovery of blood flow is not observed after a few minutes, small incisions can be made on the tunica using a 26-G needle to release the pressure.

7. Return the testis into abdomen and suture the incision. The whole procedure takes 15–30 min for one mouse or two testes. Recipients are maintained until analysis or to mate with females.

Protocol 22.5 Analysis

As described earlier, when donor cells are derived from transgenic mouse strains that express *lacZ* marker gene in germ cells, the donor-derived spermatogenesis can be readily visualized with incubation of recipient testes in a solution of 5-bromo-4-chloro-3-indolyl-β-D-galactoside (X-gal). To observe complete establishment of colonies of donor-derived spermatogenesis, 2 months are required when ROSA26 mice are used as donor and 3 months when Syx4 mice are used as donor (1–3, 18). Each colony can be further analyzed in histological sections.

lacZ *staining*

1. Sacrifice recipient mice and remove the testes and the epididymis. The epididymis has endogenous β-galactosidase activity, so it serves as the positive control for the staining reaction.
2. Place the testis in PBS and remove the tunica albuginea.
3. Using fine forceps, mechanically disperse the seminiferous tubules to facilitate penetration of fixative and staining solution.
4. Place the testes in cold 4% paraformaldehyde in PBS.
5. Fix the testes for 2 h at 4°C with gentle agitation.
6. Transfer the testes into *lacZ* rinse buffer.
7. Incubate the testes for 1 hr at 4°C with gentle agitation.
8. Repeat step 7 twice. The total length of the washing step is 3 h for 3 washes.
9. Place the testes in X-gal solution and incubate 8–16 h at 32°C with gentle agitation.
10. Fix the testes in 10% neutral-buffered formalin.

Analysis of *lacZ*-stained recipient testes

Beginning at 1 month after transplantation, donor-derived colonies are recognized as distinctive segments of the seminiferous tubules ("colonies") that are stained blue (figure 22.2) and the number of donor-derived colonies is readily counted (18). The size of a colony can be measured using a stereomicroscope with an eyepiece micrometer (18). Precise counting and measurement of colonies may require further dispersion of tubules after *lacZ* staining using fine forceps or a pair of 1-ml syringes with 26-G needles (Figure 22.2).

Histological examination of colonies can be made with paraffin sections. Since β-galactosidase is expressed in the cytoplasm of the cells derived from the mouse strains used in this protocol (ROSA26 and Syx4), nuclear fast red can be used for counterstaining.

References

1. Brinster, R. L., and Zimmermann, J. W. (1994). *Proc. Natl. Acad. Sci. USA*, **91**, 11298.
2. Brinster, R. L., and Avarbock, M. R. (1994). *Proc. Natl. Acad. Sci. USA*, **91**, 11303.
3. Ogawa, T., Arechaga, J. M., Avarbock, M. R., and Brinster, R. L. (1997). *Intl. J. Dev. Biol.*, **41**, 111.
4. Wing, T. Y., and Christensen, A. K. (1982). *Am. J. Anat.*, **165**, 13.

5. Sharpe, R. M. (1994). In: *The Physiology of Reproduction*, 2nd ed., E. Knobil and J. D. Neill, eds., p. 1063. Raven Press, New York.

6. Barratt, C. L. R. (1995). In: *Gametes—The Spermatozoon*, J. G. Grudzinskas and J. L. Yovich, eds., p. 250. Cambridge University Press, New York.

7. Russell, L. D., Ettlin, R. A., Sinha Hikim, A. P., and Clegg, E. D. (1990). *Histological and Histopathological Evaluation of the Testis*. Cache River Press, Clearwater, FL.

8. Byers, S., Pelletier, R.-M., and Suarez-Quian Carlos (1993). In: *The Sertoli Cell*, (eds. L. D. Russell and M. D. Griswold), p.431. Cache River Press, Clearwater.

9. Meistrich, M. L., and van Beek, M. E. A. B. (1993). In: *Cell and Molecular Biology of the Testis*, C. Desjardins and L. L. Ewing, eds., p. 266. Oxford University Press, New York.

10. Weissman, I.L. (2000). *Science*, **287**, 1442.

11. Nagano, M., and Brinster, R. L. (1998). *APMIS*, **106**, 47.

12. Schlatt, S., Rosiepen, G., Weinbauer, G. F., Rolf, C., Brook, P. F., and Nieschlag, E. (1999). *Hum. Reprod.*, **14**, 144.

13. Bucci, L. R., and Meistrich, M. L. (1987). *Mutat. Res.*, **176**, 259.

14. Ogawa, T., Dobrinski, I., Avarbock, M. R., and Brinster, R. L. (2000). *Nat. Med.*, **6**, 29.

15. Dinsmore, J. H. (1997). In: *Xenotransplantation*, D. K. C. Cooper, E. Kemp, J. L. Platt, and D. J. G. White, eds., p. 199, Springer-Verlag, Berlin.

16. Nagano, M., Brinster, C. J., Orwig, K. E., Ryu, B.-Y., Avarbock, M. R., and Brinster, R. L. (2001). *Proc. Natl. Acad. Sci. USA*, **98**, 13090.

17. Belleve, A. R., Cavicchia, J. C., Millette, C. F., O'Brien, D. A., Bhatnagar, R. M., and Dym, M. (1977). *J. Cell Biol.*, **74**, 68.

18. Nagano, M., Avarbock, M. R., and Brinster, R. L. (1999). *Biol. Reprod.*, **60**, 1429.

19. Boettger-Tong, H. L., Johnston, D. S., Russell, L. D., Griswold, M. D., and Bishop, C. E. (2000). *Biol. Reprod.*, **63**, 1185.

20. Dobrinski, I., Ogawa, T., Avarbock, M. R., and Brinster, R. L. (1999). *Mol. Reprod. Dev.*, **53**, 142.

21. Parreira, G. G., Ogawa, T., Avarbock, M. R., Franca, L. R., Brinster, R. L., and Russell, L. D. (1998). *Biol. Reprod.*, **59**, 1360.

22. Shinohara, T., Orwig, K. E., Avarbock, M. R., and Brinster, R. L. (2001). *Proc. Natl. Acad. Sci. USA*, **98**, 6186.

23. Shinohara, T., Avarbock, M. R., and Brinster, R. L. (2000). *Dev. Biol.*, **220**, 401.

24. Shinohara, T., Avarbock, M. R., and Brinster, R. L. (1999). *Proc. Natl. Acad. Sci. USA*, **96**, 5504.

25. Shinohara, T., Orwig, K. E., Avarbock, M. R., and Brinster, R. L. (2000). *Proc. Natl. Acad. Sci. USA*, **97**, 8346.

26. Franca, L. R., Ogawa, T., Avarbock, M. R., Brinster, R. L., and Russell, L. D. (1998). *Biol. Reprod.*, **59**, 1371.

27. Clouthier, D. E., Avarbock, M. R., Maika, S. D., Hammer, R. E., and Brinster, R. L. (1996). *Nature*, **381**, 418.

28. Dobrinski, I., Avarbock, M. R., and Brinster, R. L. (1999). *Biol. Reprod.*, **61**, 1331.

29. Dobrinski, I., Avarbock, M. R., and Brinster, R. L. (2000). *Mol. Reprod. Dev.*, **57**, 270.

30. Nagano, M., McCarrey, J. R., and Brinster, R. L. (2001). *Biol. Reprod.*, **64**, 1409.

31. Beamer, W., Cunliffe-Beamer, T. L., Shultz, K. L., Langley, S. H., and Roderick, T. H. (1988). *Biol. Reprod.*, **38**, 899.

32. Mahato, D., Goulding, E. H., Korach, K. S., and Eddy, E. M. (2000). *Endocrinology*, **141**, 1273.

33. Johnston, D. S., Russell, L. D., Friel, P. J., and Griswold, M. D. (2001). *Endocrinology*, **142**, 2405.

34. Chapman, R. M., Sutcliffe, S. B., and Malpas, J. S. (1979). *JAMA*, **242**, 1877.

35. Schilsky, R. L., Lewis, B. J., Sherins, R. J., and Young, R. C. (1980). *Ann. Intern. Med.*, **93**, 109.

36. Avarbock, M. R., Brinster, C. J., and Brinster, R. L. (1996). *Nat. Med.*, **2**, 693.

37. Nagano, M., Avarbock, M. R., Leonida, E. B., Brinster, C. J., and Brinster, R. L. (1998). *Tissue Cell*, **30**, 389.

38. Nagano, M., Shinohara, T., Avarbock, M. R., and Brinster, R. L. (2000). *FEBS Lett.*, **475**, 7.

39. Zambrowicz, B. P., Imamoto, A., Fiering, S., Herzenberg, L. A. A., Kerr, W. G., and Soriano, P. (1997). *Proc. Natl. Acad. Sci. USA*, **94**, 3789.

40. Tajima, Y., Onoue, H., Kitamura, Y., and Nishimune, Y. (1991). *Development*, **113**, 1031.
41. Ogawa, T., Dobrinski, I., Avarbock, M. R., and Brinster, R. L. (1998). *Tissue Cell*, **30**, 583.
42. Dobrinski, I., Ogawa, T., Avarbock, M. R., and Brinster, R. L. (2001). *Tissue Cell*, **33**, 200.
43. McCune, J. M., Namikawa, R., Kaneshima, H., Shultz, L. D., Lieberman, M., and Weissman, I. L. (1988). *Science*, **241**, 1632.
44. Fraser, C. C., Kaneshima, H., Hansteen, G., Kilpatrick, M., Hoffman, R., and Chen, B. P. (1995). *Blood*, **86**, 1680.

23

PAUL DE SOUSA
WILLIAM RITCHIE
ANDRAS DINNYES
TIM KING
LESLEY PATERSON
IAN WILMUT

Somatic Cell Nuclear Transfer

1. NUCLEAR TRANSFER

The methods of nuclear transfer in mammals have created many new opportunities in biology, medicine, and agriculture. Potential uses of these technologies include production of animal organs and tissues for transplantation to human patients (xenotransplantation), improvement of livestock, provision of human cells for treatment of degenerative diseases (stem cell therapy), and analysis of specific gene product function and the mechanisms that regulate gene expression. These applications are discussed in detail elsewhere (1, 2). Many of these nuclear transfer applications introduce genetic changes in cells before their use as nuclear donors in order to genetically modify the offspring. For example, research is focusing on producing genetically modified pigs so that the organs and tissues will not be rejected by human patients following xenotransplantation.

The purpose of this chapter is to describe, in detail, the Roslin sheep nuclear transfer protocol. First, however, we begin with a brief review of what has been achieved in somatic cell nuclear transfer in mammals to date, the limitations of the present procedures, and a summary of the factors that influence development of cloned embryos.

1.1 Achievements and Limitations in Nuclear Transfer Technology

The currently used nuclear transfer techniques are repeatable, as shown by their success in many independent laboratories. Viable offspring have been produced using somatic cell nuclear transfer in at least five species: including sheep, mouse, cow, goat, and pig. Despite these successes, it should not be assumed that the methods will be effective in all animals, as offspring have not been obtained in at least 3 other species (rat, dog, and rhesus monkey), despite considerable effort by the same laboratories that have been successful. A great variety of different cell types have been used in nuclear transfer, including representatives of all three lineages, and failure to obtain offspring has been reported in relatively few cell types (Table 23.1).

352

Table 23.1. A Variety of Different Cell Types Have Been Used as Nuclear Donors in the Five Different Species in Which Somatic Cell Nuclear Transfer Has Been Successful

Donor Cell Age	Donor Cell Type	Offspring Produced	No Offspring Produced
Adult	Cumulus	Mouse, Cattle, Goat	
	Oviduct	Cattle	
	Uterine	Cattle	
	Granulosa	Cattle, Pig, Goat	
	Mammary Gland	Cattle, Sheep	
	Muscle	Cattle	
	Fibroblasts	Mouse, Cattle	
	Sertoli		Mouse
	Spleen		Mouse
	Macrophages		Mouse
	Thymus		Mouse
Newborn	Fibroblasts	Cattle	
	Liver	Cattle	
	Testis		Cattle
	Sertoli	Mouse	
	Fibroblasts	Mouse, Cattle, Sheep, Pig, Goat	
	Germ Cells	Cattle	
Fetal	Liver		Cattle
	Gonad	Mouse	
ES cells		Mouse	

Regardless of the species or the particular procedure used, nuclear transfer is an inefficient process, and typically between 1–4% of reconstructed embryos become live offspring. This very low rate is the cumulative result of the death of embryos, fetuses, and offspring at all stages of development. Following activation and fusion, only a certain percentage of embryos are able to develop (between 10% and 50%). Once implanted, many different abnormalities of fetal and placental development can occur, and the rate of miscarriage can be high. In the offspring that do reach term, many die during or after birth due to lung maturation failure, kidney malfunction, and cardiovascular abnormalities. Furthermore, it should not be assumed that all surviving clones are normal; despite apparent physical well-being, they may still have genetic or epigenetic abnormalities. Epigenetic effects on gene expression may have been responsible for the death of an apparently healthy calf 7 weeks after birth, whose immune system failed to develop (3). Other postnatal deaths have been reported, including one in a goat cloning study, in which a cloned kid died after 1 month and another after 3 months (4).

1.2 Factors Affecting Development

Several factors influence the development of cloned embryos, of which the need to maintain normal ploidy is the most important. Ploidy is influenced by coordinating the cell cycles of donor cells and recipient oocytes and by allowing or preventing polar body extrusions, depending on the circumstances (see section 2.2.1). The biological mechanisms underlying this effect have been described elsewhere (5).

There are two main strategies to maintain normal ploidy. If the oocyte is at metaphase II of meiosis and so expected to have a high level of maturation promoting factor (MPF) activity (which causes the nucleus to breakdown, followed by DNA replication), then the donor cell nucleus should be awaiting DNA replication (therefore in the G_0 or G_1). Alternatively, the oocyte may be preactivated and allowed to enter S phase and thus inducing low levels of MPF activity. In this case, ploidy is maintained, regardless of the donor cell cycle, and therefore such preactivated oocytes are called "universal recipients."

There are also other effects of cell cycle beyond the simple maintenance of ploidy. There is a window of opportunity during the donor cell cycle including G_2/M, G_1, and G_0 that is more appropriate for transfer. It is possible that during this window there are changes in chromatin structure that allow access to those factors in the oocyte cytoplasm that reprogram gene expression to be appropriate for normal development. Historically, the first offspring to be produced by nuclear transfer from somatic cells were obtained when the donor cells had been induced to leave the growth cycle and become quiescent (6). Similarly, in most of the subsequent studies the donor cells were quiescent, although others have claimed that they used cells in G_1 (7), but the method used to characterize cell-cycle stage was inadequate (8). More recently, offspring have been produced from mouse embryonic stem cells judged to be in G_2 phase (9). Earlier studies in mouse using blastomeres as donors showed an advantage of the G_1 phase (presumed) in comparison to presumed S phase or G_2 stage (10). However, there is still no satisfactory comparison of which cell-cycle stage is the most advantageous with respect to somatic cell nuclear transfer.

Other aspects of the transfer procedure are also important and include the timing of activation after nuclear transfer and the recipient cell. In some species, there is a benefit in delaying oocyte activation after nuclear transfer (from minutes up to several hours). This was reported to be essential in mice (11) and to be advantageous in cattle (12), but to offer no benefit in sheep (13). The mechanism is not understood.

Nuclear transfer was successful in the production of mice (14) and pig (15) clones using serial nuclear transfer. This technique consists of an initial donor nuclear transfer into an oocyte. After a period of time (15 min to hours), the donor nucleus is transferred to an enucleated zygote. The authors in both studies suggested that the benefit of using a zygote, as opposed to an unfertilized oocyte, may reflect more effective activation by the initial fertilization or expression of transcripts from the pronucleus before enucleation.

There was a brief report of a dramatic improvement in nuclear transfer efficiency (> 20% of the reconstructed embryos became live offspring) in mice after inducing chemical enucleation (using demecolcine) of an activated oocyte (16). The positive effect of chemically removing the nucleus over mechanical removal remains to be extended to other species.

2. DESCRIPTION OF TECHNIQUES

Production of cloned offspring also depends on several enabling technologies, including those for oocyte recovery or maturation, enucleation, cell fusion or nuclear injection (see Figure 23.1 for summary), embryo culture, storage, and transfer. The number of species for which all these techniques have been established is very small.

2.1 Oocytes as Recipient Cytoplasts

2.1.1 In Vitro-Matured Versus Ovulated Oocytes

Selection of the most appropriate source of mature oocytes is influenced by the state of the art in reproductive technologies for a given species and the costs associated with each one. Oocytes to be matured in vitro can be obtained from slaughterhouse ovaries (in the case of livestock) or by harvesting from females at an appropriate stage in their estrous cycle. Otherwise, mature oocytes can be obtained from superovulated females. In the mouse, large numbers of uniform metaphase II oocytes (15–20 per female, depending on the strain) can be superovulated using established protocols at minimal expense (17). As a result, most nuclear transfer studies in mice have used superovulated oocytes (11, 18). In contrast, decades of research into optimizing methods for the in vitro maturation, fertilization, and culture of bovine embryos (reviewed by Thompson [19]) have yielded systems that have effectively replaced superovulation as a source of cytoplasts for nuclear transfer (7, 12, 20, 21, 22). However, these in vitro systems still yield a more heterogeneous population of oocytes with variable developmental competence (23–25). With respect to other species, somatic cell nuclear transfer has produced viable piglets and goat herds from both superovulated and in vitro-matured oocytes (15, 26, 27). However, primates have only been cloned (with embryonic blastomeres) using the former (28). All of the original studies describing the production of viable embryonic or somatic nuclear transfer sheep also relied on superovulatated oocytes (6, 13, 29).

2.1.2 In Vitro Handling of Oocytes for Nuclear Transfer

Optimal handling of metaphase II oocytes entails processing oocytes as quickly as possible. Time constraints can depend on the nature of the nuclear transfer approach selected as well as on the species. As oocytes age (relative to their time of ovulation or maturation), they become more prone to spontaneous activation (30–32). This appears to be related to time-dependent cell cycle and cytoplasmic changes that result in a partially activated state that can make it more difficult to coordinate cell cycle activities between an oocyte cytoplast and a transplanted karyoplast (33), the importance of which is discussed below. Despite this, aged cattle oocytes are still able to support embryonic karyoplasts and can yield viable offspring provided they are used as universal recipients (21). Also, in pigs the optimal time to electrically activate in vitro-matured or ovulated oocytes for the best yield of diploid parthenogenetic blastocysts is confined to a limited window of time lasting 2–3 h (34–36).

Before their use as cytoplasts for nuclear transfer, oocytes or fertilized zygotes need to be stripped of their cumulus investment. Because cumulus cells contribute to oocyte developmental competence (37, 38), they are removed immediately before enucleation, either mechanically by pipetting or enzymatically using hyaluronidase. Whereas pipetting is sufficient and preferred when cumulus is sparse, hyaluronidase is an effective means to process large numbers of oocytes or zygotes, especially when their cumulus investment is great.

2.1.3 Enucleation

Oocyte or zygotic enucleation is generally achieved mechanically by aspiration using glass pipettes. Frequently, this entails pretreatment of oocytes or zygotes with cytochalasin B or D to destabilize cortical microfilaments and thus facilitate the disruption of the plasma membrane. A piezo microinjection system, designed to minimize physical damage of the oocyte during enucleation and nuclear transfer, has been used successfully for somatic nuclear transfer in mice and pigs (11, 27). However, this system is not absolutely essential for somatic cloning in these species (26, 39).

The enucleation of meiotic spindles or pronuclei from species such as mice whose oocytes and zygotes are translucent can largely be achieved using plane-polarized light, specifically by the method of Differential Interference Contrast (DIC) microscopy (40). But this method is of limited value when working with the lipid-rich and thus optically dense eggs of large domestic animals.

Traditional enucleation of optically dense nonrodent eggs has relied on the use of DNA fluorescent dyes such as Hoechst 33248 to visualize the metaphase chromosomes (Sigma, Dorsett, Poole, UK) (41). To minimize harmful irradiation of oocytes, the position of the meiotic spindle in a metaphase II oocyte is first estimated by the position of the first polar body, without the use of UV. Following mechanical enucleation with a pipette, success is confirmed by UV illumination of the pipette alone (see Protocol 23.4

for details; see also figure 23.5). However, the orientation of the first polar body with respect to the spindle can become uncoupled with oocyte aging or in vitro handling, further reinforcing the importance of micromanipulating in a timely fashion (42, 43).

2.2 Somatic Nuclei as Karyoplasts for Nuclear Transfer

2.2.1 Nuclear Donor Preparation

The optimum method for preparing cells for nuclear transfer has been the subject of debate. Induction of a quiescent (G_0) state in nuclear donor cells by serum deprivation continues to be routine in our laboratory. The importance of cell cycle coordination between donated nuclei and recipient egg cytoplasm, to maintain DNA integrity and ploidy, is well established. In cloning experiments using metaphase II-arrested eggs as cytoplasts, the DNA content of nuclear reconstructed embryos can also be regulated by controlling polar body extrusion after egg activation (see Figure 23.1). Traditionally, this has been achieved using cytochalasin to depolymerize cortical microfilaments (18, 44, 45). Alternatively, nuclear retention and structure after activation can be influenced by treatments with the serine threonine kinase inhibitor 6-dimethylaminopurine (DMAP) (46) or microtubule depolymerizing agents such as nocodozole (35). Thus, as has been best exemplified in the mouse, the DNA content of an unreplicated diploid nucleus (2n, 2C) in G_0/G_1 can be preserved after transfer by suppression of polar body extrusion. Conversely, the DNA content of a replicated diploid nucleus (2n, 4C) in G_2 or M phase can be returned to normal ploidy by allowing polar body extrusion after transfer (11, 18, 47).

2.3 Methods of Nuclear Transfer

The donor nucleus is transferred to the enucleated oocyte by direct injection or cellular fusion. The latter technique is perhaps the most common. Experiments in mice have used the fusogenic properties of viruses by precoating cells with inactivated Sendai virus with or without phytohemagglutinin, the latter intended to make cell surfaces stickier (reviewed by Barton and Suvani [40]. In large domestic animals, fusion by electroporation of closely apposed membranes has been the method of choice (reviewed in Smith [48] and has been used successfully for somatic nuclear transfer in sheep, cattle, goats, and pigs (6, 7, 15, 49). Recently, electrofusion of somatic cells has also been reported to yield viable mice (50). Arguably, the advantage provided by injection is to deliver a karyoplast relatively devoid of donor cytoplasm that therefore may be able to integrate better with the oocyte cytoplasm. However, this approach requires greater micromanipulation skill and access to costly specialized systems (e.g., piezo).

2.4 Activation

Although a broad range of physical and chemical stimuli are known to parthenogenetically activate mammalian eggs (reviewed by Machaty and Prather [51]), four have been proven successful in the cloning of viable animals following somatic nuclear transfer. These include the use of electrical activation for sheep and pigs (6, 27); strontium chloride ($SrCl_2$) for mice (11); ethanol activation in goats (49); calcium ionophore (e.g., ionomycin) followed by DMAP in the cloning of cattle and pigs (7, 26).

2.5 Postactivation Handling

The choice of strategy for embryo culture and transfer varies with species and depends on the effect of the hole in the zona needed for nuclear transfer, the normal number of young in that species, the efficiency of culture methods, and the approach to embryo transfer. Surgical transfer of mouse embryos may occur at the two-cell stage up to blastocyst stages,

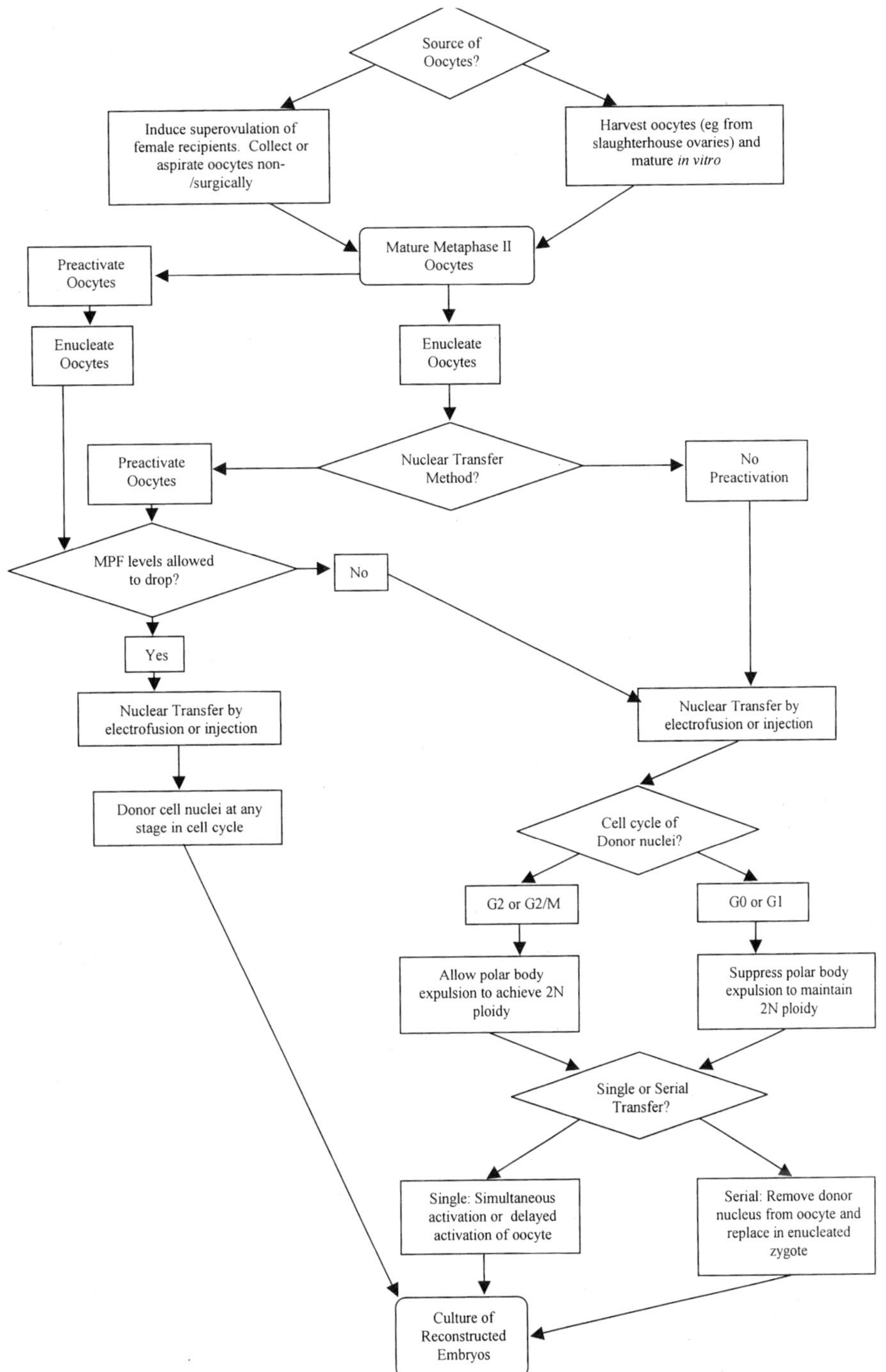

Figure 23.1. Summary diagram for nuclear transfer.

to oviduct or uterus as appropriate (52). In cattle, one or two in vitro-cultured blastocysts are nonsurgically transferred to the uterus (7). In sheep, as culture in the laboratory is less efficient at present, embryos have usually been cultured in the ligated oviduct of a temporary recipient before a second transfer at the blastocyst stage (6). The pig requires a minimum of four viable conceptuses at the time of maternal recognition of pregnancy for that pregnancy to be maintained (53). In light of the high rate of developmental failure associated with somatic nuclear transfer, pregnancy can be maintained either by the transferring vast numbers of reconstructed embryos or fertilized embryos or by hormonal supplementation of the recipient (15, 27). A novel alternative is the co-transfer of parthenogenetic embryos, known to have a finite fetal life span (35).

3. PROTOCOLS FOR SHEEP NUCLEAR TRANSFER

We provided information that illustrates a chronological, step-by-step account of the protocols, standard operating procedures (SOPs), instruments, and materials required to perform somatic cell nuclear transfer in sheep (Protocols 23.1–23.11).

4. EXAMPLES OF THE TYPE OF DATA TO BE COLLECTED AND THEIR ANALYSES

Monitoring the success of the nuclear transfer procedure is usually based on technical and pre- and postimplantation developmental parameters. The nuclear transfer procedure is monitored by collecting data on morphology and nuclear status of oocytes; rate of oocytes suitable for enucleation; rate of oocytes successfully enucleated; fusion rates of oocytes (in case of fusion methods); and lyses of oocytes during the manipulation steps.

4.1 Preimplantation Period

In the preimplantation period, including in vitro or temporary in vivo culture, early developmental parameters such as cleavage rates, blastocyst rates, cell numbers, and proportion of inner cell mass/trophoblast cells are informative (55). Data can be collected on a variety of parameters, reflecting the quality of the embryos, or characterizing the reprogramming procedure. Here we list only a few of the potential parameters to be recorded: nuclear swelling (56), metabolic assays (57), ploidy of the embryos and nucleolar protein changes (58), gene expression and methylation changes of certain genes (59), and telomerase activation (60).

4.2 Postimplantation Period

In the postimplantation period, nuclear transfer procedures often result in abnormalities. Careful monitoring of the pregnancy is important to provide data to improve the procedures and to enable veterinary interventions for the welfare of the recipient animals and the developing fetuses. Technical development, especially in the area of ultrasonography, is expected to contribute greatly to the better monitoring of nuclear transfer pregnancies. Data can be collected on pregnancy rates and time of pregnancy losses; fetal developmental parameters and abnormalities (ultrasonography or recovered fetuses [61]); length of pregnancy; birth weight; perinatal abnormalities (see more details in section 5); postnatal weight gain and behavioral studies (62); reproductive potential of cloned animals; mitochondrial inheritance studies (63); and telomere length studies (64–66).

5. TROUBLESHOOTING AND COMMON PROBLEMS

Nuclear transfer is a complex procedure, involving numerous technical steps, some of which require high level of skills to perform. The quality of the equipment and tools can strongly influence the outcome. Furthermore, this technique is based on biological resources varying greatly in quality among or even within experiments. In this section, we address some of the most common problems. However, troubleshooting can be a difficult issue in nuclear transfer, and the variety of interactions among the elements of the complex procedures can sometimes puzzle even the most experienced researchers, resulting in mysterious declines and increases in efficiency over a period of time.

5.1 Biological Resources

5.1.1 Recipient Oocytes

In vivo oocytes can be collected from the oviduct of recently ovulated donors. The quality of the oocytes for nuclear transfer can vary tremendously among donor individuals or even within the same donor, especially in farm animals. Other factors that could influence oocyte quality and/or number are breed or strain, age, and nutritional and reproductive history of the donor. Oocytes from certain individuals can have much more fragile zona pellucida, sticky or very dark cytoplasm, and can be more prone to lysis or cytoplasmic fragmentation during manipulation or after activation. Seasonal differences are clear, especially in certain species or breeds, and most probably even occasional meteorological events can influence the oocyte quality and numbers by changing the donor animals hormonal system. Superovulation protocols profoundly affect the number, synchrony, and the quality of the oocytes. The age of oocytes is also an important parameter and can affect the localization of the metaphase plate compared to the polar body (42, 43). Electrofusion of the oocyte with the newly introduced cell is less efficient in aged oocytes; however, in most species older oocytes can also be activated more readily (67). In other species, such as pigs, however, the optimal window of age for activation is narrower (35).

In vitro-matured oocytes, obtained from slaughterhouse ovaries or by ovum pickup, can be used for nuclear transfer with good success. Slaughterhouse material is usually much less expensive; however, the efficiency of the in vitro maturation methods varies among species and in general results in lower quality oocytes.

5.1.2 Nuclear Donor Cells

A large variety of cells have been used successfully for nuclear transfer experiments, whereas other cell types have been less successful or failed entirely (see table 23.1). The cell-cycle status of the donor cell has a major influence on the outcome of the nuclear transfer (5). However, the optimal method for preparing the chosen cells for nuclear transfer is not established. Size and morphology of the cells varies even from a fairly homogenous cell culture, affecting the efficiency of the nuclear transfer embryo reconstruction procedure. Larger cells usually electrofuse better with the cytoplast, due to a larger membrane surface in contact. However, in fibroblast cell cultures, smaller cells are more likely to be in G_0/G_1 stage and are often selected for nuclear transfer. Fresh cumulus cells are smaller than fibroblast cells and more homogenous in size. The morphology of the membrane of the donor cells may vary depending on the preparation of the cell suspension and could influence fusion rates.

Cell lines can vary in their capability to produce offspring after nuclear transfer. Furthermore, cell culture, depending on the number of cell doublings and culture conditions, can result in epigenetic changes. Monitoring the cultured cells for ploidy changes and major translocations by staining methods allows gross abnormalities to be detected and prevents the use of abnormal cell lines in nuclear transfer. However, minor changes are difficult to detect, and it is impossible to guarantee that the cell chosen for nuclear transfer is genetically suitable to develop into a healthy offspring. Targeting experiments requires longer culture periods, and this may increase the chances for abnormalities. Cell senescence might result in apoptosis and degradation of DNA, although it is not known how early apoptotic events might influence the outcome of nuclear transfer.

In conclusion, it is advisable to use early passage cells and monitor their ploidy. However, the only reliable test for the suitability of a cell line for nuclear transfer is the production of a healthy cloned offspring.

5.2 Technical Steps of Nuclear Transfer

5.2.1 General Setup

Nuclear transfer requires a high level of attention and control of technical parameters. The less time the oocytes are exposed to suboptimal conditions (e.g., during micromanipulation, outside of the incubator) the better the results. An environment free of disturbances, arranged to accommodate the researcher's personal work preferences is essential. Noises, strong airflow, and temperature changes can result in suboptimal conditions.

5.2.2 Microscope

The quality of the manipulation microscope is very important, and the proper visualisation of the metaphase plate during enucleation is essential. In certain species it is possible to observe the metaphase plate with DIC or Normarski optics (in the mouse, and to some extent in sheep and rabbit), or it has to be visualised with fluorescent staining and UV exposure of the oocyte or the removed cytoplasm. New technical developments, such as high-sensitivity video cameras to reduce the intensity of UV exposure (68), or new, more sensitive dyes (69) might improve the results.

5.2.3 Temperature Control

Oocytes are sensitive to low temperature, and, depending on the species, this may be an important factor during micromanipulation. Porcine oocytes are sensitive and the use of heated microscope stages or warming up the manipulation room is advisable. Temperature of the micromanipulation can affect membrane flexibility and lead to lysis of the oocytes after enucleation or injection of cells.

5.2.4 Vibration Attenuation

Vibrations from the floor or the walls can be transmitted to the microscope, disturbing the micromanipulation work. The detachment of the micromanipulator bench from the walls and the use of antivibration tables or microscope stands is important. This equipment is available commercially or can be constructed easily using squash balls under heavy metal plates.

5.2.5 Micromanipulation

Various systems are available on the market, often modified by the researcher to accommodate personal preferences. In general, the success of the enucleation or embryo reconstruction process mostly depends on the control of the vacuum in the system. The sensitivity of the microinjector syringe is crucial, and the commercially available models are often not fine enough. Sensitivity can be modified by varying the size of the syringe, the tubing, the filling (air, different-density oils, or fluorinert), and the parameters of the enucleation capillary. Loss of control is often due to air bubbles entering the oil-filled system. Purging the system of air bubbles and refilling with oil is time consuming and frustrating but difficult to avoid. Capillary parameters are very important, and commercially available ones are not always suitable for the given manipulation process. Elements of the micromanipulation system are interactive; for example, changes in the filling of the tubing might require changes in the size of the holder pipette. Piezo devices used for cloning require a more steady manipulator-arm system and rigorous control of technical parameters, including the filling and shape of the capillary and the strength of the piezo pulses. Use of mercury in the capillary can improve the results but represents a potential health hazard.

The manipulation solution must maintain its pH and osmotic parameters during the entire process. In general, it is important to avoid changes in osmolarity during the micromanipulation. Usually the solution contains agents that modify the flexibility of the cytoplasmic membrane. Cytochalasin B is the most commonly used, but it has side effects, including interfering with glucose transport mechanisms. Cytochalasin D does not have this latter side effect and a lower concentration is required for the same effect, on the membrane.

The manipulation chamber can vary depending on the process and personal preferences. In general, the optical quality of the chamber and its suitability for DIC optics is better with the use of glass slides and with the reduction of the overlaying oil. However, plastic Petri dishes with microdrops are also suitable and often used. The manipulation chamber, the design of the capillaries, and the position of the micromanipulator arms must be fitted together for a successful process.

5.2.6 Glass Microtools

While making the enucleation pipette, avoid overheating the glass capillary when breaking it on the microforge, as this can shorten and thicken the glass at the break. If the broken end of the microtool is thick, it will take longer to grind the tool on the grinding wheel and may also cause the tool to have a constriction, which makes the pipette difficult to use. Make sure that the pipette is properly ground because otherwise it is difficult to put a spike on a pipette.

5.2.7 Enucleation, Electrofusion, and Activation

To prevent harmful irradiation, the oocytes must be removed from the field of view before exposing the enucleation pipette to the UV light.

A variety of electrofusion devices are available on the market. It is critical that at the point where oocytes and donor cell are in contact, the membrane is perpendicular to the electric field. Application of a short AC pulse before the fusion pulse can improve the contact between the membranes and the positioning of the two cells. However, in nuclear transfer with small somatic cells, the effect is not very marked, and manual correcting of the positioning between the AC and DC pulse is beneficial. It is important to reduce changes in the liquid, so frequently changing the fusion solution is beneficial, depending on the drop size. Introduction of electrolytes into a nonelectrolyte fusion solution can result in the lyses of the oocytes and must be avoided by proper washes between the solutions. In case of delayed activation methods, the fusion solution must be free of Ca^{2+} ions. Accuracy of fusion rates can be compromised by the occasional lyses of the newly introduced, yet not fused cell; the presence of cytoplasmic or polar body fragments under the zona pellucida, similar in size to the donor cell; or by late fusion events, after the time of the evolution.

5.3 Embryo Culture, Fetal Development, and Birth

5.3.1 In Vivo Culture

Several factors influence the choice of culture system and time of embryo transfer. Effective in vitro culture methods are only available for some species (e.g., mouse and cow), and in other species in vivo culture is beneficial at present (e.g., sheep). Nonsurgical transfer to the uterus is possible in cattle if the embryos have been cultured to the blastocyst stage. In some species, embryos are destroyed after transfer to the oviduct after a hole has been made in the zona pellucida for nuclear transfer. This appears to be the case in sheep and cattle, but it is not in mice or pigs. To protect sheep embryos during transfer, in vivo agar chips can be made around the embryos (6); however, extreme care is required to control the temperature of the agar chip. The transfer and the recovery of the embryos from the temporary recipient introduces additional surgical procedures, which can result in embryo losses. Care during removal of hatching or hatched blastocysts from agar chips is necessary to avoid damage to the embryos without the protective zona. The optimal stage of the recipient's cycle compared to the embryo age is not well defined in the case of nuclear transfer embryos. When transferring the nuclear transfer oocytes to oviduct of the temporary recipient ewe, care must be taken not to damage the oviduct with the transfer pipette.

5.3.2 In Vitro Culture

In vitro culture allows the temporary recipient step to be omitted and provides more accurate data on the dynamics of early development after nuclear transfer. However, in vitro culture systems often result in embryos with inferior viability compared to in vivo-cultured ones. Furthermore, in vitro systems including serum can result in complication such as large offspring syndrome in ruminants (70). Evaluation of the cultured embryos before final transfer without nuclear staining is difficult due to the presence of cytoplasmic fragments.

5.3.3 Fetal Development

Nuclear transfer embryos are prone to losses during the entire period of gestation. The reasons for the high rates of losses are not fully understood. Nuclear transfer pregnancies must be treated as problematic, with frequent monitoring and special precautions. Elective caesarean sections, oxygen application after birth, or hydrocortisone pretreatments before birth to mature the respiratory system results in better survival of the animals (12).

5.3.4 Postnatal Losses

Live birth of a cloned animal does not necessarily guarantee a healthy offspring. Respiratory problems can be resolved soon after birth; however, late effects might cause further losses. Lethal immunodeficiency in a calf developed several weeks after birth (3) and abnormal obesity of cloned mice several weeks after birth (62) have been reported.

6. CONCLUSION

Somatic cell nuclear transfer is a repeatable procedure that has been successful in several species but is severely limited in its applications at present because the production rate of cloned healthy offspring is very low. The nuclear transfer protocol provided here has been successfully used in sheep and provides excellent guiding principle for nuclear transfer in all mammals. Yet the exact procedures used for oocyte recovery, nuclear transfer, timing of activation, activation method, enucleation method, serial or single transfer, in vitro versus in vivo embryo culture, and means of embryo transfer will depend on the species, and techniques should be changed accordingly. Unfortunately, there have been few comparisons between different methods. As these experiments are carried out and new ideas are tried, we should expect the efficiency to increase.

Acknowledgments   This work was supported by the Ministry of Agriculture, Food and Fisheries, the Biotechnology and Biological Sciences Research Council, the biotechnology program of the European Economic Community, Roslin Biomed, and Geron Biomed.

Protocol 23.1. Synchronization and superovulation of ewes for oocyte recovery

The following sections provide details on obtaining mature oocytes for nuclear transfer recipients from superovulated ewes.

Materials and solutions

Ewes in reasonable body condition (condition score > 2)
Ovagen: ovine FSH (oFSH), 1 U (9 mg), 8, 2 ml doses injected subcutaneously (from Immunochemicals Ltd., New Zealand).
Veramix sponges (Upjohn Ltd.)
GnRH receptal, 2 ml/ewe injected intramuscularly (Hoechst UK, Ltd.)
PMSG: Folligon 150 IU/ewe injected intramuscularly (Intervet)
Suture: Dexon (Davis and Geck Ltd)
Thiopentone sodium B.P. (Sodium Intraval)

Donor ewe treatment schedule

Procedure	Time	Day
Sponge in	—	0
oFSH	8 AM	10
oFSH + PMSG	5 PM	10
oFSH	8 AM	11
oFSH	5 PM	11
oFSH	8 AM	12
oFSH	5 PM	12
oFSH[a]	7 AM	13
oFSH	5 PM	13
Sponge out[b]	9 PM	13
Heat + GnRH	7 AM	14
Starve	PM	14
Oocyte recovery	AM	15

[a]And sponge out for Scottish Black Face.
[b]For Poll Dorset.

Protocol 23.2. Recovery of oocytes from superovulated donor ewes

Materials and Solutions

Catheter: a short piece of plastic tubing of suitable diameter pushed into the oviduct of the ewe.
Flushing medium: PBS + 1% FCS PBS—Dulbecco a, Oxoid BR 14a).

Oocyte recovery

1. Anesthetize a superovulated donor ewe using sodium intraval.
2. Perform a midline laparotomy.
3. Move the uterus outside the body cavity to expose the ovaries and oviducts.
4. Place a small catheter in the fimbria and hold in place, preventing leakage past the side of the canula.
5. Using a blunt 18-G needle introduced close to the uterotubule junction, flush 20 ml of warm flushing medium through each oviduct.
6. Collect the flushings in a sterile collection tube.
7. Search for the cumulus oocyte complexes.

Protocol 23.3. Instruments and materials required for nuclear transfer

Microscopes and manipulators

For embryo manipulation, a Nikon TE300 inverted microscope with DIC (differential interference contrast) and an epifluorescent light source is used. The microscope is fitted with two Narishige MO-Joystick Hydraulic Micromanipulators and IM-16 Micrometer type microinjectors. The microinjectors are fitted with 100, 250, or 500 μl gas tight syringes (Sigma, Dorset, UK) in addition to a three-way tap (Vigon VG1) and syringe to allow easier filling of the gas-tight syringe. The system is filled with Fluorinert FC 77 (Sigma, F-4758).

The manipulation pipettes are from Clark Electromedical Instruments. The pipettes are pulled by a Campden Instruments moving coil microelectrode puller (model 753). A Research Instruments MF1 microforge is used, underneath which a homemade grinder is fitted. This has a rotating disc of 4 cm diameter, which can be fitted with aluminium oxide paper in sizes of 0.3, 1.0, and 3.0 μm (3M), and which rotates at speeds of 0–60 rpm. An additional tool holder with a length of tubing and a syringe is used as an air pump when grinding the pipettes.

A BLS CF 150/B impulse generator produces cell fusion. The fusion chamber is constructed by gluing two platinum wires onto the bottom of a glass Petri dish, at a distance of 200 μm using araldite epoxy resin. Embryos are handled using Leica MZ 75 and MZ 125 dissecting microscopes, fitted with Linkam, MTG minitub and homemade warm stages. In addition to the above equipment, the following instruments are also required: an electronic thermometer (PTM1; Petracourt Ltd.); Drummond model 105 and 325 pipettes; automatic pipits (Gilson), 10, 20, 100, 200 and 1000 μl; and glass chips measuring 4 × 25 mm cut from 2-mm thick glass.

Embryos are cultured in Heraeus Instruments series 6060 triple gas incubators for long-term culture (Kendro Laboratory Products), Hera Cell CO_2 incubators for short-term culture, and Merck mini incubators for keeping oocytes warm on the bench.

Medium

Both SOFaaBSA and hSOF media are required and need to be made with and without calcium and with or without the addition of 10% FCS (54). See Tables 23.2 and 23.3 for compositions.

Table 23.2. SOFaaBSA Composition

Component	Sigma Cat. N.	g/1000 ml
NaCl	S-5886	6.29
KCl	P-5405	0.534
Kh_2PO_4	P-5655	0.162
$MgSO_4H_2O$	M-1880	0.182
Sodium lactate *	L-7900	0.6 ml
Penicillin	P-4687	0.06
$NaHCO_3$	S-5761	2.1
Phenol Red	P-5530	0.01
Na pyruvate	P-4562	0.0357
$CaCl_2H_2O$	C-7902	0.262
L-Glutamine	G-5763	0.3
BME	B-6766	20 ml
MEM	M-7145	10 ml
BSA	A-6003	4

Table 23.3. hSOF Composition

Component	Sigma Cat. N.	g/1000 ml
NaCl	S-5886	6.29
KCl	P-5405	0.534
Kh_2PO_4	P-5655	0.162
$MgSO_4H_2O$	M-1880	0.182
Sodium lactate *	L-7900	0.6 ml
Penicillin	P-4687	0.06
$NaHCO_3$	S-5761	0.42
Phenol Red	P-5530	0.01
Na pyruvate	P-4562	0.0357
$CaCl_2H_2O$	C-7902	0.262
Hepes	H-3784	5.208
BSA	A-6003	4

Holding pipettes

1. Pull glass capillaries (GC10-100) by hand over a small flame to give a diameter of 100–150 µm.
2. Mount a drawn capillary in the microforge and make the first bend close to the start of the pulled part at 45°.
3. Make a second bend 10 mm from the first to straighten the capillary and make parallel to the rest of the capillary.
4. Make a third bend 20 mm from the last bend at 45° and a fourth bend, close to the last bend, making the pipette parallel to the pipette again.
5. Cut the glass using a diamond pencil 2 mm from the last bend, making sure that the cut is at 90° to the glass.
6. Fire polish the end of the pipette over the filament of the forge to close the end to diameter of approximately 20 µm.

Enucleation pipettes

1. Pull glass capillaries (GC100T 15) using a moving coil microelectrode puller to give a diameter of slightly more than the required diameter, and a second taper which is almost parallel.
2. Mount a capillary in the microforge and break this at the required diameter by fusing the glass onto the glass bead on the microforge and turning off the heat while drawing it away. Do not overheat the capillary, as this may lead to distortion of the pipette and thickening of the glass.
3. Mount the pipette in an instrument holder and attach the instrument to a pump so that air is blown through the pipette (this will help prevent dirt from entering the pipette). Use a turntable mounted on the microforge with an aluminium oxide disc with various sizes of grit of 0.3, 1.0, and 3.0 µm (depending on the diameter of the pipette being ground) to grind the pipette to an angle of 35–45°. Care should be taken to grind the pipette completely because problems can arise if grinding is incomplete.
4. Mount the ground pipette in a tool holder attached to a syringe and tubing.
5. Clean the pipette by immersing it in 20% hydrofluoric acid for 30–60 s to dissolve any glass particles and makes the glass smooth.
6. Wash the pipette with sterile distilled water to remove the acid.
7. Mount the pipette horizontally in the microforge so that the hole is visible, and touch the end of the pipette onto the glass bead at the minimum temperature to melt the glass. Pull the tip into a spike, which will help the pipette pierce the zona pellucida more easily.

Note: enucleation pipettes have also been made by Humagen Fertility Diagnostics Inc. (Charlottesville, A).

Materials and solutions for preparing the manipulation chamber

Petroleum jelly/wax mixture. Mix 9 g petroleum jelly and 1 g hard paraffin wax in a beaker and heat gently until the mixture is melted. Decant into 10-ml disposable syringes and allow to cool.

Cytochalasin B stock solution. Add DMSO to cytochalasin (Sigma) to make a 5 mg/ml solution.

Siliconized glass slides. Wash good-quality slides thoroughly and rinse five times in distilled water. Dip the washed slides in Sigmacote (Sigma) and drain off excess fluid, dry well, and wash again with distilled water and 70% alcohol.

Preparation manipulation chamber (Figures P23.1 and P23.2)

1. Apply a thin line of the petroleum jelly mixture 22 mm in length along both edges of an alcohol-sterilized and siliconized slide.
2. Attach a cleaned and alcohol-sterilized glass chip to each line of the mixture.
3. Apply a thin line of petroleum jelly mixture to the top of the glass chips.
4. Pipette 300 ml of manipulation medium (HEPES-buffered SOF minus calcium with the addition of 10% FCS and 7.5 μg/ml cytochalasin B) into the middle of the chamber.
5. Clean a glass coverslip with alcohol and place it over the manipulation medium.
6. Seal the ends of the chamber with Dow-Corning silicone fluid (BDH). This will prevent the osmolarity from changing and also allow the entry of the glass microtools.
7. Place the prepared manipulation chamber on the microscope stage.
8. Attach the holding pipette to the left tool holder, ensuring that any air bubbles are removed from the tubing tool holder and pipette.
9. Move the holding pipette into the chamber, keeping it parallel to the microscope stage. Draw a small quantity of the manipulation mixture into the pipette.
10. Wash the enucleation pipette with FCS to prevent the karyoplast from sticking to the glass during enucleation.
11. Attach the enucleation pipette into the right holding tool, ensuring that all air bubbles have been expelled.

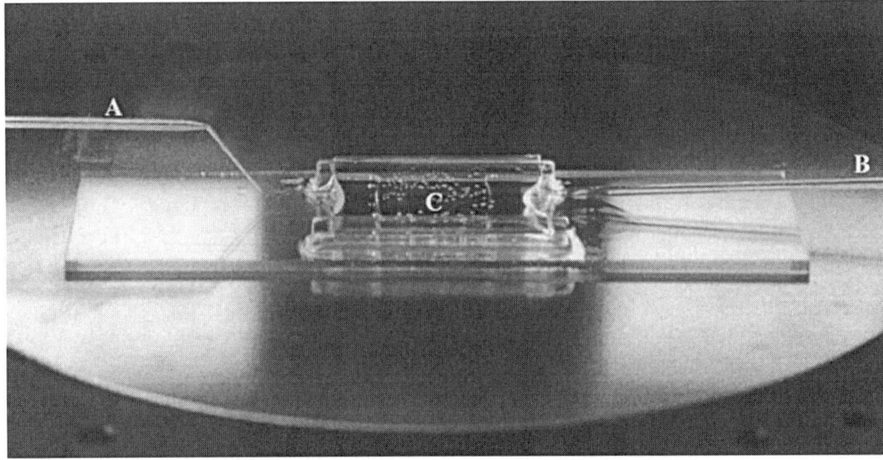

Figure P23.1. Manipulation chamber showing positions of the holding pipette (A) and enucleation pipette (B) in reference to the manipulation chamber (C). (Photograph by W. A. Ritchie.)

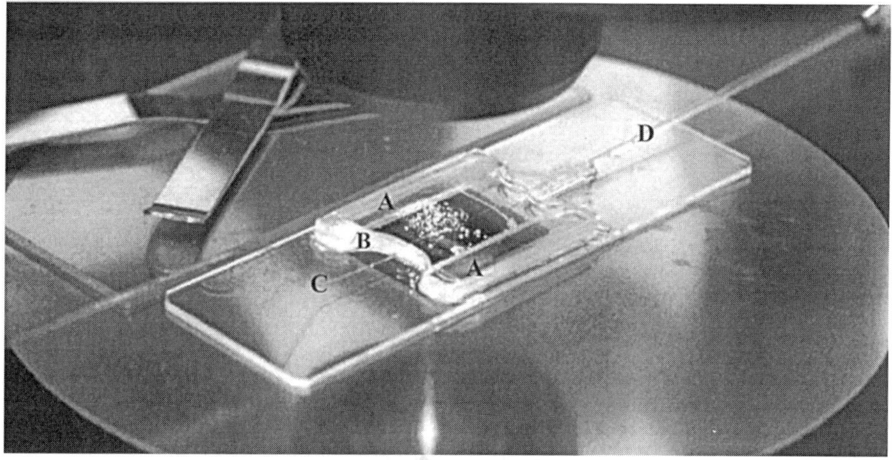

Figure P23.2. Manipulation chamber set-up. The manipulation chamber contains the manipulation fluid and is covered with a coverslip which rests on the glass strips (A), previously coated in petroleum jelly mix. The open edges are sealed with Dow Corning silicone fluid (B). Once on the stage, the holding pipette (C) and eunucleation pipette (D) are inserted through the silicone fluid seal into the manipulation chamber. (Photograph by W.A. Ritchie.)

12. Move the enucleation pipette into the manipulation chamber, tilting the pipette at a slight angle to allow the end of the pipette to touch the bottom of the slide.
13. Rotate the pipette until it is seen in profile with the ground edge seen as a straight line.

Protocol 23.4. Enucleation of oocytes

Materials and solutions

Manipulation chamber and microtools
Metaphase II oocytes (see Protocol 23.2 on recovery of mature oocytes from superovulated ewes)
Hyaluronidase solution, aliquotted in PBS in suitable volumes to give a final concentration of 300 IU/ml (store at –20°C)
hSOF medium without calcium plus 10% FCS
Cytochalasin B (store in the dark at 4°C)
Bisbenzamide (Hoechst 33342, Sigma; store in aliquots at –20°C)
hSOF medium with 10% FCS

Enucleation method (see Figures P23.1 and P23.2)

1. Remove the cumulus cells from the oocytes in hSOF minus calcium with the addition of hyaluronidase to give a concentration of 300 IU/ml. Incubate for up to 10 min at 38°C while pipetting gently with a Gilson pipettor. All of the cumulus cells should be removed.
2. Place batches of 10–20 oocytes in hSOF minus calcium with cytochalasin B and Hoechst at 38°C for 15 min.
3. After incubation, transfer the oocytes to the manipulation chamber.
4. With 40× magnification, pick up an oocyte on the holding pipette.
5. Move the microscope stage to a position where the other oocytes are not in view.
6. Move the lens to give a magnification of 200× DIC.
7. Use the enucleation pipette to turn the oocyte so the first polar body is in a position at either 4 or 2 o'clock (depending on which way the enucleation pipette is facing).

8. Suck the oocyte firmly onto the holding pipette so that it does not move.
9. Insert the enucleation pipette through the zona pellucida and aspirate the polar body and area adjacent to it into the pipette (Figure P23.3).
10. Remove the oocyte from the field of view, turn off the light, and expose the enucleation pipette plus the fragment of cytoplasm to UV light using a UV-2A filter block. The polar body fluoresces bright blue and the metaphase plate is a less bright blue (figure P23.3). The metaphase II plate may appear as a line or a disk or anything in between depending on the orientation of the plate.
11. If the enucleation was successful, move the lens back to the 40× magnification and then move the microscope stage so the enucleated oocyte is deposited at the right side of the chamber.
12. Deposit the enucleated oocyte in a position that makes it easy to remove the oocytes from the chamber and then discard the fragment from the enucleation pipette.
13. The process can be repeated if the metaphase chromosomes have not been removed.
14. Remove batches of enucleated oocytes from the chamber and store in hSOF + 10% FCS.

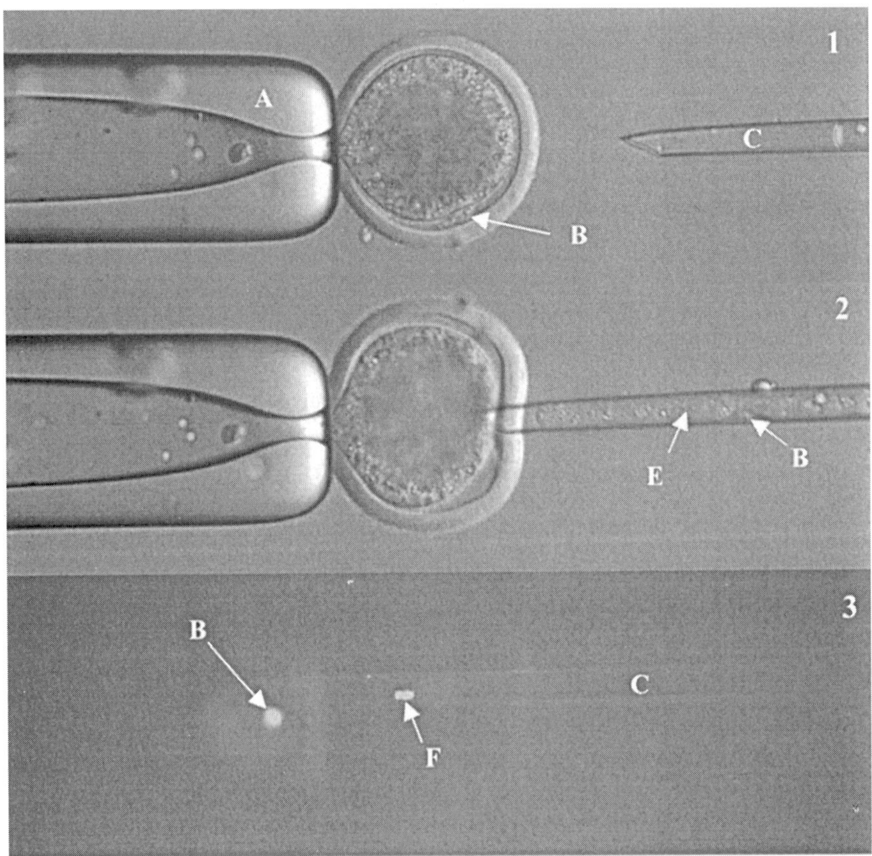

Figure P23.3. Enucleation. The holding pipette (A) secures the Hoescht-stained oocyte in place. The polar body (B) and metaphase chromosomes (usually located next to the polar body) are removed by the enucleation pipette (C) by piercing the zona pellucida (D) and then sucking out the chromosome, the polar body (B), and some cytoplasm (E). The eunucleation pipette (C) is irradiated with UV light to determine that both polar body (B) and metaphase chromosomes (F) have been removed. (Photograph by W. Ritchie.)

Protocol 23.5. Embryo reconstruction

Materials and solutions

Manipulation chamber and microtools
Cumulus-free, enucleated oocytes
Serum-starved, quiescent donor cells
Fusion medium (0.3 M manitol with 0.1 mM $MgSO_4$ and 0.05 mM $CaCl_2$ in twice-distilled water, 280 mOsm.
Fusion chamber, consisting of two 100-μm diameter platinum wires glued to the bottom of a 4.5-cm glass Petri dish 200 μm apart (see figure P23.4).
hSOF with 10% FCS

Nuclear transfer

1. Prepare the manipulation chamber as for the enucleation procedure, containing hSOF + 10% FCS.
2. Deposit a single-cell suspension of serum-starved cells in the top right side of the manipulation chamber.
3. Place a group of 10 enucleated oocytes into the center of the manipulation chamber.
4. Pick up an oocyte with the holding pipette and move the chamber to the right side of the slide.
5. Focus the microscope on the bottom of the chamber and select a suitable cell.
6. Aspirate the cell into the enucleation pipette.
7. Refocus the microscope onto the enucleated oocyte and move the enucleation pipette up to the level of the oocyte.
8. Push the pipette through the zona pellucida into the perivitelline space.
9. Expel the cell into the perivitelline space (to make a couplet), making sure that it remains in contact with the oocyte cytoplasm.
10. On completion of the batch of 10 oocytes, carry out electrofusion.

Protocol 23.6. Electrofusion

Materials and solutions

Electrofusion chamber with platinum electrodes 200 μm apart (see figure P23.4)
BSL CF-150/B impulse generator for cell fusion
Manitol solution with calcium and magnesium
hSOF + 10 % FCS

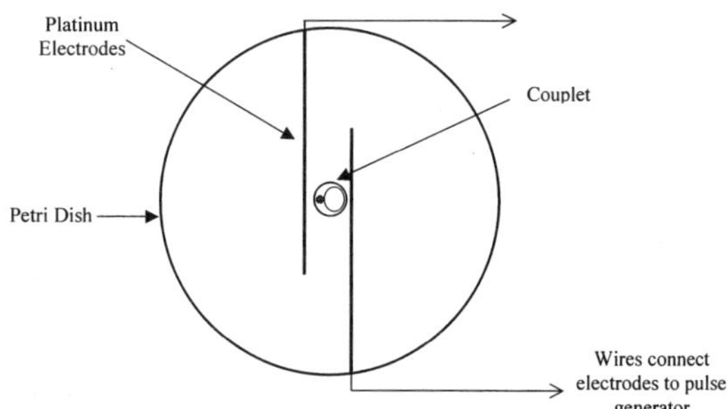

Figure P23.4. Fusion chamber. Two platinum electrode wires are secured to the Petri dish. The oocyte, containing the donor cell in the perivitelline space (the couplet), is placed in a drop of medium, between the electrodes. An electric pulse is supplied by the pulse generator to fuse the oocyte and donor cell together.

Electrofusion

1. Wash the couplets (enucleated oocytes containing the inserted donor cell in the perivitelline space) through a drop of manitol solution to wash off any manipulation medium.
2. Place 300 µl of the manitol solution into the fusion chamber spanning the electrodes.
3. Pipette the couplets into the fusion chamber, to one side of the electrodes.
4. Place a single couplet between the electrodes and manually align the cells so that the plane of contact of the cell and cytoplast are parallel to the electrodes. Precise orientation is necessary for fusion to take place.
5. Apply a 0.25 kV/cm AC electric pulse for a few seconds followed by 3×1.25 kV/cm DC electric pulses in rapid succession for 80 µs (for each pulse).
6. Remove each fusing couplet from the chamber and place in hSOF + 10% FCS.
7. Transfer the fusing couplets into SOF for 1 h and culture at 38°C in a $5\%CO_2{:}5\%O_2{:}90\%\ N_2$ gas mixture.
8. Assess for fusion after 1 h by turning the embryos around and confirming the absence of the cell in the perivitelline space.
9. Transfer fused couplets into Nunc 4-well plates with 500 µl of SOFaaBSA/well, in batches of 25–40, and overlay with Sigma mineral oil for long-term culture or culture overnight.

Protocol 23.7. Culture of reconstructed embryos for temporary transfer into recipient ewes

Materials and solutions

Molten agar (1.2%) dissolved in PBS (Dulbecco a) and heated until dissolved.
Recipient ewe
Reconstructed embryos

In vitro embryo culture

1. Place the molten agar into a suitable dish and allow it to cool to 39°C, using an electronic thermometer to monitor the temperature.
2. When the agar is at 39°C, transfer the dish to the microscope warm stage and drop one or two embryos into the warm agar using a 5-µl Drummond pipette.
3. Suck a small quantity of agar containing the embryos into the pipette, taking care that the embryos are embedded completely in the agar and are not too close to the edge of the agar cylinder.
4. Remove the pipette from the agar and allow the agar to solidify before expelling the agar cylinder and embryos into hSOF.
5. Trim any excess agar from the ends of the cylinders.
6. Transfer the agar cylinders into the double-ligated oviduct of a temporary recipient ewe using a 25-µl Drummond pipette. The ligatures should be placed close to the utero-tubal junction.
7. Transfer the reconstructed embryos into 4-well Nunc dishes containing SOFaaBSA and culture under oil, in a 5:5:90 gas mixture, for 6 days at 38°C.

Protocol 23.8. Preparing temporary recipient ewes for embryo transfer

Materials and solutions

Ewes in reasonable body condition (condition score > 2.5)
Veramix sponges
PMSG: Folligon, 400 Iu/ewe injected intramuscularly
Suture: Dexon
Thiopentone sodium B.P. (Sodium Intraval)

Temporary recipient ewe schedule

Procedure	Time	Day
Sponge in	—	0
Sponge out + PMSG	—	12
Heat	—	14
Starve	PM	15
Embryo transfer	—	16
Starve	PM	21
Embryo recovery	—	22

Protocol 23.9. Transfer and recovery of agar chips from temporary recipient ewes

Materials and solutions

Synchronized ewes
Embryos embedded in agar
Drummond pipettor
hSOF

Transfer and recovery of agar chips

1. Anesthetize a pseudopregnant ewe using sodium intraval and carry out a mid-ventral laparotomy and expose the reproductive tract.
2. Place the agar-embedded embryos into hSOF at 38°C.
3. Double ligate the oviducts at the uterotubual junction.
4. Pick up the agar cylinders using a Drummond pipettor.
5. Insert the end of the pipette into the fimbria end of the ligated oviduct.
6. Push the pipette as far down the oviduct as possible using fine forceps.
7. Expel the agar cylinders into the oviduct and carefully remove the pipette.
8. Suture the incision and allow the ewe to recover.
9. At day 7 after embryo reconstruction, euthanize the ewe and recover the oviducts.
10. Dissect out the oviducts and remove the ligatures.
11. Flush the agar cylinders from the oviducts using a blunt 18-G needle and 10 ml of medium.
12. Examine the flushings under the dissecting microscope and remove the agar cylinders.
13. Place in fresh medium and assess embryo development.
15. Dissect out the embryos, which have successfully developed to the compacted morula or blastocyst stage, using two 25-G × 0.5-inch needles attached to two 1-ml syringe barrels.

Protocol 23.10. Preparing final recipient ewes for embryo transfer

Materials and solutions

Ewes in reasonable body condition (condition score > 3)
Veramix sponges
Suture: Dexon
Thipentone sodium HB.P. (Sodium Intraval)

Final recipient ewe schedule

Procedure	Time	Day
Sponge in	—	0
Sponge out	—	13
Heat	—	15
Starve	PM	21
Embryo transfer	—	22

Protocol 22.11. Transfer of embryos to final recipient ewes

Embryo transfer

1. Anesthetize a day 7 pseudopregnant recipient ewe and perform a midventral laparotomy.
2. Pick up one, two, or three of the recovered compacted morula or blastocyst stage embryos in a 3.5 Fr G Tom Cat catheter 11.5 cm long.
3. Make a small hole in the wall of the uterus using a blunt 18-G needle.
4. Insert the tip of the catheter into the lumen of the uterus and expel the embryos into the animal.
5. Suture the incision and allow the animal to recover.
6. Examine the animal with ultrasound from day 21 to determine pregnancy.

References

1. Trounson, A. (2001). *Reprod. Fertil. Dev.*, **13**, 31.
2. Wilmut, I., Young, L., and Campbell, K. H. (1998). *Reprod. Fertil. Dev.*, **10**, 639.
3. Renard, J. P., Chastant, S., Chesne, P., Richard, C., Marchal, J., Cordonnier, N., Chavatte, P., and Vignon, X. (1999). *Lancet*, **353**, 1489.
4. Keefer, C. L., Baldassarre, H., Keyston, R., Wang, B., Bhatia, B., Bilodeau, A. S., Zhou, J. F., Leduc, M., Downey, B. R., Lazaris, A., and Karatzas, C. N. (2001). *Biol. Reprod.*, **64**, 849.
5. Campbell, K. H., Loi, P., Otaegui, P. J., and Wilmut, I. (1996). *Rev. Reprod.*, **1**, 40.
6. Wilmut, I., Schnieke, A. E., McWhir, J., Kind, A. J., and Campbell, K. H. S. (1997). *Nature*, **385**, 810.
7. Cibelli, J. B., Stice, S. L., Golueke, P. J., Kane, J. J., Jerry, J., Blackwell, C., de Leon, F. A. P., and Robl, J.M. (1998). *Science*, **280**, 1256.
8. Wilmut, I., and Campbell, K. H. S. (1998). *Science*, **281**, 1611.
9. Zhou, Q., Jouneau, A., Brochard, V., Adenot, P., and Renard, J. P. (2001). *Biol. Reprod.*, **65**, 412.
10. Cheong, H. T., Takahashi, Y., and Kanagawa, H. (1993). *Biol. Reprod.*, **48**, 958.
11. Wakayama, T., Perry, A. C. F., Zuccotti, M., Johnson, K. R., and Yanagimachi, R. (1998). *Nature*, **394**, 369.
12. Wells, D. N., Misica, P. M., and Tervit, H. R. (1999). *Biol. Reprod.*, **60**, 996.
13. Campbell, K. H. S., McWhir, J., Ritchie, W. A., and Wilmut, I. (1996). *Nature*, **380**, 64.
14. Ono, Y., Shimozawa, N., Ito, M., and Kono, T. (2001). *Biol. Reprod.*, **64**, 44.
15. Polejaeva, I. A., Chen, S. H., Vaught, T. D., Page, R. L., Mullins, J., Ball, S., Dai, Y. F., Boone, J., Walker, S., Ayares, D. L., Colman, A., and Campbell, K. H. S. (2000). *Nature*, **407**, 86.
16. Baguisi, A., and Overstrom, E. W. (2000). *Theriogenology*, **53**, 209.
17. Hogan, B., Constantini, F., and Lacy, E., eds. (1986). *Manipulating the Mouse Embryo.* Cold Spring Harbor Laboratory Press, Cold Spring Harbor, NY.
18. Kono, T., Kwon, O. Y., Watanabe, T., and Nakahara, T. (1992). *J. Reprod. Fertil.*, **94**, 481.
19. Thompson, J. G. (2000). *Anim. Reprod. Sci.*, **60–61**, 263.
20. Prather, R. S., Barnes, F. L., SIMS, M. M., Robl, J. M., Eyestone, W. H., and First, N. L. (1987). *Biol. Reprod.*, **37**, 859.
21. Barnes, F., Endebrock, M., Looney, C., Powell, R., Westhusin, M., and Bondioli, K. (1993). *J. Reprod. Fertil.*, **97**, 317.
22. Kubota, C., Yamakuchi, H., Todoroki, J., Mizoshita, K., Tabara, N., Barber, M., and Yang, X.Z. (2000). *Proc. Natl. Acad. Sci. USA*, **97**, 990.
23. Pavlok, A., Lucas-Hahn, A., and Niemann, H. (1992). *Mol. Reprod. Dev.*, **31**, 63.
24. Fair, T., Hyttel, P., and Greve, T. (1995). *Mol. Reprod. Dev.*, **42**, 437.
25. De Sousa, P. A., Westhusin, M. E., and Watson, A. J. (1998). *Mol. Reprod. Dev.*, **49**, 119.
26. Betthauser, J., Forsberg, E., Augenstein, M., Childs, L., Eilertsen, K., Enos, J., Forsythe, T., Golueke, P., Jurgella, G., Koppang, R., Lesmeister, T., Mallon, K., Mell, G., Misica, P., Pace, M., Pfister-Genskow, M., Strelchenko, N., Voelker, G., Watt, S., Thompson, S., and Bishop, M. (2000). *Nature Biotech.*, **18**, 1055.
27. Onishi, A., Iwamoto, M., Akita, T., Mikawa, S., Takeda, K., Awata, T., Hanada, H., and Perry, A. C. F. (2000). *Science*, **289**, 1188.

28. Meng, L., Ely, J. J., Stouffer, R. L., and Wolf, D. P. (1997). *Biol. Reprod.*, **57**, 454.
29. Willadsen, S. M. (1986). *Nature*, **320**, 63.
30. Nagai, T. (1987). *Gamete Res.*, **16**, 243.
31. Ware, C. B., Barnes, F. L., Maiki-Laurila, M., and First, N. L. (1989). *Gamete Res.*, **22**, 265.
32. Pinyopummin, A., Takahashi, Y., Hishinuma, M., and Kanagawa, H. (1993). *Jpn. J. Vet. Res.*, **41**, 81.
33. Xu, Z., Abbott, A., Kopf, G. S., Schultz, R. M., and Ducibella, T. (1997). *Biol. Reprod.*, **57**, 743.
34. De Sousa, P. A., King, T., Wilmut, I, and Zhu, J. (1999). Porcine oocytes with improved developmental competence. U.S. Patent No. 6, 548, 741.
35. De Sousa, P. A., Dobrinsky, J. R., Zhu, J., Archibald, A. L., Ainslie, A., Bosma, W., Bowering, J., Bracken, J., Ferrier, P. M., Fletcher, J., Gasparrini, B., Harkness, L., Johnston, P., Ritchie, M., Ritchie, W. A., Travers, A., Albertini, D., Dinnyes, A., King, T. J., and Wilmut, I. (2002). *Biol. Reprod.*, **66**, 642.
36. Zhu, J., Telfer, E. E., Fletcher, J., Springbett, A., Dobrinsky, J., De Sousa, P. A., and Wilmut, I. (2002). *Biol. Reprod.*, **66**, 635.
37. Zhang, L., Jiang, S., Wozniak, P. J., Yang, X., and Godke, R. A. (1995). *Mol. Reprod. Dev.*, **40**, 338.
38. Carabatsos, M. J., Sellitto, C., Goodenough, D. A., and Albertini, D. F. (2000). *Dev. Biol.*, **226**, 167.
39. Zhou, Q., Boulanger, L., and Renard, J. P. (2000). *Cloning*, **2**, 35.
40. Barton, S. C. and Surani, M. A. (1993). *Methods Enzymol.*, **205**, 732.
41. Tsunoda, Y., and Kato, Y. (1998). *J. Reprod. Fertil.*, **113**, 181.
42. Kim, N. H., Moon, S. J., Prather, R. S., and Day, B. N. (1996). *Mol. Reprod. Dev.*, **43**, 513.
43. Webb, M., Howlett, S. K., and Maro, B. (1986). *J. Embryol. Exp. Morphol.*, **95**, 131.
44. Smith, L. C., and Wilmut, I. (1989). *Biol. Reprod.*, **40**, 1027.
45. Meng, L., Ely, J. J., Stouffer, R. L., and Wolf, D. P. (1997). *Biol. Reprod.*, **57**, 454.
46. Loi, P., Ledda, S., Fulka, J., Cappai, P., and Moor, R. M. (1998). *Biol. Reprod.*, **58**, 1177.
47. Kwon, O. Y., and Kono, T. (1996). *Proc. Natl. Acad. Sci. USA*, **93**, 13010.
48. Smith, L. C. (1992). In: *Guide to Electroporation and Electrofusion*, D. C. Chang, B. M. Chassy, A. E. Saunders, and A. E. Sowers, eds., p. 371, Academic Press, San Diego, CA.
49. Baguisi, A., Behboodi, E., Melican, D. T., Pollock, J. S., Destrempes, M. M., Cammuso, C., Williams, J. L., Nims, S. D., Porter, C. A., Midura, P., Palacios, M. J., Ayres, S. L., Denniston, R. S., Hayes, M. L., Ziomek, C. A., Meade, H. M., Godke, R. A., Gavin, W. G., Overstrom, E. W., and Echelard, Y. (1999). *Nature Biotech.*, **17**, 456.
50. Ogura, A., Inoue, K., Takano, K., Wakayama, T., and Yanagimachi, R. (2000). *Mol. Reprod. Dev.*, **57**, 55.
51. Machaty, Z., and Prather, R. S. (1998). *Reprod. Fertil. Dev.*, **10**, 599.
52. Wakayama, T., and Yanagimachi, R. (2001). *Mol. Reprod. Dev.*, **58**, 376.
53. Polge, C., Rowson, L. E., and Chang, M. C. (1966). *J. Reprod. Fertil.*, **12**, 395.
54. Walker, S. K., Hill, J. L., Kleemann, D. O., and Nancarrow, C. D. (1996). *Biol. Reprod.*, **55**, 703.
55. Van, D., V, Liu, L., Bols, P. E., Ysebaert, M. T., and Yang, X. (1999). *Mol. Reprod. Dev.*, **54**, 57.
56. Collas, P., and Robl, J. M. (1991). *Biol. Reprod.*, **45**, 455.
57. Rieger, D., Luciano, A. M., Modina, S., Pocar, P., Lauria, A., and Gandolfi, F. (1998). *J. Reprod. Fertil.*, **112**, 123.
58. Hyttel, P., Viuff, D., Fair, T., Laurincik, J., Thomsen, P. D., Callesen, H., Vos, P. L., Hendriksen, P.J., Dieleman, S. J., Schellander, K., Besenfelder, U., and Greve, T. (2001). *Reproduction*, **122**, 21.
59. Wrenzycki, C., Wells, D., Herrmann, D., Miller, A., Oliver, J., Tervit, R., and Niemann, H. (2001). *Biol. Reprod.*, **65**, 309.
60. Xu, J., and Yang, X. (2000). *Biol. Reprod.*, **63**, 1124.
61. De Sousa, P. A., King, T., Harkness, L., Young, L. E., Walker, S. K., and Wilmut, I. (2001). *Biol. Reprod.*, **65**, 23.
62. Tamashiro, K. L. K., Wakayama, T., Blanchard, R. J., Blanchard, D. C., and Yanagimachi, R. (2000). *Biol. Reprod.* **63**, 328.
63. Steinborn, R., Schinogl, P., Zakhartchenko, V., Achmann, R., Schernthaner, W., Stojkovic, M., Wolf, E., Muller, M., and Brem, G. (2000). *Nat. Genet.*, **25**, 255.

64. Shiels, P. G., Kind, A. J., Campbell, K. H., Waddington, D., Wilmut, I., Colman, A., and Schnieke, A. E. (1999). *Nature*, **399**, 316.
65. Lanza, R. P., Cibelli, J. B., Blackwell, C., Cristofalo, V. J., Francis, M. K., Baerlocher, G. M., Mak, J., Schertzer, M., Chavez, E. A., Sawyer, N., Lansdorp, P. M., and West, M.D. (2000). *Science*, **288**, 665.
66. Tian, X. C., Xu, J., and Yang, X. (2000). *Nat. Genet.*, **26**, 272.
67. Suzuki, H., Liu, L., and Yang, X. (1999). *Reprod. Fertil. Dev.*, **11**, 159.
68. Debey, P., Renard, J.P., Geze, M., and Adenot, P. (1989). *C. R. Acad. Sci. III*, **309**, 197.
69. Dominko, T., Chan, A., Simerly, C., Luetjens, C. M., Hewitson, L., Martinovich, C., and Schatten, G. (2000). *Biol. Reprod.*, **62**, 150.
70. Young, L. E., Sinclair, K. D., and Wilmut, I. (1998). *Rev. Reprod.*, **3**, 155.

24

TETSUNORI MUKAIDA
MAGOSABURO KASAI

Cryobiology: Slow Freezing and Vitrification of Embryos

1. CRYOPRESERVATION

To cryopreserve cells alive without decreasing survival with the duration of storage, they must be preserved at temperatures below the glass transition temperature, which is around $-130°C$. In practice, liquid nitrogen ($-196°C$) is used for maintaining such a temperature. To date, various methods for embryo cryopreservation have been developed, and embryos of more than 20 mammalian species have been successfully cryopreserved (1). Cryopreservation, coupled with embryo transfer, is used to preserve genetic variants in laboratory animals (mainly mice), for breeding and reproduction in farm animals (mainly cattle), and for treatment of infertility in humans. Furthermore, embryo cryopreservation is expected to be used for the conservation of wild species. Here we give a brief outline of embryo cryobiology and describe some reliable protocols for embryo cryopreservation.

1.1 Historical View

In 1949, Polge et al. (2) happened to discover that glycerol enhanced the survival of frozen-thawed fowl spermatozoa. This discovery of a cryoprotectant made it possible to cryopreserve various types of cells. However, protocols developed for small cells, such as spermatozoa, were not effective because embryos have quite a large amount of cytoplasm. In larger cells, ice is more likely to form in the cytoplasm, which is a major cause of cell injury in cryopreservation.

In 1972, Whittingham et al. (3) reported the successful deep freezing of mouse embryos, resulting in the production of viable young. They cooled embryos with approximately 1 M cryoprotectant very slowly ($0.3–0.5°C$ /min) to $-80°C$, stored them in liquid nitrogen, and then warmed them slowly ($4°–25°C$/min) before embryo recovery. The original slow-freezing method proved effective for embryos of other mammalian species, including humans (4–6). However, it required a long time for cooling and an elaborate device to control the cooling rate. Subsequently, it was found that slowly cooled samples could be plunged into liquid nitrogen at around −30°C if the samples were thawed rapidly (360°C/min) (7). This conventional slow-freezing method is now widely used for the cryopreservation of mammalian embryos.

In 1985, Rall and Fahy (8) devised a rapid method called vitrification, in which embryos suspended in a highly concentrated solution are directly plunged into liquid nitrogen. Vitrification is the complete solidification of a solution without crystallization, and thus without the formation of ice. This approach eliminated the slow-cooling process as well as the need for elaborate equipment. Another advantage of this approach is the high levels of viability of embryos if conditions are optimized because of the absence of extracellular ice.

More recently, modified vitrification methods have been developed, which enable ultrarapid cooling and warming by minimizing the volume of the vitrification solution (9–11). This approach is expected to enable cryopreservation of embryos that are sensitive to cryo-induced injuries such as damage from intracellular ice and chilling.

1.2 Mechanisms of Cell Injury

If mammalian embryos are frozen in a physiological solution, they will be injured by intracellular ice or perhaps crushed by the extracellular ice because the embryo is a large cell mass (12). To prevent this, the inclusion of a cryoprotectant in the suspending solution is essential. Although the cryoprotectant allows the embryo to survive, intracellular ice is still a major cause of cell injury. Furthermore, the cryoprotectant can be toxic to the cell, and removal of the permeated cryoprotectant from the cell at recovery may cause osmotic injuries. In addition, cryopreserved embryos are at risk of chilling injury and fracture damage. All of these obstacles must be circumvented for embryos to survive cryopreservation (Figure 24.1).

1.2.1 Chilling Injury

At certain stages, embryos of some species (e.g., pig embryos before the peri-hatching stage) are sensitive only to cooling to below +20°C. Embryos with chilling sensitivity appear dark with cytoplasmic lipid droplets. Possible strategies to circumvent this injury are to remove the droplets (13), to use embryos at later stages when the amount of droplets has decreased (14), or to adopt ultrarapid cooling and warming in which the critical temperature can be passed rapidly (9). Embryos of rodents and humans appear bright and are not susceptible to chilling.

1.2.2 Intracellular Ice Formation

With slow freezing, embryos are at high risk of forming intracellular ice from seeding by extracellular ice. To prevent this, embryos must be loaded with cryoprotectant, and the suspending solution must be seeded and cooled very slowly so that cellular contents become concentrated by gradual dehydration in response to the concentration of the extracellular unfrozen fraction during the growth of extracellular ice (15). After sufficient concentration, the embryos, together with the extracellular unfrozen fraction, will be vitrified in liquid nitrogen (Figure 24.2). With the vitrification method, in contrast, the chance of intracellular ice forming is small because there is no ice outside the cells during cooling. In a solution with a reduced amount of cryoprotectant, ice may form in the solution during warming, a process known as devitrification. Devitrification is not harmful as long as the ice remains outside the cell. In any case, however, intracellular ice can form even in vitrification if the concentration of cryoprotectant in the cell is not high enough.

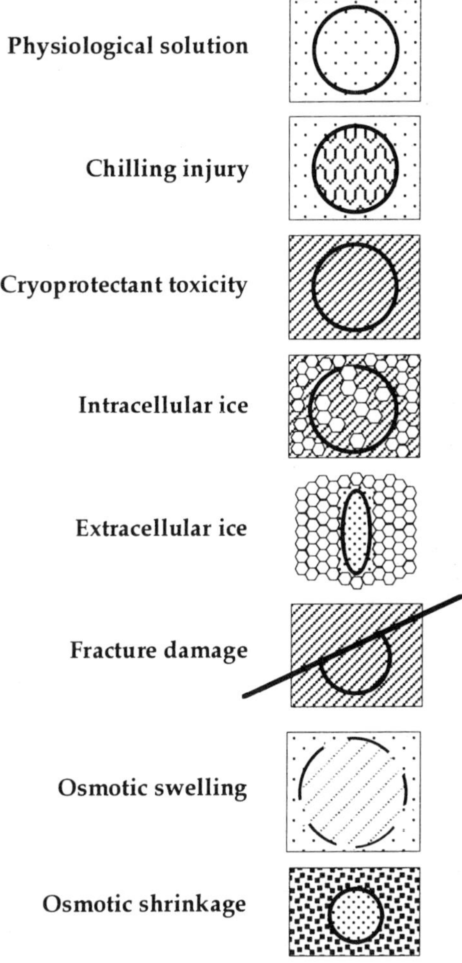

Figure 24.1. Schematic of injuries in cryopreserved embryos.

1.2.3 Toxicity of Cryoprotectants

The cryoprotectant used for embryo cryopreservation is almost always glycerol, ethylene glycol, DMSO, or propylene glycol. The mechanism of the protective action of these permeating agents is considered the same, but their toxicities are different. For slow freezing, the concentration is limited to 1–2 M, and the toxicity is relatively low. In vitrification, however, the concentration can be as high as 8 M, and selection of a low-toxicity agent is more important. Ethylene glycol and glycerol are considered less toxic than propylene glycol and acetamide (16).

1.2.4 Fracture Damage

When cryopreserved embryos are recovered, they are occasionally cracked. This physical injury, called fracture damage, is thought to be caused by nonuniform changes in the volume of the liquid and solid phases of the medium during rapid phase changes. Fracture damage can be reduced by reducing the cooling and warming velocities during passage through the temperature range where the phase change would occur (around

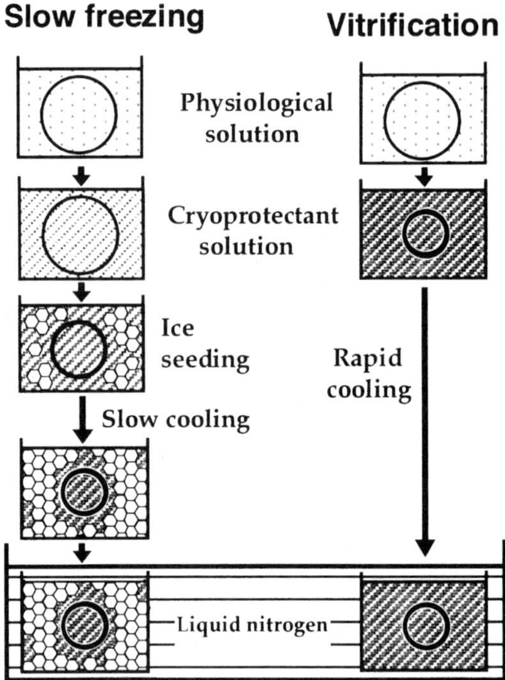

Figure 24.2. Schematic showing the intra- and extracellular states of embryos during cooling by slow freezing and vitrification.

−130°C) (17). Fracture damage is also related to the container for the embryos, with more flexible containers resulting in less damage (18). In conventional vitrification using straws, fracture damage can be prevented completely by cooling and warming to the phase transition temperature in a gas phase (19).

1.2.5 Osmotic Swelling

Just after warming, cryopreserved embryos contain a permeated cryoprotectant, which has to be removed. If the embryos are directly recovered in an isotonic solution, they risk injury from osmotic swelling because water diffuses in far more rapidly than the cryoprotectant diffuses out. The sensitivity of embryos to osmotic swelling differs with the stage of development, but cryopreserved embryos just after warming are more sensitive than fresh ones at all stages (20). The most common strategy for preventing this injury is to dilute embryos with a hypertonic solution containing sucrose as a non-permeating solute to counteract the inflow of excess water (21, 22). Including sucrose in the cryopreservation solution may help prevent cells from swelling in a sucrose solution (23, 24), as small saccharides such as sucrose promote cell shrinkage before dilution.

1.2.6 Osmotic Shrinkage

When the permeated cryoprotectant diffuses out of the cells when using a sucrose solution, embryos remain shrunken. This hypertonic shrinkage may also be injurious to the cells (25). As in the case of hypotonic stress, cryopreserved cells are sensitive to this injury just after warming. Therefore, after diffusion of the permeated cryoprotectant out of the cell in a hypertonic solution, the embryos should be transferred into a less hypertonic solution, and finally into an isotonic solution.

1.3 Permeability of Embryos to Cryoprotectant

The permeability of the cell membrane of embryos to cryoprotectant is an important factor in determining the conditions for cryopreservation (26). The permeability can be analyzed from the change in volume during suspension in the cryoprotectant solution (27). Upon suspension, embryos shrink quickly, losing water in response to the extracellular osmolality. Then, as the cryoprotectant permeates, water reenters the cell to maintain intracellular osmotic equilibrium, and the volume is slowly regained. The permeating property differs not only among embryos (species and developmental stages) but also with the cryoprotectant. For instance, the permeability of mouse embryos generally increases as development proceeds to the compacted morula. However, at the one–cell stage, ethylene glycol is less permeating than propylene glycol, whereas in morulae, ethylene glycol is far more permeating than other cryoprotectants (28). This is also true for bovine embryos produced in vitro (29).

Generally, rapidly permeating agents are favored because the exposure time before cooling can be shortened and because osmotic swelling during removal of the cryoprotectant can be minimized. In general, the difference in the permeability of the cell membrane among embryos of various species seems to be much smaller than the difference among embryos at various developmental stages. However, human blastocysts may be an exception because Mukaida et al. (30) suggest that human blastocysts are much less permeable not only to cryoprotectant but also to water.

2. METHODS OF CRYOPRESERVATION

2.1 Conventional Slow Freezing

Slow freezing is widely used for the cryopreservation of mouse, bovine, and human embryos. Embryos are first suspended in 1–2 M cryoprotectant dissolved in a physiological solution (modified PBS) at room temperature, then allowed to equilibrate with the cryoprotectant for full permeation and loaded in a container (a cryostraw or a cryotube) (Figure 24.3). At around –5°C, the formation of ice is induced (seeded), and the sample is cooled very slowly (0.3°–0.5°C/min) to around –30°C and then stored in liquid nitrogen. A slow cooling stage is necessary to prevent intracellular ice from forming. In liquid nitrogen, the cytoplasm vitrifies in a state of supercooling. However, because a supercooled solution can devitrify during warming at temperatures between –90° and –40°C, samples must be warmed rapidly during passage through this temperature range (17). The warmed sample is diluted with a hypertonic solution containing nonpermeating saccharide (usually sucrose) before the embryos are recovered in a physiological solution. Although this method is likely to produce more consistent results than vitrification, it has the disadvantage of a long, controlled slow cooling.

Although the mechanism of cryoprotection is the same, various protocols have been used. For instance, DMSO has been used as cryoprotectant for mouse embryos, glycerol or ethylene glycol for bovine embryos, and propylene glycol with sucrose for human embryos. The temperature at which slow cooling is terminated is also variable, ranging from 25° to 40°C. These variations are based not necessarily on the characteristics of the embryo but on previous successful reports.

2.2 Conventional Vitrification

Vitrification, with recent improvements, has become a reliable strategy, not only because it is simple but also because it can lead to high survival. To induce vitrification in liquid nitrogen, the solution must contain a very high concentration of cryoprotectant. As it is a less toxic cryoprotectant, ethylene glycol is widely used. In addition, a small

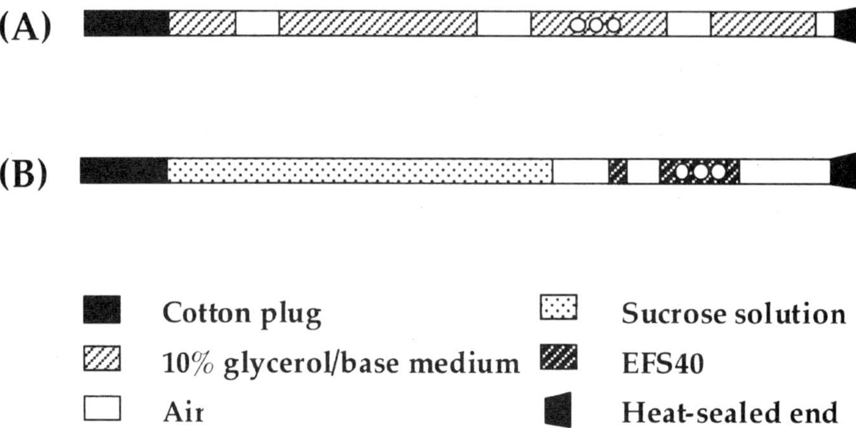

Figure 24.3. Configuration of the straw for (A) slow freezing and (B) conventional vitrification.

saccharide (e.g., sucrose) and a macromolecule (e.g., Ficoll 70, BSA, or PVP) are frequently included in vitrification solutions. These nonpermeating agents are much less toxic and are known to promote vitrification of the solution (8). Therefore, their inclusion can reduce the toxicity of the solution by decreasing the concentration of the permeating agent required for vitrification. In addition, inclusion of a saccharide promotes shrinkage of embryos and thus reduces the amount of intracellular cryoprotectant, which will also reduce the toxic effect of the permeating cryoprotectant (31). At the same time, the osmotic action of saccharide plays an important role in minimizing the swelling of embryos during dilution, since a quick dilution is necessary to prevent the toxic effect of the solution.

The basic procedure for vitrification is simple. Embryos are suspended in a vitrification solution and then plunged in liquid nitrogen. The sample is warmed rapidly and diluted quickly with a sucrose solution. The most important process is the exposure of embryos to the vitrification solution before cooling. To prevent intracellular ice from forming, a longer period of exposure is desirable. However, if the exposure is too long, cells suffer from the toxicity of the solution. Therefore, the optimal exposure time for successful vitrification must be a compromise between preventing the formation of intracellular ice and preventing toxic injury. Actually, however, embryos may be injured by the toxicity of the cryoprotectant before enough cryoprotectant can permeate the cells. Therefore, a two-step procedure is commonly adopted, in which embryos are first equilibrated in a dilute (e.g., 10%) cryoprotectant solution, followed by a brief (30–60 s) exposure to a vitrification solution before the sample is cooled with liquid nitrogen (32, 33). The optimal exposure time in the vitrification solution depends not only on the cryoprotectant solution but also on the temperature, as both the permeability of embryos and the toxicity of the cryoprotectant are largely influenced by the temperature.

2.3 Ultrarapid Vitrification

In ultrarapid vitrification, essentially, embryos are treated by a procedure similar to conventional vitrification. However, the volume of the vitrification solution is reduced to a minimum using various tools. For instance, embryos are stuck on an EM grid (9) or on a minute loop at the tip of a stick placed in a cryotube (a cryoloop) (11), aspirated in a thin, open capillary (an open pulled straw) (10), or vitrified without any container (a microdrop) (34). The aim of this approach is to cool and warm embryos literally in

Table 24.1. Solutions for Ultrarapid Vitrification with a Cryoloop

Base medium	Hepes-buffered modified HTF containing 5 mg/ml HSA*
Solution A	To 10 ml base medium, add 0.1 g Ficoll 70 and 3.4 g sucrose
Cryoprotectant solution I	To 840 ml base medium, add 75 ml DMSO and 75 ml ethylene glycol
Cryoprotectant solution II	To 840 ml Solution A, add 175 ml DMSO and 175 ml ethylene glycol
Solution B	To 10 ml base medium, add 3.4 g sucrose
Dilution solution I	To 840 ml base medium, add 420 ml solution B
Dilution solution II	To 840 ml base medium, add 210 ml solution B

*When 25% HSA solution is used, add 2% (v/v) HAS solution to base medium.

a moment by minimizing the volume needed for the vitrification. This new approach is especially effective for embryos in which the permeability of the surface membrane is low and thus preventing the formation of intracellular ice is not easy. An example is the human blastocyst. It has been observed that human blastocysts are dehydrated and concentrated more slowly than earlier stage embryos, and a few trials to cryopreserve human blastocysts by conventional vitrification using straws resulted in disappointing rates of survival (30). However, the cryoloop method (Table 24.1) has proven effective for human blastocysts (30). Ultrarapid vitrification has also been shown to improve the postwarming survival of embryos and oocytes that have chilling sensitivity because the critical temperature can be passed rapidly before the cells are injured (9).

With this method careful handling of the embryos and solution is required because the concentration of the vitrification solution surrounding each embryo is liable to change. For instance, technical skill and even the bore size of the pipette can make a big difference, and a small volume of the vitrification solution can become concentrated by evaporation in a short period. In addition, the vitrification solution is usually exposed to liquid nitrogen, and contamination with pathogens cannot be excluded (35). Further optimization of the protocol for each type of embryo will make this approach more reliable and practical.

3. PROTOCOLS

In Protocols 24.1–24.3, we describe three methods of cryopreservation: conventional slow freezing, conventional vitrification, and ultrarapid vitrification. Materials common to all the protocols are listed below.

- Instruments: dissecting microscope with transmitted light, liquid nitrogen storage tank, CO_2 incubator.
- Tools: Dewar flask (or styrofoam vessel), Pasteur pipette and mouth piece for embryo handling, 35-mm sterile plastic tissue-culture dish, thermometer, forceps, indelible marker pens, timer, filter unit for sterilization.
- Reagent and material: paraffin oil (or mineral oil), liquid nitrogen.
- Solutions: all solutions should be filter sterilized and stored in tightly capped bottles or tubes at 4°C. Keeping the solutions in sterile glass ampoules is best for long-term storage.

4. EXAMPLES OF THE TYPE OF DATA TO BE COLLECTED

At cryopreservation, the following should be recorded:

- Sample number
- Date, time, room temperature

- Embryo donor: species, strain, and parents
- Number of embryos in the sample
- Age, developmental stage, origin, and quality of embryos
- Type of sample: straw or cryovial
- Mark on the sample
- Cryoprotectant solution and method of cryopreservation
- Compartment in the storage tank.

At warming, the following should be recorded:

- Sample number
- Date, time, room temperature
- Number of embryos recovered
- Morphology of embryos at recovery in physiological solution
- Morphology of embryos after 1–3 h of culture
- Morphology and development of embryos after ~ 1 day of culture.

For a summary diagram, see Table 24.2.

5. TROUBLESHOOTING

5.1 Conventional Slow Freezing

For mouse embryos, the protocol is effective at various developmental stages, although
1.5 M DMSO is used as the cryoprotectant and the 20% calf serum in base medium is
replaced by BSA. Glycerol is unsuitable for two-cell embryos because it is less permeating in embryos at early cleavage stages.

Essentially, the same protocol is effective in human embryos at the two- to eight-cell
stages, except that the cryoprotectant is 1.5 M propylene glycol + 0.1 M sucrose rather
than 10% glycerol (24), and the base medium is modified HTF, in which serum is replaced by HSA (36).

5.2 Conventional vitrification

The temperature for handling embryos in a cryoprotectant solution has significant effects. When the room temperature is higher than 25°C, the duration of exposure of
embryos to EFS40 must be shortened and vice versa (37). Elevating the temperature is
preferable for promoting the permeation of the cryoprotectant, but the optimal range of
the exposure time decreases because the toxicity of the cryoprotectant increases con-
siderably. For the protocol described (at 25°C), keep the lights off so as not to elevate
the temperature of the microscope stage when embryos are not being manipulated under
the microscope. Avoid warming the EFS40 column of the straw with fingers.

Table 24.2. Summary of the Three Protocols

	Slow Freezing	Conventional Vitrification	Ultrarapid Vitrification
Concentration of permeating CPA	1.4 M	7.2 M	4.7 M
Temperature of treatment	25°C	25°C	37°C
Period of CPA pretreatment	0 min	2 min	2 min
Time in final CPA solution	15-20 min	1 min	30 sec
Container	Straw	Straw	Cryovial
Time required for cooling	90 min	3 min	< 0.1 sec

CPA: cryoprotectant

If only a small Dewar flask is available, immerse half of the straw containing the embryos in liquid nitrogen first, and then cool the rest of the straw slowly in liquid nitrogen vapor. By this method, a small percentage of vitrified embryos may suffer fracture damage. If the whole straw is immersed in liquid nitrogen in one step, it may rupture because of the rapid increase in the volume of the freezing sucrose solution.

Compact morulae have the advantage of affording the highest survival after vitrification among mouse embryos at various stages (38). Furthermore, morulae can be directly transferred to the EFS40 column in the straw without pretreatment in EFS20; the optimal exposure time in EFS40 (at 25°C) is 30–60 s (37). In this case, embryos must be loaded at the tip of a fine pipette to minimize the volume of base medium introduced into EFS40.

For mouse blastocysts with a large blastocoel, base medium containing 10% (v/v) ethylene glycol can be substituted for EFS20. In this case, suspend blastocysts in the cryoprotectant solution for 5 min at 25°C, and then suspend the embryos in EFS40 in the straw for 30 s before cooling in liquid nitrogen vapor (33).

For human embryos at two- to eight-cell stages, the protocol for two-cell mouse embryos is effective, except that BSA in base medium is substituted with HSA.

For bovine blastocysts, embryos pretreated with EFS20 should be washed quickly with EFS40 outside the straw, using a pipette loaded with EFS40, before transferring the embryos to EFS40 in the straw. The total exposure time in EFS40 should be restricted to 35–40 s.

The present protocol, with slight modification, has proven effective in embryos of other species such as rabbit (39), horse (40), sheep (41), mastomys (multimammate rat) (42), and Mongolian gerbil (43). Notably, rabbit embryos at the morula stage can be cryopreserved efficiently; direct exposure to EFS40 for 2 min at 20°C before cooling results in quite high survival (39).

5.3 Ultrarapid Vitrification: Cryoloop

Instead of warming solutions after preparation in a four-well multidish, it is possible to warm the solutions and then prepare them in the dish. Because the amount of vitrification solution on the cryoloop is very small, the solution is liable to be concentrated by evaporation, which increases its toxicity. To prevent this, the cryoloop should be loaded with the vitrification solution just before the loading embryo(s) on the loop. The amount of vitrification solution on the loop need not be minimal, as long as the drop is kept on the loop by surface tension. A Dewar flask containing liquid nitrogen should be placed nearby to enable quick cooling.

At warming, the loop containing embryo(s) should be soaked in the sucrose solution as quickly as possible, making sure the sample is never held in air. Again, liquid nitrogen and the sucrose solution should be placed nearby. When the loop is dipped in the sucrose solution, the steel pipe portion should not be immersed to prevent bubbling.

At recovery in an isotonic solution, the blastocoel of the blastocyst will be collapsed, and the embryo may look like a "morula." However, the blastocoel will reform within 1–3 h of culture.

Data on the survival of human blastocysts from the HART Clinic (44) are presented in Table 24.3.

Table 24.3. Survival Rate of Human Blastocysts

No. embryos vitrified/recovered	725
No. embryos that survived	583 (80%)
No. transfers	207
No. clinical pregnancies	76 (37%)
No. ongoing/delivered	55 (27%)

6. CONCLUSION

Numerous protocols for the cryopreservation of mammalian embryos have been reported. The protocols can be classified into four methods, original slow freezing, conventional slow freezing, conventional vitrification, and ultrarapid vitrification. Although the principle of cryopreservation is the same, strategies to circumvent various injuries (especially from the formation of intracellular ice) are different. The most suitable protocol should be adopted for each case. For certain types of embryos such as human blastocysts and bovine embryos at earlier stages, ultrarapid vitrification would be a preferable choice because the survival rates of embryos cryopreserved by other methods have been low. For other embryos such as mouse embryos, bovine blastocysts, and human embryos at two- to eight-cell stages, both slow freezing and conventional vitrification have proven effective. Slow freezing will bring more consistent results because embryos are less likely to be injured by the toxicity of the cryoprotectant and thus can be handled under less strict conditions. However, vitrification has a potential advantage in that higher survival can be obtained if conditions, such as temperature and period of exposure of embryos to the cryoprotectant, as well as the skill of pipetting, is optimized.

Acknowledgments   We thank David K. Gardner and Michelle Lane (Colorado Center for Reproductive Medicine, Englewood, Colorado, U.S.A.) for teaching us the cryoloop technique. We also thank Katsuhiko Takahashi, (Hiroshima HART Clinic, Hiroshima, Japan), Tatsuhiro Tomiyama (Osaka HART Clinic, Osaka, Japan), and Chikahiro Oka (Tokyo HART Clinic, Tokyo, Japan) for establishing a clinical program for vitrification of human embryo.

Protocol 24.1. Conventional slow freezing

A protocol for bovine blastocysts used at the National Livestock Breeding Center, Japan, which was modified from that reported by Takeda et al. (45) is described. Essentially, the same protocol is available for embryos of other species (see "Troubleshooting" section in text).

Cryopreservation of embryos

Materials

- Instruments: programmable freezer, heat sealer (heated pliers or straw powder can be substituted)
- Tools: 0.25-ml insemination straw (e.g., IMV, AA201, L'Aigle, France; straws can be sterilized by ethylene oxide gas), 1-ml syringe, straw connector (2-cm-long silicone tube or a shortened yellow tip)
- Reagents: glycerol, sucrose, calf serum
- Solutions: base medium (Dulbecco's PBS containing 20% [v/v] calf serum), cryoprotectant solution (base medium containing 10% (v/v) glycerol).

Methods

1. Label the 0.25-ml straw with a finely tipped marker pen (marker pens of various colors are useful).
2. Connect the straw and 1-ml syringe with the connector.
3. Suspend embryos in the cryoprotectant solution at room temperature (~ 25°C).
4. Allow them to equilibrate in the solution for 10–15 min. Equilibrium can be confirmed by the recovery of the volume following initial shrinkage.
5. By aspirating the syringe, load ~15 mm of cryoprotectant solution, ~ 8 mm of air, ~30 mm of cryoprotectant solution and ~ 8 mm of air. Under a dissecting microscope, aspirate embryos with ~20 mm of cryoprotectant solution, followed

by ~ 8 mm of air, ~10 mm of cryoprotectant solution and air (till the end), successively, into the straw as shown in figure 24.3A. If the straw is to be sealed with the straw powder, the end must be loaded with cryoprotectant solution without air.

6. Seal the open end of the straw with the heat sealer. If the sealing is incomplete, liquid nitrogen will penetrate the straw, causing it to explode and consequently leading to the loss of the embryos.

7. At 15–20 min after suspending the embryos in cryoprotectant solution, transfer the straw to the alcohol bath at –7°C in the programmed freezer.

8. After 5–10 min, touch the surface of the straw with forceps precooled in liquid nitrogen to induce the formation of ice (seeding).

9. After 10 min, confirm that the entire cryoprotectant solution has seeded. Then, start cooling the sample at 0.3°C/min in the programmed freezer.

10. At -30°C, stop cooling and plunge the straw into liquid nitrogen.

11. Preserve the straw in a canister in a liquid nitrogen storage tank.

Recovery of cryopreserved embryos

Materials

- Tools: water bath maintained at 38°C, embryological watch glass (or culture dish).
- Solutions: base medium (Dulbecco's PBS containing 20% calf serum), dilution solution I (base medium containing 6% [v/v] glycerol and 0.3 M sucrose), dilution solution II (base medium containing 3% [v/v] glycerol and 0.3 M sucrose), sucrose solution (base medium containing 0.3M sucrose).

Methods

1. Prepare 100-µl drops of dilution solution I, dilution solution II, sucrose solution, and base medium under paraffin oil in culture dishes.

2. Remove the straw from the liquid nitrogen using forceps and start the timer. After keeping it in air (~25°C) for 10 s, immerse the straw in the water bath for 10 s until the ice has completely melted.

3. Remove the straw and wipe the water with a paper towel.

4. Cut off both ends of the straw and recover the content in a watch glass.

5. Under a dissecting microscope, recover the embryos in a pipette and transfer them successively to dilution solution I, dilution solution II, the sucrose solution, and the base medium at 10-min intervals.

6. The embryos can be transferred to recipients immediately or after a certain period of culture.

Protocol 24.2. Conventional vitrification

A protocol for mouse two-cell embryos (46) is described. Essentially, the same protocol is effective in mouse embryos at other stages of development (33, 37, 38), for human embryos at the two- to eight-cell stage (47), and for bovine blastocysts (48) with slight modification (see "Troubleshooting" section in text).

Cryopreservation of embryos

Materials

- Instruments: heat sealer (heated pliers or straw powder can be used instead)
- Tools: 200-ml flask, 1-ml disposable syringe, 18-G needle, 10-ml disposable syringe, tightly stoppered 10-ml sterile plastic test tube, 0.25-ml insemination straw (e.g., AA201, IMV, L'Aigle, France), straw connector (silicone tube or a short-

ened yellow tip), Styrofoam plate (5–10 mm thick, one on which can be placed 123-mm-long straws), embryological watch glass (or culture dish).

- Reagents: ethylene glycol, Ficoll 70 (mol wt ~70,000, Amersham Pharmacia Biotech, Uppsala, Sweden), sucrose, BSA
- Solutions: base medium (Dulbecco's PBS modified by addition of 5.56 mM glucose, 0.33 mM pyruvate, 100 IU/ml penicillin, and 3mg/ml BSA [49]; other physiological solutions [e.g., M2 medium and HEPES-buffered HTF] can be substituted), EFS40 and EFS20 (see below), sucrose solution (base medium containing 0.5 M sucrose).

Methods

1. To make EFS40 and EFS20, prepare FS solution, which is base medium containing 30% (w/v) Ficoll and 0.5 M sucrose. To a 200-ml flask containing 35.1 ml of base medium (omitting the BSA), add 15.0 g of Ficoll 70, and leave it to dissolve (it takes some time). Add 8.56 g of sucrose and shake to dissolve. Then, add 105 mg of BSA and dissolve. For a small volume of sample, one-tenth of the FS solution can be made in a 10-ml test tube with a tight stopper.

2. Using 1-ml disposable syringes (with an 18-G needle), add 4 ml of ethylene glycol and 6 ml of FS solution together in a 10-ml test tube with a tight stopper to make EFS40, which is base medium containing 40% (v/v) ethylene glycol, 18% (w/v) Ficoll, and 0.3 M sucrose (31). Syringes are used to measure solution volumes, as the viscosity of the liquids makes accurate pipetting difficult.

3. Likewise, add 2 ml of ethylene glycol and 8 ml of FS solution together in a 10-ml test tube with a tight stopper to make EFS20, which is base medium containing 20% (v/v) ethylene glycol, 24% (w/v) Ficoll and 0.4 M sucrose.

4. Repeatedly tilt the tubes hundreds of times until mixed completely. Filter sterilize the 10 ml of EFS solution with a 0.45 mm filter unit because EFS solutions are viscous. Preserve filtered EFS40 and EFS20 in tight-stoppered tubes at 4C. Keeping the solutions in sterile ampoules is best for long time storage. For vitrification, load EFS40 and EFS20 (~1 ml each) in 1 ml syringes attached with an 18-G needle.

5. Adjust the room temperature on the bench to 25°C and equilibrate all the solutions and instruments at this temperature (for other temperatures, see "Troubleshooting" in text).

6. Pour liquid nitrogen into a Dewar flask and float a Styrofoam disk on it.

7. Label a 0.25-ml straw with a finely tipped marker pen (various colors are useful). Connect the straw and a 1-ml syringe with a connector.

8. Pour 2–3 ml S-PB1 medium into a culture dish, and drop a few drops of EFS40 from the syringe containing EFS40 into a lid of a culture dish. Replace with fresh EFS40 for each sample (evaporation of the solution can concentrate the cryoprotectant). By aspirating the syringe with a connector carefully, load ~60 mm of sucrose solution, ~15 mm of air, ~4 mm of EFS40, ~5 mm of air, and ~13 mm of EFS40 successively into the straw as shown in Figure 24.3B. Place it horizontally near the edge of the bench. The first small EFS column removes the sucrose solution sticking to the inner surface of the straw. The EFS40 is confined to a rather small portion of the straw to increase the dilution rate with the sucrose solution during perfusion after warming.

9. Drop a few drops of EFS20 into a lid of a culture dish. Replace with fresh EFS20 for each sample (evaporation of the solution can concentrate the cryoprotectant).

10. Prepare two pipettes; one aspirated with base medium, the other with EFS20.

11. Under a dissecting microscope, pick up embryos at the tip of the pipette with base medium and suspend them in EFS20 and start the timer. The embryos shrink instantly and move to the surface of EFS20.

12. Aspirate the embryos into the other pipette containing EFS20, and then transfer the embryos to the EFS20 in the watch glass to wash them.

13. Pick up the embryos in the pipette, and, at 2 min, transfer them into the 13-mm EFS40 column in the straw with a minimal volume of EFS20.

14. Aspirate the syringe with air until the first larger column of sucrose solution is aspirated into the cotton plug. If the straw is to be sealed with the straw powder, the end must be loaded with sucrose solution separated by air.

15. Pull the straw out of the connector and completely seal the open end with the heat sealer.

16. At 60 s after exposure of the embryos to EFS40, place the straw on the Styrofoam in liquid nitrogen to allow it to cool moderately in the vapor. For cooling in a small Dewar flask, see "Troubleshooting" in text.

17. After 3 min or more, store the straw in a canister in the liquid nitrogen storage tank.

Recovery of cryopreserved embryos

Materials

- Tools: scissors, 1-ml disposable syringe, 18-G needle, embryological watch glass, water bath maintained at ~25°C.
- Solutions: base medium, sucrose solution, culture medium (e.g., M16 modified by adding 10 mM EDTA and 1 mM glutamine).

Methods

1. Prepare two pipettes: one aspirated with the sucrose solution and the other with the base medium.

2. Prepare 100-ml drops of the sucrose solution and base medium under paraffin oil in a culture dish.

3. Fill a 1-ml syringe, with an 18-G needle attached, with 1 ml of the sucrose solution.

4. Remove the straw from the liquid nitrogen using forceps and start the timer. Keep the straw in air at room temperature for 10 s, then immerse it into the water bath and shake gently. During the 10-s suspension in air, the sample passes through −130°C rather slowly, which prevents fracture damage. However, further slow warming is injurious because the cytoplasm can devitrify.

5. In about 8 s, when the sucrose solution begins to melt, remove the straw and wipe the water quickly with a paper towel. Immediately cut off the sealed end and cottonwool plug, tilt the straw slightly with EFS40 side down, and perfuse the straw slowly with 1 ml of the sucrose solution in the syringe into a watch glass. Start the timer and shake the watch glass gently to dilute the EFS40 quickly. After perfusion, leave the sucrose solution in the straw and keep the straw horizontal.

6. Using a pipette containing the sucrose solution, recover the embryos under a dissecting microscope and then transfer them to the sucrose solution in a culture dish. If all the embryos are not recovered, decant the residual sucrose solution left in the straw into the watch glass and examine.

7. At about 5 min after perfusion, transfer the embryos into the base medium and wash them in the medium using a pipette containing base medium.

8. Transfer the embryos into a drop of culture medium under paraffin oil in a culture dish, which has been equilibrated in a CO_2 incubator at 37°C overnight, and culture them in the incubator until transfer to recipients. The apparent integrity of blastomeres is observable after ~1 h of culture, and cleavage is observable the next day.

Protocol 24.3. Ultrarapid vitrification: cryoloop

A protocol effective for vitrification of human blastocysts using a cryoloop technique (30) is described. The protocol is adopted from the report by Lane et al. (11, 50) with slight modification. This new technique may also be available for other embryos; the first successful cryopreservation of hamster embryos was accomplished by this method (50).

Cryopreservation of embryos

Materials

- Instrument: microscope stage warmer.
- Tools: cryovial with a cryoloop (a minute nylon loop, (20 mm wide; 0.5–0.7 mm in diameter) mounted on a stainless-steel pipe inserted into the lid of a cryovial; a metal rod is inserted in the lid for handling [Crystalwand, Hampton Research, Laguna Niguel, CA] (50), handling stainless-steel rod with a small magnet (Hampton Research), four-well mutiplate dish, storage cane
- Reagents: Ficoll 70 (Pharmacia Biotech), sucrose, HSA (25% solution can be used), ethylene glycol, DMSO
- Solutions (see table 24.1): base medium (HEPES-buffered modified HTF containing 5 mg/ml of HSA; other buffered medium can be substituted), solution A (0.83 M sucrose in base medium containing 1% [w/v] Ficoll 70 and 5 mg/ml of HSA), cryoprotectant solution I (base medium containing 7.6% [v/v] DMSO and 7.6% [v/v] ethylene glycol), cryoprotectant solution II (solution A containing 14.7% [v/v] DMSO, 14.7% [v/v] ethylene glycol; the final concentration of sucrose is 0.58 M)

Methods

1. In the 4-well multidish, prepare ~1 ml of base medium, cryoprotectant solution I, and cryoprotectant solution II in each of the wells (see table 24.1).
2. Place the dish in an incubator with no CO_2 at 37°C for ~30 min.
3. Attach an open cryovial to the cane and immerse it in liquid nitrogen in a Dewar flask.
4. Take the warmed 4–well dish out of the incubator and place it on the stage warmer at 37°C.
5. Into each of two prewarmed pipettes, aspirate base medium and cryoprotectant solution II.
6. Suspend an embryo (or embryos) in cryoprotectant solution I and start the timer.
7. Dip the cryoloop into cryoprotectant solution II to create a thin, filmy layer of solution, by surface tension, on the nylon loop.
8. Place two 20-ml drops of cryoprotectant solution II on the lid of a Petri dish.
9. After 100–120 s of suspension of the embryo in cryoprotectant I, pick up the embryo in a pipette containing cryoprotectant solution II and transfer it into the 20-ml drops of the solution.
10. Immediately wash the embryo in the solution by pipetting up and down, and transfer the embryo onto the filmy layer on the nylon loop using the pipette under a dissecting microscope.
11. Within 25–30 s of exposure of the embryo to cryoprotectant solution II, plunge the loop into liquid nitrogen.
12. Screw the lid (with the loop) into the cryovial using the stainless-steel rod, with the loop containing the embryo being submerged in the liquid nitrogen. Store the cane with the cryovial in liquid nitrogen.

Recovery of cryopreserved embryos

Materials

- Tools: 4-well multidish, stainless-steel rod with a small magnet (Hampton Research).

- Reagent: sucrose.
- Solutions: base medium, solution B (base medium containing 0.82 M sucrose), dilution solution I (base medium containing 0.27 M sucrose), dilution solution II (base medium containing 0.16 M sucrose), culture medium (e.g., G2.2).

Methods

1. In the 4-well multidish, place ~1 ml of dilution solution I, dilution solution II, and solution A in each of the wells.
2. Place the dish in an incubator at 37°C for ~30 min.
3. Take out the warmed 4-well dish and place it on the stage warmer at 37°C.
4. Submerge the cryovial attached to the cane in liquid nitrogen in a Dewar flask.
5. Remove the lid (with a loop) of the cryovial with the stainless-steel rod under liquid nitrogen.
6. In one movement, dip the loop into dilution solution I. The vitrified embryo will immediately fall from the loop into the solution.
7. After 2 min, pick up the embryo under a dissecting microscope, and transfer it to dilution solution II.
8. After 3 min, transfer the embryo into base medium and wash it twice.
9. After 5 min, transfer the embryo to culture medium, which has been equilibrated in a CO_2 incubator at 37°C overnight, for further culture until transfer.
10. About 2 h after warming, observe the appearance of the embryo under a dissecting microscope to assess the survival.

References

1. Rall, W. F. (2001). In: *Assisted Fertilization and Nuclear Transfer in Mammals*, D. P. Wolf and M. Zelinski-Wooten, eds., p. 173. Humana Press, Totowa, NJ.
2. Polge, C., Smith, A. U., and Parkes, A. S. (1949). *Nature*, **164**, 666.
3. Whittingham, D. G., Leibo, S. P., and Mazur, P. (1972). *Science*, **178**, 411.
4. Wilmut, I., and Rowson, L. E. A. (1973). *Vet. Rec.*, **92**, 686.
5. Bank, H., Maurer, R. R. (1974). *Exp. Cell Res.*, **89**, 188.
6. Trounson, A., and Mohr, L. (1983). *Nature*, **305**, 305.
7. Willadsen, S. M. (1977). In: *The Freezing of Mammalian Embryos*, Ciba Foundation Symposium 52, K. Elliott, and J. Whelan, eds. p. 175. Elsevier, Amsterdam.
8. Rall, W. F., and Fahy, G. M. (1985). *Nature*, **313**, 573.
9. Martino, A., Songsasen, N., and Leibo, S. P. (1996). *Biol. Reprod.*, **54**, 1059.
10. Vajta, G., Booth, P. J., Holm, P., Greve, Y., and Callesen, H. (1997). *Cryo-Lett.*, **18**, 191.
11. Lane, M., Schoolcraft, W. B., and Gardner, D. K. (1999). *Fertil. Steril.*, **72**, 1073.
12. Schneider, U., and Mazur, P. (1987). *Cryobiology*, **24**, 17.
13. Nagashima, H., Kashiwazaki, N., Ashman, R. J., Grujpen, C. G., and Nottle, M. B. (1995). *Nature*, 374, 416.
14. Hayashi, S., Kobayashi, K., Mizuno, J., Saitoh, K., and Hirano, S. (1989). *Vet. Rec.*, **125**, 43.
15. Mazur, P. (1990). *Cell Biophys.*, **17**, 53.
16. Kasai, M. (1994). In: *Perspectives on Assisted Reproduction*, T. Mori, T. Aono, T. Tominaga, and M. Hiroi, eds., *Frontiers in Endocrinology*, vol. **4**, p. 481. Ares-Serono Symposia Publications, Rome.
17. Rall, W. F., Reid, D. S., and Polge, C. (1984). *Cryobiology*, **21**, 106.
18. Landa, V. (1982). *Folia Biol.*, **28**, 350.
19. Kasai, M., Zhu, S. E., Pedro, P. B., Nakamura, K., Sakurai, T., and Edashige, K. (1996). *Cryobiology*, **33**, 459.
20. Pedro, P. B., Zhu, S. E., Makino, N., Sakurai, T., Edashige, K., and Kasai, M. (1997). *Cryobiology*, **35**, 150.
21. Kasai, M., Niwa, K., and Iritani, A. (1980). *J. Reprod. Fertil.*, **59**, 51.
22. Leibo, S. P. (1983). *Cryo-Lett.*, **4**, 387.
23. Kasai, M., Iritani, A., and Chang, M. C. (1979). *Biol. Reprod.*, **21**, 839.
24. Lassalle, B., Testart, J., and Renard, J. P. (1985). *Fertil. Steril.*, **44**, 645.

25. Pedro, P. B., Sakurai, T., Edashige, K., and Kasai, M. (1997). *J. Mamm. Ova Res.*, **14**, 66.

26. Leibo, S. P. (1986). In: *Genetic Engineering of Animals*, J. W. Evans and A. Hollaender, eds., vol. **37**, p. 251. Plenum Press, New York.

27. Jackowski, S., Leibo, S. P., and Mazur, P. (1980). *J. Exp. Zool.*, **212**, 329.

28. Kasai, M. (1996). *Anim. Reprod. Sci.*, **42**, 67.

29. Pedro, P. B., Kasai, M., Mammaru, Y., Yokoyama, E., and Edashige, K. (1996). *Proc. Int. Congr. Anim. Reprod.*, **13**, 15.

30. Mukaida, T., Nakamura, S., Tomiyama, T., Wada, S., Kasai, M., and Takahashi, K. (2001). *Fertil. Steril.*, **76**, 618.

31. Kasai, M., Komi, J. H., Takakamo, A., Tsudera, H., Sakurai, T., and Machida, T. (1990). J. *Reprod. Fertil.*, **89**, 91.

32. Scheffen, B., Van Der Zwalmen, P., and Massip, A. (1986). *Cryo-Lett.*, **7**, 260.

33. Zhu, S. E., Kasai, M., Otoge, H., Sakurai, T., and Machida, T. (1993). *J. Reprod. Fertil.*, **98**, 139.

34. Landa, V., and Tepla, O. (1990). *Folia Biol.*, **36**, 153.

35. Bielanski, A., Nadin-Davis, S., Sapp, T., and Lutze-Wallace, C. (2000). *Cryobiology*, **40**, 110.

36. Ashwood-Smith, M. J., Hollands, P., and Edwards, R. G. (1989). *Hum. Reprod.*, **4**, 702.

37. Kasai, M., Nishimori, M., Zhu, S. E., Sakurai, T., and Machida, T. (1992). *Biol. Reprod.*, **47**, 1134.

38. Miyake, T., Kasai, M., Zhu, S. E., Sakurai, T., and Machida, T. (1993). *Theriogenology*, **40**, 121.

39. Kasai, M., Hamaguchi, Y., Zhu, S. E., Miyake, T., Sakurai, T., and Machida, T. (1992). *Biol. Reprod.*, **46**, 1042.

40. Hochi, S., Fujimoto, T., Braun, J., and Oguri, N. (1994). *Theriogenology*, **42**, 483.

41. Martinez, A. G., Furnus, C. C., Matkovic, M., de Matos, D. G. (1997). *Theriogenology*, **47**, 351.

42. Mochida, K., Matsuda, J., Noguchi, Y., Yamamoto, Y., Nakayama, K., Takano, K., Suzuki, O., and Ogura, A. (1998). *Biol. Reprod.*, **58**, Suppl 1, 180.

43. Mochida, K., Wakayama, T., Takano, K., Noguchi, Y., Yamamoto, Y., Suzuki, O., Ogura, A., and Matsuda, J. (1999). *Theriogenology*, **51**, 171.

44. Mukaida, T., Nakamura, S., Tomiyama, T., Wada, S., Oka, C., Kasai, M., and Takahashi, K. (2003). *Hum. Reprod.*, **18**, 384.

45. Takeda, T., Henderson, W. B., and Hasler, J. F. (1987). *Theriogenology*, **27**, 285.

46. Edashige, K., Asano, A., An, T. Z., and Kasai, M. (1999). *Cryobiology*, **38**, 273.

47. Mukaida, T., Wada, S., Takahashi, K., Pedro, P. B., An, T. Z., and Kasai, M. (1998). *Hum. Reprod.*, **13**, 2874.

48. Mahmoudzaddeh, A. R., Van Soom, A., Bols, P., Ysebaert, M., and de Kruif, A. (1995). *J. Reprod. Fertil.*, **103**, 33.

49. Whittingham, D. G. (1971). *Nature*, **233**, 125.

50. Lane, M., Bavister, B. D., Lyons, E. A., and Forest, K. (1999). *Nature Biotech.*, **17**, 1234.

Index